Open School-Conference of NIS Countries Ultrafine Grained and Nanostructured Materials (UFGNM-2018)

IOP Conference Series: Materials Science and Engineering Volume 447

Ufa, Russia
1 – 5 October 2018

ISBN: 978-1-5108-7660-6
ISSN: 1757-8981

Printed from e-media with permission by:

Curran Associates, Inc.
57 Morehouse Lane
Red Hook, NY 12571

Printed by Curran Associates, Inc. (2019)

For permission requests, please contact the Institute of Physics
at the address below.

Institute of Physics
Dirac House, Temple Back
Bristol BS1 6BE UK

Phone: 44 1 17 929 7481
Fax: 44 1 17 920 0979

techtracking@iop.org

Additional copies of this publication are available from:

Curran Associates, Inc.
57 Morehouse Lane
Red Hook, NY 12571 USA
Phone: 845-758-0400
Fax: 845-758-2633
Email: curran@proceedings.com
Web: www.proceedings.com

TABLE OF CONTENTS

IOP Conf. Series: Materials Science and Engineering **447** (2018) 011001 doi:10.1088/1757-899X/447/1/011001

PREFACE

Proceedings of the Open School-Conference of NIS Countries «Ultrafine Grained and Nanostructured Materials – 2018» (UFGNM- 2018)

Radik R. Mulyukov, Ayrat A. Nazarov, Renat M. Imayev

The Open School-Conference of NIS Countries «Ultrafine Grained and Nanostructured Materials – 2018» (UFGNM - 2018) was successfully held in Ufa, Bashkortostan Republic, Russia, from October 1 till October 5, 2018. The UFGNM serial conference, which is held biennially since 2008, is becoming one of the leading materials science conferences in the territory of the former USSR, and aims to address fundamental problems and application prospects of advanced nanostructured and ultrafine grained materials. The conference scope includes but is not limited to various topics of this interdisciplinary field such as nanoparticles, nanowires, nanofilms and coatings, nanostructured composite materials, carbon nanostructures, bulk ultrafine grained metals and alloys, processing methods, structure and properties, interfaces in materials, superplasticity and related phenomena, solid state joining, theory and simulation etc.

UFGNM-2018 was organized by the Institute for Metals Superplasticity Problems of the Russian Academy of Sciences in cooperation with Bashkir State University and was supported by the Ministry of Science and Higher Education of the Russian Federation, Russian Foundation for Basic Research (grant # 18-02-20122), sponsored by LLC "TESCAN" (S.-Petersburg, Russia). The organizers also acknowledge donations from the Bashkortostan Regional Department of the Russian Engineering Union and LLC "Gazprom Tranzgas Ufa".

The proceedings contain 88 papers presented on the conference. All papers have been peer reviewed and passed several stages of edition procedure. We express our sincere gratitude to all people who contributed to this process and did their best to enhance the scientific merit and quality of the Proceedings. We also are grateful to the authors for their contributions, to the speakers and participants of the school-conference UFGNM-2018. We hope to welcome them along with new participants to the next school-conference UFGNM-2020.

Ufa, Bashkortostan, Russia

COMMITTEES

of the Open School-Conference of NIS Countries «Ultrafine Grained and Nanostructured Materials – 2018» (UFGNM- 2018)

International Organizing Commitee

Chairman

Dr. Sci., Prof., Corresponding Member of RAS R.R. Mulyukov, Institute for Metals Superplasticity Problems of the Russian Academy of Sciences, Ufa, Russia

Co-Chairman

Dr. Sci., Prof. N.D. Morozkin, Bashkir State Univerisity, Ufa, Russia

Committee Members

Dr. Sci., Prof., Corresponding Member of RAS M.I. Alymov, Merzhanov Institute of Structural Macrokinetics and Materials Science of the Russian Academy of Sciences, Chernogolovka, Russia
Dr. Sci., Prof., Corresponding Member of the National Academy of Sciences of Ukraine V.N. Varyukhin, Donetsk, Ukraine
Dr. Sci., Prof., Corresponding Member of the Russian Academy of Sciences A.A. Inozemstev, JSC "Aviadvigatel", Perm', Russia
Dr. Sci., Prof., Corresponding Member of the Russian Academy of Sciences S.G. Psakhie, Institute of Strength Physics and Materials Science of the Russian Academy of Sciences, Tomsk, Russia
Dr. Sci., Prof., Corresponding Member of the National Academy of Sciences of Belarus V.V. Rubanik, Institute of Technical Acoustics of the National Academy of Sciences of Belarus, Vitebsk, Belarus
Dr. Sci., Prof., Corresponding Member of the Russian Academy of Sciences V.V. Rybin, Peter the Great Saint-Petersburg Polytechnic University, S.-Petersburg, Russia
Dr. Sci., Prof., Corresponding Member of the Russian Academy of Sciences V.V. Sagaradze, M.N. Mikheev Institute of Metals Physics of Ural Branch of the Russian Academy of Sciences, Ekaterinburg, Russia
Dr. Sci., Prof. of the Russian Academy of Sciences Yu.A. Baimova, Institute for Metals Superplasticity Problems of the Russian Academy of Sciences, Ufa, Russia
Dr. Sci., Prof. A.M. Glezer, Moscow National University of Science and Technology MISIS, Moscow, Russia
Dr. Sci. R.M. Imayev, Institute for Metals Superplasticity Problems of the Russian Academy of Sciences, Ufa, Russia
Dr. Sci. N.K. Kasmamytov, Institute of Physico-Tehcnical Problems and Materials Science of the National Academy of Sciences of Kyrgyzstan, Bishkek, Kyrgyzstan
Dr. Sci., Prof. N.A. Koneva, Tomsk State University of Architecture and Building, Tomsk, Russia
Dr. Sci. G.F. Korznikova, Institute for Metals Superplasticity Problems of the Russian Academy of Sciences, Ufa, Russia
Dr. Sci., Prof. K.A. Kuterbekov, L.N. Gumilyov Eurasian National University, Astana, Kazakhstan
Dr. Sci. M.V. Markushev, Institute for Metals Superplasticity Problems of the Russian Academy of Sciences, Ufa, Russia
Dr. Sci., Prof. D.L. Merson, Togliatti State University, Togliatti, Russia
Dr. Sci., Prof. M.M. Myshlyaev, Baykov Metallurgy and Materials Institute of the Russian Academy of Sciences, Moscow, Russia
K.S. Nazarov, Institute for Metals Superplasticity Problems of the Russian Academy of Sciences, Ufa, Russia
Dr. Sci., Prof. S.P. Pavlinich, JST "Kuznetsov", Samara, Russia
Dr. Sci., Prof. V.V. Perevezentsev, Federal Research Center Institute of Applied Physics, Nizhniy Novgorod, Russia

IOP Conf. Series: Materials Science and Engineering **447** (2018) 011002 doi:10.1088/1757-899X/447/1/011002

Dr. Sci., Prof. S.V. Smirnov, Institute of Engineering Science of Ural Branch of the Russian Academy of Sciences, Ekaterinburg, Russia
Dr. Sci., Prof. M.D. Starostenkov, Polzunov Altai State Technical University, Barnaul, Russia
Dr. Sci., Prof. R.A. Yakshibaev, Bashkir State Univerisity, Ufa, Russia

Local Organizing Committee

Chairman

Dr. Sci. A.A. Nazarov, Institute for Metals Superplasticity Problems of the Russian Academy of Sciences, Ufa, Russia

Co-Chairman

Dr. Sci., Prof. V.P. Zakharov, Bashkir State University, Ufa, Russia

Secretary

Dr. V.M. Valitova, Institute for Metals Superplasticity Problems of the Russian Academy of Sciences, Ufa, Russia

IOP Conf. Series: Materials Science and Engineering **447** (2018) 011003 doi:10.1088/1757-899X/447/1/011003

Peer review statement

All papers published in this volume of *IOP Conference Series: Materials Science and Engineering* have been peer reviewed through processes administered by the proceedings Editors. Reviews were conducted by expert referees to the professional and scientific standards expected of a proceedings journal published by IOP Publishing.

IOP Conf. Series: Materials Science and Engineering **447** (2018) 012001 doi:10.1088/1757-899X/447/1/012001

Ion sputtering rate of nanostructured FCC, BCC and HCP metals processed by severe plastic deformation

R Kh Khisamov[1], K S Nazarov[1], S N Sergeev[1], R U Shayakhmetov[1], J A Baimova[1,2], Y M Yumaguzin[1,2] and R R Mulyukov[1,2]

[1] Institute for Metals Superplasticity Problems of the Russian Academy of Sciences, 39 Stepan Khalturin St., Ufa, 450001, Russia
[2] Bashkir State University, 32 Zaki Validi St., Ufa, 450074, Russia

E-mail: r.khisamov@mail.ru

Abstract. The effect of nanostructuring on the ion sputtering rate of metals and the glow discharge current is studied. Nanostructured samples were processed by severe plastic deformation using high pressure torsion. For comparative measurements, coarse-grained samples were obtained by annealing of nanostructured samples. The sputtering rates of nanostructured and coarse-grained nickel, iron and zirconium were measured after ion sputtering in glow gas-discharge. It has been found that nanostructuring of metals leads to an increase in the ion sputtering rate and the glow discharge current. Possible causes of changes in the ion sputtering rate due to nanostructuring are analyzed.

1. Introduction

Functional elements of ion sources and similar devices, such as walls, cold cathodes etc. suffer from ion sputtering, which results in degradation of these elements with time. Therefore, the elements are fabricated of metals with a low value of the ion sputtering yield. From another point of view, ion sputtering can also play a positive role, for instance, in the case of sputtering targets, which are widely used in nano- and microelectronics. Target materials must meet certain requirements, such as high purity and a lack of pores, homogeneity of the grain size and crystallographic texture.

Nanostructured metals, which are polycrystals with an average grain size of the order of 100 nm, potentially can be used for a fabrication of elements of ion sources or sputtering targets. One of the attractive methods for obtaining nanostructured metals is deformation nanostructuring, often referred to as severe plastic deformation [1]. To date, various physical properties of nanostructured metals have been studied. It has been found that these materials have significantly modified values of magnetic, thermal, mechanical, electrical, emission properties as compared to their coarse-grained counterparts [1-5]. However, there are some fields that are poorly investigated, for example, properties associated with ion sputtering of nanostructured metals. In this regard, it is important to study the effect of nanostructuring on the rate of ion sputtering of metals.

2. Materials and experimental procedure

Pure nickel, iron and zirconium were chosen as the FCC, BCC and HCP metals for the investigation. To obtain a nanosized microstructure, the samples were processed by severe plastic deformation using high pressure torsion [6,7] in Bridgman anvils under pressure P = 6.5 GPa (10 anvil rotations at a rotation rate of 2 revolutions per minute). For comparative measurements, coarse-grained samples were obtained by annealing of the nanostructured samples at 873-1073 K in vacuum for 1 hour. All samples had the form of discs with a diameter of 10 mm and a thickness of 0.5 mm. The microstructure of the samples was investigated by a Mira 3 LHM scanning electron microscope (Tescan, Czech Republic).

Ion sputtering of the samples was carried out in a glow discharge plasma in a gas discharge setup. The gas discharge setup (similar to the one described in [8]) comprises a glass bulb with an anode and six sputtering targets placed inside. The distance between the anode and the targets was equal to d = 15 cm. The nanostructured and coarse-grained samples of nickel, iron and zirconium (six samples in total) were used as the sputtering targets. The simultaneous use of six targets provides the same conditions of sputtering for different samples. Before sputtering, all samples were subjected to mechanical grinding to obtain flat surfaces. The samples were attached to T- shaped metal holders. To measure the gas discharge current, the ammeters were connected to the metal holders. The gas discharge bulb was placed in a vacuum chamber. The experiment was performed at a vacuum level in a chamber of about 10^{-5} Torr. To achieve the glow discharge conditions, the bulb was filled with Ar at a pressure of P ~ 0.1-1 Torr. The discharge voltage was equal to U = 800 V. The discharge current varied for different samples. Ion sputtering was performed for 2 hours. The ion sputtering rate was determined by evaluating the difference in weight of the samples before and after ion sputtering on the precision analytical balance of MV 210-A (Sartogosm, Russia).

3. Results and discussion

Electron backscatter diffraction (EBSD) was used to study the microstructure of nanostructured nickel and iron. Corresponding EBSD maps are presented in figure 1. Analysis of these maps shows that the average grain size of nanostructured nickel and iron after high pressure torsion is almost the same and equal to 180 nm. Analysis of the spectrum of grain boundary misorientations shows that the majority of grain boundaries (up to 80%) are high angle. The microstructure of nanostructured zirconium is studied using the back-scattered electron (BSE) mode, which showed that the grain size was equal to 100-300 nm.

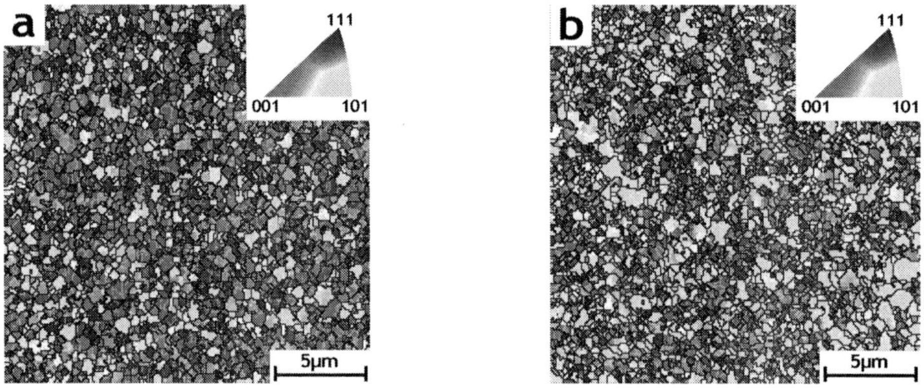

Figure 1. EBSD maps of nanostructured Ni (a) and Fe (b).

Figure 2 shows the BSE images of the microstructure of coarse-grained nickel, iron and zirconium, i.e. after annealing. It was observed that the grain size of coarse-grained samples is larger than 10 μm.

The values of the average glow discharge current for all investigated samples for U =800 V are collected in figure 3a. This figure shows that the glow discharge current J of nanostructured samples is higher than that of coarse-grained samples ($J_{NS-Ni}>J_{CG-Ni}$, $J_{NS-Zr}>J_{CG-Zr}$, $J_{NS-Fe}>J_{CG-Fe}$) and the nickel sample has the highest value ($J_{Ni}>J_{Zr}>J_{Fe}$). The glow discharge current depends on the ion-induced electron emission yield γ of metal as $J \sim \gamma$ [9].

The ion sputtering rate V can be determined from the measurements of the sample mass before (m_1) and after (m_2) ion sputtering, as

$$V = \frac{x}{t} = \frac{\Delta m}{\rho S t},$$

where x is the thickness of the sputtered layer, t is the total sputtering time, Δm is the mass change of the sample $\Delta m = m_1 - m_2$, ρ is the density of the metal, S is the sample area. Ion sputtering rates for nanostructured and coarse-grained nickel, iron and zirconium for J = 1.0 mA/cm^2 are presented in figure 3b. These data show that the ion sputtering rate of nanostructured samples is again higher than that of coarse-grained samples: $V_{NS\text{-}Ni} > V_{CG\text{-}Ni}$, $V_{NS\text{-}Zr} > V_{CG\text{-}Zr}$, $V_{NS\text{-}Fe} > V_{CG\text{-}Fe}$ and $V_{CG\text{-}Ni} > V_{CG\text{-}Zr} > V_{CG\text{-}Fe}$.

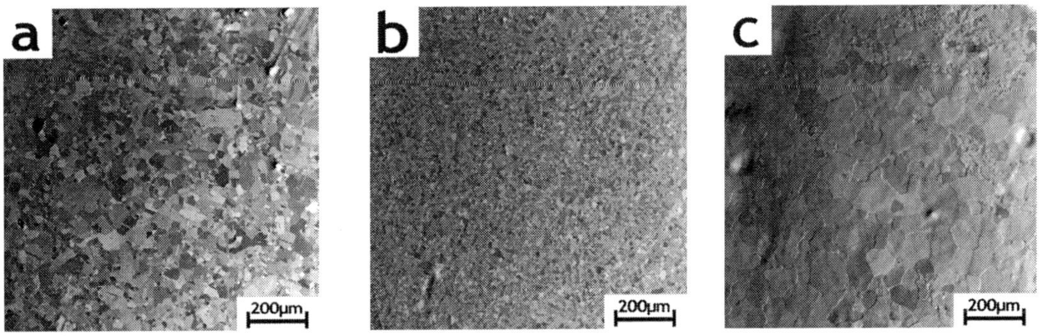

Figure 2. BSE images of coarse-grained nickel (a), iron (b) and zirconium (c).

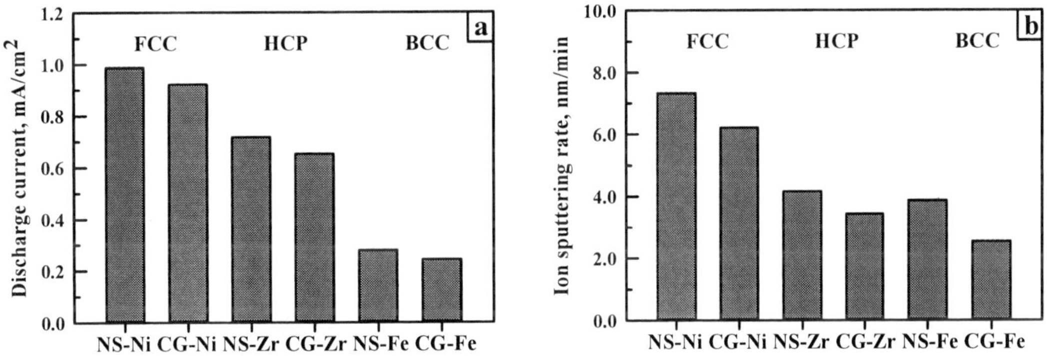

Figure 3. The average glow discharge current density (a) and the calculated ion sputtering rates at J=1.0 mA/cm^2 (b) for the nanostructured (NS) and coarse-grained (CG) FCC, BCC and HCP metals for discharge voltage U=800 V, P~0.1-1 Torr, d=15 cm.

It is known that ion sputtering of a metal depends on its microstructure. For instance, the smaller the grain size, the higher the ion sputtering rate of metals [10-12]. The increase in the ion sputtering rate due to nanostructuring can be explained as follows. The ion sputtering yield Y depends on the binding energy of the atom E on the surface as Y ~ 1/E. The presence of defects reduces the binding energy. Therefore, the ion sputtering yield Y of the grain boundary is higher than that of the grain [10]. Consequently, one can assume that grain boundaries are sputtered faster than grains. Nanostructured metals have a larger volume fraction of grain boundaries, which may explain the higher values of their ion sputtering rates as compared to coarse-grained metals. Thus, the increase in the ion sputtering rate due to nanostructuring is a result of the increase in the volume fraction of grain boundaries.

4. Summary

The effect of deformation nanostructuring on the rate of ion sputtering of nickel, iron and zirconium in a glow gas-discharge has been studied. It has been shown that the ion sputtering rate and the glow discharge current of nanostructured samples are higher than those of coarse-grained samples for all investigated metals having FCC, HCP and BCC lattices. The enhanced values of both quantities are the result of a decrease in the grain size and, correspondingly, an increase in the specific grain boundary area in the materials. Further studies are required for better elucidation of the effect of nanostructuring on the ion sputtering process.

Acknowledgments

The reported study was funded by RFBR according to the research project № 18-32-00754. Fabrication of the gas discharge setup was accomplished according to the state assignment of IMSP RAS № AAAA-A17-117041310213-0. Electron microscopic studies were carried out on the facilities of shared services center of the Institute for Metals Superplasticity Problems of Russian Academy of Sciences «Structural and Physical-Mechanical Studies of Materials».

References

[1] Nazarov A A and Mulyukov R R 2002 *Handbook of NanoScience, Engineering and Technology* (CRC Press, Boca Raton) p. 22-1-22-41
[2] Mulyukov R R 2006 *J. Vac. Sci. Technol. B* **24** 1061
[3] Khisamov R Kh, Safarov I M, Mulyukov R R, Yumaguzin Yu M, Zubairov L R and Nazarov K S 2011 *Techn. Phys.* **56** 1661
[4] Khisamov R Kh, Safarov I M, Mulyukov R R andYumaguzin Yu M 2013 *Physics of the Solid State* **55** 1
[5] Nazarov K S, Khisamov R Kh, Yumaguzin Yu M, Mulyukov R R 2015 *Techn. Phys. Lett.* **41** 16
[6] Valeev I Sh, Valeeva A Kh, Mulyukov R R and Khisamov R Kh 2016 *Letters on Materials* **6** 347
[7] Mukhametgalina A A, Samigullina A A, Sergeev S N, Zhilyaev A P, Nazarov A A, Zagidullina Yu R, Parkhimovich N Yu, Rubanik V V and Tsarenko Yu V 2017 *Letters on Materials* **7** 85
[8] Khisamov R Kh, Nazarov K S, Safarov I M, Musabirov I I, Yumaguzin Yu M and Mulyukov R R 2017 *Mater. Phys. Mech.* **33** 161
[9] Raizer Yu P 2009 *Gas Discharge Physics* (Intellect, Dolgoprudny) 736 p. (In Russian)
[10] Behrisch R and Eckstein W 2007 *Sputtering by Particle Bombardment* (Springer-Verlag Berlin Heidelberg) 509 p.
[11] Michaluk C A 2002 *J. Electron. Mater.* **31** 2
[12] Reza M, Sajuri Z, Yunas J and Syarif J 2016 *IOP Conf. Ser.: Mater. Sci. Eng.* 114 012116

Effect of severe plastic deformation on the saturation magnetization of ferromagnetic DO_3-type ordered Fe_3Al alloy

A M Glezer[1,2], I ATimshin[1], M A Libman[2], I V Shchetinin[1], E S Savchenko[1], A A Tomchuk[2], R V Sundeev[3]

[1] National University of Science and Technology "MISIS", 4 Lenin av., Moscow, 119049 Russia
[2] I.P. Bardin Science Institute for Ferrous Metallurgy, 23/9 Radio st., Moscow, 105005 Russia
[3]MIREA - Russian Technological University, 78 Vernadsky av., Moscow, 119454 Russia

E-mail: a.glezer@mail.ru

Abstract. The effect of severe plastic deformation in Bridgman anvils at room temperature on the specific saturation magnetization (SSM) of the ferromagnetic DO_3 ordered Fe–24 at.% Al alloy was studied by X-Ray diffraction analysis and magnetic measurements. High pressure torsion at certain deformation parameters was shown to fully suppress the long-range ordering in the alloy and simultaneously increase its SSM by 11% relative to that in the equilibrium state. A theoretical model which qualitatively explains this phenomenon is proposed.

1. Introduction

At present, methods of a substantial increase in physical and mechanical properties of metallic materials based on severe plastic deformation (SPD) attract a great attention of researchers [1]. One of the most commonly used SPD methods is high pressure torsion (HPT) in Bridgman anvils [2]. In the literature, much attention is paid to the effect of SPD on the mechanical properties of many metals and alloys [3]. Unfortunately, only a few papers are devoted to the effect of SPD on the magnetic properties of ferromagnetic alloys. One can mention only some publications, in which the structure-dependent magnetic properties (coercivity, permeability, frequency characteristics) of crystalline Nd(Pr)-Fe-B [4], Fe-C [5], α-Fe, FeCo, Ni_3Fe [6], Fe-Ni [7,8], Fe-Si [8], Fe-Co [9], Ni_3Al [10] and amorphous Fe-Ni-B, Fe-Cr-B, Fe-Si-B, Fe-Si-B-Nb-Cu, Ni-Fe-Co-Si-B [11,12] ferromagnetic metals and alloys are considered as functions of SPD. However, in some studies, SPD by HPT in Bridgman anvils has been found to noticeably affect the saturation magnetization of crystalline [6,7] and amorphous [11-13] alloys. As is known [14], this characteristic is not a structure-dependent parameter of ferromagnetic materials and is normally determined only by chemical and phase compositions of a ferromagnetic material regardless of its structural state. No reasonable explanation for this phenomenon has been found yet. Only assumptions related to the densification of the crystal lattice [6] or a change in the degree of short-range order [7] upon SPD have been put forward.

The aim of this paper is to analyze in detail the nature of the possible effect of SPD (by the HPT method) on the fundamental characteristic of ferromagnetic materials (specific saturation magnetization σ) in relation to the DO_3–type Fe–24 at. % Al alloy of the composition close to the stoichiometric A_3B one. In addition to the specific saturation magnetization σ, the lattice parameter (a)

of the BCC Fe-24 at % Al solid solution and the degree of the DO_3 long-range order (η) in the first coordination sphere have been analyzed as functions of HPT.

2. Experimental

The Fe-24 at.% Al alloy was melted in a vacuum furnace in the form of rods with a diameter of 10 mm. Then discs with a thickness of 0.3 mm were cut from the rods and annealed in an argon atmosphere at a temperature of 500 °C for 50 hours to attain an equilibrium state corresponding to the DO_3 atomic order [15]. Then the disc samples were subjected to plastic deformation in Bridgman anvils at room temperature to a certain fraction / number of full revolutions N (¼, ½, 1, 2, 3, 4, and 6) of the movable anvil under quasi-hydrostatic pressure P = 4 GPa at a movable anvil rotation velocity of 1 rev/min.

The specific saturation magnetization σ was measured with a VSM-250 vibrating magnetometer after room-temperature HPT to N = 0.25-6. X-ray structure analysis of the samples before and after SPD was carried out with a RigakuUltima IV diffractometer using Co K_α radiation and graphite monochromator.

3. Results and discussion

Figure 1 shows the measured magnetic hysteresis curves for the Fe-24 at.% Al alloy samples in the initial state and after room-temperature HPT to N = 1. One can see a noticeable increase in σ from 160 emu/g (without HPT processing) to 177 emu/g after HPT, i.eby about approximately 11%.

Figure 2 shows the measured specific saturation magnetization σ (a), the lattice parameter a (b), and the long-range order parameter η (c) of the alloy as functions of the degree of deformation (number of revolutions N) at room temperature.

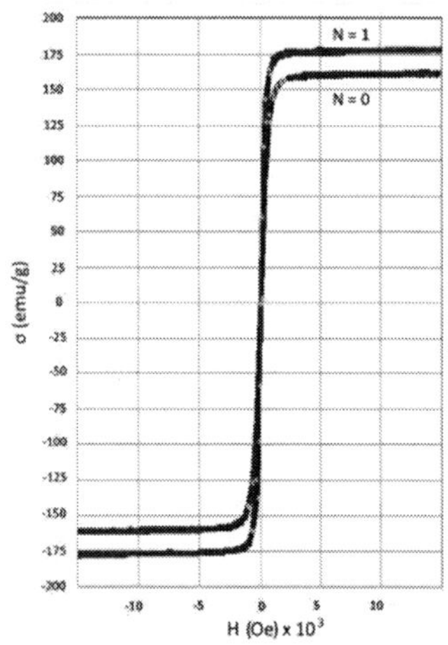

Figure 1. Magnetic hysteresis curves for the Fe-24 at.% Al alloy in the initial state (N=0) and after HPT (N=1).

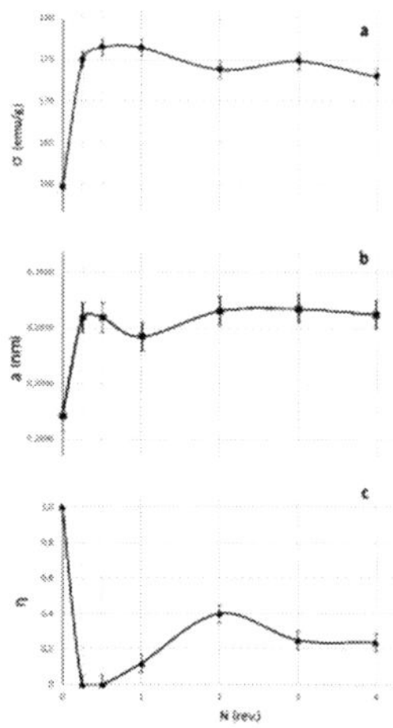

Figure 2. Effect of the degree of deformation (N) on the specific saturation magnetization σ (a), lattice parameter a (b), and long-range order parameter η (c).

At $N \leq 1$, there is a noticeable increase in both σ and a, while η is substantially reduced to almost zero, and the alloy undergoes deformation-induced disordering. Then, at $N > 1$, the long-range order is partially restored (the η parameter increases). There is also a stabilization of a and a weak decrease in σ, which, however, remains substantially higher than that in the initial state prior to the HPT processing.

The nature of the unusual change in the long-range order parameter η (figure 2 c) was analyzed in detail in our work [16]. The earlier unknown phenomenon of a partial recovery of a completely destroyed $D0_3$ long-range order has been found to occur upon room-temperature HPT deformation in the Fe_3Al (Fe - 24 at.% Al) super lattice. The established effect was attributed [16] to low-temperature dynamic recrystallization, at which a high-temperature two-phase A2+B2 state is formed in the grains dynamically recrystallized upon SPD. At high degrees of deformation by HPT ($N \geq 2$), a stationary state which includes a "two-phase mixture" of deformation fragments and dynamically recrystallized grains is formed in the alloy in accordance with the laws of nonequilibrium thermodynamics. In addition, note that we apparently for the first time succeeded to get the Fe-24 at. % Al alloy with a composition close to the one of stoichiometric Fe_3Al alloy in an unordered state using HPT method at N <1. It is known [17] that even melt quenching at a rate of about 10^6K/s cannot suppress the long-range ordering, which occurs in the stoichiometric Fe_3Al alloy at relatively high temperatures as a second-order phase transition (the T_C temperature is 730 °C [15]).

We assume that a significant increase in the magnetic parameter σ in figure 2 a is partially related only to the suppression of the long-range ordering in the ferromagnetic BCC Fe-Al solid solution. To check this assumption, we constructed the σ - η parametric dependence (figure 3), at which each point corresponds to σ and η for all considered HPT processing regimes including the initial (undeformed) state. Extrapolating the experimental data from figure 3 as a linear correlation function, we obtain a very high correlation coefficient K = -0.85. This, of course, means that the degree of long-range order in the first coordination sphere actually strongly affects the specific saturation magnetization of the alloy, although the character of such dependence can be nonlinear.

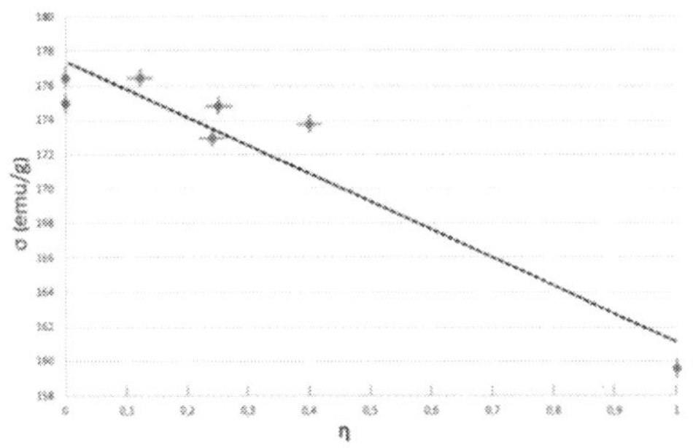

Figure 3. Linear correlation between the specific saturation magnetization σ and the degree of long-range order η corresponding to a constant degree of HPT deformation N (K = - 0.85).

The experimental results presented above allow one to conclude that a decrease in the degree of atomic order η leads to an increase in the specific saturation magnetization σ and, therefore, to an increase in the degree of magnetic ordering of the Fe-24 at.% Al ferromagnetic alloy. In addition, the change in lattice parameter upon HPT (figure 2 b) is mainly due to the order-disorder transformations in the Fe-Al system [17]. It was shown experimentally [18] that a change in the lattice parameter in ordered iron-based alloys can affect the saturation magnetization. Therefore, the structure changes occurring under HPT can be conventionally considered as a sequence of order-disorder phase transitions accompanied by the appearance of transformation-induced deformation $\varepsilon = \Delta a/a$ at increased N. To establish an analytical relationship between the long-range order parameters in the magnetic and atomic subsystems, we use, according to [19], the following equation for the thermodynamic potential of the spatially inhomogeneous system (G), in which the phase transition characterized by the φ parameter is accompanied by the appearance of the transformation-induced deformation $\hat{\boldsymbol{\varepsilon}}$:

$$G = \int_V \{g(\varphi, \hat{\boldsymbol{\varepsilon}}) + \nabla\varphi\hat{\boldsymbol{\alpha}}\nabla\varphi\}dV \approx \int_V \left\{g(\varphi, \widehat{\boldsymbol{\varepsilon}_0}) + \nabla\varphi\hat{\boldsymbol{\alpha}}\nabla\varphi + \frac{1}{2}(\hat{\boldsymbol{\varepsilon}} - \widehat{\boldsymbol{\varepsilon}_0})\hat{\boldsymbol{E}}(\hat{\varepsilon} - \widehat{\boldsymbol{\varepsilon}_0})\right\} dV. \quad (1)$$

Here g (φ, $\hat{\boldsymbol{\varepsilon}}$) is the specific thermodynamic potential; $\hat{\boldsymbol{\varepsilon}}$ is the total deformation characterizing the lattice distortion at a certain point; $\hat{\varepsilon}_0(\varphi)$ is the transformation-induced deformation corresponding to the parameter φ; $\hat{\boldsymbol{E}}$is the tensor of elastic moduli; $\hat{\boldsymbol{\alpha}}$ is the parameter, which characterizes the interaction attempting to keep the system in a uniform state; and V is the volume.

We consider a spatially uniform system, in which the processes of changes in long-range order simultaneously occur in the magnetic and atomic subsystems and the interaction between such processes occurs through an elastic field caused by the transformation-induced deformation in the crystal. In this case, the formula for the thermodynamic potential takes the following form:

$$G = g(\eta;\xi;\widehat{\boldsymbol{\varepsilon}_0})V = -\frac{N}{2}W(\widehat{\boldsymbol{\varepsilon}_0})\eta^2 - \frac{N}{2}A(\widehat{\boldsymbol{\varepsilon}_0})\xi^2 + \frac{1}{2}\widehat{\boldsymbol{\varepsilon}_0}\widehat{\boldsymbol{E}}\widehat{\boldsymbol{\varepsilon}_0}V - TS(\eta;\xi), \quad (2)$$

whereg $(\xi,\eta,\widehat{\boldsymbol{\varepsilon}_0})$ is the specific thermodynamic potential of a spatially homogeneous system; $\xi = \sigma/\sigma_0, \eta, \widehat{\boldsymbol{\varepsilon}_0}$ are the parameters of magnetic and atomic order and the transformation-induced deformation, respectively; σ_0 is the specific saturation magnetization at absolute zero. N_0 is the number of atoms in the alloy; $W(\hat{\boldsymbol{\varepsilon}}_0)$ is the mixing energy; $A(\hat{\boldsymbol{\varepsilon}}_0)$ is the exchange interaction energy; $S(\eta,\xi)$ is the entropy; and T is the absolute temperature.

The dependences of the mixing energies and exchange interaction can be represented in the form of expansions in powers of the transformation-induced deformation. Due to the small degree of the transformation-induced deformation $\varepsilon = \Delta a/a << 1$, we restrict ourselves by the linear terms of the formula and have:

$$W(\widehat{\boldsymbol{\varepsilon}_0}) = W_0(1 + \widehat{\boldsymbol{\beta}_1\boldsymbol{\varepsilon}_0}),$$
$$A(\widehat{\boldsymbol{\varepsilon}_0}) = A_0(1 + \widehat{\boldsymbol{\beta}_2\boldsymbol{\varepsilon}_0}), \quad (3)$$

where $\widehat{\boldsymbol{\beta}_{1,2}}$ are the coefficients of the expansion at the linear members of the transformation-induced deformation.

Taking into account the cubic symmetry of the crystal lattice of the alloy and the dilatational character of the transformation-induced deformation, we have:

$$\widehat{\boldsymbol{\varepsilon}_0} = (\varepsilon_0, \varepsilon_0, \varepsilon_0);\ \widehat{\boldsymbol{\beta}_{1,2}} = (1/3\beta_{1,2},\ 1/3\beta_{1,2},\ 1/3\beta_{1,2}).$$

From the conditions of the minimum of thermodynamic potential

$$\frac{\partial G}{\partial \varepsilon_0} = \frac{\partial G}{\partial \eta} = \frac{\partial G}{\partial \xi} = 0, \quad (4)$$

we obtain the following equations for determining the equilibrium degrees of the transformation-induced deformation (5) and the parameters of the atomic (6) and magnetic (7) order:

$$\varepsilon_0 = \frac{N}{2VE}(W_0\beta_1\eta^2 + A_0\beta_2\xi^2), \quad (5)$$
$$\frac{1}{4}kTln\frac{(1+3\eta)(3+\eta)}{(1-\eta)^2} = W_0\left(1 + \frac{N}{2V}\frac{A_0}{E}\beta_1\beta_2\xi^2\right)\eta + \frac{N}{2V}\frac{{W_0}^2}{E}{\beta_1}^2\eta^3, \quad (6)$$
$$\frac{1}{2}kTln\frac{1+\xi}{1-\xi} = A_0\left(1 + \frac{N}{2V}\frac{W_0}{E}\beta_1\beta_2\eta^2\right)\xi + \frac{N}{2V}\frac{{A_0}^2}{E}{\beta_2}^2\xi^3, \quad (7)$$

where $E = 3(C_{11} + 2C_{12})$; C_{11}, C_{12} are the components of the elastic modulus tensor of the cubic lattice in the Voigt notation.

A qualitative analysis of Eqs. (6) and (7) leads to the conclusion that the character of the interaction of the order parameters in the magnetic and atomic subsystems within the proposed model is determined by the sign of the product of coefficients β_1 and β_2. Upon atomic ordering, the magnetic order parameter and, consequently, the saturation magnetization increase at $\beta_1\beta_2 > 0$ and decrease at $\beta_1\beta_2 < 0$.

The proposed theoretical model explains the experimentally observed (figure 2) effect of the DO_3 atomic order on the long-range ordering in the magnetic subsystem. Since a decrease in the order in the atomic subsystem leads to an increase in magnetization (figure 2), then, in accordance with the proposed theoretical model, we find that, in the Fe-24 at.% Al alloy, the product $\beta_1\beta_2 < 0$. To determine the sign of each coefficient, we use Eq. (5) and obtain:

$$\beta_1 = \frac{EV}{W_0 N\eta}\frac{\partial \varepsilon_0}{\partial \eta} \sim \frac{EV}{W_0 N\eta}\frac{\partial}{\partial \eta}\left(\frac{\Delta a}{a}\right) < 0$$
$$\beta_2 = \frac{EV}{A_0 N\xi}\frac{\partial \varepsilon_0}{\partial \xi} \sim \frac{EV}{A_0 N\xi}\frac{\partial}{\partial \xi}\left(\frac{\Delta a}{a}\right) > 0, \quad (8)$$

Since, in accordance with figure 2, the lattice parameter in the alloy is increased by magnetic ordering and decreased by atomic ordering, it follows from Eqs. (8) that $\beta_1 < 0$ and $\beta_2 > 0$. Experiments in [15] show that exactly this situation occurs in binary Fe-Al alloys.

Our experiments in the Bridgman anvils allowed us to obtain a wide range of atomic order parameters η(figure 2c). In particular, for the first time it was possible to obtain the alloy of near-stoichiometric composition Fe_3Al in a fully disordered state ($\eta \cong 0$). Accordingly, this allowed us to obtain an extremely high magnetic order parameter ξ of the Fe-24 at.% Al alloy under the action of HPT and, therefore, the maximum possible specific saturation magnetization σ, which is noticeably (by about 11%) higher than the initial σ of the alloy in the equilibrium state.

4. Conclusions
The effect of severe plastic deformation in the Bridgman anvils at room temperature on the specific saturation magnetization of the ferromagnetic DO_3 ordered Fe–24 at.% Al alloy was discovered. Torsion under high pressure at certain deformation parameters was shown to fully suppress the long-range ordering in the alloy and simultaneously increase the specific saturation magnetization by 11% relative to that of the equilibrium state. It was shown that a strong correlation occurred between the specific saturation magnetization and the degree of DO_3-type long-range order in the ordered states. It was shown that a theoretical model, including consideration of transformation-induced lattice deformation, can qualitatively explain this phenomenon.
It was shown that a theoretical model, including consideration of lattice deformation caused by the transformation, can qualitatively explain this phenomenon.

Acknowledgments
The work was supported by State Task no. 2017/113 of Ministry of Education and Science of the Russian Federation., as well as partially by the RFBR, project no. 18-02-00398.R.V.S. is grateful for the financial assistance to Russian Science Foundation project no. 18-72-00026 in X-ray analysis.

References
[1] Vinogradov A and Estrin Yu 2018 *Progr. Mater. Sci.* **95** 172
[2] Edalati K and Horita Z 2016 *Mater. Sci. Eng. A* **652** 325
[3] GlezerA M, KozlovE V, Koneva N A, Popova N A and Kurzina I A 2017 *Plastic Deformation of Nanostructured Materials* (Taylor & Francis Group, N-Y)
[4] StolyarovV V, GunderovD V , Popov A G, Gaviko V S and Ermolenko A S 1998 *J. Alloys Compd.* **281** 69
[5] Straumal B B, Mazilkin A A, Protasova S G, Dobatkin S V, Rodin A O, Baretzky B, Goll D and Schütz G 2009 *Mater. Sci. Eng. A* **503** 185
[6] HosokawaA, OhtsukaH, Li T, Li S and Tsuchiya K 2014 *Mater. Trans.* **55** 1286
[7] GlezerA M, TomchukA A,. Savchenko A G, Korovushkin V V and Shchetinin I V 2016 *Bull. Russ. Acad. Sci.: Phys.***80** 1021
[8] Scheriau S, Rumpf K, Kieber S and Pippan R 2008 *Mater. Sci. Forum* **584-586** 923
[9] Grossinger R, Mehboob N and Zehetbauer M 2014 *IOP Conf. Seres: Mater. Sci. Eng.***60** 012019
[10] Kazantseva N V, Pilugin V P, Zavalishin V A, Rigmant M B, Davidov D I and Stepanova N N 2013 *Lett. Mater.***3** 16
[11] Glezer AM, PlotnikovaM R, ShalimovaA V and Dobatkin S V 2009 *Bull. Russ. Acad. Sci.:*

Phys. **73** 1233
[12] Abrosimova G, Aronin A, Matveev D and Pershina E 2013 *Mater. Lett.* **97** 15
[13] Zhilyaev A P and Langdon T G 2008 *Prog. Mater. Sci.* **53** 893
[14] Kakehashi Y 2012 *Modern Theory of Magnetism in Metals and Alloys* (Springer, Heidelberg)
[15] Swann P R, Duff W R and Fisher R M 1972 *Metal. Trans.* **3** 413
[16] Glezer A M, Timshin I A, Shchetinin I V, Gorshenkov M V, Sundeev R V and Ezhova A G 2018 *J. Alloys Compd.* **744** 791
[17] GlezerA M and Permyakova I E 2013 *Melt-Quenched Nanocrystals* (N-Y, Taylor & Francis Group)
[18] Shiga M 1974 *AIP Conf. Proc.* **18** 463
[19] Khachaturyan A G 2008 *Theory of Structural Transformations in Solids* (N-Y, Dover)

IOP Conf. Series: Materials Science and Engineering **447** (2018) 012003 doi:10.1088/1757-899X/447/1/012003

Molecular dynamics simulation of the effect of cyclic stresses on nanocrystals with nonequilibrium grain boundaries: the role of the grain size

A A Nazarov

Institute for Metals Superplasticity Problems, Russian Academy of Sciences, 39 Khalturin st., Ufa 450001, Russia

E-mail: aanazarov@imsp.ru

Abstract. Grain boundaries (GBs) in bulk nanostructured materials processed by severe plastic deformation (SPD) have a nonequilibrium structure caused by extrinsic grain boundary dislocations (EGBDs) absorbed during deformation. Under external influences (annealing or cyclic straining) these GBs relax towards equilibrium resulting in a release of the excess energy of nanomaterials. Using our earlier developed method, atomic models of nanocrystals with nonequilibrium GBs having grain sizes of 10, 15, and 20 nm are constructed and relaxed by molecular dynamics. Then, the effect of oscillating tension-compression stresses with amplitudes of 2, 3 and 4 GPa on the systems thus obtained is simulated. Relaxation of nonequilibrium GBs by dislocation emission is found and the effect is shown to be independent of the grain size.

1. Introduction

The interaction of grain boundaries (GBs) with lattice dislocations results in the formation of extrinsic grain boundary dislocations (EGBDs) and a specific structure of GBs, which is commonly referred to as a nonequilibrium structure [1]. The nonequilibrium GBs (NGBs) induce long-range internal stress fields and possess an excess energy as compared to equilibrium GBs of the same geometry [2]. The nonequilibrium GB structure was shown to be typical for bulk nanostructured materials processed by severe plastic deformation (SPD) and to largely determine their properties [3]. For better understanding of the effect of NGBs on the properties of materials, it is necessary to know their atomic structure.

Recently, we have developed a method for the construction of initial structures for molecular dynamics (MD) simulations of nanocrystals with NGBs and studied the atomic structure, and energy of a model nickel [112] columnar nanocrystal with NGBs [4,5]. Also, the behavior under cyclic stresses was studied for the case of a nanocrystal with the grain size of $d = 15$ nm.

The present paper aims to study the role of grain size in the behavior of a [112] columnar nanocrystal with NGBs under cyclic stresses.

2. Model description and simulation procedure

The geometry of a computation cell containing four columnar grains with crystallographic axes [112] parallel to the z direction of a coordinate frame is presented in figure 1a. The grains have a regular hexagon-shape cross section with an edge a. The maximum size of the hexagons, $d = 2a$, will be taken as the grain size. Grains 1 and 2 have orientations of slip planes $(11\bar{1})$ parallel to *xOz* and *yOz* planes, respectively. Taking the orientation of grain 1 as a reference, grains 3 and 4 are rotated to angles $\pm\alpha = \pm 25°$ to form a symmetric high-angle tilt GB between them. To initialize the structures with NGBs, the construction method proposed in [4] is used. This method is based on an artificial shear strain of selected grains of the nanocrystal to introduce EGBDs into its boundaries. In order to form EGBDs on the boundaries of the selected grain, it is assumed to be removed from its place, freely deformed by plastic shear γ, then strained back elastically to $-\gamma$ and embedded into the polycrystal. In figure 1 a, grains 3 and 4 are deformed in this way. Then the system with EGBD precursors is subjected to energy minimization with special precautions preventing the EGBDs' annihilation. As an example, presented in figure 1 b is the initial structure of a nanocrystal constructed for the grain size $d = 5$ nm and shear strain $\gamma = 0.15$ of grains 3 and 4.

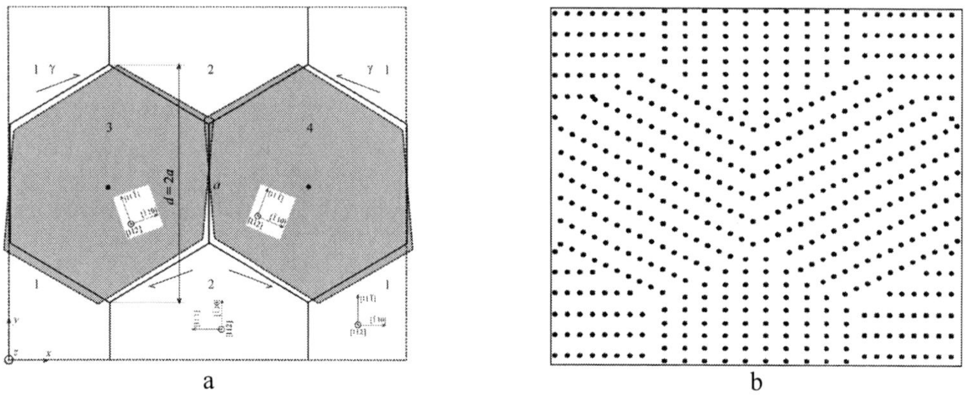

Figure 1. Computational cell construction for [112] columnar nanocrystals with NGBs (a) and initial structure of a nanocrystal with grain size $d = 5$ nm and shear strain $\gamma = 0.15$ of grains 3 and 4 (b).

Using the method described above, the initial structures of nanocrystals with grain sizes $d = 10$, 15 and 20 nm and shear strain $\gamma = 0.10$ of grains 3 and 4 were constructed for further studies. Simulations were carried out for nickel using embedded atom method (EAM) potential developed in [6]. The corresponding lattice parameter is equal to $a_0 = 0.352$ nm and the potential energy per atom $E_{a0} = -4.45$ eV. The period of the computational cell along the *z*- axis was fixed equal to $H_z = |3a_0/2[112]| = 1.293$ nm for $T = 0$ K and scaled by a factor of 1.0044 at $T = 300$ K to account for thermal expansion. Periodic boundary conditions with a constant pressure were applied along the other two directions (zero along *x* and zero or *p* along *y*). To obtain relaxed configurations of the systems, an energy minimization was carried out first keeping all atoms having energies $\varepsilon > -4.40$ eV (i.e. atoms on and near GBs) fixed to their positions that allowed one to avoid the annihilation of EGBD precursors. Then, a full relaxation at $T = 0$ K followed by equilibration at $T = 300$ K was carried out.

The effect of cyclic straining was simulated at $T = 300$ K, applying a periodic tension-compression stress with a period of 160 ps along the *y*- axis. For this, the period was divided into 40 intervals, in each of which the value of the stress changed from $p_{k-1} = p_0 \sin[2\pi(k-1)/40)]$ to $p_k = p_0 \sin 2\pi k/40)$ in each of the *k*- th interval ($k = 1, 2, \ldots, 40$) and for each interval of the stress 2000 steps of MD run with time step 2 fs were done. Three values of the stress amplitude, $p_0 = 2$, 3 and 4 GPa were taken for simulations.

Simulations were carried out using the LAMMPS MD code [7], visualization and common-neighbor analysis (CNA) [8] of the structures were done by OVITO [9]. Events of lattice dislocation emission, changes in the sizes of computational cells H_x and H_y, and of the potential energy of atoms E_a with time were analyzed.

3. Simulation results and discussion

Figures 1 a,b present the results of the CNA of the atomic structures of a nanocrystal with the grain size d = 10 nm with the nonequilibrium structure of GBs at 0 K and after ten periods of cyclic deformation with the stress amplitude of 4 GPa at 300 K.

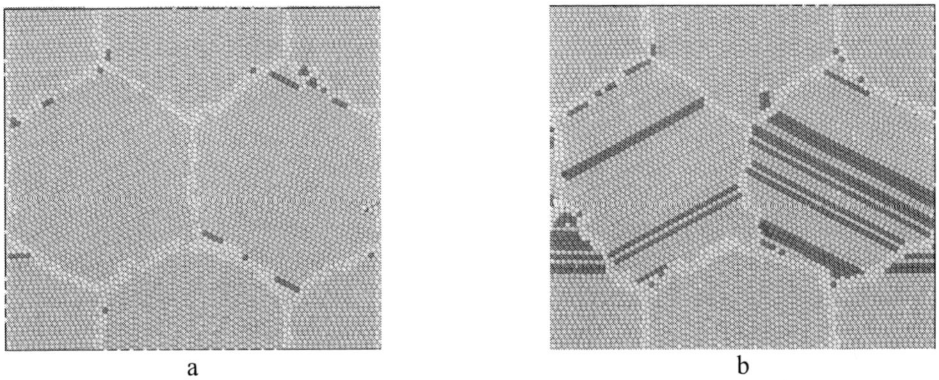

Figure 2. Results of a CNA of the atomic structures of a nanocrystal with the grain size d = 10 nm: (a) with nonequilibrium GBs at 0 K; (b) after ten cycles of straining with the stress amplitude of 4 GPa at 300 K. Green atoms have a fcc environment, red ones an hcp environment, and white and blue atoms have other types of coordination.

In the figures, red atoms having an hcp environment visualize stacking faults, which are formed due to the glide of partial lattice dislocations across the grains. Therefore, the comparison of the figures shows that under the effect of cyclic stresses GBs generate lattice dislocations. Analysis shows that the latter are emitted when the applied stress has the same sign as the internal stresses induced by EGBDs, and, therefore, these dislocations result in a relaxation of the stress fields of NGBs.

The fact that the generation of dislocations results in relaxation of NGBs is clearly seen from figures 3 a-c, which demonstrate the change of atomic potential energies with time expressed in terms of the number of stress cycles N_{cycle} for the three grain sizes studied. One can see that the potential energy decreases with the number of cycles. At the stress amplitude of 2 GPa, there is nearly no change of the average energy, since lattice dislocations are not emitted by NGBs. At the amplitude of 3 GPa, dislocation activity occurs in the grains and at 4 GPa it becomes higher. Accordingly, the average potential energy per atom decreases with time of stress cycling.

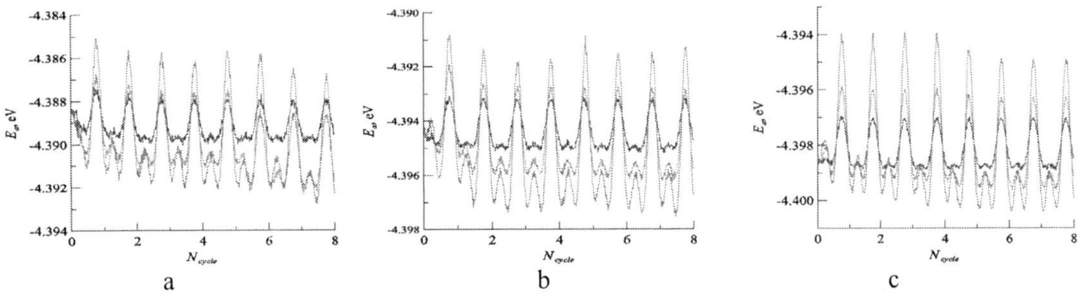

Figure 3. Dependence of atomic energies of nanocrystals with grain sizes d = 10 (a), 15 (b) and 20 (c) nm on the number of stress cycles; p_0 = 2,3,4 GPa (blue, green and red curves, respectively).

The fact that the average potential energy E_a decreases due to the elimination of long range stress fields induced by EGBDs is seen from figures 4 a,b, where atomic energy maps for a nanocrystal with the grain size of 10 nm before and after cyclic deformation are presented. One can clearly see that the atoms inside the deformed grains had an enhanced energy due to the long range elastic fields of EGBDs, and after stress cycling the atoms have energies close to that of perfect crystal atoms due to the compensation of the stress fields by the emission of dislocations.

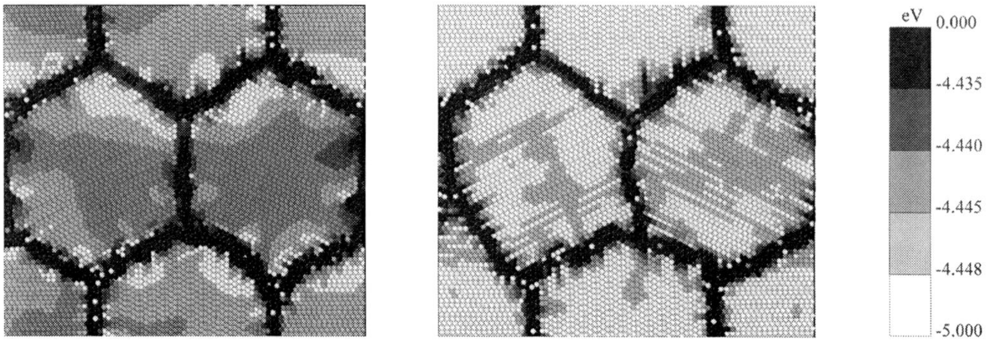

a b c

Figure 4. Atomic energy maps of nanocrystals with the grain size of $d = 10$ nm in the state with NGBs (a) and after ten cycles of straining with the stress amplitude of 4 GPa (b) and energy scale (c) to demonstrate the relaxation of long range stress fields.

Figure 3 also shows that, despite a two times difference between the minimum and maximum grain sizes of the studied nanocrystals, there is no apparent change of the stress amplitude necessary for initiating the dislocation emission by NGBs: in all three cases, the generation of dislocations occurs at an amplitude between 3 and 4 GPa. Therefore, the stresses necessary for an emission of individual dislocations from GBs, NGBs as well, depend mainly on their local atomic structure rather than on long range stress fields. However, the latter bring an asymmetry to the dislocation emission process under cyclic stresses resulting in the possibility of structure relaxation under completely symmetric stress oscillations.

4. Conclusions

Molecular dynamics simulations show that oscillating applied stresses can result in relaxation of the nonequilibrium structure of GBs in nanocrystals by emission of partial dislocations, which compensate the stress fields of EGBDs. This process occurs at the stress amplitudes above a certain lower bound, which for the present case is between 2 and 3 GPa. This lower bound seems to be independent of the grain size of nanocrystals. Therefore, the relaxation of NGBs by dislocation emission is controlled by the local structure of GBs rather than by their long range stress fields.

Acknowledgments
The present work was supported by the Russian Science Foundation (grant No. 16-19-10126).

References

[1] Valiev R Z., Gertsman V Yu and Kaibyshev O A 1986 *Phys. Status Solidi* A **97** 11
[2] Nazarov A A, Romanov A E and Valiev R Z 1993 *Acta Metall. Mater.* **41** 1033
[3] Nazarov A A and Mulyukov R R 2003 *Handbook of Nanoscience, Engineering, and Technology*, ed W Goddard, D Brenner, S Lyshevski and G Iafrate (Boca Raton: CRC Press) p 22-1
[4] Nazarov A A 2018 *Letters on Materials* **8** 5
[5] Nazarov A A and Murzaev R T 2018 *Comp. Mater. Sci.* **151** 204.
[6] Foiles S M, Baskes M I and Daw M S 1986 *Phys. Rev. B* **33** 7983
[7] Plimpton S 1995 *J. Comp. Phys.* **117** 1
[8] Honeycutt J D and Andersen H G 1987 *J. Chem. Phys.* **91** 4950
[9] Stukowski A 2010 *Modell. Simul. Mater. Sci. Eng.* **18** 015012

IOP Conf. Series: Materials Science and Engineering **447** (2018) 012004 doi:10.1088/1757-899X/447/1/012004

Threshold displacement energy in Ni, Al and B2 NiAl

N T H Trung[1,2], H S M Phuong[1,2], M D Starostenkov[1*], V V Romanenko[1], V A Popov[1]
[1]Altai State Technical University, 46 Lenin st., Barnaul 656038, Russia
[2]Nuclear Research Institute, 01 Nguyen Tu Luc st., Da Lat 670000, Vietnam

E-mail: genphys@mail.ru

Abstract. For simulation of radiation effects and defects in the Ni-Al system, a new many body potential was developed by joining the equilibrium part of the Mishin's EAM potential with the universal function of Ziegler, Biersack and Littmark at a suitable interatomic spacing and the corresponding pairwise energy from the DFT calculations. To assess the qualities of this potential, we performed molecular dynamics simulations to calculate the threshold displacement energy of pure metals Ni, Al and their NiAl B2 phase superalloy at ambient temperature. It was found that the threshold energy of displacement of nickel in nickel is higher than that in the B2 long-range ordered alloy NiAl in contrast to aluminium, which has lower threshold energy than B2 NiAl. The threshold displacement energy ranges between 14±2 eV and 189±2 eV depending on the crystallographic direction. The lowest threshold energy of two fcc metals (14±2 eV for Al and 28±2 eV for Ni) was found in the direction <101>, these values correspond to the mean value in the easiest directions of atomic displacements and they are in good agreement with many reports from the literature.

1. Introduction

The threshold displacement energy (TDE) E_d is the minimum kinetic energy that must be given to an atom in a solid in order to displace it permanently from its lattice site to the defect position. TDE is an important factor, which determines the defect production, particularly, the number of Frenkel pairs in NRT and Kinchin-Pease displacement models of primary radiation damage in a material. It is also often used in binary collision approximation computer codes such as SRIM. In a crystalline solid, the TDE magnitude is a function of the crystallographic direction $E_d(\theta,\phi) = E_d([hkl])$, temperature, applied strain and many other factors. One distinguishes the minimum threshold displacement energy, $E_{d,min} = \min (E_d(\theta,\phi))$, and the average or effective one, $E_{d,ave} = \text{ave} (E_d(\theta,\phi))$, averaged over all lattice directions [1].

The calculation of the threshold displacement energy is not a trivial problem, because the cascade displacement under irradiation is a complex multi-body collision process, as has been shown by various research groups, in which several methods have been used [2, 3, 4]. In the case of Ni, Al and B2 NiAl alloy, many studies, both simulation and experiment, have been conducted to find their threshold displacement energy in several crystallographic directions at different temperatures, however the results are significantly different between these research groups [5]. Discrepancies may come from different method and potential applied. For this reason, we performed several classical molecular dynamics (MD) simulations to investigate the threshold displacement energies of pure fcc metals Ni, Al and their B2 phase NiAl superalloy in some crystallographic directions at ambient temperature (300 K). The authors of this work would like to provide another perspective to this classical problem of radiation damage.

Table 1. Recommended values of the effective displacement energy for use in displacement calculation [1].

Metal	Lattice(c/a)	$E_{d,min}$(eV)	E_d (eV)
Al	fcc	16	25
Ti	hcp (1.59)	19	30
V	bcc	-	40
Cr	bcc	28	40
Mn	bcc	-	40
Fe	bcc	20	40
Co	fcc	22	40
Ni	fcc	23	40
Cu	fcc	19	30
Zr	hcp	21	40
Nb	bcc	36	60
Mo	bcc	33	60
Ta	bcc	34	90
W	bcc	40	90
Pb	fcc	14	25
Stainless steel	fcc	-	40

2. Potential for simulation of radiation damage

Many potentials have been developed for the Ni-Al system, including the Finnis–Sinclair potential of Yan and the EAM potentials by Foiles and Daw, Voter and Chen [6]. In this work, to investigate the TDE in Ni, Al and B2 NiAl we chose the embedded-atom-method (EAM) potential developed by Mishin et al, [6]. This potential was developed in 2009 using an empirical fitting method. Within the EAM description, the total energy of atom *i* is given by:

$$E_i = F_\alpha\left(\sum_{i\#j}\rho_\beta\left(r_{ij}\right)\right)+\frac{1}{2}\sum_{i\#j}\phi_{\alpha\beta}\left(r_{ij}\right)$$

where r_{ij} is the distance between atoms i and j, $\phi_{\alpha\beta}$ is the pair-wise potential function, ρ_β is the contribution to the electron charge density from atom j of type β at the location of atom i, and F_α is the embedding function that represents the energy required to place atom i of type α into the electron cloud.

However, the Mishin's potential does not meet the requirements and criteria associated with the simulations of radiation damage. Like the aforementioned potentials based on the EAM formalism, it is too "soft" and not suitable for simulating radiation damage at a very short interatomic distance. At such small atomic separations, the force fields fitted to the equilibrium properties tend to significantly underestimate the potential energy of the colliding dimer [7]. To enable molecular dynamics simulations of high-energy collisions, the Mishin EAM potential was modified following the procedure in the work of Stoller et al [7] that involves the use of *ab initio* calculations to determine the magnitude and spatial dependence of pair interactions at intermediate distances, along with systematic criteria for choosing the joining parameters.

There is no generally accepted procedure for choosing the inner and outer cutoff distances to join the equilibrium region of the potential with the higher energy region [4, 7, 8, 9]. In order to test its effect, some MD runs were performed with different inner and outer cutoffs to the spline used to connect the many-body term and the universal function of Ziegler, Biersack and Littmark (ZBL potential) [10] to the Mishin potential. The most appropriate cutoffs are presented in table 2.

Table 2. Electron density and pair-wise interaction term cutoffs for the modification of the Mishin potential

Interaction	Ni-Ni	Ni-Al	Al-Al
r_{inner}(Å)	0.2	1	1.7
r_{outer} (Å)	2	1.6	2.3
Electron density cutoff (Å)		1.5	

Modifications are made to the many-body term of the EAM, namely the density and embedding functions. The goal is to have the repulsive pair interaction term of the EAM play the major role in describing the resulting energetics as the distance between two atoms decreases. To ensure this, the embedding energy term has to become negligible by either becoming constant or very small [7]. A spline interpolation was used to create a smooth transition from the original density function at $r = 0.15$ nm to zero at $r = 0$. This modification reduces the contribution of the many-body embedding energy term as the atom closely approaches its neighbor [7].

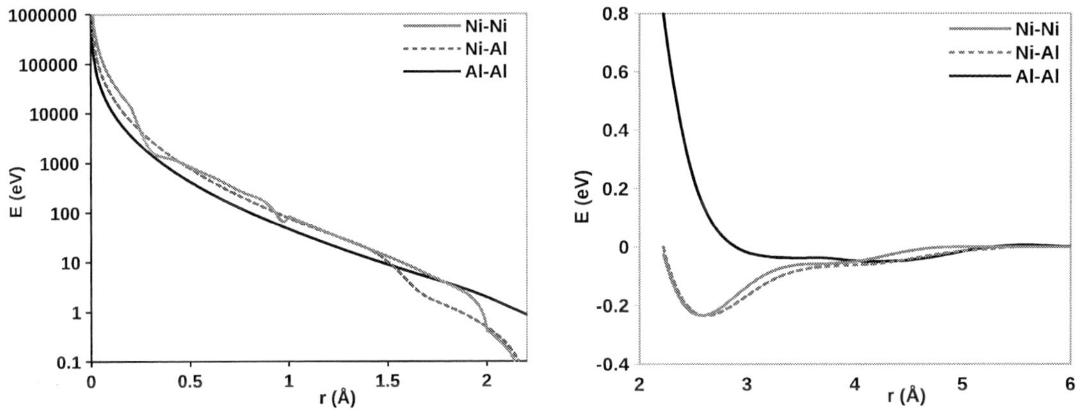

Figure 1. Pair-interaction functions of the modified potential in this study. The right panel is a linear scale plot of the pairwise energy near the equilibrium distance ($r > 2.2$ Å), while the left panel is a log-scale plot of the pairwise energy at a shorter distance ($r < 2.2$ Å).

All two-body terms, pair interaction terms as described in figure 1 need to be modified in two regions: at a very short distance $0 < r < r_{inner}$ the total energy must match the ZBL potential, at an intermediate distance $r_{inner} < r < r_{outer}$ it must match target values for the dimer energy in crystalline Ni-Al obtained from DFT calculations [7]. In the equilibrium region, where $r_{outer} < r < r_{cutoff}$ the force field needs to be equal to the original Mishin potential in order to preserve all its equilibrium properties. All regions were joined together using polynomial regression to maintain the continuity of the second derivative at the joining points, particularly when the Verlet method is used to integrate the equations of motion. After that, the *atsim.potential* package [11] was used for the specification and tabulation of pair-terms and many-body terms to EAM setfl file format, which is directly supported by the most popular classical MD code LAMMPS [12].

The modified potential correctly reproduced the equilibrium lattice parameters, cohesive energies, lattice constants and others equilibrium properties of pure Ni, Al and their alloys B2 NiAl, L12 Ni_3Al.

IOP Conf. Series: Materials Science and Engineering **447** (2018) 012004 doi:10.1088/1757-899X/447/1/012004

To assess the qualities of this potential in higher energy collision simulation, we performed several classical MD simulations using LAMMPS to calculate the threshold displacement energies of Ni, Al and B2 NiAl in some crystallographic directions at ambient temperature.

3. Threshold displacement energy simulation

The LAMMPS MD with GPU and Voronoi package [12, 13, 14] was used to simulate the displacement cascades in a simulation box containing 30×30×30 unit cells with periodic boundary conditions that are applied in three directions.

Our simulations were run with a two layer wide Langevin bath at 300 K to evacuate the extra heat. Initially, the temperature was thermostated at 300 K, and the pressure was barostated to zero in 10 ps. After the primary knock-on atom (PKA) was given the first kinetic energy of 100 eV in some of the basic crystallographic directions, the MD run was carried out at a constant NVE. A variable time-step of at least 0.2 ps was used to limit the distance traveled by atoms between time-steps to ensure numerical accuracy and simulation efficiency. After 0.2 ps with a variable time-step for a primary collision, the simulation time-step resets to 0.001 ps and the system is equilibrated for 10 ps. The number of stable Frenkel pairs is calculated by the Voronoi and Wigner-Seitz cell analysis [14, 15, 16]. This process continues with the binary search algorithm until at least one stable Frenkel pair remains after the simulation. For each direction, total 10 simulations were performed to obtain a confident estimate of the average TDE. The average results of the TDE in each direction are presented in table 3.

Table 3. Average values of the TDE in corresponding direction

	Threshold displacement energy (E_d±2 eV)			
Direction [hkl]	Ni	Al	Ni in B2 NiAl	Al in B2 NiAl
100	48	14	17	14
010	54	20	17	17
001	45	20	24	20
110	32	14	29	20
101	28	14	17	17
011	28	14	60	39
111	98	24	83	36
1 0.5 0.5	132	45	92	36
0.5 1 0.5	142	39	51	54
0.5 0.5 1	64	54	58	76
135	189	36	101	60
720	45	23	20	45
210	113	33	26	39
211	132	45	92	36
520	72	20	23	39
441	114	14	23	40
322	189	30	92	54
311	130	45	117	23

From the results shown in table 3, it is seen that the threshold energy of nickel in pure metal nickel is higher than its value in the B2 long-range ordered NiAl alloy in contrast to aluminum with the threshold energy in B2 NiAl larger than its value in pure metal Al. The threshold displacement energy is ranging between 14±2 eV – 189±2 eV, depending on the crystallographic direction. There are no experimental data available for the highest threshold displacement energy $E_{d,max}$ and the threshold displacement energy in the B2 NiAl alloy [2]. The lowest threshold energy $E_{d,min}$ of two fcc metals (14±2 eV for Al and 28±2 eV for Ni) was found in the direction <101>, these values correspond to the mean value in the easiest directions of atomic displacement and they are in good agreement with many reports from the literature [1, 2, 3, 5 , 17, 18]. However, our simulations do not provide a consistent value of the effective threshold displacement energy. This is the expected behavior, because the modified potential in this work is an empirical potential, in which the effects of the ion-electron interaction (electronic stopping power) and the electron-phonon interaction were neglected. Moreover, in order to obtain the value of the effective threshold displacement energy, we have to execute sufficiently larger simulations in a sufficiently larger crystallographic direction.

4. Conclusions

To enable molecular dynamics simulation of high energy collisions, a new many body potential was developed by joining the equilibrium part of the Mishin EAM potential for the Ni-Al system with the universal function of Ziegler, Biersack and Littmark at a suitable interatomic spacing and the corresponding pairwise energy from DFT calculation. To assess the qualities of this potential, we performed MD simulations to calculate the threshold displacement energy of pure metal Ni, Al and their NiAl B2 phase superalloy at ambient temperature. Many authors from the literature used the linear search method to calculate the TDE with the time complexity O(n). In this work, we applied the binary search method with the time complexity O(log(n)), a faster approach to simulation of the threshold displacement energy. The simulation results are in good agreement with other data from the literature, sufficient to prove the accuracy of this potential. Therefore, it is suitable for simulating radiation damage with the higher energies collision.

References

[1] Was G S 2007 *Fundamentals of Radiation Materials Science: Metals and Alloys* (Berlin: Springer).
[2] Dimitrov C, Sitaud B and Dimitrov O 1994 *J. Nucl. Mater.* **208** 53
[3] Caro A, Victoria M and Averback R. S *1990 J. Mater. Res.* **5** 1409
[4] Béland L K, et al 2016 *J. Appl. Phys.* **119** 085901
[5] *Atomic Defects in Metals* 1991 Landolt-Börnstein - Group III Condensed Matter, Atomic Defects in Metals, vol 25, ed H. Ullmaier (Springer-Verlag Berlin Heidelberg)
[6] Purja Pun G P and Mishin Y 2009 *Phil. Mag.* **89** 3245
[7] Stoller R E et al. 2016 *J. Chem. Theory Comput.* **12(6)** 2871
[8] Doan N V and Tietze H 1995 *Nucl. Instr. Methods Phys.* **102** 58
[9] Nordlund K and Dudarev S 2008 *C. R. Physique* **9** 343
[10] Ziegler J F, Biersack J P and Littmark U 1985 *The Stopping and Range of Ions in Matter* vol 1, In: Bromley D.A. (eds) Treatise on Heavy-Ion Science. (Springer, Boston) pp 93-129
[11] M. Rushton, http://atsimpotentials.readthedocs.io/en/latest/
[12] Plimpton S 1995 *J Comp Phys* **117** 1
[13] Brown W M, Kohlmeyer A, Plimpton S J and Tharrington A N 2013 *Comp. Phys. Comm.* **184** 449
[14] Rycroft C H 2009 *Chaos* **19** 041111
[15] Stukowski A 2012 *Modell. Simul. Mater. Sci. Eng.* **20** 045021
[16] Stukowski A 2010 *Modelling Simul. Mater. Sci. Eng.* **18** 015012
[17] Konobeyev A Yu, Fischer U, Korovin Yu A and Simakov S P 2017 *Nucl. En. Techn.* **3** 169
[18] Kenworthy H M and Neely H H 1956 *Threshold Displacement Energy in Nickel* (Canoga Park, California: Atomics International) p 19

IOP Conf. Series: Materials Science and Engineering **447** (2018) 012005 doi:10.1088/1757-899X/447/1/012005

Structure and electronic properties of 5-7 graphene

M E Belenkov [1], V M Chernov [1], E A Belenkov [1], V M Morilova [2]

[1] Physics Department, Chelyabinsk State University, Chelyabinsk 454001, Russia
[2]Snezhinsk Physics and Technology Institute of National Research Nuclear University MEPhI, Snezhinsk, Russia

E-mail: me.belenkov@gmail.com

Abstract. The electronic properties and the geometrically optimized structure of the 5-7 graphene layer were calculated by the methods of the density functional theory in the generalized gradient approximations. The L_{5-7} graphene should have metallic properties, since the density of its electronic states at the Fermi level is different from zero. The sublimation energy is 7.49 eV, which is less than the sublimation energy for the hexagonal graphene layer, but higher than the sublimation energy of such basic polymorphous varieties of graphene as L_{4-8}, L_{3-12} and L_{4-6-12}.

1. Introduction

Polymorphic varieties of graphene differ from hexagonal graphene in that the three-coordinated (sp^2 hybridized) atoms that make up the graphene layers are located not at the vertices of hexagons, but at the vertices of the polygons having a different number of sides [1,2]. In the structure of the basic polymorphous varieties of graphene, all the atoms are in equivalent structural positions. There can be only four main polymorphs of graphene: L_6, L_{4-8}, L_{3-12}, L_{4-6-12} [2,3]. Existence of an unlimited variety of polymorphic variations of graphene where the atoms are located in more than one structural position is possible [4]. Interest in polymorphous varieties of graphene was caused by the fact that some of them should have semiconductor properties, rather than metallic properties. Formally, graphene polymorphs can be considered as monolayers consisting of topological defects 4, 5, 7, 8, and so on, which replace the hexagons of the usual graphene layer [5]. The stability of defects and the layers consisting of them increase as the difference between the number of sides in the defect and the number of sides in the hexagon diminishes. Therefore, the most stable defects should combine 5-7 topological defects and graphene layers constructed from such defects. In this paper, the structure and electronic properties of one of the structural varieties of the L_{5-7} graphene are studied.

2. The methods of model calculations

Calculations of the geometrically optimized crystal structure, the density of electronic states, and the band structure of the L_{5-7} graphene were performed by the methods of the density functional theory (DFT) [6] in generalized gradient approximations, (GGA) [7]. Calculations of the electronic properties and geometrically optimized structure were performed using the Quantum ESPRESSO software package [8]. An oblique unit cell of a minimal area containing 8 atoms and a rectangular expanded unit cell containing 16 atoms were observed in the calculations (figure 1). The volume distribution of the electron density was calculated for three-dimensional unit cells corresponding to the stacks of graphene layers with an interlayer distance of 10 Å. This distance ensured the absence of influence of adjacent layers on each other, therefore the calculated structure of the layers and their properties corresponded to the values characteristic for isolated layers. Densities of electronic states were calculated using a set of k-points: 12 × 12 × 12. The wave functions were decomposed along a truncated basis set of plane waves, and the dimension of the set of basic functions was limited to the value E_{cutoff} = 1 keV.

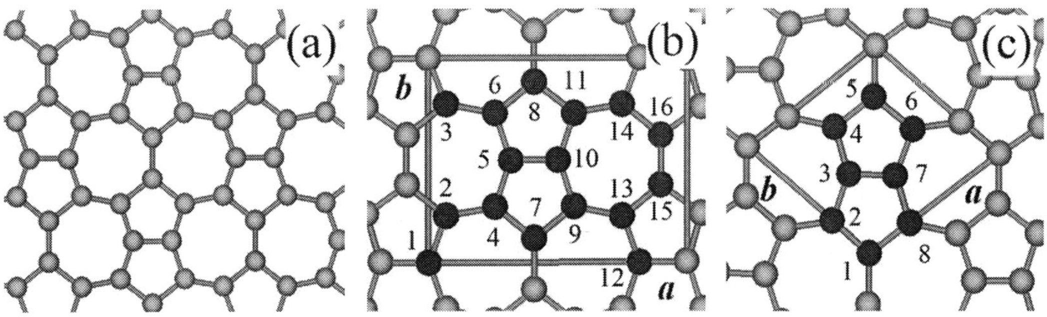

Figure 1. Geometrically optimized structure of the 5-7 graphene layer: (a) crystal lattice; (b) oblique unit cell; (c) rectangular unit cell.

3. Results and discussion

As a result of the calculations, a geometrically optimized structure of the 5-7 graphene layer was found (figure 1a). The choice of various unit cells did not affect the values of the structural parameters. In $L_{5\text{-}7}$ graphene there are three structural positions corresponding to atoms number 4, 5 and 7, designated in a rectangular unit cell or atoms number 2, 3 and 1 in an oblique unit cell (figure 1b, 1c). This is one of the main differences between the $L_{5\text{-}7}$ graphene and the graphene layers of the main polymorphic varieties L_6, $L_{4\text{-}8}$, $L_{3\text{-}12}$ and $L_{4\text{-}6\text{-}12}$ where all the atoms are in the same structural positions [1-3]. Therefore, 5-7 graphene layer has five different lengths of interatomic bonds (r_{ij}) and seven different angles between bonds (φ_{ijk}) characterizing three structural positions of atoms. The calculated values of these structural parameters are given in table 1.

Table 1. The lengths of the interatomic bonds (r_{ij}) and the angles between the bonds (φ_{ijk}) in the $L_{7\text{-}5}$ graphene (i, j, k are the atomic numbers in the rectangular unit cell)

r_{ij} (Å)		ϕ_{ijk} (°)	
$r_{4.7}$	1.492	$\phi_{2.4.7}$	130.44
$r_{4.5}$	1.413	$\phi_{2.4.5}$	122.50
$r_{5.10}$	1.399	$\phi_{5.4.7}$	107.06
$r_{2.4}$	1.406	$\phi_{4.5.10}$	110.32
$r_{15.16}$	1.426	$\phi_{4.5.6}$	139.36
		$\phi_{4.7.9}$	105.30
		$\phi_{13.15.16}$	127.35

The lengths of the bonds and the angles between them differ from the values characteristic for hexagonal graphene. This indicates a deformation of the $L_{5\text{-}7}$ structure compared to the L_6 structure. The calculation of the total energy per atom showed that for 5-7 graphene, this value is -157.05 eV/atom. The sublimation energy, which was found as the difference between the energy of an atom in a graphene layer and the energy of an isolated carbon atom, was 7.49 eV/atom. This value is less than the sublimation energy of hexagonal graphene (7.78 eV/atom), but higher than the sublimation energy of all other major polymorphic graphene varieties described in [3]. This indicates a lower stability of the L_{5-7} graphene compared to hexagonal graphene and a higher stability of the structure of the L_{5-7} graphene compared to other polymorphous graphene varieties.

The values of the unit cell parameters for the rectangular cell are a = 7.511 Å, b = 5.893 Å, and for oblique cells a = b = 4.804 Å. The number of atoms in a rectangular unit cell is 16, the number of atoms in an oblique unit cell is 8. The values of the coordinates of atoms in the unit cells are given in fractions of the vectors of elementary translations and are shown in table 2.

Table 2. The coordinates of the atoms in the unit cells of the L_{5-7} graphene (in fractions of the vectors of elementary translations)

№	x (a)	y (b)	№	x (a)	y (b)
		Oblique unit cell of the L_{5-7} graphene			
1	0.046	0.060	9	0.612	0.336
2	0.112	0.285	10	0.547	0.560
3	0.112	0.836	11	0.612	0.785
4	0.295	0.336	12	0.860	0.060
5	0.360	0.560	13	0.795	0.285
6	0.295	0.785	14	0.795	0.836
7	0.453	0.181	15	0.954	0.439
8	0.453	0.940	16	0.954	0.681
		Rectangular unit cell of the L_{5-7} graphene			
1	0.121	0.121	5	0.879	0.879
2	0.116	0.434	6	0.884	0.566
3	0.407	0.593	7	0.593	0.407
4	0.566	0.884	8	0.434	0.116

The results of calculations of the band structure and density of electronic states of the L_{5-7} graphene are shown in figures 2 and 3, respectively. The choice of the unit cell did not have a significant effect on the results - the dependence of the density of electronic states on energy is practically identical for orthorhombic and monoclinic unit cells. This indicates the correctness of the results obtained. It can be seen from the graphs that, at the Fermi energy level, the density of electronic states is different from zero - there is no band gap and, therefore, the 5-7 graphene layer should have metallic properties. The density of electronic states at the level of the Fermi energy in the L_{5-7} graphene turned out to be non-zero, so that there is a significant overlap of the valence and conduction bands, and not only their tangency in hexagonal graphene.

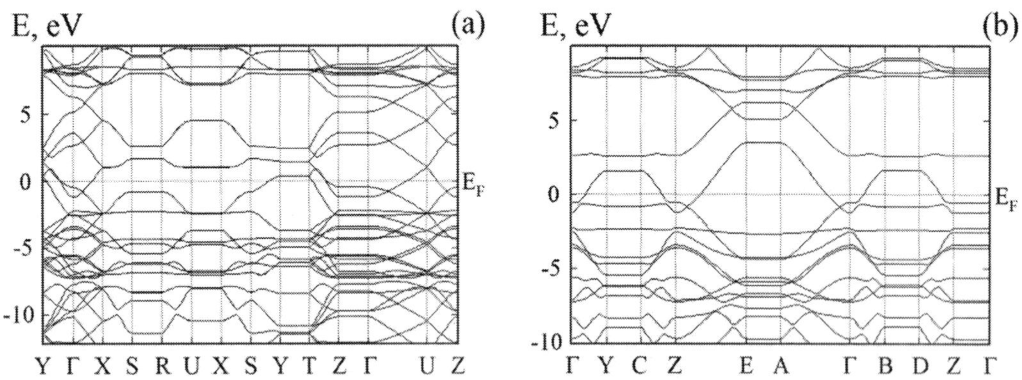

Figure 2. Band structure of the L_{5-7} graphene layer: (a) orthorhombic unit cell (b) monoclinic unit cell.

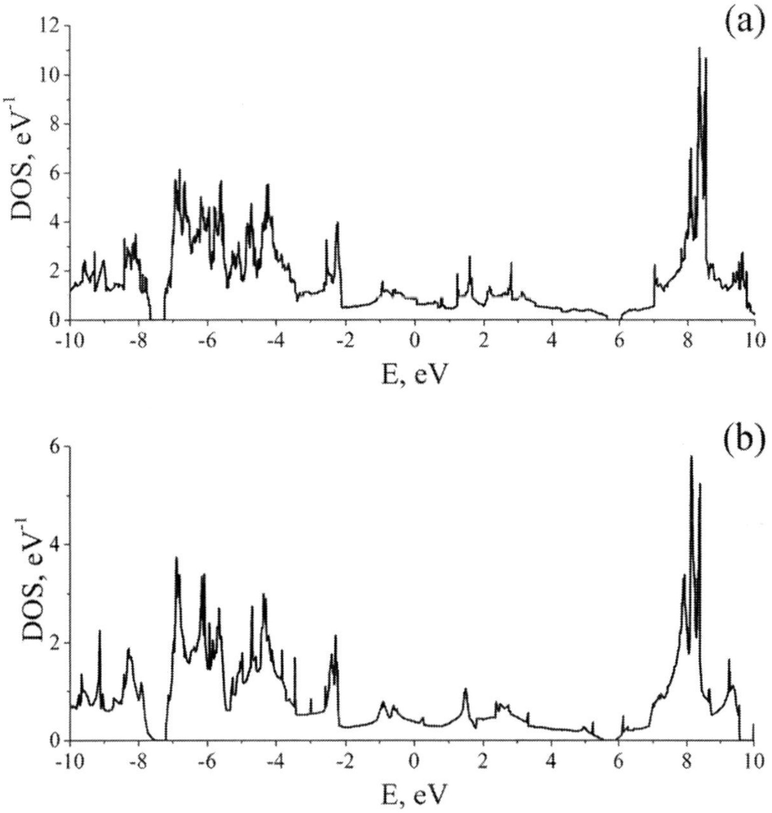

Figure 3. Density of electronic states of the L_{5-7} graphene: (a) orthorhombic unit cell (b) monoclinic unit cell.

4. Conclusions

Thus, as a result of calculations of the structure and electronic properties of the new polymorphic variety of 5-7 graphene, it was established that the sublimation energy of this structure turned out to be lower than that of hexagonal L_6 graphene, but higher than the sublimation energy of all other major polymorphic varieties of graphene. This indicates a high degree of stability of this structural variety and the possibility of its experimental synthesis. 5-7 graphene does not have a band gap at the Fermi energy level and must have metallic properties.

References

[1] Enyashin A N, Ivanovskii A L 2011 *Phys. Status Solidi B* **248** 1879
[2] Belenkov E A and Greshnyakov V A 2013 *New Carbon Mater.* **28** 273
[3] Belenkov E A and Kochengin A E 2015 *Phys. Solid State* **57** 2126
[4] Kochengin A E and Belenkov E A 2017 *Perspektivnye Materialy* **2** 5 (In Russian)
[5] Belenkov E A and Zinatulina Yu A 2005 *Bulletin of Chelyabinsk State University* No.25 32
[6] Koch W A, Holthausen M C 2001 *Chemist's Guide to Density Functional Theory. 2nd edition.* (Wiley-VCH Verlag GmbH)
[7] Perdew J P, Chevary J A, Vosko S H, Jackson K A, Pederson R A, Singh D J and Fiolhais C 1992 *Phys. Rev. B.* **46** 6671
[8] Giannozzi P et al 2009 *J. Phys.: Condens. Matter.* **21** 395502

Microstructure and mechanical properties of titanium alloy based composites fabricated in situ by casting and subjected to hot forging

V M Imayev, R A Gaisin, R M Imayev

Institute for Metals Superplasticity Problems, Russian Academy of Sciences
39 Khalturin st, Ufa 450001, Russia

E-mail: vimayev@mail.ru

Abstract. The microstructure and mechanical properties of discontinuously reinforced Ti/TiB and Ti/(TiB+TiC) based composite materials fabricated in situ by casting were studied. High temperature α+β and near-α titanium alloys were taken as matrix materials. The composites in as-cast conditions contained randomly oriented TiB whiskers or TiB whiskers and TiC particles. Different hot forging procedures followed by heat treatment providing most creep resistant matrix conditions were applied. The produced composite materials based on Ti/TiB demonstrated appreciably higher yield stress and creep resistance in comparison with those of the matrix alloy, while retaining quite reasonable ductility. For the composite materials based on Ti/TiB and Ti/TiB+TiC, the load-bearing capacity of the TiB whiskers mainly contributed to the enhancement of strength and creep resistance, whereas the role of TiC particles was found to be negligible. At the same time, the reinforcement with TiC particles strongly reduced the room temperature ductility.

1. Introduction

Titanium matrix composites (TMCs), discontinuously reinforced with ceramic whiskers/particles and fabricated in situ, have recently received considerable attention due to their superior high strength, elastic modulus, creep and fatigue resistances [1-12]. Among various ceramic reinforcements, TiB and TiC are the most suitable, because these compounds possess the most appropriate balance of chemical stability, high strength and elasticity modulus, thermal expansion coefficient similar to that of titanium and low solubility of boron and carbon in α and β titanium [1,2]. In contrast to TMCs, obtained by ex-situ methods, in situ techniques provide little contamination and low cost. In situ synthesis techniques include mechanical alloying, self-propagating high-temperature synthesis, methods based on conventional powder metallurgy, such as reaction hot pressing, spark plasma sintering and hot isostatic pressing [1-6]. An alternative way to fabricate in situ TMCs discontinuously reinforced with ceramic whiskers/particles is common casting [7-12]. For instance, using boron and graphite as additives to the titanium alloy matrix TiB whiskers and TiC particles are formed during casting. The casting route is one of the cost-effective ways to produce TMCs based on Ti/TiB and Ti/(TiB+TiC). In the casting route, mainly homogeneously distributed reinforcements can be obtained. The size of whiskers and particles in as-cast TMCs depends on the amounts of boron and carbon additives, whereas the shape (length/diameter ratio) and orientation of reinforcements can be varied by hot working of as-cast TMCs. The present work was aimed to study the effect of two- (2D) and three-directional (3D) hot forging in the upper part of the α+β phase field followed by heat treatment on the microstructure and mechanical properties of TMCs based on Ti/TiB and Ti/(TiB+TiC). 2D hot forging was applied to align TiB whiskers along the third direction and to retain a high aspect ratio of whiskers. 3D forging was applied to obtain TMCs with uniformly distributed refined reinforcements. Carbon was added to the Ti/TiB based composites, which demonstrated the most attractive mechanical properties.

2. Experimental

As initial components, α+β, near-α titanium alloys and boron and graphite powders were taken. The composite materials were prepared as 100-gram ingots using a laboratory arc-melting furnace in an argon atmosphere. For comparison, the matrix alloys were also prepared as 100-gram ingots using the same facility. The compositions of the materials under study are represented in table 1. The Ti-6.8Al-2.1Sn-2Zr-3.5Mo-0.8W-0.2Si and Ti-6.8Al-4Zr-2.5Sn-1Nb-0.7Mo-0.15Si alloys are known as Russian titanium alloys VT25U and VT18U, respectively, and for the sake of simplicity, these abbreviations have been used throughout the text. Thereafter, the composites are designated as VT25U/TiB, VT25U/(TiB+TiC), VT18U/TiB and VT18U/(TiB+TiC). The volume fractions of TiB whiskers and TiC particles are also shown in table 1.

Table 1. Chemical compositions of the materials under study.

Material abbreviation	Composition in wt.%									
	Ti	Al	Zr	Sn	Mo	W	Nb	Si	B	C
VT25U	Bal.	6.8	2	2.1	3.5	0.8	-	0.2	-	-
VT25U/8 vol.%TiB	Bal.	6.8	2	2.1	3.5	0.8	-	0.2	1.5	-
VT25U/(8 vol.%TiB+2 vol.%TiC)	Bal.	6.8	2	2.1	3.5	0.8	-	0.2	1.5	0.5
VT18U	Bal.	6.8	4	2.5	0.7	-	1	0.15	-	-
VT18U/6.5 vol.%TiB	Bal.	6.8	4	2.5	0.7	-	1	0.15	1.2	-
VT18U/(6.5 vol.%TiB+1.9 vol.%TiC)	Bal.	6.8	4	2.5	0.7	-	1	0.15	1.2	0.5

The materials were subjected to 2D or 3D hot forging in the upper part of the α+β phase field, followed by heat treatment. The post-forging heat treatments included high temperature anneals in the β or the upper part of the α+β temperature fields, followed by two-step annealing at lower temperatures. The post-forging high temperature anneals were carried out to reach tailored mechanical properties, including creep resistance that is unfeasible in as-forged conditions. Thus, the composite materials and the alloys under study were subjected to the same hot forging and nearly the same heat treatment, which allowed comparing the mechanical properties of the composite materials and the matrix alloys. Detailed information on processing of the materials under study can be found elsewhere [10-12].

A microstructural study was performed using scanning electron microscopy (SEM) in the secondary electron (SE) or backscattering electron (BSE) mode. Tensile tests were carried out at T=20-700 °C with an initial strain rate of $\dot{\varepsilon}=8.3\times10^{-4}$ s^{-1}. The ultimate tensile strength, σ_{UTS}, the yield stress, $\sigma_{0.2}$, and the elongation to failure, δ, were determined from the tests. Creep tests were performed at T=500-600 °C in air. All specimens for mechanical testing were prepared by electrospark cutting followed by fine grinding of work surfaces.

3. Results and discussion

Initial as-cast material

Figure 1 represents microstructural images of the matrix alloys and the composite materials reinforced with TiB and TiB+TiC in as-cast conditions. The microstructures of the base alloys are characterized by coarse primary β-grains and α/β colonies with a size of d_β~1000 and d_c~100 μm, respectively (figure 1 a, d). The presence of TiB whiskers and TiC particles led to the refinement of matrix microstructures (figure 1 b, c, e, f). XRD analysis confirmed the presence of the TiB and TiC phases in the composite materials along with titanium α and β phases [10-12]. Carbon was partially dissolved in the titanium matrix and being a strong α-stabilizer reduced the β phase content in VT25U/(TiB+TiC) as compared with that in VT25U and VT25U/TiB. The β phase content in the alloy and the VT25U/TiB composite was about 10 vol.%, whereas in VT25U/(TiB+TiC) the β phase content was reduced down to 5-6 vol. %. The same effect was observed in VT18U/(TiB+TiC). The β phase content in the VT18U alloy and the VT18U/TiB composite was found to be 3-4 vol. %, whereas in VT18U/(TiB+TiC) the β phase content was reduced down to 1-2 vol. %. In the case of VT25U/(TiB+TiC), the addition of carbon resulted in the formation of coarser borides in contrast to VT25U/TiB. The size of TiB whiskers was in the range of $L\times D$=(10-200)×(0.5-10) μm in VT25U/TiB and of $L\times D$=(10-400)×(0.5-20) μm in VT25U/(TiB+TiC), the size of TiC particles was d_p=1-10 μm. In the case of the VT18U based composites, the size of TiB whiskers was in the range of $L\times D$=(10-200)×(0.5-5) μm, the size of TiC particles was d_p=1-6 μm. The volume fractions of reinforcements in VT25U/(TiB+TiC) and VT18U/(TiB+TiC) were about 10 and 8.4 vol.%, respectively.

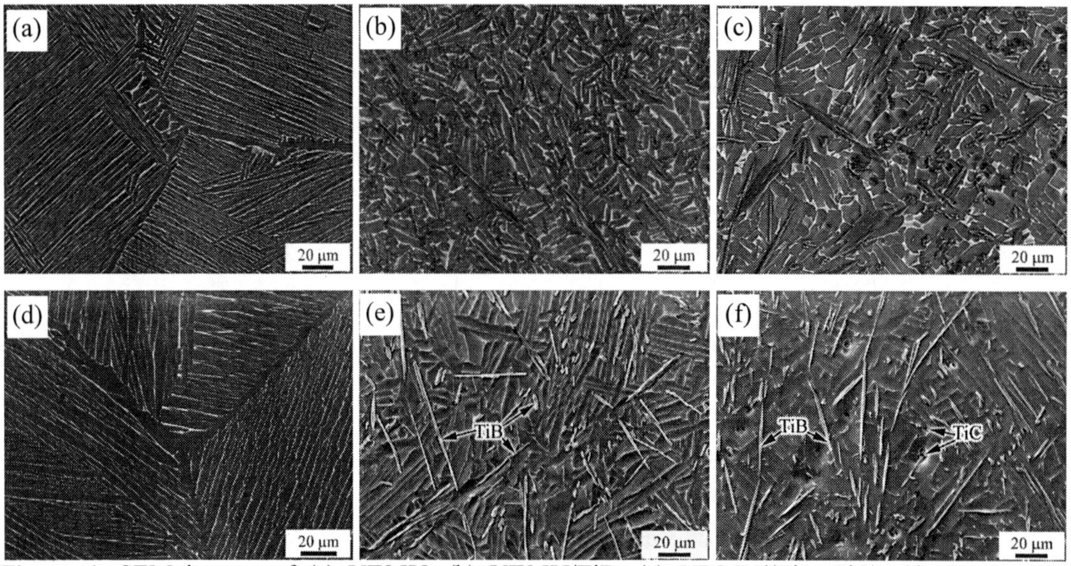

Figure 1. SEM images of (a) VT25U, (b) VT25U/TiB, (c) VT25U/(TiB+TiC), (d) VT18U, (e) VT18U/TiB, (f) VT18U/(TiB+TiC) in as-cast conditions.

Mechanical properties after hot forging and heat treatment
Hot forging led to alignment and a higher aspect ratio of the TiB whiskers after 2D forging and refinement and random orientation of the whiskers after 3D forging [10-12]. The post-forging heat treatment provided nearly the same matrix microstructures in VT25U based composites as compared to the VT25U alloy, as well as in VT18U based composites as compared to the VT18U alloy. As mentioned, this allowed comparing the mechanical properties of the composite materials and the matrix alloys. Figures 2 and 3 represent the tensile mechanical properties of the composite materials as compared to the base alloys. One can see that the presence of reinforcements provided a significant strengthening effect both at room and elevated temperatures. Table 2 shows the yield stress increment obtained in the composite materials as compared with the base alloys. One can see that this value varied in the range of 16.5-50%. The addition of carbon did not give appreciable strengthening, especially at elevated temperatures or even led to a slight softening of the composite material as in VT25U/(TiB+TiC). This can be ascribed to the fact that the addition of carbon being a strong α-stabilizer led to softening of the matrix due to reducing the β phase content and due to coarsening of the TiB whiskers as in the case of VT25U/(TiB+TiC).

IOP Conf. Series: Materials Science and Engineering **447** (2018) 012006 doi:10.1088/1757-899X/447/1/012006

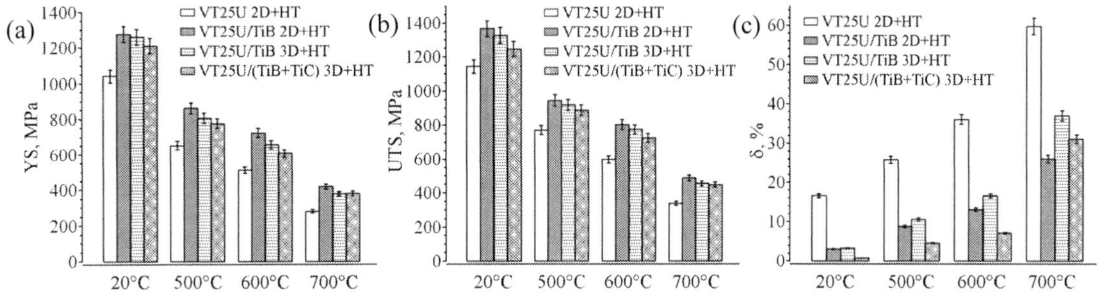

Figure 2. Temperature dependencies of (a) yield stress, $\sigma_{0.2}$, (b) ultimate tensile strength, σ_{UTS}, and (c) elongation to failure, δ, obtained for VT25U, VT25U/TiB and VT25U/(TiB+TiC) subjected to 2D and 3D forging followed by heat treatment (HT – heat treatment).

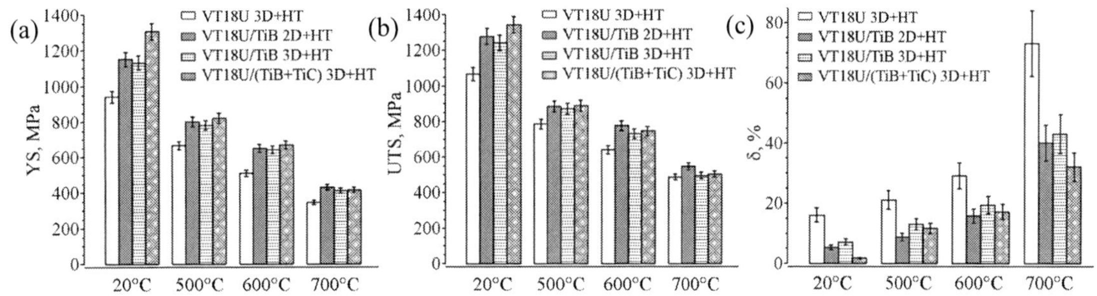

Fig. 3. Temperature dependencies of (a) yield stress, $\sigma_{0.2}$, (b) ultimate tensile strength, σ_{UTS}, and (c) elongation to failure, δ, obtained for VT18U, VT18U/TiB and VT18U/(TiB+TiC) after 2D and 3D hot forging followed by heat treatment (HT – heat treatment).

Table 2. The yield stress increment obtained in the VT25U and VT18U based composite materials as compared with the base alloys.

Type of the matrix titanium alloy	Material / Processing	$\Delta YS^a/YS_m^b \times 100$, %			
		20	500	600	700
α+β	VT25U/TiB / 2D forging + HT[c]	22	32	41	50
	VT25U/TiB / 3D forging + HT	20	23.5	28.5	36.5
	VT25U/(TiB+TiC) / 3D forging + HT	16.5	19	21	37.5
Near-α	VT18U/TiB / 2D forging + HT	22.3	20.3	28	24.6
	VT18U/TiB / 3D forging + HT	20.2	17.5	26.5	18.8
	VT18U/(TiB+TiC) / 3D forging + HT	39	23.3	31.7	20.3

[a]ΔYS – the increase in the yield stress due to the reinforcements; [b]YS_m – the yield stress of the matrix alloy, [c]HT – heat treatment.

The use of 2D forging gave a slight increase in the yield stress as compared to 3D forging. This is due to the higher load-bearing capacity of the TiB whiskers having a higher aspect ratio and is aligned along the tensile direction. However, the difference in the yield stress of the composites reinforced with TiB and subjected to 2D and 3D forging, followed by the same heat treatment, is small. This suggests that the yield stress in VT25U/TiB also depended on the spacing between the TiB whiskers. In other words, the mechanism of precipitation hardening, apparently, worked if the aspect ratio of the whiskers decreased. As is known, precipitation hardening is inversely proportional to the spacing between the precipitates. Breaking of the borides during 3D forging led to a decrease of the aspect ratio as compared to the material subjected to 2D forging. As mentioned, this reduced the spacings between the TiB whiskers and provided a more uniform distribution of reinforcement that most likely compensated the effect of reducing the aspect ratio and the random orientation of the TiB whiskers. The composite materials under study showed appreciably lower ductility than the base alloys both at room and elevated temperatures. However, in the case of the composite materials reinforced only with TiB, quite reasonable room temperature elongations (δ=5.2-7% in VT25U/TiB and 3-3.3% in VT18U/TiB) were attained.

Creep tests revealed that the composite materials demonstrated appreciably higher creep resistance as compared to that of the base alloys (figure 4). In the case of VT25U based composites, the highest creep resistance was obtained in VT25U/TiB after 2D forging, followed by heat treatment. 3D forging followed by heat treatment gave nearly the same creep resistance in VT25U/TiB and VT25U/(TiB+TiC). In the case of the VT18U based composites nearly the same creep resistance was obtained in VT18U/TiB and VT18U/(TiB+TiC) irrespective of the processing route. Thus, one can conclude that the presence of the TiB whiskers always gave a significant improvement in the creep resistance compared to that of the base alloys. The aligned TiB whiskers with a higher aspect ratio additionally slightly improved the creep resistance comparing with the refined and randomly oriented TiB whiskers. The presence of TiC particles resulted only in a negligible increase of creep resistance both in VT25U/(TiB+TiC) and VT18U/(TiB+TiC) as compared to VT25U/TiB and VT18U/TiB after the same processing route. Thus, reinforcement with TiB whiskers was quite effective from the viewpoint of creep resistance irrespective of the processing route, whereas the TiC particles exerted only a negligible positive influence on the creep resistance.

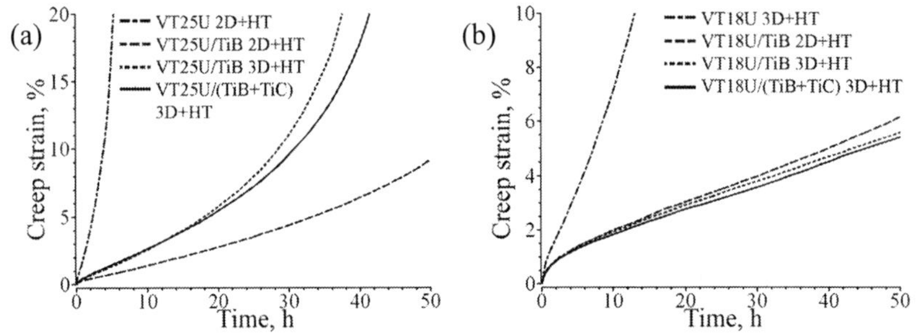

Figure 4. Creep curves obtained for (a) VT25U based composites and (b) VT18U based composites in comparison with the base alloys (HT – heat treatment). The test temperature and a loading were T=600 °C and P=300 MPa, respectively.

Fracture behavior

The investigation of the fracture behavior showed that the load-bearing capacity of the TiB whiskers was retained at elevated temperatures up to 700 °C and mainly contributed to the enhancement in strength and creep resistance (figure 5).

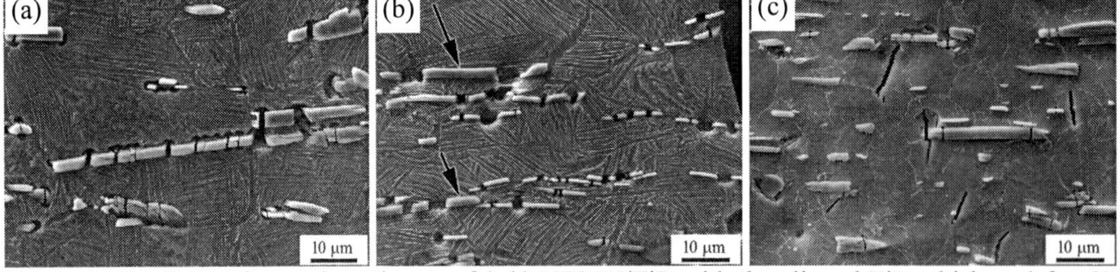

Figure 5. The flat surfaces of specimens of (a,b) VT25U/TiB with the aligned TiB whiskers (after 2D forging and heat treatment) and (c) VT18U/(TiB+TiC) tensile tested (a,c) at T=20 °C and (b) at T=600

°C. Arrows show unbroken boride fragments, along of which interfacial debonding is not observed. The pictures were obtained near the fracture zones, the tensile axis is horizontal.

The presence of TiC particles did not lead to any appreciable improvement in strength and creep resistance, but had a negative influence on the room temperature ductility, provoking the formation of brittle cracks (figure 5). Fracture surface observations (not presented here) showed that the main mechanism of failure of the composites at T=20-700 °C was the brittle fracture of the TiB whiskers and the TiC particles, followed by ductile failure of the matrix.

4. Conclusions

The composites based on Ti/TiB and Ti/(TiB+TiC) were prepared by casting using the α+β titanium alloy Ti-6.8Al-2.1Sn-2Zr-3.5Mo-0.8W-0.2Si and near-α titanium alloy Ti-6.8Al-4Zr-2.5Sn-1Nb-0.7Mo-0.15Si as matrix materials and boron and carbon additives. 2D and 3D forging, followed by heat treatment, was applied for the composites and the reference matrix alloys. The forged and heat treated composite materials demonstrated significantly higher strength and creep resistance, while retaining reasonable ductility as compared to the base alloys after the same processing history. The load-bearing capacity of the TiB whiskers contributed mainly to the enhancement of strength and creep resistance. The main mechanism of failure of the composites both at room and elevated temperatures was the brittle fracture of the TiB whiskers and the TiC particles, followed by ductile failure of the matrix.

Acknowledgements

The work was supported by the program of fundamental scientific researches of Governement Academy of Sciences No. AAAA-A17-117041310215-4. The work was performed using the facilities of the shared services center «Structural and Physical-Mechanical Studies of Materials» at the Institute for Metals Superplasticity Problems of Russian Academy of Sciences.

References

[1] Chandran K S R, Panda K B and Sahay S S 2004 *JOM* **56** 42
[2] Abkowitz S, Abkowitz S M, Fisher H and Schwartz P J 2004 *JOM* **56** 37
[3] Huang L J, Geng L and Peng H X 2015 *Prog. Mater. Sci.* **71** 93
[4] Wang B, Huang L J, Geng L and Yu Z S 2017 *J. Alloys Comp.* **690** 424
[5] Wang B, Huang L J and Geng L 2012 *Mater. Sci. Eng. A.* **558** 663
[6] Hu H T, Huang L J, Geng L, Sun J F and Tian H 2016 *J. Alloys Comp.* **688** 958
[7] Zhang C J, Kong F T, Hiao S L, Zhao E T, Xu L J and Chen Y Y 2012 *Mater. Sci. Eng. A* **548** 152
[8] Qu J, Zhang C, Zhang S, Han J, Chai L, Chen Z and Chen Y 2017 *Mater. Sci. Eng. A* **701** 16
[9] Ma F, Lu S, Liu P, Li W, Liu X, Chen X, Zhang K, Pan D, Lu W and Zhang D 2017 *J. Alloys Comp.* **695** 1515
[10] Gaisin R A, Imayev V M and Imayev R M 2017 *Lett. Mater.* **7(2)** 186 (in Russian)
[11] Gaisin R A, Imayev V M and Imayev R M 2017 *J. Alloys Comp.* **723** 385
[12] Imayev V M, Gaisin R A and Imayev R M 2018 *J. Alloys Comp.* **762** 555

IOP Conf. Series: Materials Science and Engineering **447** (2018) 012007 doi:10.1088/1757-899X/447/1/012007

Effect of high-speed dynamic channel angular pressing and aging on the microstructure and properties of Cu-Cr-Zr alloys

I V Khomskaya[1], V I Zel'dovich[1], N Yu Frolova[1], A E Kheifets[1], E V Shorokhov[2] and D N Abdullina[1]

[1]Miheev Institute of Metal Physics, Ural Branch, Russian Academy of Sciences, 18 S. Kovalevskaya st., Ekaterinburg 620108, Russia
[2] Russian Federal Nuclear Center–Zababakhin All-Russian Research Institute of Technical Physics, PO box 245, Snezhinsk 456770, Russia

E-mail: khomskaya@imp.uran.ru

Abstract. The microstructure, mechanical and functional properties, including electrical and wear resistances have been investigated for the Cu-0.09%Cr-0.08%Zr and Cu-0.14%Cr-0.04%Zr alloys subjected to high-speed dynamic channel angular pressing (DCAP) followed by annealing (aging). It was shown that DCAP led to the formation of a submicrocrystalline (SMC) structure (*d*=200–400 nm) with second-phase precipitates up to 5 nm in a size. The sequence of decomposition of a copper based α solid solution, accompanied by recrystallization and precipitation of second-phase nanoparticles during annealing (aging), was determined. DCAP and aging at 400-450 °C led to a decrease in the electrical resistance of the Cu-Cr-Zr alloys from 4.2 to 2.1 μΩ·cm and to a drastic increase in their strength properties (σ_u from 197 to 542 MPa, $\sigma_{0.2}$ from 94 to 464 MPa), while retaining reasonable ductility. It was revealed that the SMC structure processed by DCAP and annealing (aging) at 400 °C can be additionally strengthened after sliding friction tests due to the formation of a nanocrystalline structure in the surface layer.

1. Introduction

Copper alloys with small additions of chromium and zirconium belong to the group of precipitation-hardenable low-alloyed alloys characterized by a good combination of strength and electrical conductivity [1-3]. They are widely used in the electrical and atomic-energy industries. To obtain a submicrocrystalline (SMC) structure in copper alloys, methods of severe plastic deformation (SPD), such as high-pressure torsion, equal-channel angular pressing (ECAP), multidirectional forging have been used [1-5]. In this study, we used a new SPD method, namely, dynamic channel-angular pressing (DCAP) [6], which is a high-rate ($\sim 10^5$ s^{-1}) version of ECAP [5]. The advantage of the DCAP method consists in the use of the energy of gun-powder gases instead of energy-consuming pressing equipment and the short duration ($\sim 5 \cdot 10^{-4}$ s) of one pressing cycle. In the case of the DCAP, the deformation by simple shear is a high-strain-rate process. In addition, upon DCAP, high-speed shock-wave compressive strain takes place, which creates an additional source of deformation-induced hardening [7, 8]. The aim of the work was to study the changes in the microstructure and properties of Cu-Cr-Zr alloys processed by DCAP and the thermal stability of the produced SMC structures.

2. Experimental

The alloys Cu-0.09% Cr-0.08% Zr and Cu-0.14% Cr-0.04% Zr (wt. %) were selected for the study.

The alloys were melted from pure components. The as-cast rods were quenched from 1000 °C in icy water, and DCAP samples with a size of Ø16×65 mm were cut from the quenched material. The samples were accelerated in a gun to a velocity of 230 m/s and were directed to a die with channels with a diameter of 16 and 14 mm, which intersect each other at an angle of 90°. The strain rate was $\sim 10^5$ s^{-1}, the time for one pass was $5 \cdot 10^{-4}$ s, and the number of passes was n=4. The samples were analyzed in the as-quenched and DCAP states and after annealing (aging) at 300-700 °C for 1 hour. The TEM observations were performed using a JEM-200CX microscope. The electrical resistance, ρ, was measured by the potentiometric (compensation) method on samples of 0.3×2.0×15 mm, and the error did not exceed $\Delta\rho/\rho$=1%. Microhardness was determined at a load of 0.49 N. Uniaxial tensile tests were carried out at room temperature with an initial strain rate of 10^{-2} s^{-1}. Tribological tests by sliding friction were carried out without lubrication at a normal load of 196 N and a sliding velocity of 0.014 m/s, the number of cycles was 1000, and the sliding distance was 9 m. The friction force was continuously measured and registered. The friction coefficient f was determined as the ratio of the integral values of the friction force to the normal load.

3. Results and discussion

3.1. Microstructural changes after DCAP and annealing (aging)

The grain size of the alloys in the initial quenched state was in the range of d=200-400 μm. A fragmented structure consisting of subgrains 200-400 nm in a size with an internal dislocation structure was formed indicating occurrence of dynamic polygonization during DCAP. Second-phase precipitates with a size up to 5 nm are visible at the boundaries and in the interior of subgrains (figure 1 a). One can conclude that the coarse-grained structure of the alloys was transformed into a SMC structure due to high-rate processes of fragmentation, dynamic polygonization and strain aging.

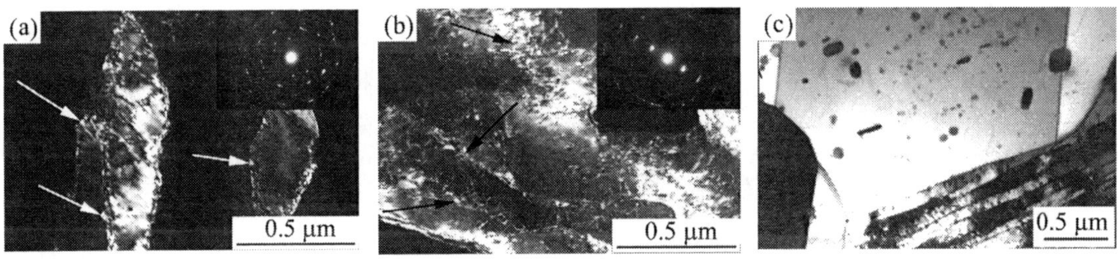

Figure. 1. TEM images of the Cu-0.14Cr-0.04Zr (a, b) and Cu-0.09Cr-0.08Zr (c) alloys after: DCAP (a), DCAP and aging at: 400 °C (b) and 600 °C (c). Note second-phase precipitates (arrowed) at boundaries and in the interior of subgrains.

The microstructure of the alloys obtained after DCAP remained almost the same after annealing at 300 °C. Annealing (aging) at 350-400 °C caused the decomposition of the supersaturated α solid solution of copper, which led to the precipitation of Cu_5Zr nanoparticles (5-10 nm) (figure 1 b). Cu_5Zr nanoparticles precipitated at dislocations (and subgrain boundaries) and promoted their pinning. This impeded the formation of recrystallization centers. An increase in the annealing (aging) temperature to 450-500 °C caused further decomposition of the supersaturated α solid solution and Cr-containing particles precipitated in the Cu-Cr-Zr alloys along with Cu_5Zr particles. The recrystallization of the alloys started at 500 °C and ended at 700 °C. After annealing (aging) at 600 °C, Cr-containing particles grew up to 100 nm in size and often had an elongated shape. Cu_5Zr globular precipitates, grouped in stripes, grew more slowly and gained around 10 nm (figure 1c). As a result, a disperse microstructure, which consisted of grains with a size of d=1-5 μm containing a large number of Cr-containing and Cu_5Zr nanoparticles was obtained in the Cu-Cr-Zr alloys. After annealing (aging) at 700 °C the grain size did not exceed 5-10 μm [9].

3.2. Mechanical properties after DCAP and annealing (aging)

The microstructural changes affected the properties of the alloys. The microhardness of the investigated alloys in the initial coarse-grained state was 700 MPa (figure 2). After DCAP, the microhardness increased to 1600 MPa (in 2.3 times) as compared to the initial values. This is associated with a substantial microstructure refinement (by three orders of magnitude) and the strain aging of the alloys during DCAP. Annealing of the deformed alloys at 300 °C did not change their microhardness. After annealing (aging) at 400-450 °C, the microhardness increased up to 1,800 MPa in contrast to the deformed state (figure 2). This behavior is related to the decomposition of the supersaturated α solid solution of copper [1-3]. An increase in the annealing (aging) temperature to 500-600 °C led to a decrease in the microhardness, which is explained by the development of recrystallization. Annealing (aging) at 650-700 °C did not cause further softening of the investigated alloys because no appreciable grain growth occured.

As expected, the high-speed DCAP led to a significant increase (in 2.6-3.3 times) in the strength properties of the Cu-0.14Cr-0.04Zr alloy (figure 3). It should be noted that the strength properties of copper after the same treatment increased only in 1.2-1.4 times [8]. The strength properties of the deformed alloy increased even more significantly (in 2.8-5.1 times) after annealing (aging) at 400 °C (figure 3). Thus, using the combined treatment, DCAP (n=4 passes) and aging at 400 °C, we can fabricate an alloy with superior microhardness (1800 MPa) and strength properties (σ_u=542 MPa, $\sigma_{0.2}$=464 MPa) while retaining reasonable ductility (δ=11%). Note that ECAP and n=8-12 passes were required to achieve near the same level of strength properties in the Cu-0.44%Cr-0.2%Zr alloy [1].

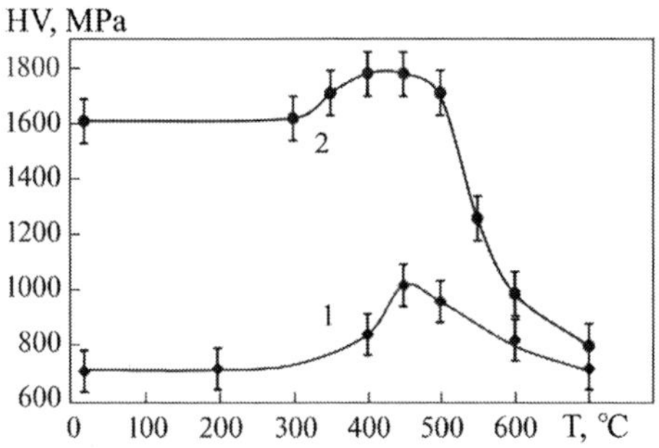

Figure 2. Dependences of the microhardness of the Cu-0.14Cr-0.04Zr alloy samples on the annealing (aging) temperature after (1) quenching, (2) DCAP

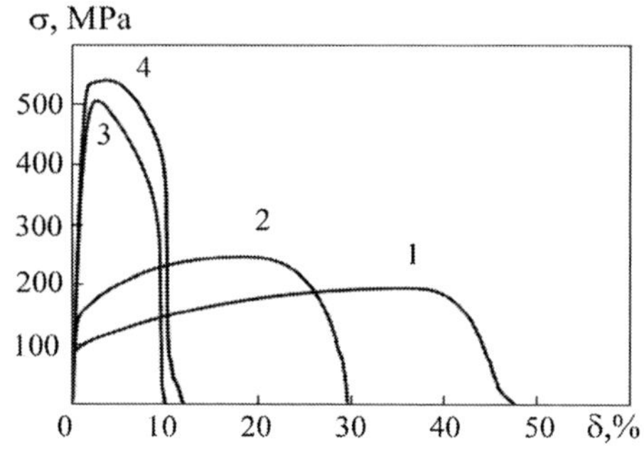

Figure 3. Stress-strain curves obtained as a result of tensile tests of the Cu-0.14Cr-0.04Zr alloy samples after: (1) quenching, (2) quenching and annealing (aging) at 450 °C, (3) DCAP, (4) DCAP and annealing (aging) at 400 °C

3.3. Electrical resistance and wear resistance after DCAP and aging

The electrical resistance (ρ) of the Cu-0.14Cr-0.04Zr alloy in the quenched state was $\rho \approx 4.3$ μΩ·cm (figure 4). The DCAP resulted in a decrease of the ρ value as compared to that of the quenched state (figure 4). This is associated with a partial decomposition of the supersaturated α solid solution of copper after DCAP, which is confirmed by the microstructural observations and the measured properties (figures 1 b, 2 and 3). The ρ value of the quenched alloy decreased significantly with increasing the annealing (ageing) temperature in the range of 350-500 °C (figure 4), which was caused by the decomposition of the α solid solution of copper. The ρ value of the deformed alloy began to decrease in the range of 200-300 °C, which is related to the recovery processes, and decreased significantly in the range of 350-450 °C (figure 4). The latter can be ascribed to the extensive precipitation of the second-phase particles. It should be noted that the decrease in the ρ value after DCAP took place at a lower temperature than after quenching, since the decomposition of the solid solution after DCAP was more extensive than after quenching. Note that the increase in microhardness started at higher temperatures than the decrease in the ρ value (figures 2 and 4). This finding suggests that the depletion of the solid solution by the alloying elements occurred earlier than the precipitates were formed, which led to increasing the microhardness. It is interesting that the ρ value of the alloy after DCAP and aging at 500 °C decreased to 2.1 μΩ·cm. This should be ascribed to the almost complete decomposition of the α solid solution. Note that the ρ value of the alloy after quenching and aging at 500 °C was 2.5 μΩ·cm, which indicated an incomplete decomposition of the α solid solution.

The samples of the Cu-0.09Cr-0.08Zr alloy were subjected to sliding friction tests after different treatments. In the quenched state, the wear rate (*Ih*) and the coefficient of friction (*f*) of the samples were $3.1\cdot10^{-7}$ and 0.5, respectively. After DCAP, the *Ih* value decreased to $2.3\cdot10^{-7}$, and the *f* value increased to 0.62 (figure 5). The microhardness of the deformed by DCAP alloy after testing for sliding friction increased from 1600 to 3200 MPa. As was earlier shown [9], this was the result of the formation of a nanocrystalline microstructure with a grain size of *d*=40-50 nm in the surface layer of the tested samples.

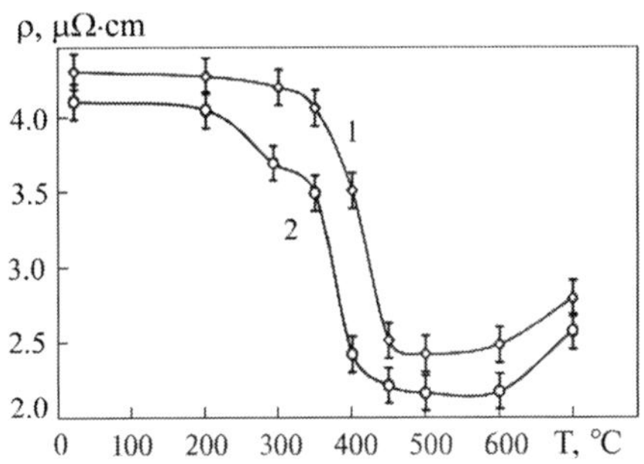

Figure. 4. Dependences of electrical resistance on the aging temperature obtained for samples of the Cu-0.14Cr-0.04Zr alloy after (1) quenching and (2) DCAP

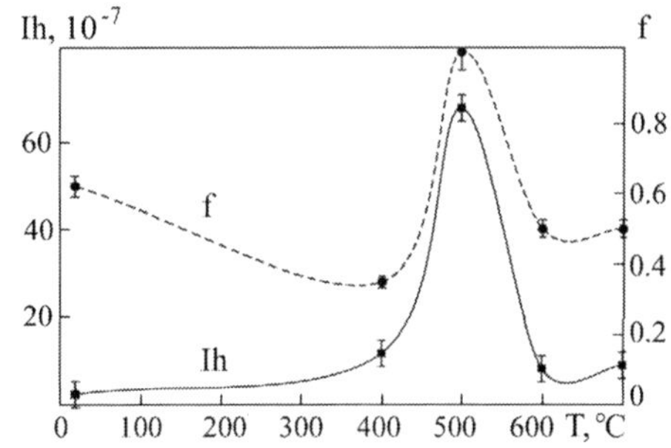

Figure 5. Dependences of the wear rate, *Ih,* and friction coefficient, *f,* on the aging temperature after DCAP obtained for samples of the Cu-0.09Cr-0.08Zr alloy

Annealing (aging) at 400 °C led to an insignificant increase in the *Ih* value as compared to the DCAP state. When this happens, the *f* value decreased and reached its minimum value of 0.35, the microhardness increased to 3350 MPa. This was the result of microstructural refinement down to 15-30 nm [8] and decomposition of a supersaturated α solid solution of copper followed by precipitation of particles of the second-phase. The elevation of the aging temperature to 500 °C resulted in a significant growth (in near 30 times) of the *Ih* value and an increase of the *f* value from 0.35 to 1.0. At higher aging temperatures, 600 and 700 °C, the *Ih* and f values returned to the values close to those obtained after DCAP and DCAP and annealing (aging) at 400 °C (figure 5).

4. Summary

In the present work, two Cu-Cr-Zr alloys of near the same composition have been studied. It was shown that DCAP and annealing (aging) of the Cu-Cr-Zr alloys under study led to a refinement of the microstructure and to the decomposition of the α solid solution followed by precipitation of nanosized particles. This resulted in a significant improvement of their mechanical and functional properties. DCAP and aging at 400 °C of the alloys provided an increase in the ultimate strength in 2.6-3.3 times and the yield stress in 2.8-5.1 times as compared to those in the initial coarse-grained state, while retaining reasonable ductility.

DCAP and aging at 400-500 °C led to a decrease in the electrical resistance of the Cu-Cr-Zr alloys to 2.1 μΩ·cm. Tests for sliding friction of samples subjected to DCAP and aging at 400, 600, and 700 °C gave rise to the formation of the nanocrystalline microstructure in the surface layer. That provided a high level of the microhardness and acceptable tribological properties.

IOP Conf. Series: Materials Science and Engineering **447** (2018) 012007 doi:10.1088/1757-899X/447/1/012007

Acknowledgments
The work was supported within a State Assignment (theme «Structure»AAAA-A18-118020190116-6).

References
[1] Vinogradov A, Patlan V, Suzuki Y, Kitagawa K and Kopylov V I 2002 *Acta Mater.* **50** 1639
[2] Purcek G, Yanar H, Shangina D V, Demirtas M, Bochvar N R and Dobatkin S V 2018 *J. Alloys Comp.* **742** 325
[3] Zhilyaev A P, Morozova A, Cabrera J M, Kaibyshev R and Langdon T G 2017 *J. Mater. Sci.* **52** 305
[4] Mulyukov R R, Imayev R M and Nazarov A A 2008 *J. Mater. Sci.* **43** 7257
[5] Valiev R Z and Aleksandrov I V 2007 *Bulk Nanostructural Metal Materials: Production, Structure and Properties* (Moscow: Akademkniga)
[6] Shorokhov E V, Zhgilev I N and Valiev R Z 2006 Russian Federation Patent No. 2283717
[7] Zel'dovich V, Shorokhov E, Frolova N, Zhgiliev I, Kheifetz A and Khomskaya I 2009 *Int. J. Mater. Res.* **100** 830
[8] Khomskaya I V, Shorokhov E V, Zel'dovich V I, Kheifets A E, Frolova N Yu, Nasonov P A, Ushakov A A and Zhgilev I N 2011 *Phys. Metals Metallogr.* **111** 612
[9] Kheifets A E, Khomskaya I V, Korshunov L G, Zel'dovich V I and Frolova N Yu 2018 *Phys. Metals Metallogr.* **119** 420

Effect of aging on structure and strength of high pressure torsion processed 2024 aluminum alloy

M V Markushev, E V Avtokratova, R R Ilyasov, S V Krymskiy, A A Khazgalieva and O Sh Sitdikov

Institute for Metals Superplasticity Problems, Russian Academy of Sciences, 39 Khalturin st., Ufa 450001, Russia.

E-mail: mvmark@imsp.ru

Abstract. Pre-quenched samples with a diameter of 20 mm and a thickness of 1 mm, cut from commercial hot-pressed rod of the 2024 aluminum alloy, were severely deformed by high-pressure torsion (HPT) with 10 revolutions under a pressure of 6 GPa at room temperature and then annealed for up to 50 hours in the temperature range of 100-190 °C. It was found that the post-deformation annealing at relatively low temperatures and/or short times can result in the alloy aging and 10-15% increased hardness. However, the conventional regimes of artificial aging of the alloy according to T6 heat treatment led to a sharp drop in its strength due to recovery and recrystallization of the matrix and simultaneous overaging. The nature of the revealed structural and property relations is discussed.

1. Introduction

Data published over the last decades have shown a high potential for enhancement the service properties of conventional metallic materials due to novel methods of their structure/phase control through nanostructuring (NS) - imparting the size less than 100 nm to the main structural components. Owing to the comparative simplicity and applicability for many metals and alloys, NS based on severe plastic deformation (SPD) to effective strains of e>1 [1] is of particular interest. To date, a number of SPD techniques have been developed, a wide range of nanomaterials have been processed and several industrial applications have been claimed. However, despite the huge number of studies, there are many unexplored points, starting with routes of cost affordable NS technologies. In addition, the mechanisms and factors responsible for the effective formation of nanostructures and improved service properties are still not so clear. As for wrought aluminum alloys, these points are narrowed to the effect of precipitates, which, judging by literature data, is rather ambiguous. For instance, in [2,3] it was established that the formation and/or increase in the volume fraction of disperse phases before SPD led to a greater proportion of nanograins of less size in the deformed structure. In contrast, in [4,5] it was shown that the formation of high densities of nanosized precipitates in high-strength alloys can completely suppress recrystallization of the matrix under SPD.

One of the topical issues for the age-hardenable Al alloys is the effectiveness of post-SPD aging. Along with the deformation strengthening owing to formation of new grains and subgrains, and an increase in the density of dislocations, the mechanical behavior of the SPD-ed alloy with further aging will be determined by changes in the impacts of the disperse hardening and aluminum solid solution strengthening. Though the contribution of the former to the strength of SPD materials is comparatively small, alloying elements can provide noticeable post-deformation disperse hardening of the matrix with formation of nanosized precipitates. However, their nucleation and growth in severely deformed matrix will be more heterogeneous and intense than in a non-deformed one, and may occur in a non-conventional sequence of phase transformations [6]. This behavior is commonly explained by the increased fraction of new grain boundaries and triple junctions, the high defectiveness of structure and

IOP Conf. Series: Materials Science and Engineering **447** (2018) 012008 doi:10.1088/1757-899X/447/1/012008

the level of stored energy, as well as the intensity of recovery and recrystallization [7-11]. Therefore, the development of the optimal aging regimes for SPD materials is a rather important task, which requires an understanding of the nature of structure transformations, ensuring hardening or softening.

The aim of the study is to analyze the influence of post-SPD aging on the structure and strength of a medium strength age-hardenable aluminum alloy subjected to cold high pressure torsion (HPT) in a pre-quenched condition.

2. Material and procedure

A commercial hot-pressed rod of the 2024 alloy of standard chemical composition (Al-4.4Cu-1.4Mg-0.7Mn, wt.%) was used as the research material. Disk-shaped samples with a diameter of a 20 mm and a thickness of 2 mm, cut out from the rod across its axis, were solution treated and quenched in water from a temperature of 505 °C. Further SPD was implemented immediately through the HPT technique with 10 revolutions under a pressure of 6 GPa at room temperature. The final artificial aging of the SPD-processed alloy was carried out in the temperature range of 100-190 °C with an exposure time up to 50 hours.

The structure of the alloy was analyzed using optical metallography (OM) and transmission electron microscopy (TEM). The objects for analysis were prepared by standard procedures with electropolishing at -28°C in 20% solution of HNO_3 in CH_3OH. Microhardness was determined at a 15-second load of 1N. The hardness and structure of the alloy were examined in the middle of the radius of the SPD-ed disk.

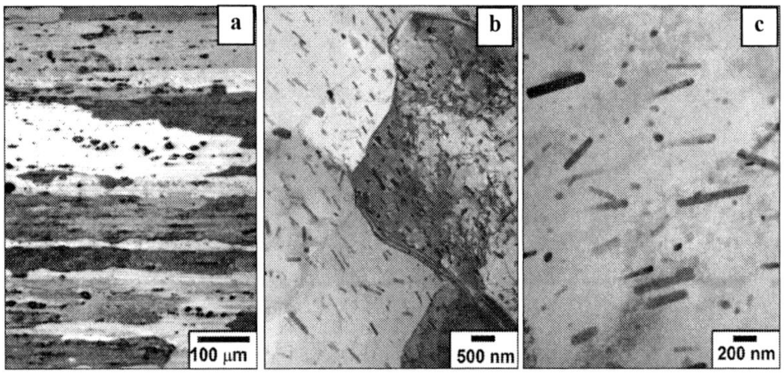

Figure 1. Structure of the pre-quenched 2024 alloy rod. a – OM, b and c – TEM.

3. Results and discussions

The initial pre-quenched alloy had a predominantly coarse-fibered structure with a fiber thickness of ~100-200 μm (figure 1). The grain and subgrain boundaries were decorated with coarse particles of excess and impurity phases (figure 1a), and their interiors contained elongated in the pressing direction precipitates of secondary T-phase ($Al_{20}Cu_2Mn_3$) with the sizes of 330×70 nm and densities about 3-4×10^2 μm^{-3} (figures 1b, c). With the subsequent HPT the structure of the matrix – aluminum solid solution supersaturated with Cu and Mg, transformed into a non-equilibrium nano-fragmented one with a fragment / (sub)grain size of about 75 nm (figures 2a, e). Judging by the TEM analysis, the deformation structure of the alloy was not so homogeneous: in particular, it was easy to find out the band components, indicating the development of a microshear banding under HPT. Such heterogeneity was virtually caused by the non-uniform distribution of primary and secondary phases, involving T-phase precipitates, and by their interaction with dislocations. Thus, frequent observations of these precipitates at non-equilibrium boundaries of fragments mean their active participation in the dynamic recrystallization of the matrix [7].

As a result of HPT, the alloy was considerably strengthened and its hardness increased from ~130 up to 270 Hv. Providing the almost maximum hardness, recorded for HPT aluminum alloys [9], was caused by deformation strengthening, resulting from nanostructural hardening and maintaining the solid solution strengthening. Subsequent annealing at 100°C led to a further increase in the hardness of

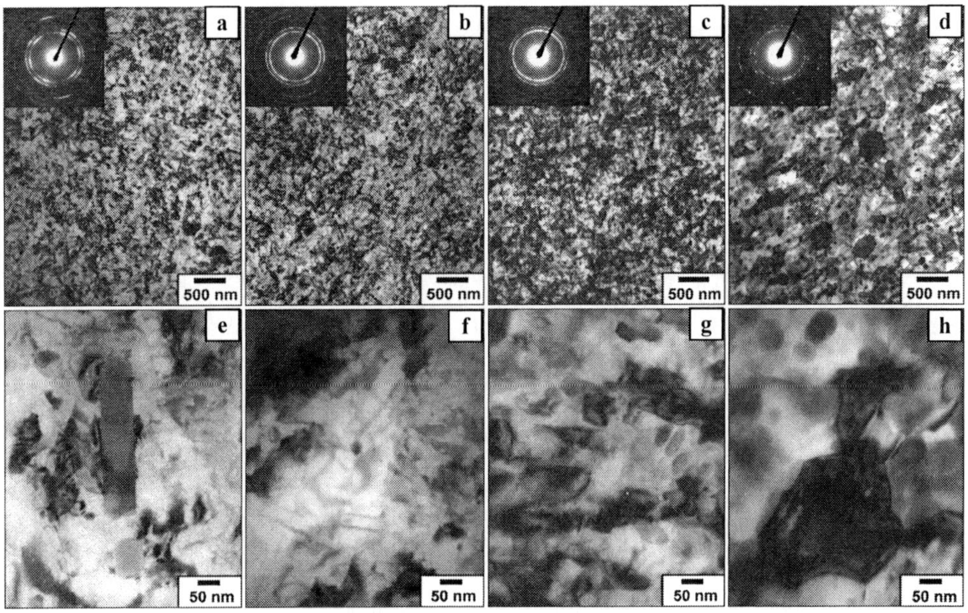

Figure 2. TEM structures of the HPT processed (a, e) and post-HPT aged for 48 hours at 100 (b, f), 140 (c, g) and 190 °C (d, h) alloy.

the alloy, and its gain proceeded up to 50 hours (figure 3). This fact testified that the additional post-deformation hardening of the alloy could be caused by the effect of disperse strengthening (artificial aging) only. The signs of the latter in the form of lamellar precipitates of the main strengthening phase(s) (S and/or Θ) were hardly distinguished and found only in separate fragments (figure 2f) after transformation of their structure into a more equilibrium one due to recovery. The morphology of these phases had no obvious differences from those commonly observed, except for their extremely small length, limited by the size of a fragment. Thus, it is possible to claim that during aging of the alloy at 100 °C, along with the decomposition of aluminum solid solution and the predominant formation of Guinier-Preston-Bagaryatsky zones, the structure of the fragments underwent static recovery. Therewith, losses in the strength of the alloy due to elimination of crystal defects, as well as from a decrease in the level of alloying of aluminum solid solution were compensated and overcame by disperse strengthening with an even more than 10% increase in alloy. First of all, two maxima on the dependence of hardness on annealing time have been observed (figure 3). The first one was probably conditioned by hardening at the above mentioned stage of zone aging, and the second one – at the stage of phase aging. It should be noted that the latter was found at much lower temperatures and shorter times than those under conventional T6 route (artificial aging to maximum strength is usually realized in the pre-quenched alloy under annealing at T = 185-195 °C during 11-13 hours). This finding was not surprising, since the decomposition of an aluminum solid solution should occur more rapidly in a severely deformed alloy, which suggested a shift of the maximum on the kinetic hardness curves to lower temperatures and times. By the way, a similar result was obtained for highly work hardened conventional sheets, showing the maximum of hardness after 7-9 hours of aging at conventional temperatures.

At the same time, these data testify to a low thermal stability of the obtained nanostructure and a lowered interval of working temperatures for HPT-processed nanostructured alloy. Judging by the TEM data (figures 2c, g), static recovery and continuous static recrystallization were the main reasons for the softening behavior of the alloy during annealing at 140 °C. Recrystallization occurred through in-situ mechanism and was followed by intense migration of fragment/grain boundaries. The latter resulted in simultaneous changes in the morphology of the secondary phases, giving them a compact shape (figure 2g) due to the early violation of their coherency by continuous changes in the lattice of the surrounding matrix.

The overaging stage and intensive decrease in the hardness of the alloy due to the loss of disperse strengthening were observed at the annealing times of beyond 10 hours).

During annealing at 190 °C the zone stage of aging was not fixed and a short phase hardening was followed by a strong hardness drop due to the activation of both discontinuous and continuous static recrystallization and strengthening phase coagulation (figure 2). As a result, a bimodal recrystallized structure was formed with spherical precipitates rather evenly distributed (figure 3). Though a considerably decreased strength was observed, the alloy did not lose its nanostructured nature, since the sizes of all phase components, including statically recrystallized grains, still retained in the nano range.

Besides, it should be noted that the detected behavior of the alloy was virtually caused by processes at the level of interaction of mobile dislocations with clusters containing atoms of Cu and Mg – fragments of zones and metastable phases formed during processing. Such a point of view was formulated and proved in [12], and the results of this study act in its support.

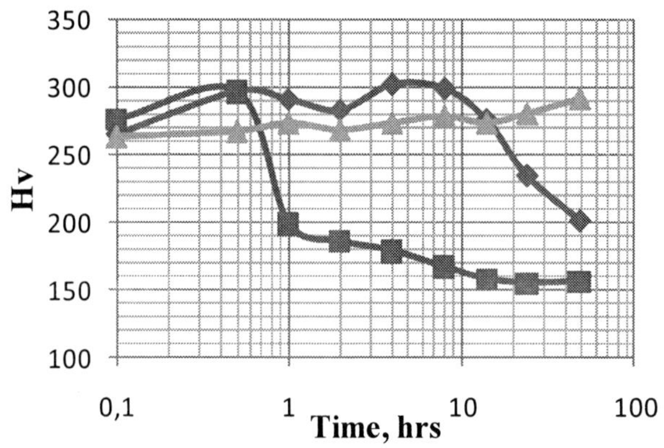

Figure 3. HPT alloy hardness vs time of aging at 100 (▲), 140 (♦) and 190 °C (■).

4. Conclusions

Few (nano)structured states of the 2024 aluminum alloy, having a hardness in the range of 150-300 Hv, were processed by thermomechanical treatment using HPT and artificial aging. It is shown that the ambient temperature strength of the nanostructured alloy is determined not only by grain boundary strengthening, as is traditionally believed, but also strongly depends on the dislocation density, nature, morphology and distribution of the second phases. The data obtained testify to low thermal stability of the HPT-processed nanostructured alloy and to decreased interval of its working temperatures. The main reasons for this behavior of the alloy are intensive recovery and recrystallization of the matrix with the simultaneous effect of overaging during the post-SPD annealing.

Acknowledgments

The work was performed under the grant No. 16-19-10152 of Russian Science Foundation on the basis of Collective center of IPSM RAS "Structural and physics-mechanical analyses of materials".

References

[1] Valiev R Z, Estrin Y, Horita Z, Langdon T G, Zehetbauer M J and Zhu Y 2016 *JOM* **68** 1216
[2] Rabinovich M Kh, Markushev M V and Murashkin M Yu 1997 *Mater. Sci. Forum* **243** 591
[3] Barlow C Y, Hansen N and Liu Y L 2002 *Acta Mater.* **50** 171
[4] Krymskiy S V, Sitdikov O Sh and Markushev M V 2016 *AIP Conf. Proc.* **1785** 040030
[5] Markushev M V, Burdastykh Yu L, Krymskiy S V and Sitdikov O Sh 2017 *Lett. Mater.* **7** 101

[6] Sha G, Wang Y B, Liao X Z, Duan Z C, Ringer S P and Langdon T G 2009 *Acta Mater.* **57** 3123
[7] Humphreys F J, Hatherly M 2004 *Recrystallization and Related Annealing Phenomena* (Oxford: Elsevier) p 658
[8] Krymskiy S V, Sitdikov O Sh, Avtokratova E V, Murashkin M Yu and Markushev M V 2012 *Rev. Adv. Mater. Sci.* **31** 145
[9] Markushev M V, Avtokratova E V and Sitdikov O Sh 2017 *Lett. Mater.* **7** 465
[10] Markushev M V 2009 *Phys. Metals Metallogr.* **108** 161
[11] Markushev M V, Avtokratova E V, Krymskiy S V and Sitdikov O Sh 2018 *J. Alloys Comp.* **743** 773
[12] Chen Y, Gao N, Sha G, Ringer S P and Starink M J 2015 *Mater. Sci. Eng. A* **627** 10

Analysis of the distribution of alloying elements in ultrafine-grained steel 05G2MFBT by atom-probe tomography

S N Sergeev [1], I M Safarov [1], R M Galeyev [1], S V Gladkovsky [2], D I Dvoynikov [2], S V Rogozhkin [3,4], A A Iskandarov [3], G F Korznikova [1]

[1]Institute for Metals Superplasticity Problems of Russian Academy of Sciences, 39 Khalturin st., Ufa 450001, Russia
[2]Institute of Engineering Science of the Urals Branch of RAS, 34 Komsomolskaya st., Yekaterinburg 620049, Russia
[3]A.I. Alikhanov Institute of Theoretical and Experimental Physics of National Research Centre Kurchatov Institute, 25 Cheremushkinskaya st., Moscow 117218, Russia
[4]National Research Nuclear University MEPhI (Moscow Engineering Physics Institute), 31 Kashirskoye Shosse, Moscow 115409, Russia

E-mail: nicocem17@gmail.com

Abstract. It is shown that in low carbon steel, the structure after the controlled rolling comprises grains of ferrite and carbides. The average grain size of ferrite was 2.5 μm, the carbide size varied from 1 to 5 μm. Additional warm rolling led to the formation of a fibrous ultrafine-grained (UFG) structure with an average grain/subgrain size about 0.3 μm, the carbide size did not exceed 100 nm. The mechanical tests carried out under the uniaxial tension scheme at room temperature showed that the ultimate strength after the controlled rolling is 600-700 MPa, with a relative elongation of 21% while the impact strength at room temperature is 3.15 MJ/m^2. In the state after warm rolling, the ultimate strength increased to 1000-1050 MPa, the elongation decreased from 21% to 17%, the impact strength at room temperature increased from 3.15 to 3.38 MJ/m^2. One of the important factors leading to an increase in the mechanical properties of alloys is solid-solution hardening. To confirm this, a tomographic atomic-probe analysis of steel samples in both structural states was carried out, which made it possible to detect an increase in the concentration of carbon in steel ferrite by a factor of two in the state after warm rolling, compared with the state after the controlled rolling.

1. Introduction

In materials subjected to deformation to high degrees, significant structural changes occur, accompanied not only by a sharp increase in the level of mechanical properties [1, 2], but, in some cases, by the manifestation of anomalies in the change of properties. In [3], for example, an anomalous temperature dependence of the toughness in rolled low-carbon steel was found. To understand the nature of such effects, traditional techniques for structure analysis, such as optical metallography, X-ray diffraction analysis, and electron microscope studies, are sometimes not sufficient and it is necessary to handle new high-resolution methods of analysis. One of such progressive techniques is the method of atom-probe tomography, which allows to carry out chemical analysis and evaluate local inhomogeneities at the atomic level.

In the present work, the microstructure and mechanical properties of pipe steel 05G2MFBT were studied after the controlled rolling with accelerated cooling and after additional warm rolling.

2. Experimental

Material of the study was low-carbon steel 05G2MFBT (Fe-1.61% Mn-0.03% Cu-0.192% Mo-0.26% Si-0.236% Cr-0.025% V-0.22% Ni-0.054% C-0.059% Nb-0.009% P-0.0012% S) in two different structural states. The first one was obtained by the controlled rolling at an accelerated cooling rate, and the second one was subjected additionally to plastic deformation by means of intensive warm rolling. Rolling was carried out at a temperature of 550 °C by several passes to the final cross-section of 10 mm x 10 mm rods with a relative narrowing of e = 10-15% in each pass. The accumulated degree of deformation was equal to e = 2.7.

Microstructure analysis was conducted using a Tescan Mira 3LMH scanning electron microscope (SEM) with a back scattered electron (BSE) detector.

Tensile tests of flat samples were carried out on an "INSTRON-1185" universal dynamometer at room temperature with an initial strain rate of 10^{-3} s^{-1}.

Impact tests of standard size specimens with a V-shaped notch according to GOST 9454 were carried out on an instrumented pendulum tester "Tinius Olsen IT542M" in the temperature range from +20 to -80 °C. Tomographic studies were performed on an atomic-probe tomography (APT) with laser evaporation (PAZL-3D), using the equipment of the CAMIX Center for Collective Use (http://kamiks.itep.ru/) of the Scientific Research Center "Kurchatov Institute" developed at ITEP [4].

3. Results and discussion

Structural studies of steel after the controlled rolling revealed a grained structure, the average grain size of ferrite was 2.5 μm, the carbide size varied from 1 to 5 μm (figure 1). Warm rolling led to the formation of a fibrous ultrafine-grained (UFG) structure with an average grain/subgrain size of 0.3 μm, while the carbide size did not exceed 100 nm (figures 2,3).

So, the toughness increased from 3.15 MJ/m^2 to 3.38 MJ/m^2. Mechanical tests carried out under the uniaxial tension scheme at room temperature showed that the ultimate strength after the controlled rolling is 600-700 MPa, with a relative elongation of 21%. After warm rolling, the strength properties increased to 1000-1050 MPa, while elongation decreased from 21% to 17%. The results of mechanical tests are shown in table 1.

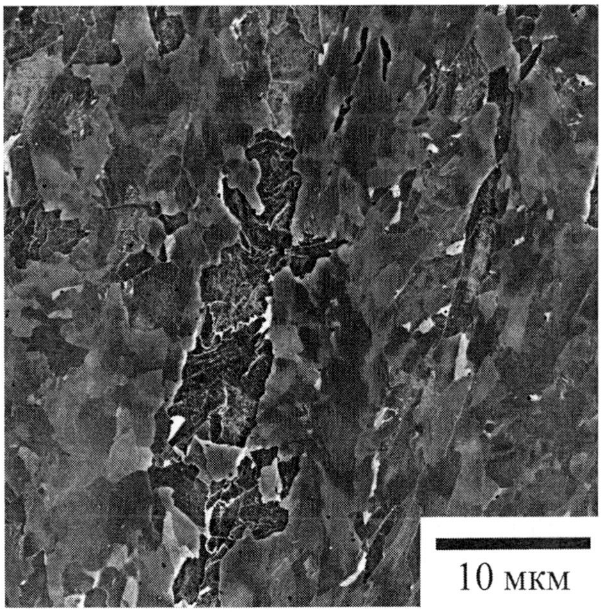

Figure 1. Microstructure of steel 05G2MFBT subjected to the controlled rolling with accelerated cooling

Figure 2. Microstructure of steel 05G2MFBT subjected to warm rolling in transverse section.

Figure 3. Microstructure of steel 05G2MFBT subjected to warm rolling in longitudinal section.

Table 1. Mechanical characteristics of steel 05G2MFBT.

Treatment conditions	$\sigma_{0.2}$, MPa	σ_u, MPa	δ, %	KCV, MJ/m^2
Controlled rolling	612	685	21	3.15
Warm rolled	1017	1080	17	3.38

Tomographic analysis of low-carbon steel after two different deformation-thermal treatments showed that after the controlled rolling, the distribution of alloying elements in the ferrite matrix is uniform, while warm rolling resulted in carbides grinding, and an increase in the concentration of carbon by two times as compared to the state after the controlled rolling. There is also an increase in the concentration of manganese and nickel at the interface between the matrix of iron and carbides, while the distribution of chromium remains homogeneous (figure 4). It is important to note that if carbon and nickel in the particle are uniformly distributed, then the concentration of manganese increases twice exactly on the boundary by substituting iron, wherein in the center of the particle the content of manganese corresponds to its content in the matrix. The revealed differences in the supersaturation of the solid solution and the increase in the concentration of the alloying elements at the carbide boundary contribute to an increase in the strength properties and impact strength of steel 05G2MFBT after warm rolling compared to the state after the controlled rolling followed by accelerated cooling.

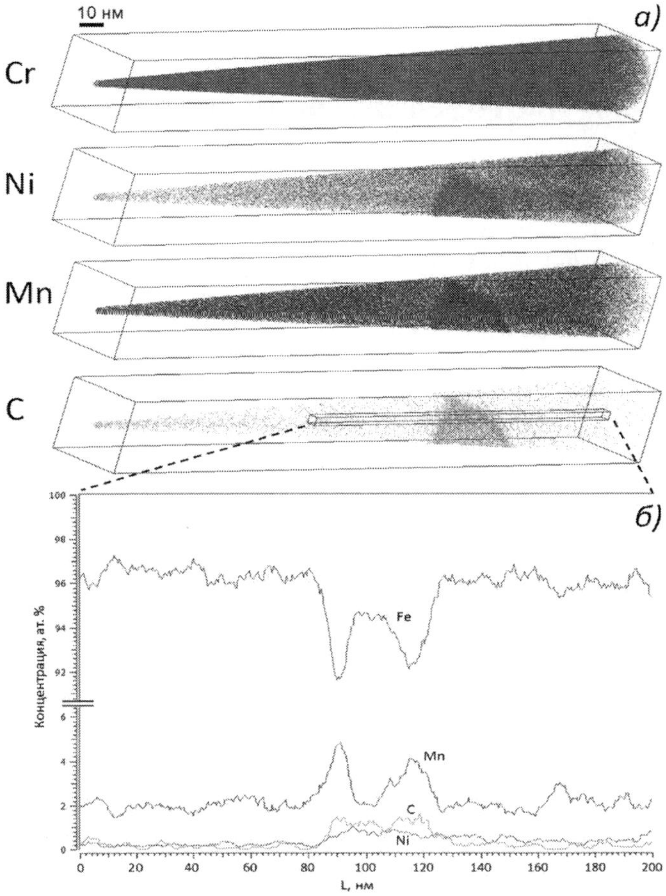

Figure 4. a) Representation of a three-dimensional reconstruction of the atomic map of UFG steel with distribution of Cr, Ni, Mn and C atoms in the investigated volume of steel 05G2MFB K65 subjected to warm rolling; b) linear concentration profiles along the direction of research.

4. Summary

As a result, one of the factors leading to an increase in the impact and tensile strength is an increase in the degree of supersaturation of the solid solution of steel ferrite and an increase in the concentration of the content of alloying elements on the boundary of carbides after deformation, as confirmed by atom-probe tomography.

References

[1] Safarov I M, Korznikov A V, Galeyev R M, Sergeev S N, Gladkovskii S V, Borodin E M and Pyshmintsev I Yu 2014 *Phys. Metals Metallogr.* **115** 295

[2] Astafurova E G, Maier G G, Koshovkina V S, Melnikov E V, Naydenkin E V, Smirnov M A, Bataev V A, Odessky P D and Dobatkin S V 2015 *Letters on Materials* **5** 432

[3] Safarov I M, Korznikov A V, Galeyev R M, Sergeev S N, Gladkovskii S V and Pyshmintsev I Yu 2016 *Doklady Phys.* **61** 15

[4] Rogozhkin S V, Aleev A A, Lukyanchuk A A, Shutov A S, Raznitsyn O A and Kirillov S E 2017 *Instr. Exper. Techn.* **60** 428.

IOP Conf. Series: Materials Science and Engineering **447** (2018) 012010 doi:10.1088/1757-899X/447/1/012010

Investigation of the effect of hydrogen on the properties of fcc nanowires

M D Starostenkov, O V Yashin, A V Yashin

Altai State Technical University, Lenin Str. 46, 656038 Barnaul, Russia

E-mail: frs22@mail.ru

Abstract. In the present work, the features of the processes of structure-energy transformations observed during deformation of fcc Ni nanowires containing hydrogen atoms in octahedral and tetrahedral pores are studied. As a result of the investigation at temperatures of 50, 100, 300, 600, 900 and 1200 K, the effect of the concentration of hydrogen atoms on the strength characteristics of fcc Ni nanowires having a cylindrical shape was revealed.

1. Introduction

The scenario of the strain-induced structure-energy transformations of fcc Ni nanowires containing hydrogen atoms was studied by means of the molecular dynamics method. The axis of the tensile direction corresponded to the <100> direction. Ni nanowires had a cylindrical shape (figure 1a). The interaction of atoms was calculated using the Morse pair potential, adopted from [1], approved in [2].

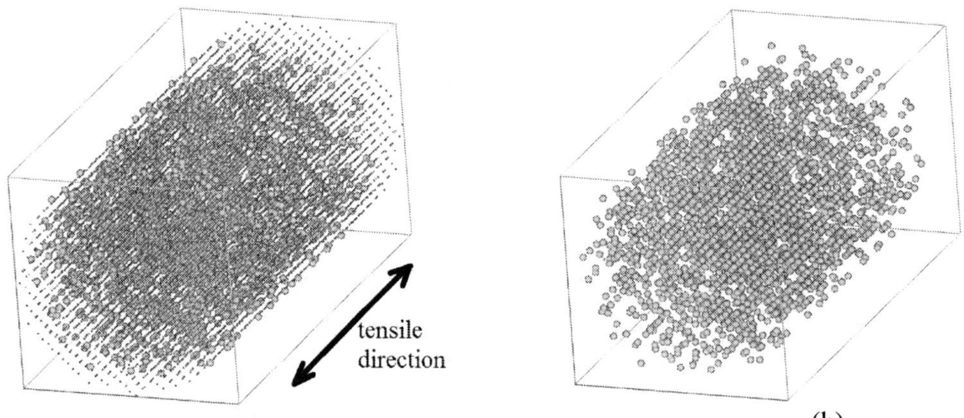

(a) (b)

Figure 1. Cylindrical <100> oriented Ni nanowires with 40% of pores filled with hydrogen. (a) atomic structure of the wire before the deformation begins (small dots stand for Ni, large ones for H), (b) hydrogen sub-lattice

Hydrogen atoms of were placed in octahedral and tetrahedral pores of the wire (figure 1b). In the present work 0, 20, 40, 60 and 80% of the pores of the nanowires were filled with hydrogen, one hydrogen atom in each pore. The ratio of the number of hydrogen atoms in the octahedral and tetrahedral pores of the nanowires corresponded to the ratio of the total volume of the octahedron and tetrahedral pores in the unit cell.

A periodically repeated translational displacement of the atoms constituting absolutely rigid grips along the axis of stretching of the nanowires in opposite directions from each other was performed to simulate the straining of the nanowires. The total velocity of the gripper movement was 20 m / s and corresponded to the strain rate of the order of · 10^9 s^{-1}. The numerical experiment was performed at temperatures of 50 K, 300 K, 600 K, 900 K and 1200 K. At the beginning of the computer experiment, the temperature was set through the initial velocities of the atoms. During deformation, thermal stabilization was performed in accordance with the Berendsen algorithm [3]. The described model of the experiment was approved earlier in studies of the deformation dynamics of nanowires of such alloys as CuAu [4], Ni_3Al [5,6] and Ni_3Fe [7].

2. Results and discussion

As a result of the study of structure-energy transformations of fcc Ni with 0, 20, 40, 60, and 80% of pores filled with hydrogen, four basic stages of deformation were revealed: quasi-elastic (I), plastic (II), flows (III) and fracture (IV). The deformation stages are clearly seen on the graphs of the dependence of the stored deformation energy on the experiment time (figure 2).

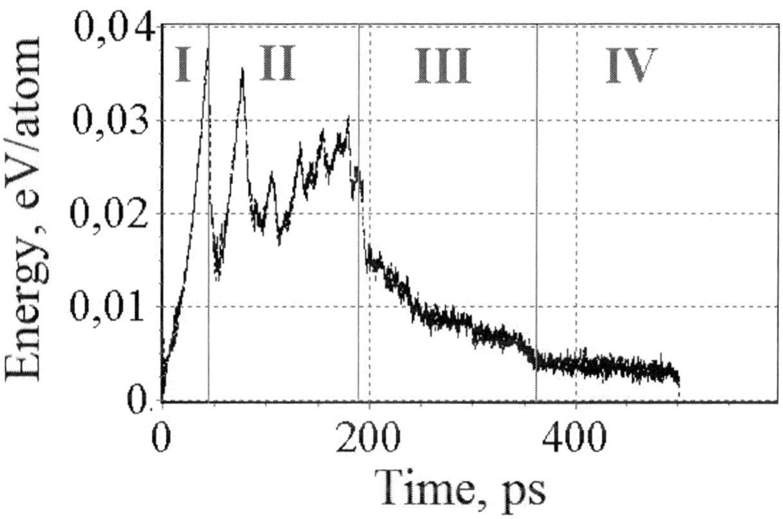

Figure 2. Dependence of stored energy of deformation per 1 atom on the experiment time, obtained at T=300 K for nanowires <100> Ni with 40% of pores filled with hydrogen

At the same time, the behavior of this graph is reflected in the graph of the dependence "stress on the grips / time of the experiment", where the deformation stages can be indicated as well (figure 3). At high temperatures (900 and 1200 K) for <100> Ni nanowires with a large proportion of hydrogen filled pores (more than 40%), it was impossible to distinguish the deformation stages due to the fact that the material acquired an amorphous state.

In the course of the study, the points of maximum stress on the grips, which were reached at the end of the first stage (figure 3), were of the greatest interest

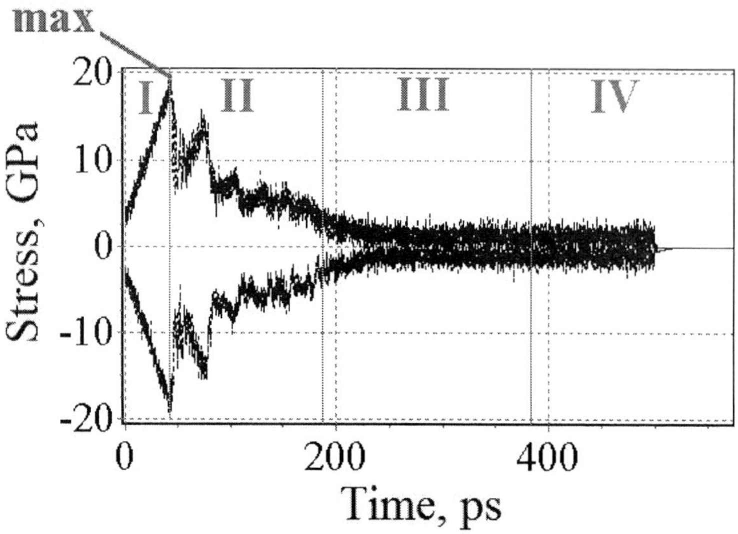

Figure 3. Dependence "stress on the grips / the experiment time", obtained at T=300 K for nanowires <100> Ni with 40% of pores filled with hydrogen. The stress applied to the left grip is indicated as negative for better display on the graph.

The next graph was build using the obtained values of the maximum stress on the grips for nanowires of cylindrical shape <100> Ni at different temperatures. The dependence shows the change in the value of the maximum stress on the grips from the temperature for nanowires with 0, 20, 40, 60 and 80% of pores filled with hydrogen.

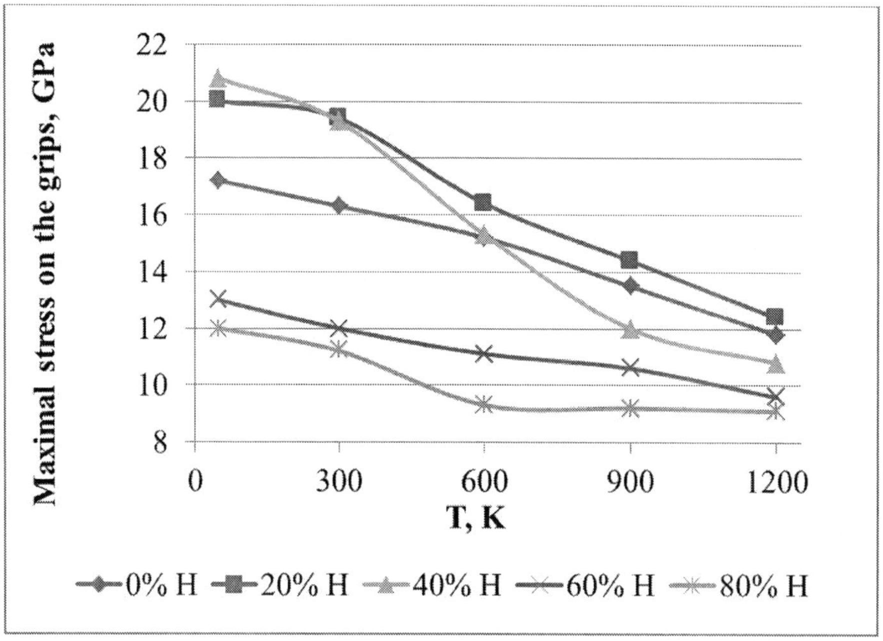

Figure 4. Dependence of maximal stress value on the grips at different temperatures in nanowires with 0, 20, 40, 60 and 80% of pores filled with hydrogen

IOP Conf. Series: Materials Science and Engineering **447** (2018) 012010 doi:10.1088/1757-899X/447/1/012010

The obtained results show that when hydrogen filled up to 20% of the pores, the strength of nanowires at different temperatures is higher than in nanowires without hydrogen. At the same time the filling of more than 40% of pores with hydrogen results in the degradation of the strength properties of nanowires. This fact is associated with the formation of clusters of hydrogen atoms in the form of globules at the stage of plastic deformation. The obtained results are in accordance with the data presented in [9-14].

References

[1] Poletaev G, Starostenkov M and Dmitriev S 2016 *Mater. Phys. Mech.* **27** 53
[2] Kulabukhova N, Poletaev G, Starostenkov M, Kulagina V and Potekaev A 2012 *Russ. Phys J.* **54** 1394
[3] Berendsen H, Postma J, van Gunsteren W, DiNola A and Haak J 1984 *J. Chem. Phys.* **81** 3684
[4] Starostenkov M, Yashin A, Sinica N 2014 *Key Eng. Mater.* **592-593** 51
[5] Potekaev A, Starostenkov M, Sinitsa N, Yashin A, Kharina E and Kulagina V 2011 *Russ. Phys J.* **53** 818
[6] Potekaev A, Starostenkov M, Sinitsa N, Yashin A, Kharina E and Kulagina V 2011 *Russ. Phys J.* **54** 180
[7] Starostenkov M and Aish M 2014 *Adv. Mater. Res.* **1013** 242
[8] Starostenkov M, Yashin O and Yashin A 2017 *Nano Hyb. Compos.* **15** 21
[9] Zheng Y, He M, Cheng G, Zhang Z, Xuan F and Wang Z 2018 *Molec. Simul.* **44** 173
[10] Poletaev G, Martynov A and Starostenkov M 2010 *Basic Problems of Material Science* **7** 27 (In Russian)
[11] Sannikov A, Poletaev G, Soskov A and Starostenkov M 2014 *Basic Problems of Material Science* **11** 317 (in Russian)
[12] Martynov A, Poletaev G and Starostenkov M 2011 *Letters on materials* **1** 43
[13] Poletaev G, Starostenkov M and Patseva Y 2004 *Basic Problems of Material Science* **1** 147 (In Russian)
[14] Rakitin R, Poletaev G, Aksenov M and Starostenkov M 2005 *Basic Problems of Material Science* **2** 124 (in Russian)

IOP Conf. Series: Materials Science and Engineering **447** (2018) 012011 doi:10.1088/1757-899X/447/1/012011

Dehydrogenation of graphane by external driving

J A Baimova[1,2], I Evazzade[3], L A Kalyakin[4], S V Dmitriev[1,5] and K Zhou[6]

[1]Institute for Metals Superplasticity Problems, Russian Academy of Sciences, 39 Khalturin st., Ufa 450001, Russia
[2]Bashkir State University, 32 Zaki Validy St., Ufa 450076, Russia
[3]Department of Physics, Faculty of Science, Ferdowsi University of Mashhad, Mashhad, Iran
[4]Institute of Mathematics, Ufa Research Center, Russian Academy of Sciences, 112 Chernyshevsky St., Ufa 450008, Russia
[5]National Research Tomsk State University, 36 Lenin Ave., Tomsk 634050, Russia
[6]Nanyang Technological University, 50 Nanyang Ave., Singapore 639798, Singapore

E-mail: julia.a.baimova@gmail.com

Abstract. The dehydrogenation of graphane due to external harmonic driving is studied by molecular dynamics simulation. It is found that the dehydrogenation can be initiated by harmonic driving of carbon atoms with as small amplitude as 0.02 Å, if the driving frequency is somewhat above the small-amplitude vibration frequency of the C-H bond. Thus, a novel mechanism for the dehydrogenation of graphane is reported.

1. Introduction

Graphene, a zero band gap two dimensional material where carbon atoms are arranged in a honeycomb lattice, presents outstanding properties by itself and when combined with other organic molecules. Its chemical [1], electrical [2], mechanical [3, 4, 5] and optical [6] characteristics have attracted a tremendous attention of researchers over the last decade. Graphene can be effectively modified via the surface functionalization, in which the carbon atoms are converted from sp^2 to sp^3 hybridization to bond with the adatoms. Hydrogenation has been widely accepted as an efficient way to manipulate the electronic [7], magnetic [8], and mechanical [9] properties of graphene. Moreover, hydrogenated graphene is a very promising structure for hydrogen storage. Previously, it was shown that a gravimetric density of 8% could be achieved for corrugated graphene with satisfactory reversibility [10]. In the case of carbon nanofibers and carbon nanotubes, hydrogen uptake of 1.5 wt.% per 1000 $m^2\ g^{-1}$ was observed under ambient temperature and pressure conditions [11].

About a decade ago, hydrogenation of a graphene sheet has been successfully achieved, and a hydrocarbon often referred to as "graphane" was experimentally fabricated [12]. In this structure, each hydrogen (H) atom is bonded to a carbon atom (C) alternately on each side of the graphene sheet. Each C atom in this conformation is pulled out of the plane by an H atom at a small distance, forming a not perfectly flat structure in contrast to graphene. Previously, in [13, 14, 15, 16, 17, 18], the role of nonlinear localized modes or discrete breathers (DBs) in the dehydrogenation process was discussed. DBs can affect various properties of crystals, and they were found in different carbon structures [19, 20]: in graphene [21, 22, 23], diamond [24] and carbon nanotubes [25, 26]. The existence of DBs in graphane was shown for the first time by molecular dynamics (MD) simulation [15] and was proved later by ab-initio simulation [18]. It was shown that spontaneously exited at elevated temperatures DBs can lead to a detachment of a hydrogen atom from a graphene sheet.

Recently the effect of external harmonic driving on the nonlinear dynamics of graphene has been analyzed, and it has been observed that DBs can assist energy transfer from driven atoms to a graphene nanoribbon [27]. In the present work, MD simulation of harmonically driven graphane is studied to analyze the possibility of dehydrogenation under such conditions.

2. Model description and simulation procedure

In figure 1a the structure of graphane is presented with black and white circles, showing carbon and hydrogen atoms, respectively. The computational cell used in the present simulation included 12×12 primitive translational cells and was subjected to periodic boundary conditions. According to the considered stoichiometry, each primitive cell includes two C and two H atoms. The mass of hydrogen atom is equal to 1 g/mole or $1.66\ 10^{-27}$ kg.

The simulations are performed using the LAMMPS package with the adaptive intermolecular reactive empirical bond order (AIREBO) potential [28] to describe the interatomic interaction in graphane. The AIREBO potential was successfully used to study various properties of hydrocarbon systems, e.g., in [13, 14, 15]. The equations of motion for atoms were integrated numerically using the fourth-order Verlet method with a time step of 0.1 fs.

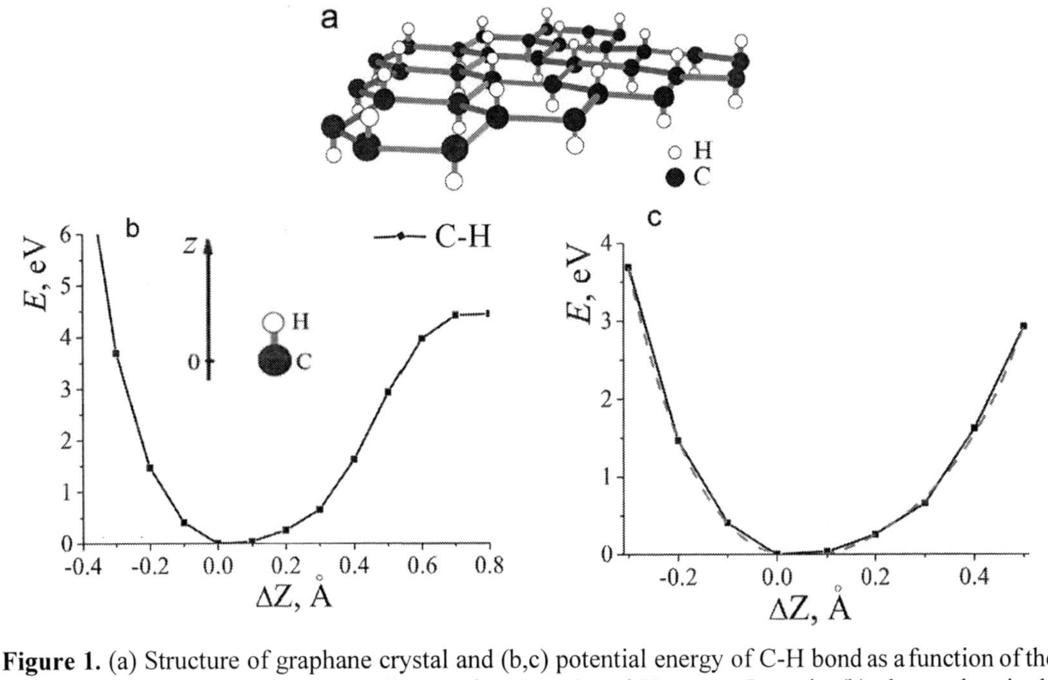

Figure 1. (a) Structure of graphane crystal and (b,c) potential energy of C-H bond as a function of the displacement from the equilibrium distance between C and H atoms. Inset in (b) shows the single degree of freedom model. In (c) polynomial fit (2) of C-H bond energy is shown by the red dashed line.

All carbon atoms are forced to move in accordance with the harmonic law

$$\Delta Z_C(t) = A\sin(\omega t), \tag{1}$$

where ΔZ_C is the displacement of C atoms from their initial positions in the direction normal to the graphane sheet, A and ω are the driving amplitude and frequency, respectively. At $t = 0$ all H atoms have zero initial displacements from equilibrium positions and zero initial velocities.

Since all C atoms move synchronously, assuming that the hydrogen atoms also move synchronously, the problem under consideration can be reduced to analyzing a single degree of freedom model, schematically shown in the inset of figure 1b. The model includes single C atom moving according to the law (1) and H atom attached to it by the C-H bond. It is instructive to calculate the interaction energy E between H and C atoms. This was done by a quasistatic change of the distance ΔZ between C and H atoms and calculating the potential energy of their interaction. The result is shown in figure 1b. Close to the energy minimum, a potential function can be approximated by the polynomial function

$$E = a_0 + a_1\Delta Z + a_2(\Delta Z)^2 + a_3(\Delta Z)^3 + a_4(\Delta Z)^4, \quad (2)$$

with the coefficients a_0=0, a_1=-1.586278, a_2=19.52419, a_3=-37.08165, a_4=55.90332. This function is plotted in figure 1c by a red dashed line together with numerical data (black symbols and a black line). We have also found that the frequency of the small amplitude oscillations of the C-H bond is about ω^* = 80 THz.

3. Simulation results and discussion

Figure 2 shows the energy of the C-H bond as a function of time for the driving amplitude A=0.04 Å and three driving frequencies: 80, 86, and 90 THz in panels a to c, respectively. As it can be seen, for 80 THz the bond is quasiperiodically pumped with energy up to a value of 2.2 eV, and then its energy decreases to zero. This regime is always observed at sufficiently small driving amplitudes and frequencies not close to the resonance value ω^*. At high driving amplitudes or driving frequency close to ω^*, the H atom separates from the C atom. From figures 2 b, c it is seen that the disorption energy is approximately equal to 5.3 eV.

Figure 3 shows the maximal energy of the C-H bond as a function of the driving frequency for four driving amplitudes. For the driving amplitude A=0.01 Å the C-H bond does not break. For higher driving amplitudes, bond breaking takes place for driving frequencies around 90 THz. The higher the driving amplitude, the wider is the driving frequency range where bond breaking occurs.

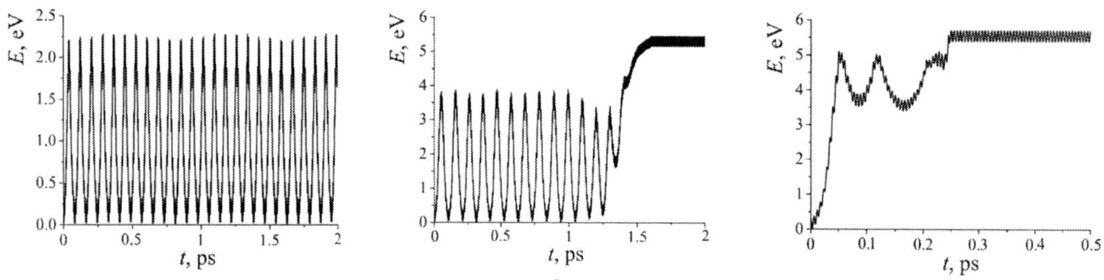

a b c

Figure 2. Energy of the C-H bond as a function of time for driving amplitude A=0.04 Å and for driving frequencies (a) 80 THz, (b) 86 THz, and (c) 90 THz.

4. Conclusions

Molecular dynamics simulations confirmed that dehydrogenation of graphane can be initiated by external driving at frequencies slightly above the frequency of small-amplitude vibration of the C-H bond. The presented work describes a new mechanism for the dehydrogenation of graphene which may be important for solving the problem of transportation and storage of hydrogen.

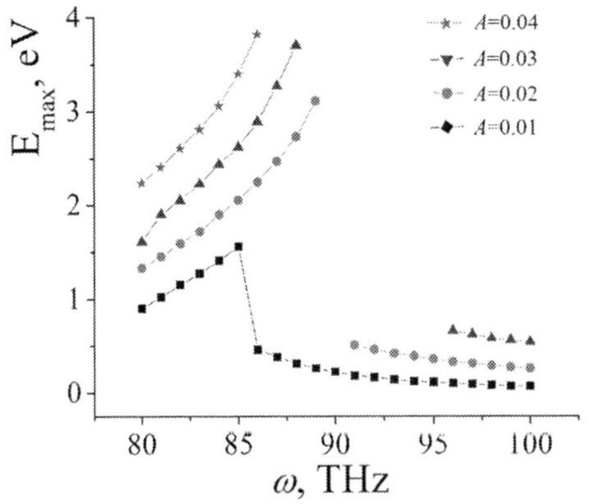

Figure 3. Maximal energy of C-H bond as the function of driving frequency for fourdriving amplitudes.

Acknowledgments
Research was designed by D.S.V. supported by the Russian Foundation for Basic Research, grant no. 17-02-00984-a. Work of J.A.B. (writing the paper) was supported by the grant of the President of the Russian Federation for state support of young Russian scientists - doctors of sciences MD-1651.2018.2. The work was partly supported by the State Assignment of IMSP RAS.

References

[1] Salehi-Khojin A, Estrada D, Lin K Y et al 2012 *Appl. Phys. Lett.* **100** 033111
[2] Zhang Y, Tan Y W, Stormer H L and Kim P 2005 *Nature* **438** 201–204
[3] Lee C, Wei X, Kysar J W and Hone J 2008 *Science* **321** 385–388
[4] Dmitriev S V, Baimova J A, Savin A V and Kivshar Y S 2012 *Comp. Mater. Sci.* **53** 194–203
[5] Baimova J A, Dmitriev S V, Zhou K and Savin A V 2012 *Phys. Rev. B* **86** 035427
[6] Boukhvalov D W, Katsnelson M I and Lichtenstein A I 2008 *Phys. Rev. B* **77** 035427
[7] Sofo J O, Chaudhari A S and Barber G D 2007 *Phys. Rev. B* **75** 153401
[8] Zhou J, Wang Q, Sun Q, Chen X S, Kawazoe Y and Jena P 2009 *Nano Letters* **9** 3867–3870
[9] Li Y, Datta D, Li Z and Shenoy V B 2014 *Comp. Mater. Sci.* **83** 212–216
[10] Tozzini V and Pellegrini V 2011 *J. Phys. Chem. C* **115** 25523–25528
[11] Zttel A, Sudan P and et al P M 2002 *Int. J. Hydrogen Energy* **27** 203–212
[12] Elias D C, Nair R R, Mohiuddin T M G, et al 2009 *Science* **323** 610–613
[13] Baimova J, Murzaev R and Rudskoy A 2017 *Phys. Lett. A* **381** 3049–3053
[14] Baimova J A, Murzaev R T, Lobzenko I P, Dmitriev S V and Zhou K 2016 *J. Exp. Theor. Phys.* **122** 869–873
[15] Liu B, Baimova J A, Dmitriev S V, Wang X, Zhu H and Zhou K 2013 *J. Phys. D* **46** 305302
[16] Lobzenko I P 2016 *Letters on Materials* **6** 73–76
[17] Korznikova EA Baimova JA Lobzenko IP Liu B Dmitriev SV and Zhou K 2018 *Mater. Phys. Mech.* **35** 71–79
[18] Chechin G M, Dmitriev S V, Lobzenko I P and Ryabov D S 2014 *Phys. Rev. B* **90** 045432
[19] Dmitriev S V 2016 *Letters on Materials* **6** 86–89
[20] Baimova JA Korznikova EA Dmitriev SV Lobzenko I P 2015 *Rev. Adv. Mater. Sci.* **42** 68–82
[21] Yamayose Y, Kinoshita Y, Doi Y, Nakatani A and Kitamura T 2007 *Europhys. Lett.* **80** 40008
[22] Baimova J A, Dmitriev S V and Zhou K 2012 *Europhys. Lett.* **100** 36005

[23] Korznikova E A, Baimova Y A, Dmitriev S V, Mulyukov R R, Savin A V 2012 *JETP Lett.* **96** 222–226
[24] Murzaev R, Bachurin D, Korznikova E and Dmitriev S 2017 *Phys. Lett. A* **381** 1003–1008
[25] Kinoshita Y, Yamayose Y, Doi Y, Nakatani A and Kitamura T 2008 *Phys. Rev. B* **77** 024307
[26] Savin A V and Kivshar Y S 2008 *Europhys. Lett.* **82** 66002
[27] Evazzade I, Lobzenko I P, Korznikova E A, Ovidko I A, Roknabadi M R and Dmitriev S V 2017 *Phys. Rev. B* **95** 035423
[28] Stuart S J, Tutein A B and Harrison J A 2000 *J. Chem. Phys.* **112** 6472–6486

IOP Conf. Series: Materials Science and Engineering **447** (2018) 012012 doi:10.1088/1757-899X/447/1/012012

Modification of the atomic structure of metallic materials after strong external exposures

V A Ivchenko

Yeltsin Ural Federal University, 19 Mira st. Yekaterinburg, 620002, Russia
Institute of Electrophysics, Ural Branch, Russian Academy of Sciences, 106 Amundsen st., 620016, Yekaterinburg, Russia

e-mail: ivchenko2008@mail.ru

Abstract. The paper presents the application of direct methods (field ion microscopy and atom probe techniques) to study the modification of the crystal structure of various materials at the atomic spatial level after strong external exposures. The effects of amorphization of subsurface volumes of alloys, deformation effects in pure metals, and structural phase transformations in subsurface volumes of solid solutions were detected during radiation exposure to charged particle beams with an energy of 20 keV. The original results show the possibility of three-dimensional reconstruction of the elemental composition of the materials investigated with atomic resolution to study the modification of the atomic structure. It is established that the nature of the crystalline structure of the boundary region of planar defects directly depends on the type of external influence. Experimental data on the elemental composition of interphase interfaces in mechanically alloyed compounds and doped alloys have been obtained. It is established that the width of the boundary regions at the metal interfaces varies within 0.8–1.5 nm, depending on the type of intensive external action to which they were subjected.

1. Introduction

Among the most efficient and advanced methods to study conducting and semiconducting materials, using which the lattice of solids can be directly studied with the atomic spatial resolution, are field ion microscopy (FIM) and various modifications of atom probes of a field ion microscope.

The objective of this paper is to demonstrate the efficient use of atom probe methods for studying nanostructured states arising in the bulk of metals and alloys during modification of the atomic structure of various materials by strong external exposures. It is shown that, along with the lattice distortions typical of thermo- mechanical, thermal, strong deformational, and other perturbations (dot, linear, dislocation, and other lattice defects), the studied materials contain such radiation defects as vacancy clusters, amorphized regions, and nano segregations of atoms of one of the components in ordered alloys. The microscope field size (~100 nm) and analysis of the sample volume by controlled layer by layer evaporation of the surface atoms make it possible to determine whether the material under study is nanostructured or not. The possibility of studying the sample in a volume down to micron sizes makes it possible to obtain macro scale experimental data.

FIM is the only microscopy technique capable of providing direct observation of individual atoms as elements composing the structure of a sample during an ordinary experiment. The use of FIM makes it possible to select an individual atom in the image for mass spectrometric identification (atom probe FIM methods), to perform in situ experiments with individual atoms deposited on the surface, and to reconstruct the structural and chemical compositions in the sample volume by controlled layer by layer removal of surface atoms by an electric field at cryogenic temperatures. Exactly the latter situation distinguishes direct FIM methods from other structurally sensitive ones with atomic resolution, but being indirect methods of microscopic study of materials.

2. Experimental

The samples to be studied were prepared as tip emitters with an apex curvature radius of 30–50 nm from materials preliminarily processed by various strong external exposures, i.e., electrochemically polished wires or rods.

The objects of interaction with accelerated beams of gas ions (E = 20 keV, implantation dose D = 10^{13}–10^{18} ion/cm^2, and current density j = 100–340 μA/cm^2) were various metal materials. Field emitters certified for ion implantation were characterized by the atomically smooth surface of the apex prepared in situ by field evaporation of surface atoms.

Figure 1 shows the ion image of Pt after certification. On the one hand, the micropattern represents the contrast of the tip end surface from atoms at the adges of the step; on the other hand, it reflects the stereographic projection of the crystal under study. Circular contour lines in the ion image are the edges of the corresponding sets of crystallographic planes of certain directions. Adjacent rings (from any set of concentric rings) represent images of parallel atomic layers. The distance between the rings corresponds to the interplanar distance for a given crystallographic direction. In terms of the ion microscopic image, this distance also refers to the lattice step height. The rings themselves in the ion microscopic image, as a rule, consist of isolated bright dots which are images of surface atoms arranged at atomic sites in step edges.

Materials such as Cu80Co20 were mechanical alloyed by high purity (99.9%) Cu and Co powders with particle sizes of ~50 μm. The powders were treated in a centrifugal planetary mill in an argon atmosphere at room temperature for 20 hours. The ratio of steel balls of the mill to powder particles was 1: 25. After mechanical alloying, the powders were compacted into pellets by compression shear at 500 MPa. Some pellets were annealed at 400 °C for 2 hours. Work pieces for future pins were rods (~0.2 × 0.2 × 10 mm^3 in volume) cut from thermally treated pellets by spark cutting. Tomographic atom probe (TAP) studies were performed with the ratio of the pulse voltage to the dc voltage of 20% and the repetition rate of 775 Hz in high vacuum (1×10^{-8} Pa). The sample temperature during measurements was 60–80 K.

3. Results and considerations

The objects of interaction with accelerated beams of gas ions were pure metals (Ir), atomically ordered alloys (Cu3Au), and solid solutions (50Pd30Cu20Ag), the decomposition of which leads to atomically ordered phase separation [1-7].

Using the FIM method, the effect of partial amorphization in the subsurface volume of atomically ordered alloys was detected due to the implantation of positive argon ions accelerated to 20 keV (D = 1018 ion/cm2 and j = 200 μA/cm2) [6]. Partial amorphization was observed at distances of at least 90 nm from the irradiated surface of materials. In the experiment, only the beam current density varied; the implantation energy and dose did not change (E = 20 keV, D = 10^{18} ion/cm^2). At j = 275 μA/cm^2, a structural phase transition occurred in the entire volume of the material, which indicated that the sample was heated to the critical phase transformation temperature (~390 °C), and the cooling rate corresponded to the conditions of the transition to a disordered state.

Figure 2 shows the ion pattern of the irradiated platinum surface with the fluence of 10^{18} ions/cm^2.

It is obvious, that on the basis of the contrast of micro patterns of the atomically smooth platinum surface when analyzing the subsurface regions of the material during the controlled elimination of atomic layers, the phase state of the metal becomes almost amorphous with an increase in the fluence of up to 10^{18} ions/cm^2.

The proof of this is the structureless arrangement of atoms in the subsurface layers. The counterpart of the observed ion pattern corresponds to the ion pattern of amorphous materials obtained by ultrafast

IOP Conf. Series: Materials Science and Engineering **447** (2018) 012012 doi:10.1088/1757-899X/447/1/012012

cooling. In our estimation, the amorphization of the pure metal (Pt) occurs in the subsurface region at a depth of 12 nm from the irradiated surface.

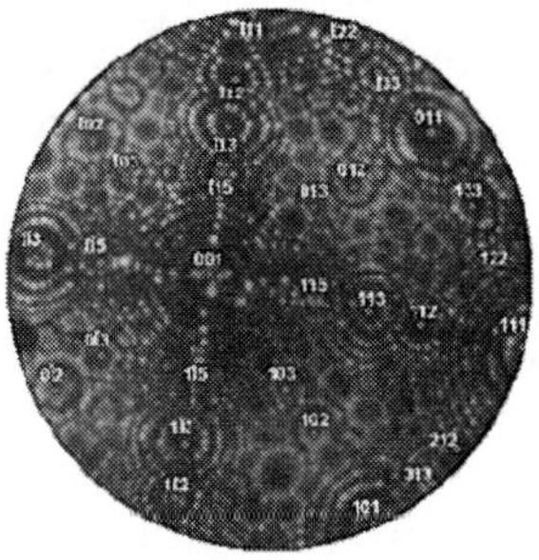

Figure 1. Neon image of Pt single crystal, U = 10 kV.

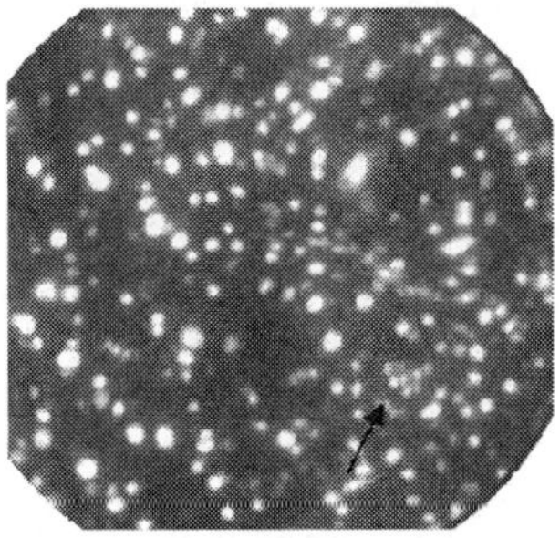

Figure 2. Ion pattern of Pt surface after irradiation by Ar^+ with D = 10^{18} ions/cm^2 (T = 300°C) (the arrow denotes the region of the crystalline state of the metal).

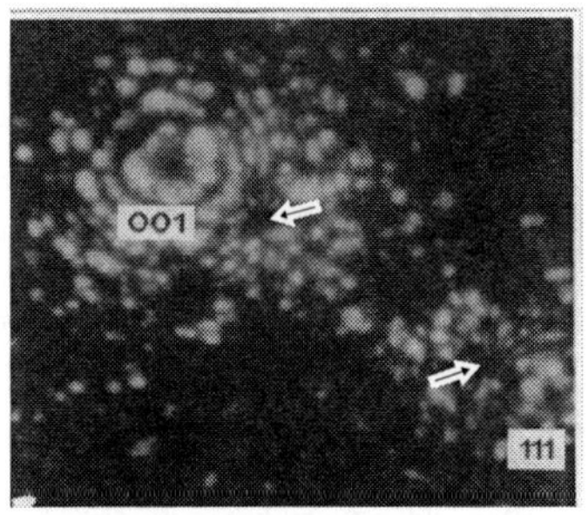

Figure 3. A region of ion contrast of ion-implanted PdCuAg after the removal of 14 atomic layers of the irradiated surface (arrows show ion contrast micropores).

The study of ion implanted PdCuAg detected a phase structural transition in its subsurface volume [4]. The size of the surface volume in which the structural phase transformation was detected was estimated as a function of the beam current density. In the subsurface volume, vacancy clusters (micropores) were detected (figure 3), and their sizes and shape were estimated. Micropores were shaped as ellipsoids with a height of 4–25 nm and a diameter of 3–12 nm. The volume fraction of micropores monotonically decreases with distance from the irradiated surface.

The study of mechanically deformed iridium samples revealed a high density of point, linear, and volume structural defects. A comparative analysis of planar defects [1], observed in pre-deformed (~90%) and irradiated iridium, showed a significant difference between their structures. Mechanically deformed iridium consisted of grains with sizes within 20–30 nm, the bodies of which were almost free of structural defects. In contrast, the ion irradiated metal (E = 20 keV, D = 10^{18} ion/cm^2, j = 300 μA/cm^2) exhibited a sub-block microstructure with a characteristic size of ~3–5 nm, block misorientation within 0.5°–1°, and various defects in the bodies of blocks.

The width of the boundary regions at the interfaces in the samples upon various external actions was, similarly to the heat treated metals and alloys, of the order of the interatomic distances.

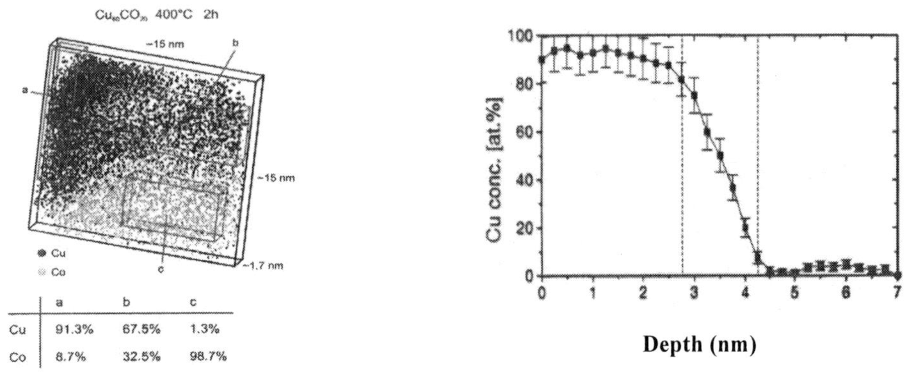

a b

Figure 4. 3D elemental map of the mechanically alloyed Cu80Co20 compound (upon annealing at 400 °C for 2h) as reconstructed by the TAP data; (b) interphase boundary width of the mechanically alloyed Cu80Co20 compound upon 2 h of annealing.

The width of the boundary region changes significantly in the case of materials obtained by mechanical alloying (figure 4) [8]. Figure 4a shows a three-dimensional (3D) reconstruction of the elemental map of the mechanically alloyed Cu80Co20 compound. The determination of the composition of the annealed material revealed three regions with different concentrations of cobalt: (a) a region containing 8.7 at.% Co, (b) an intermediate region containing 32.5 at.% Co, and (c) a region with high concentration of Co and only 1.3 at.% Cu. All these regions exhibit interphase boundaries with an effective width of about 1.5 nm (figure 4b). The plot in figure 4b shows a variation of copper content in the boundary regions of the interphase interfaces.

The FIM study of nickel samples clearly revealed the boundaries of ultradisperse blocks, which constitute the microstructure of individual grains [9]. The dimensions of blocks, estimated both on the surface FIM image and in the course of the controlled sequential removal of the surface atomic layers, varied from 1 to 10 nm. The widths of the boundary regions were of the order of the interatomic distance. In the first approximation, the obtained data are identical to the results of studying the atomic structure in submicrocrystalline (SMC) tungsten (with an average grain size of ~100 nm) upon severe plastic deformation ($e \approx 7$) [10]. The analysis of the FIM images of an SMC tungsten surface region with a grain boundary showed that its width was about 0.6–0.8 nm. It should be noted that the width of the boundary in the initial coarse-grained tungsten was 0.3–0.4 nm [9]. In addition, the use of field electron microscopy (FEM) and electron spectroscopy techniques revealed significant differences between the electron energy distributions in SMC and coarse grained tungsten [10].

4. Conclusions

Thus, the application of direct methods for studying the crystal structure of various materials at the atomic spatial level (field ion microscopy and atom probe FIM methods) after various intensive external exposures was demonstrated. For example, the effects of amorphization of subsurface volumes of alloys, deformation effects in pure metals, and structural phase transformations in subsurface volumes of solid solutions were detected during the radiation exposure to the charged particle beams with an energy of 20 keV. It is established that the nature of the crystalline structure of metal interfaces significantly depends of the type of external actions and eventually determines the physical (in particular, mechanical) properties of materials.

Experimental data on the elemental composition of interphase interfaces in mechanically alloyed compounds have been obtained. It is established that the width of the boundary regions at the metal interfaces varies within 0.8–1.5 nm, depending on the type of the intensive external action to which they were subjected.

Acknowledgments

This work was supported by RFBR Grant No. 16-08-00615.

References

[1] Ivchenko V A and Syutkin N N 1999 *Techn. Phys. Lett.* **25** 3 233
[2] Bunkin A Yu, Gavrilov N V, Ivchenko V A, Kreindel, Y E, Kuznetsova L Yu and Syutkin N N 1990 *Fizika Metallov i Metallovedenie* **69** 171 (In Russian)
[3] Bunkin A Yu Ivchenko V A, Kuznetsova L Yu and Syutkin N N 1990 *Fizika Metallov i Metallovedenie* **7** 111 (In Russian)
[4] Ivchenko V A and Syutkin N N 1995 *Surf. Sci.* **87/88** 257
[5] Ivchenko V A, Syutkin N N and Bunkin A Yu 1988 *J. Phys.* **49–C6** 379
[6] Ivchenko V A, Syutkin N N and Kuznetsova L Yu 2000 *Techn. Phys. Lett.* **26** 541
[7] Ivchenko V A, Ovchinnikov V V, Goloborodsky B Yu and Syutkin N N 1997 *Surf. Sci.* **384** 46
[8] Ivchenko V, Wanderka N, Czubayko U, Naundorf V, Ermakov AYe, Uimin M A and Wollenberg H 2000 *Mater. Sci. Forum* **343/346** 709
[9] Ivchenko V A, Efros B M, Popova E V, Efros N B, Loladze L V 2003 *Fizika i Tekhika Vysokikh Davlenii* **13** 3 159 (In Russian)
[10] Mulyukov R R, Yumaguzin Yu M, Ivchenko V A and Zubairov L R 2000 *JETP Lett.* **72** 5 257

IOP Conf. Series: Materials Science and Engineering **447** (2018) 012013 doi:10.1088/1757-899X/447/1/012013

Texture and grain-boundary misorientation distributions in Y123 ceramics deformed by torsion under pressure

M F Imayev and D B Kabirova

Institute for Metals Superplasticity Problems, Russian Academy of Sciences, 39 Khalturin st., Ufa 450001, Russia

E-mail: marcel@imsp.ru

Abstract. Texture and grain-boundary misorientation distributions in different parts of the sample of the high-T_c superconductor Y123 deformed by torsion under pressure at 1008 °C to a twist angle of 30° were investigated. The average levels of the basal plane texture (the Lotgering factor F) of undeformed and deformed samples were F = 0.25 and 0.965, respectively. In the undeformed sample, the fraction of low-angle grain boundaries (LAGBs) with misorientation angles from 2° to 10° does not exceed 5%. The deformation led to the formation of a strong, but non-uniform texture along the radius of the sample. In the center of the sample, a strong fiber texture is formed with the [001]-axis parallel to the axis of compression / torsion, and the fraction of LAGBs is about 20%. At a distance of 2 mm from the center of the sample there is a ring with a non-basal texture, where the fraction of LAGBs reaches 30%. A biaxial texture is formed at the edge of the sample: the [001] axis is parallel to the axis of compression / torsion, and the [110] axis is oriented along the radius of the sample. The fraction of LAGBs at the edge of the sample is 56%.

1. Introduction

In order to obtain high critical current density (J_c), high-temperature superconducting (HTSC) ceramics should have a strong crystallographic texture (the angles of the grain boundary misorientation should not exceed 10°) [1]. Among these materials, at present the bulk ceramics $YBa_2Cu_3O_{7-x}$ (Y123) is the closest one to practical applications. To form a strong texture in Y123, a melt texturing process has been developed, but the properties of the samples obtained by this method are not high enough. For a wide range of practical applications, the value of $J_c = 10^5$ A/cm^2 in a field of 1 T is required. The best samples of Y123 obtained by the melt texturing technique have a critical current density of the order of 10^4 A/cm^2 in 1 T field. When superconductors are used as permanent magnets (with a frozen field), another important parameter is magnetization. The magnetization of a superconductor is proportional to the critical current density J_c multiplied by the minimum size of the current loop. In a strongly textured sample, the latter is approximately equal to the size of the sample. That is, to obtain high magnetization, superconducting products must be one-piece, rather than multipart ones. At present, melt texturing techniques do not allow producing large billets, therefore large products, for example, superconducting rings in the kinetic storage [2] and rotors of superconducting motors [3], are stacked from small rectangular blocks of Y123. The production of a one-piece ring will increase the size of the current loop and, accordingly, increase the magnetization of the superconductor. In this connection, it is important to develop an alternative method for processing of Y123 ceramics, which makes it possible to obtain a high value of J_c in a large single axisymmetric product. As such a method, one can consider hot torsion under quasi-hydrostatic pressure (HTP), which allows deforming brittle ceramics to large strains.

Using this method, a strong texture and high superconducting properties were obtained in ceramics $Bi_2Sr_2CaCu_2O_{8+d}$ [4]. A strong texture can be also obtained by the HTP technique in Y123 ceramics. For the best Y123 sample, the full width at a half maximum of the (002) peak of the rocking curve was 5.2 °, and the Lotgering factor F (basal planes orientation degree) was 0.956 [5]. However, a more detailed study of the local texture on the polished sections which are perpendicular to the axis of the pellet showed that in the deformed Y123 ceramics, the texture along the radius of the samples was not homogeneous [6]. The presence of an annular region with a non-basal texture was found. In the sample with $F = 0.965$, a ring with a non-basal texture is located in the range of 0.7-3.5 mm from the center of the sample. The existence of such a ring is associated with the formation of a wavy (corrugated) structure due to the constraint of deformation during HTP. For a more detailed and accurate determination of the texture, the transverse section of the sample should be studied.

Therefore, the aim of the present paper was to investigate the local texture and distributions of grain-boundary misorientations in different parts of the polished section parallel to the axis of compression / torsion of the sample.

2. Experimental

Y123 powder was pressed to pellets of 10 mm in diameter and 2 mm in height and was sintered in air at 900 °C for 5 hours. The prepared samples were plastically deformed by HTP in the following regimes: temperature T=1008 °C, twist rate $\omega = 4\times 10^{-4}$ rpm, pressure $P = 10$ MPa, and the twist angle α varied from 0° to 60°. Details of the deformation process, the calculation of the strain, the study of the microstructure and texture are given elsewhere [5-7]. To study the distributions of grain-boundary misorientations, two samples were chosen: one undeformed and one deformed by HTP to $\alpha = 30°$. The diameter of the dense interior of the deformed sample was 14 mm. The degree of basal plane texture (the Lotgering factor) of the undeformed and deformed samples was $F = 0.25$ and 0.965, respectively [5].

The local texture and grain-boundary misorientation distributions were measured on polished sections parallel to the axis of the pellets. In the undeformed sample the measurement was made in the center of the section, and in the deformed sample it was performed at distances $r = 0.5$; 2; 4.2 and 6.3 mm from the center of the sample. The microstructure and grain-boundary misorientation distributions were examined using a TESCAN MIRA 3 scanning electron microscope equipped with an EBSD HKL Oxford attachment. For the analysis and data processing Mambo, Tango, Twist, Salsaprograms, included in the software package CHANNEL-5, were used.

3. Results and discussion

Deformation led to the formation of a strong texture (figures 1a-d). At the center of the deformed sample ($r = 0.5$ mm), the [001] fiber texture (axis [001] is parallel to the compression/torsion axis) is formed as it is evidenced by a diffuse maximum on the inverse pole figure between the poles (010) and (110) (figure 1a). At a distance of 2 mm from the center of the sample there is a ring with a non-basal texture (figure 1 b). When moving towards the edge of the sample (figures 1 b-d), the texture maximum becomes more and more localized at the pole (110), which indicates that the texture turns into a biaxial type – the axis [001] is parallel to the axis of compression/torsion and the axis [110] is oriented along the radius of the sample (figure 1d). Since the critical current density in superconducting ceramics depends on the grain-boundary misorientation angles [1], the distributions of grain-boundary misorientation has been constructed for a more complete description of the microstructure formed during deformation. The graph in figure 2a shows the distribution of grain-boundary misorientation angles in the undeformed sample. The distribution is uniform in the entire investigated range of angles from 2° to 100°, and it is dominated by high-angle boundaries with misorientations above 35°. The fraction of low-angle grain boundaries (LAGBs) with misorientation angles from 2° to 10° does not exceed 5%. The distributions of grain-boundary misorientation angles in different places of the deformed sample are shown in figures 2 b-e. The absence of boundaries with misorientation angles more than 50° is characteristic of all parts of the deformed sample, which is expected for a sample with the [001] fiber texture and a cubic crystallographic lattice. At the center of the sample (figure 2 b), the distribution is uniform, the fraction of LAGBs is about 20%. In the area with non-basal texture (figure 2 c) the distribution of misorientation angles is non-uniform, in the intervals of misorientation angles of 5-6° and 31-32° the number of boundaries increases to 6-7%. The fraction of LAGBs reaches 30%. At a distance of 4.2 mm from the center of the sample (figure 2 d), the distribution has one maximum in the angular interval of 2-3°, and then the distribution is uniform. The fraction of LAGBs is 32%. At the edge of the sample (r = 6.3 mm), the fraction of the boundaries in each interval of the range of angles from 20° to 45°, with the exception of a small maximum in the interval of 44-45°, does not exceed 1%, which is noticeably less than in other places of the sample (figure 2 e). About 31% of the grain boundaries fall within the interval of misorientation angles of 2-3°, and this value is the maximum of the entire distribution. The fraction of LAGBs at the edge of the sample is 56%. Thus, the data on the distribution of misorientation angles are consistent with the texture data.

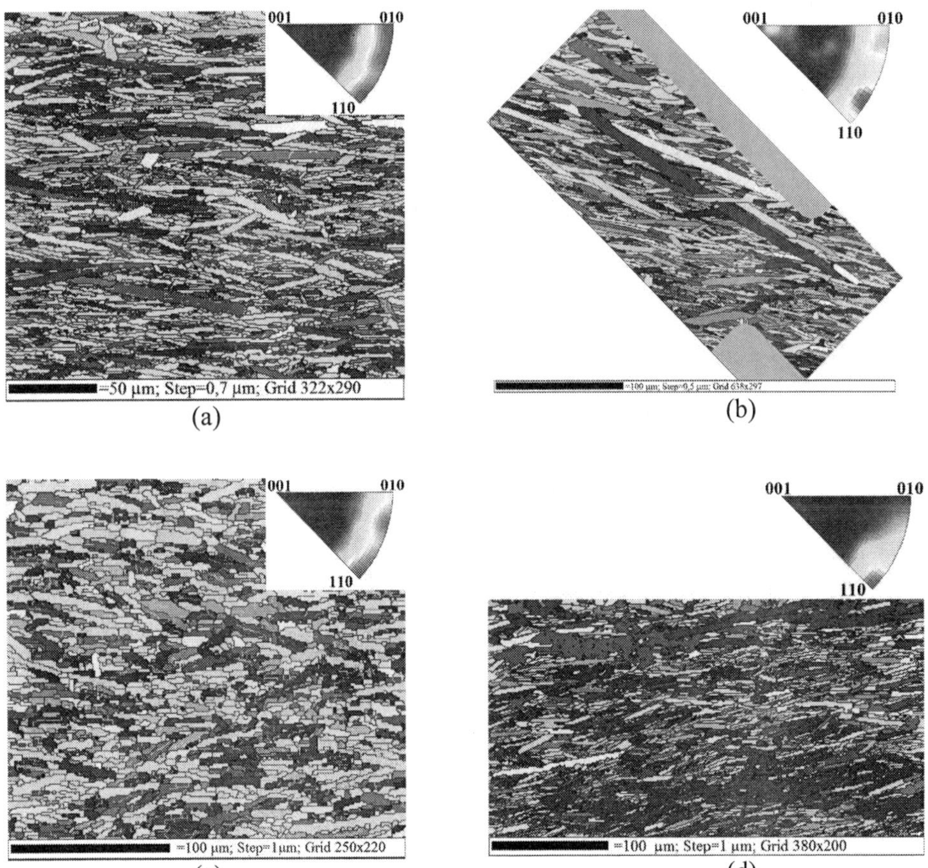

Figure 1. IPF maps and local inverse pole figures taken from the places located at a distance of (a) 0.5 mm, (b) 2 mm, (c) 4.2 mm, (d) 6.3 mm from the center of the deformed Y123 sample.

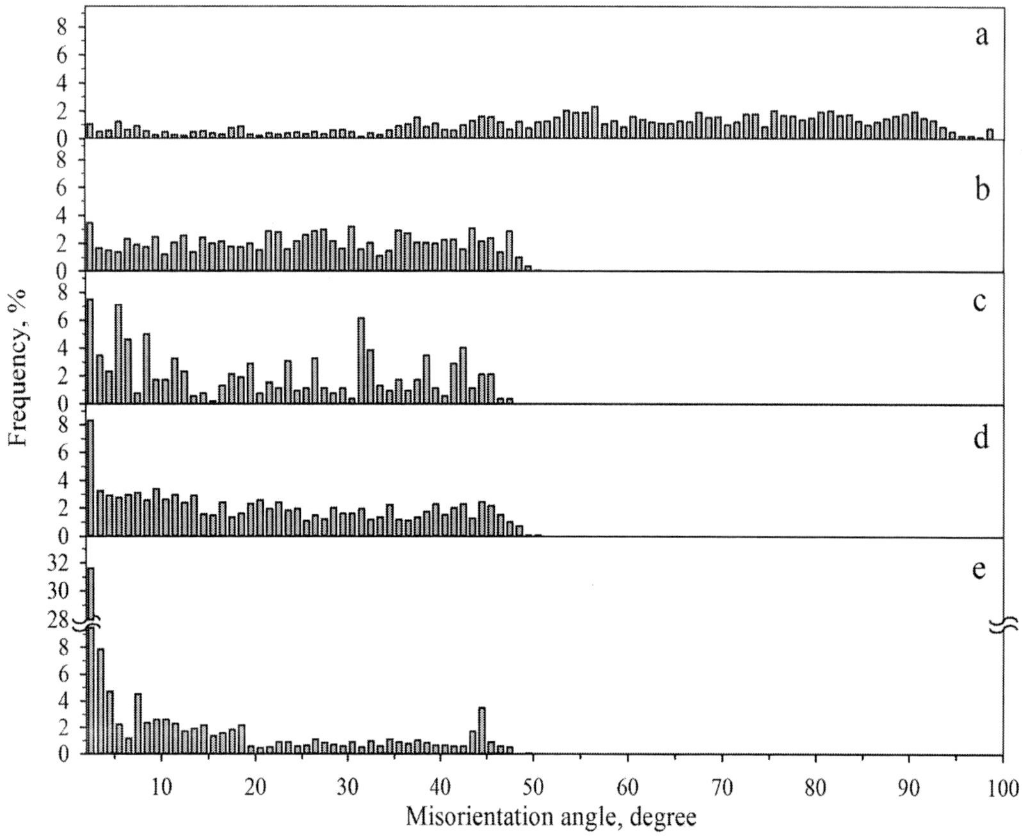

Figure 2. Distributions of grain boundary misorientation angle in (a) an undeformed sample Y123 and places of a deformed sample at a distance of (b) 0.5 mm, (c) 2 mm, (d) 4.2 mm, (e) 6.3 mm from the center of the sample.

4. Conclusions

The texture and grain-boundary misorientation distributions in different parts of the sample of the high-*Tc* superconductor Y123 deformed by torsion under pressure were studied. High critical current density should be expected at the edge of the sample, where a biaxial texture is formed, and the fraction low-angle grain boundaries exceed 50%.

Acknowledgements

The present work was accomplished according to the state assignment of IMSP RAS No AAAA-A17-117041310216-1.

References

[1] Gross R 2005 *Physica C* **432** 105
[2] Poltavets V, Kovalev K, Ilyasov R, Glazunov A, Maevsky V, Verzbitsky L, Akhmadyshev V and Shikov A 2014 *J. Phys.: Conf. Ser.* **507** 032022
[3] Levin A V, Vasich P S, Dezhin D S, Kovalev L K , Kovalev K L , Poltavets V N and Penkin V T 2012 *Phys. Procedia* **36** 747
[4] Imayev M F, Daminov R R, Reissner M, Steiner W, Makarova M V and Kazin P E 2007 *Physica C* **467** 14
[5] Imayev M F, Kabirova D B and Yakshibayeva R R 2011 *Perspektivnye Materialy. Special Issue*

No. 12 186 (in Russian)
[6] Imayev M F, Kabirova D B and Pavlova V V 2015 *Russian Phys. J.* **58** 762
[7] Imayev M F, Kabirova D B and Dementyev A V 2008 *New Research on YBCO Superconductors*, ed D M Friedman (NY: NOVA Publishers) pp 235-252

Synthesis and structure of nanoparticles in intermetallic Al_2Au

A Yu Volkov and E G Volkova

Mikheev Institute of Metal Physics, Ural Branch, Russian Academy of Sciences, 18 S. Kovalevskoi st., Ekaterinburg, 620137 Russia

E-mail: volkov@imp.uran.ru

Abstract. The intermetallic compound Al_2Au has been obtained from gold powder and aluminum particles through mechanical alloying in ball mills. The intermetallic structure was investigated by the methods of scanning transmission electron microscopy and X-ray diffraction analysis. It was found that mechanical alloying resulted in the formation of micro and nano particles of Al_2Au. The powder obtained was compacted into tablets. X-ray diffraction analysis showed intermetallic lines and one line belonging to aluminum, within the compact. The structure of compacts after annealing was investigated. It was determined that compacting promoted the agglomeration of intermetallic particles. Further annealing led to the surface depletion in aluminum, cracking and refinement of structural elements. After heat treatment, a phase enriched in gold – AlAu – is registered in the samples.

1. Introduction

The intermetallic compound Al_2Au or "purple gold" was discovered in 1892. Its bright color is first of all attractive for jewelers, but the brittleness inherent in all intermetallic compounds makes it quite difficult to use the alloy. Another obstacle consists in the formation of the intermetallic within a narrow concentration range about 1060 °C. An alternative can be found in creating Al_2Au coatings. Film samples are easier to obtain, they are not so brittle. In addition, a thin film of Al_2Au is an optical filter [1]. The synthesis of this intermetallic compound followed by the investigation of its structure and properties is of interest both for research and for practical use.

To obtain fine powders, different technologies are used, including mechanical alloying (MA) in ball mills. The use of this method is quite successful in creating materials, which are difficult to obtain in the traditional way.

Earlier, we suggested obtaining the intermetallic compound Al_2Au by MA of nanosized gold and aluminum powders [2].

The purpose of this work is to investigate the microstructure of the intermetallic compound Al_2Au obtained by MA in accordance with our proprietary technology.

2. Experimental procedure

A fine powder was obtained using the gas-phase method, which consists in evaporation of a drop of molten and condensation of vapor in an argon flow [3]. To obtain Au powder, a wire with a diameter of 0.6 mm of 99.99% pure gold was taken. The sizes of particle in the obtained powder were in the range of 30-50 nm [2]. Based on the results of previous research, it was concluded that the use of nanosized aluminum powder for MA was unreasonable due to the presence of an oxide film on its surface. That is why in this work the option of milling a mixture of nanosized gold with coarse (about 0.5 mm) aluminum filings was chosen. The test amounts of nanopowder of gold and aluminum filings were mixed in proportions corresponding to the composition of Al_2Au and dispersed by ultrasonication in toluene. MA was carried out in a planetary ball mill (PBM) Pulverisette-7 for 4 hours at a rotation speed of the milling pot of 800 rpm. The milling process was carried out in a hexane atmosphere; to ensure the most efficient running of the process, 15 balls with a diameter of 10 mm were used. The set of the planetary ball mill (milling pot and balls) was made of stainless steel. The ratio of the weights of the balls and source materials was 10:1 (the total weight of the balls was 60 g).

To carry out investigations, the powder obtained as the result of MA was compacted into tablets with a diameter of 11 mm and a thickness of 5 mm. The work included studies of compact samples after annealing in the temperature range from 300 to 600 °C. The compact was divided into 4 parts; each of them was annealed at a different temperature. Two samples with the most intense color were selected for the study. One was annealed at 350 °C for 30 minutes (mode 1). The other investigated sample was first soaked at 400 °C for 30 minutes and additionally treated at 600 °C, for 15 minutes (hereinafter – mode 2). The compacts were annealed in atmosphere of He.

The products of MA were studied using a JEM-200CX transmission electron microscope (hereinafter referred to as TEM study). The products of MA for the TEM study were placed between two copper grids. Microstructure evaluation was performed using a Philips SEM 515 scanning electron microscope (SEM). X-ray diffraction analysis was carried out using a Rigaku DMAX 2200 diffractometer.

3. Results and discussion

Figure 1 shows the results of the SEM study: it gives a general view of the synthetized powder and the elemental analysis data. Tiny particles of arbitrary shapes and agglomerations are well seen in figure 1a, b.

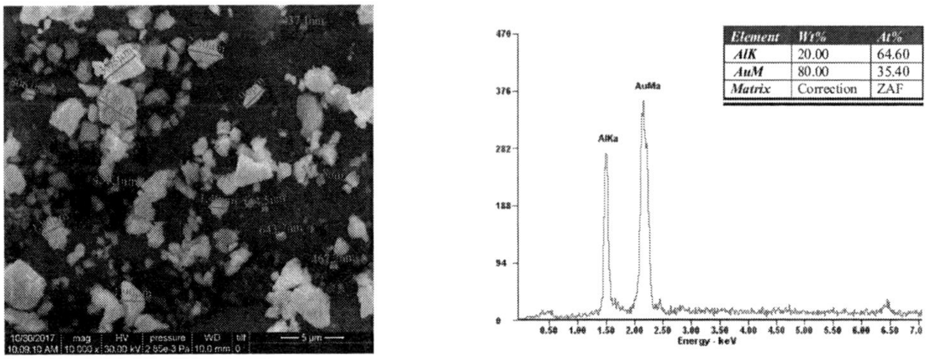

(a) (b)

Figure 1. Powder morphology (a); corresponding energy dispersive spectrum (b).

The elemental analysis (figure 1b) of the powder showed good compliance with the intermetallic compound Al_2Au (as is known, it contains 79 wt% of Au and 21 wt% of Al). The size of the structural elements of the synthetized powder, which are seen through SEM, varies from 300 nm to 3 μm. Sometimes, there are cracks and defects on the surface of particles, which can be related to their high hardness and brittleness – properties inherent in all intermetallic compounds. The TEM study of the obtained powder showed the formation of nano- and micro- particles of the intermetallic compound during MA (figure 2). Microdiffraction images are point-like for micro-particles and circular for nanograins, they correspond to the intermetallic compound Al_2Au. There were found areas of the sample where no intermetallic compound formation took place with Au grains and Al_2O_3 particles being observed. During the study, the amorphous phase was registered.

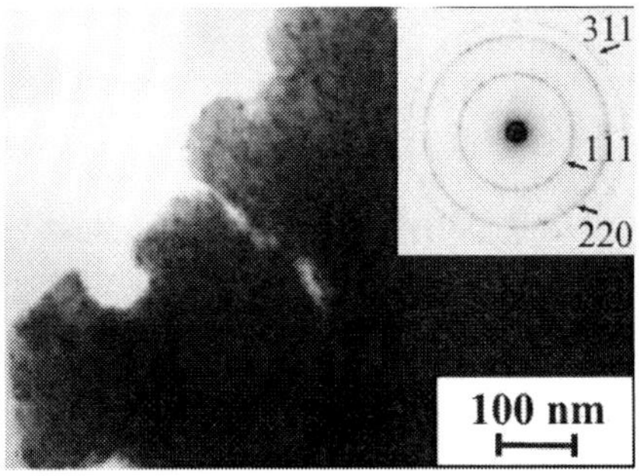

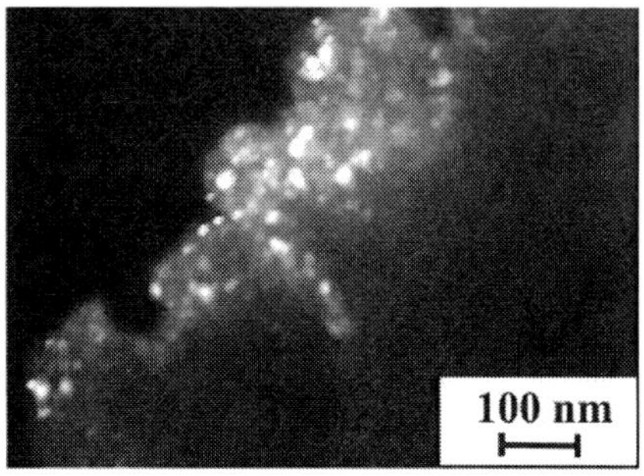

(a) (b)

Figure 2. TEM micrographs of mechanically alloyed powder Al_2Au; a – bright field image and SAD pattern, b – dark field image in 220_{Al2Au}.

To investigate the color change depending on the temperature of treatment, the synthetized powder was pressed into compacts. X-ray diffraction analysis (XDA) of a tablet pressed from MA powder defined in it the presence of intermetallic compound Al_2Au for the most part and a small amount of aluminum. The sizes of the coherent scattering regions were 25.2 ± 1.6 nm, the microstrain values: 0.29 ± 0.02 %. To study the dependence of the powder color on the treatment temperature, the compact was divided into four parts, each of which was annealed. For the studies, two samples with the most intense color were selected. After annealing, all the samples changed their color, but did not acquire the bright purple color characteristic of the intermetallic Al_2Au (figure 3).

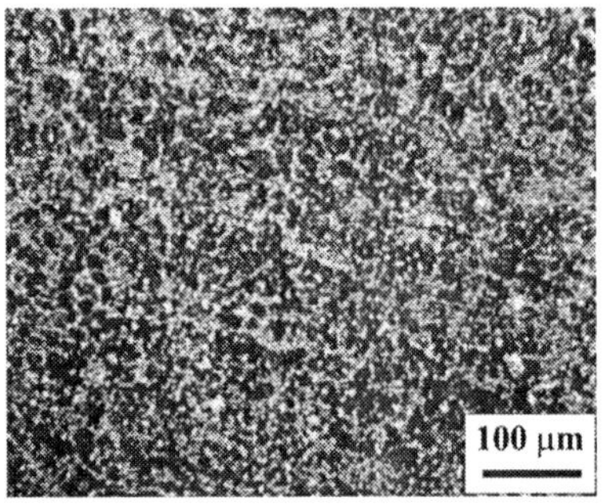

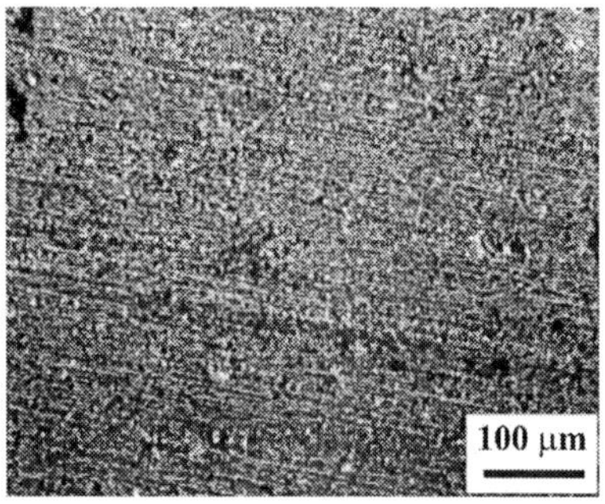

(a) (b)

Figure 3. Surface view of Al_2Au compact; a – mode 1, b – mode 2, optics.

The surfaces of the compacts, processed in two modes, were studied using SEM. Particles in heat-treated samples have arbitrary shapes; sharp edges and brittle chips are observed. Elemental analysis revealed that annealing causes a deviation in the ratio of components from the desired composition. The content of Al (13-16%) and Au (83-86%) in both samples would rather correspond to the AuAl phase. This indicates that in the process of heat treatment, the composition of the material is enriched with gold due to evaporation of some part of aluminum.

The study showed that after heat treatment in accordance with mode 1, the particle size is 0.5 ÷ 5 μm. On the surface of the particles there are multiple defects (cracks, chips). It can be assumed that during compacting and subsequent hydrostatic pressing, deformation hardening processes take place [4]. Recall that in the initial state, the synthesized powder had particle sizes from 0.3 to 3 μm (figure 1). It is obvious that agglomeration of particles occurred during this heat treatment. In turn, the particle sizes of the sample after heat treatment in mode 2 decreased to 0.2 ÷ 2 μm. Such a refinement of the grains with increasing annealing temperature is rare and requires an explanation. To obtain a more general picture of the processes taking place in compacts during annealing, X-ray diffraction studies were carried out. On the diffractograms of compacts, after annealing, lines appear from the AuAl phase along with the intermetallide Al_2Au. With an increase in the annealing temperature, the intensity of reflections from the AuAl phase goes up slightly, the number of X-ray peaks corresponding to the equiatomic intermetallic compound also increases (from two to three). In contrast to the diffractogram for the initial compact, no aluminum reflections are observed after annealing. The sizes of coherent scattering regions for mode 1 and mode 2 are: 78.5 ± 2.5 nm and 108.0 ± 4.5 nm, and the microstrain values are 0.15 ± 0.01% and 0.12 ± 0.01%, respectively. Thus, the X-ray diffraction study did not reveal an anomalous dependence of the grain size on the processing temperature. Consequently, the refinement of the structure in the compact treated in mode 2, as shown in figure 3, can be explained by the fracture of particles along cracks and defects caused by Al depletion of the surface layer of the material at a higher annealing temperature. Electron microscope studies showed that submicro- and nanocrystalline structures are formed in the sample after annealing in mode 2, the average grain size does not exceed 100 nm. Microdiffraction patterns contain rings from the Al_2Au phase. TEM study of the structure in the modes of bright and dark fields showed that a structure with high-angle boundaries is formed in the sample. In some grains of the intermetallic compound, an increased dislocation density and uneven contrast are observed.

On the basis of the data obtained, it can be assumed that the color change of the intermetallic phase of Al_2Au after annealing is caused by defects of different natures that occurred in the surface layer of the compacts as a result of heat treatment.

4. Conclusions

It is shown that it is possible to obtain a sufficiently pure powder of the intermetallic compound Al_2Au by the MA method. The formation of various intermetallic grains, ranging from nanostructures to microcrystallites, takes place. During MA, an amorphous phase is formed.

Pressing and heat treatment promote the agglomeration of intermetallic particles. Additional annealing at the temperature of 600 °C leads to cracking and refinemnt of structural elements in the surface layer. The composition of the samples after annealing changes: the AuAl phase appears.

Compacting and subsequent heat treatments lead to a change in color. The change in color is associated with the appearance of new phases, the crushing of particles and the increase in their defectiveness on the surface of the samples.

Acknowledgments

The work was carried out within the framework of the State task on the themes: Pressure № AAAA-A18-118020190104-3. Electron microscope studies were performed in the Department of Electron Microscopy of the Shared Use Center of the Institute of Metal Physics of the Urals Branch of the Russian Academy of Sciences.

References

[1] Supansomboon S, Maaroof A and Cortie M B 2008 *Gold Bull.* **41** 296
[2] Volkov A Yu, Uimin M A, Mysik A A, Novozhonov V I, Volkova E G, Shchegoleva N N, Knyazev Yu V and Kozlov K A 2011 *Inorg. Mater.* **47** 465
[3] Popov V V, Stolbovskii A V, Uimin M A and Ermakov A E 2013 *Phys. Metals Metallogr.* **114** 77
[4] Nes E 1997 *Progr. Mater. Sci.* **41** 129

IOP Conf. Series: Materials Science and Engineering **447** (2018) 012015 doi:10.1088/1757-899X/447/1/012015

Effect of deformation nanostructuring on thermal expansion and phase composition of Fe - 36% Ni alloy

K A Krylova[1], I Kh Bitkulov[1] and R R Mulyukov[1,2]

[1]Institute for Metals Superplasticity Problems of RAS, 39 Stepan Khalturin St., Ufa 450001, Russia
[2]Bashkir State University, 32 Zaki Validi St., Ufa 450076, Russia

E-mail: bukreevakarina@gmail.com

Abstract. The influence of deformation nanostructuring and subsequent annealing on the temperature dependence of the thermal expansion coefficient of the Fe-36% Ni Invar alloy is studied. The nanostructured state was obtained by severe plastic deformation via high pressure torsion in Bridgman anvils. An "anomalous" contraction of the nanostructured Fe-36% Ni alloy was detected during heating in certain temperature ranges. The phase composition of the nanostructured Invar was studied to explain the behavior of the thermal expansion coefficient. The revealed anomalies of the coefficients are explained by the formation of a bcc phase not observed in the Fe-36% Ni Invar alloy in the coarse-grained state. The formation of this phase becomes possible due to a significant increase in the diffusion coefficient as a result of nanostructuring.

1. Introduction

Invar alloys of the Fe-Ni system have found wide application in modern engineering and instrumentation due to the anomalously low value of the coefficient of thermal expansion (CTE) – α(T) ≈ 1.6·10-6 °C-1 [1]. For the first time, unique properties of Invar were found in the Fe-36% Ni alloy by Guillaume at the end of the 19th century. It is known that the chemical compositions of the Fe-36% Ni alloy and its deformation processing influence the CTE. Alloying elements or impurities commonly lead to an increase of the CTE while plastic deformation reduces the latter. Therefore, one can assume that deformation nanostructuring can reduce the CTE of the Invar alloy.

Today, there are many works [3-5] on the study of nanostructured (NS) materials, since they have properties that are not characteristic for materials in the coarse-grained (CG) state. Deformation nanostructuring also influences the phase composition of alloys, for example, leads to the suppression of martensitic [6] or polymorphic hcp-fcc [7] transformations. Accordingly, the study of the properties of the Fe-36% Ni alloy in the NS state can broaden its application and, undoubtedly, is of scientific interest.

The aim of this work is to study the CTE of the Fe-36%Ni alloy after deformation nanostructuring by high pressure torsion and upon annealing. To explain the temperature dependence of the CTE of the Invar alloy in different structural states, its phase composition was studied.

2. Experimental

The NS samples were obtained by severe plastic deformation (to a true strain of e = 7) via the high-pressure torsion method in Bridgman anvils. The average crystallite size of the processed sample was about 100 nm [8].

The dependences of CTE on temperature α (T) were measured using a Dh 1500 RHP ULVAC SINKU-RIKO dilatometer. The heating rate of the samples was 2 °C/min. Measuring the temperature dependences α(T) consisted of several cycles. The specimen was heated from room temperature to maximum temperature and then annealed at this temperature for 30 minutes, then cooled inside the device to room temperature. After that, the measurement of this sample was repeated, but it was heated to the next maximum temperature. To exclude bending deformation during heating, a sample in the form of a plate (0.1 x 4.5 x 7.4 mm3) was fixed in special slots of two quartz cylinders. The length of the plate exceeded the depth of the slits by 0.8 mm, which provides mechanical contact with the dilatometer sensor.

The phase composition of samples in a different structural state was investigated by X-ray diffraction analysis at room temperature. To obtain different structural states, the samples after deformation nanostructuring were annealed in vacuum for 1 hour at 300, 400, 420 and 480 °C.

3. Results and discussion

Figure 1 shows the temperature dependences of the CTEα(T) of the Fe - 36% Ni Invar alloy in the NS state (square symbols) and after annealing at 280 °C (circle symbols), 350 °C (rhomb symbols), 500 °C (triangle symbols) and 800 °C (the solid line) for 30 minutes. These dependences have a non-linear character and for some cases of measurements they are non-monotonic. The non-monotonic behaviour of the α (T) curves is observed during heating from room temperature to 280 °C, 350 °C and 500 °C (see curves with symbols of a circle, a rhomb and a triangle, respectively, in figure 1). All the curves show that the CTE sharply decreases when the annealing temperature of the measured sample is reached. The CTE decreases even to negative values when the specimen is heated above 350 °C, which corresponds to its annealing temperature, and after reaching above 480 °C, the value α(T) increases sharply. Such a non-monotonic temperature dependence of CTE is not observed in the CG Fe-36% Ni Invar alloy (see the dashed curve in figure 1) [9]. However, figure 1 shows that the curve α (T) for the NS Invar annealed at 800 °C (the solid line in figure 1) practically coincides with the curve of the CGFe - 36% Ni alloy (the dashed line).This indicates that the annealing of the NS Invar at 800 °C leads to the formation of a grain structure similar to the CG state.

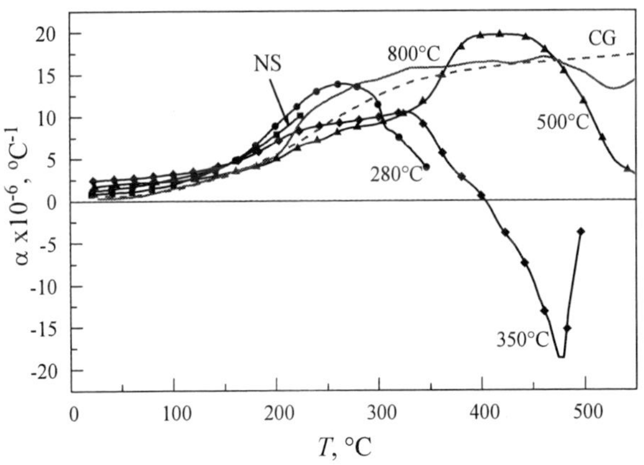

Figure 1. Temperature dependences of the CTEα (T) of Fe - 36% Ni Invar alloy in the NS state (square symbols) and after annealing at 280 °C (circle symbols), 350 °C (rhomb symbols), 500 °C (triangle symbols) and 800 °C (the solid line) [8].The dotted line shows the dependence α (T) for Invar in the CG state [9].

To explain the anomalous contraction of the NSFe-36% Ni Invar alloy during heating above 350 °C, the phase composition of the alloy in different structural states was studied by X-ray diffraction analysis. Figure 2 shows X-ray diffraction patterns of the NS Fe-36% Ni alloy in the as-processed state (a) and after annealings at 300 °C (b), 400 °C (c), 420 °C (d) and 480 °C (e). The peaks in the diffraction patterns obtained from the sample after deformation nanostructuring (see figure 2 a) correspond to the fcc lattice γ-Fe. However, the peaks have a lower intensity and a greater width than in the case of the CG sample, which is due to the structural nonequilibrium and high internal elastic stresses in the sample arising due the process of nanostructuring.

The peaks appearing in the X-ray diffraction patterns after annealing of the NS samples at 350 °C, 400 °C and 420 °C (figure 2 b-d) correspond to the bcc lattice of α-Fe. As the annealing temperature increases, the intensity of the bcc peaks increases and reaches a maximum value at 420 °C (figure 2 d). A further increase in the annealing temperature leads to the dissolution of α-Fe, and the bcc phase peaks are not observed in the diffraction pattern at 480 °C (figure 2 e). Note that at temperatures below 500 °C, ordered phases of FeNi and/or $FeNi_3$ can form in Fe-Ni alloys [10]. However, it is impossible to reliably confirm (or refute) the presence of these phases using the X-ray diffraction method, since the lattice parameters of the main γ-phase and intermetallic phases practically coincide (see figure 2).

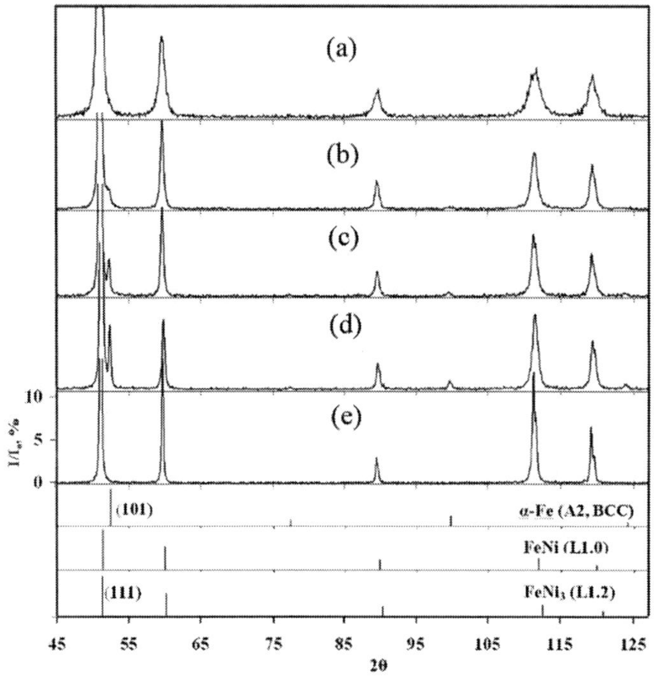

Figure 2. X-ray diffraction patterns of the NS Fe-36% Ni (a) alloy and after annealing at 300 °C (b), 400 °C (c), 420 °C (d) and 480 °C (e) for 1 hour. I/I_0 - is the relative intensity, reduced to the intensity I_0 of the (111) γ-Fe peak.

Figure 3 shows the dependences of the CTE at 20 °C $\alpha_{20°C}$ (square symbol) and the height of the diffraction peaks (101) for the bcc phase I/I0 (circle symbol) on the annealing temperature T_{an} of the NS Fe-36% Ni alloy. On the $\alpha_{20°C}$ curve one can see that the CTE increases with increasing T_{an} and after annealing at 420 °C, it reaches a value of 2.48×10^{-6} °C^{-1}. At this temperature, the intensity of the diffraction peaks of the bcc phase reaches a maximum. Thus, the degradation of the properties of invar at room temperature is related to the separation of the bcc phase during annealing of the NS Invar at temperatures above 350 °C. Similar behavior was previously found in [11, 12]. The contraction of the sample observed in the CTE temperature dependence of the Invar alloy annealed at 350 °C (curve with rhomb symbols in figure 1) is associated with the dissolution of the less dense bcc phase during heating above T_{an}. The α-phase is completely dissolve at 480 °C and the diffraction peaks, corresponding only to the main fcc phase, are observed (figure 2 e), i.e. the structure becomes completely single-phase. This leads to a sharp increase in the value of the CTE (see curve with rhomb symbols in figure 1).

The γ → α transition, usually not observed in Fe-Ni alloys with a Ni content of more than 33%, found in the annealed NS Fe-36% Ni alloy became possible due to an increase in the diffusion coefficient by several orders of magnitude after deformation nanostructuring [13, 14]. The results obtained in this work can extend the use of the Fe-36% Ni Invar alloy in various industries due to the detected influence of the annealing temperature after deformation nanostructuring on CTE. In other words, by varying the value of T_{an}, one can create an Invar specimen with a certain value of thermal expansion.

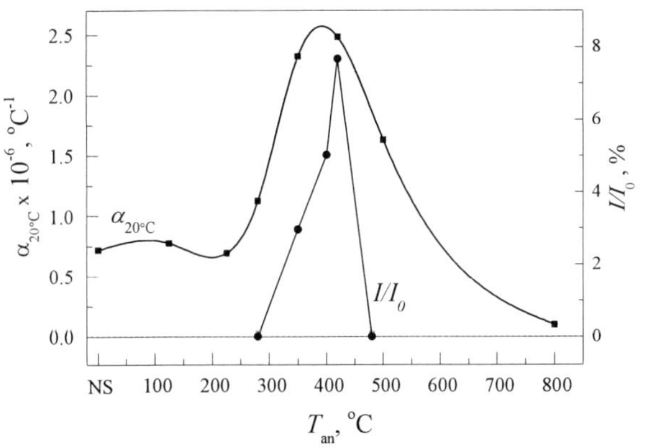

Figure 3. Dependence of the CTE at 20 °C $\alpha_{20°C}$ (square symbol) and the height of the diffraction peak (101) bcc phase I/I_0(circle symbol) on the annealing temperature T_{an} of the NS Fe - 36% Ni alloy.

4. Conclusions

The formation of a nanosized structure in the Fe-36% Ni Invar alloy using deformation by high pressure torsion and subsequent annealing leads to an increase in CTE at 20 °C at certain annealing temperatures. This behavior is associated with the separation of the bcc phase during annealing, which is not observed in the CG Fe-36% Ni alloy. In addition, the observed shortening in the NS Invar upon reheating after annealing at 350 °C is due to the dissolution of the less densely packed bcc phase. The formation of this phase becomes possible due to a significant increase in the diffusion coefficient as a result of deformation nanostructuring.

References

[1] Callister W D 2007 *Materials Science and Engineering: An Introduction* 7th edition (NY: John Wiley & Sons) p 729
[2] Scorzelli R B 1997 *Hyperfine Inter.* **110** 143
[3] Nazarov A A and Mulyukov R R 2003 *Handbook of Nanoscience, Engineering, and Technology*, Ed. Goddard W, Brenner D, Lyshevski S, Iafrate G, (Boca Raton, CRC Press) pp 22-1-41
[4] Mulyukov R R, Imayev R M and Nazarov A A 2008 *J. Mater. Sci.* **43** 7257
[5] Mulyukov R R, Sharipov I Z, Bukreeva K A and Bitkulov I Kh 2010 *Phys. Metals Metallogr.* **109** 234
[6] Musabirov I I, Safarov I M, Mulyukov R R, Sharipov I Z and Koledov V V 2014 *Letters on Materials* **4** 265
[7] Valeev I Sh, Sergeev V I and Mulyukov Kh Ya 2009 *Phys. Solid State* **51** 593
[8] Mulyukov R R, Kazantsev V A, Mulyukov Kh Ya, Burkhanov A M, Safarov I M and Bitkulov I Kh 2006 *Rev. Adv. Mater. Sci.* **11** 116
[9] Menshikov A Z 1977 *J. Magn. Magn. Mater.* **5** 188
[10] Kubaschewski O 1982 *Iron Binary Phase Diagrams* (Springer-Verlag, NY) p 175
[11] Nagayama T, Yamamoto T and Nakamura T 2016 *Electrochimica Acta* **205** 178
[12] Rodriguez R R, Valenzuela J L, Tabares J A and Pérez Alcázar G A 2014 *Hyperfine Inter.* **224** 323
[13] Valiev R Z, Korznikov A V and Mulyukov R R 1993 *Mater. Sci. Eng. A* **168** 141
[14] Würschum R, Kübler A, Gruss S, Scharwaechter P, Frank W, Valiev R Z, Mulyukov R R and Schaefer H-E 1996 *Annalles de Chimie* **21** 471

IOP Conf. Series: Materials Science and Engineering **447** (2018) 012016 doi:10.1088/1757-899X/447/1/012016

Structural varieties of carbon compounds

E A Belenkov[1], **V A Greshnyakov**[1] **and E A Belaya**[2]

[1] Physics Department, Chelyabinsk State University, Chelyabinsk 454001, Russia
[2] Chemistry Department, Chelyabinsk State University, Chelyabinsk 454001, Russia

E-mail: belenkov@csu.ru

Abstract. The existence of allotropic, polymorphic and polytypic structural varieties of carbon materials and nanostructures is possible. Allotropes differ by the atomic coordination in the structure. In polymorphs, the coordination is the same, but bond lengths and bond angles differ. In polytype structures, there is a different order of molecular layer packing. A theoretical analysis of the structures of polytypic varieties of carbon and their phase transformations is performed in this paper.

1. Introduction

Carbon materials can have different structures and properties [1,2]. With the same chemical composition, the material properties are determined by the structure, therefore, carbon materials can be described and classified according to their structural features [3]. To denote structural varieties, several terms are commonly used, such as allotropes, polymorphs and polytypes. A precise definition of these terms is necessary. In addition, there may exist different nanostructures, the features of which are nanoscale in one, two and three crystallographic directions [3]. All these structural varieties of carbon should be considered not isolated, but taking into account the possibility of their mutual structural transformations. Therefore, in this paper, a hierarchical classification of carbon structural varieties is considered. Also, the polytypic varieties and the possibilities of their mutual transformations are described in detail.

2. Structural classification of carbon compounds

Coordination of carbon atoms in the structure most strongly affects the properties of carbon compounds. Carbon atoms in compounds can be in 2-, 3-, and 4-coordinated states, that is, they can have a corresponding number of nearest neighbouring atoms. The coordination determines the hybridization of electron orbitals of the atom. Therefore, each coordinated state (2, 3, and 4) corresponds to the hybridized state (sp, sp^2, and sp^3). It is this feature of the structure that distinguishes various allotropic modifications of carbon. Typical allotropes are carbyne, graphite and diamond, in which the atoms are in three different coordinated states.

Each carbon atom in compounds tends to form four covalent bonds with neighbouring atoms, therefore allotropic varieties differ in the covalent bond order. In the diamond structure (4-coordinated atomic states), covalent bonds with neighbouring atoms are single, since each of them is formed by one pair of electrons. In graphene, four covalent bonds are formed with three neighbouring atoms, so one bond accounts for 1.33 electron pairs, which correspond to the bond order. In the third allotropic variety of carbon, carbyne, each atom has two neighbouring atoms, and the average bond order is two.

Based on this parameter, the remaining carbon compounds can be assigned to one of the three allotropic groups.

Another factor affecting the structure and properties of carbon compounds is their crystallographic dimension (nD_c). The number n in this parameter indicates the number of Cartesian coordinate axes along which the carbon structure has macroscopic dimensions. Therefore, compounds can have crystallographic dimension $0D_c$ (fullerenes), $1D_c$ (nanotubes), $2D_c$ (graphene layers), and $3D_c$ (3D-graphites and diamond-like phases).

Different combinations of the two parameters, nD_c and N_a, determine the individual structural groups of carbon compounds [3]. Allowing the possibility of changing the parameter N_a from 0 to 4, theoretically, the existence of twelve [nD_c, N_a] structural groups becomes possible [3]. The compounds of each structural group can be transformed into compounds of another group, changing the atomic coordination and crystallographic dimension of the compounds. These transformations can occur due to the formation of additional interatomic bonds or the breaking of existing bonds. This allows theoretically predict the structure of new carbon phases and nanostructures.

Within each structural group, variations in the compound structure are also possible. Such structural varieties should be called polymorphs. In polymorphic varieties, atoms have the same coordination, and the polymorphic structures have the same crystallographic dimension. Differences in the structures of polymorphs are due to the interatomic bond length and bond angles distortions, which can be quantitatively characterized by the parameters of *Str* and *Def*. As ideal structures in which there is no distortion (Str = 0 and Def = 0), polycumulene carbyne, hexagonal graphene and cubic diamond structures are adopted. For example, there may be four main varieties of graphene [3]. The differences in the structure of polymorphs from each other are less than the structural differences in allotropes, therefore the properties of polymorph differ less than in allotropes.

The next level of structural differences concerns the packing order of molecular layers in the structures of carbon compounds. Therefore, the existence of a variety of polytypes is possible for each polymorphic variety [4]. For example, in addition to hexagonal and rhombohedral graphites formed on the basis of graphene L_6 layers, polytypes based on other graphene polymorphs can exist [5]. In the structure of polytypes, the short-range order of the atom arrangement differs slightly. Significant differences are observed in distant coordination spheres. Therefore, the properties of polytypes differ slightly. It is necessary to carry out a theoretical analysis of the possible structure of polytypes of carbon compounds.

3. Structural analysis of polytypes

The theoretical analysis was carried out by enumeration of all possible combinations of molecular layers packing in unit cells of different sizes. As a result of the analysis of the graphite and diamond polytype structures, the possibility of the existence of isomeric varieties of polytypes was established. Polytypes with a period of recurrence $N > 5$ have a number of isomeric varieties with the same N, but have different hexagonality (H) (table 1). As a result, several polytypes with a different recurrence periods have a close or identical hexagonality [4]. Apparently, they can simultaneously form and sustainably exist at identical values of thermodynamic parameters, since their cohesion energies have close values [4]. Isomeric structures with the same period (N) can have different hexagonality. The formation of different polymorphic varieties probably can occur as a result of mutual phase transformations.

4. Modeling of polytype phase transformations

When modelling polytype phase transitions, it was assumed that they could be due to changes of graphene layers in the graphite crystals or molecular layers in diamond crystals. It was assumed that they can occur due to shifts of graphene layers in graphite crystals or molecular layers in diamond crystals. Calculations of the activation energy of the shifts were performed using the atomic-atomic potential method for graphite. The image of the area of the relative shift of graphene layers is shown in figure 1. The results of calculating the binding energy of neighbouring layers depending on the shift

IOP Conf. Series: Materials Science and Engineering **447** (2018) 012016 doi:10.1088/1757-899X/447/1/012016

are shown in figure 2. The height of the potential barrier that must be overcome in order to shift the layer from one local minimum to another is 65 μeV/atom. Therefore, in graphite nanocrystals, the ordered structure can be absent at room temperature, and the structural transformations of some polytypes into others can occur continuously.

Table 1. Structural characteristics of polytypes (N is the period of layer packing; M_N is the number of all possible combinations; L_N is the number of allowable combinations; P_N is the number of allowable combinations minus equivalent ones; F_N is the number of unique polytypes; H is the hexagonality of the polytype; W is the probability of polytype formation. Shown in the brackets are the equivalent combinations of those having smaller values of N).

N	M_N	L_N	P_N	F_N	H (%)	Layer alternation	W (%)
2	9	6	1	1	100	ABAB...	100
3	27	6	1	1	0	ABCABC...	100
4	81	18	3	1	50	ABAC...	66.67 (100)
5	243	30	5	1	40	ABABC...	100
6	729	66	11	2	66.7	ABABAC...	54.55 (66.67)
					33.3	ABACBC...	27.27 (33.33)
7	2187	126	21	3	57.1	ABABABC...	33.33
					57.1	ABABCAC...	33.33
					28.6	ABACABC...	33.33
8	6561	258	43	6	75	ABABABAC...	18.6 (20)
					75	ABABACAC...	9.3 (10)
					50	ABABACBC...	18.6 (20)
					25	ABABCABC...	18.6 (20)
					50	ABABCBAC...	18.6 (20)
					25	ABACBABC...	9.3 (10)

The activation energy of the diamond layer was calculated by the density functional theory method for the successive shift of two atoms (with numbers 4 and 5 in figure 3) in the ab-plane from the position ($2a/3$; $b/3$) to the position ($a/3$; $2b/3$). In diamond, the molecular layer shift energy is 2.3 eV/atom (figure 4). Such shifts of layers can cause phase transformations of the structure of any noncubic polytype into the structure of any other polytype. In addition, shear transformations can be the cause of the formation of crystals with random packing of layers [6]. Phase transitions of some polytypes into others are possible in nanodiamonds at high temperatures, whereas in macrocrystals these transitions may be absent due to the significant value of the activation energy necessary for shifts in the molecular layer.

5. Conclusions

Thus, the structural classification of carbon compounds is described. Structural varieties of carbon compounds can be divided into allotropes, polymorphs and polytypes. The structures of allotropes differ from each other in the number of atoms in the first coordination sphere. In polymorphs, the first coordination number is the same, but the bond lengths and bond angles are different. Polytype structures differ only in the packing order of the molecular layers. In addition, a theoretical analysis of the structure of carbon polytypic varieties was carried out, and the phase transformations of graphite and diamond polytypes were calculated.

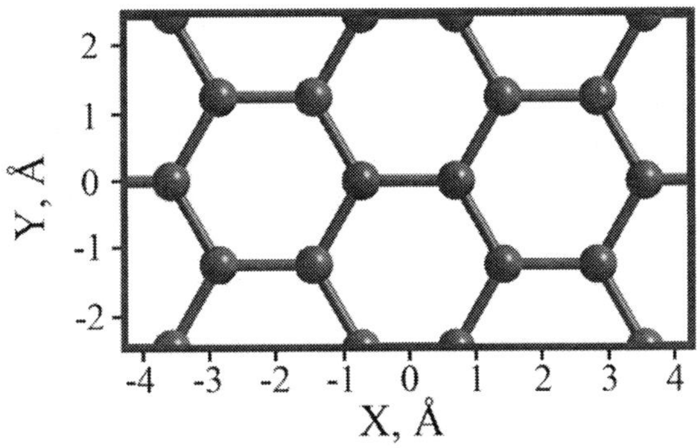

Figure 1. Graphene layer fragment within which relative shifts of adjacent layers were performed.

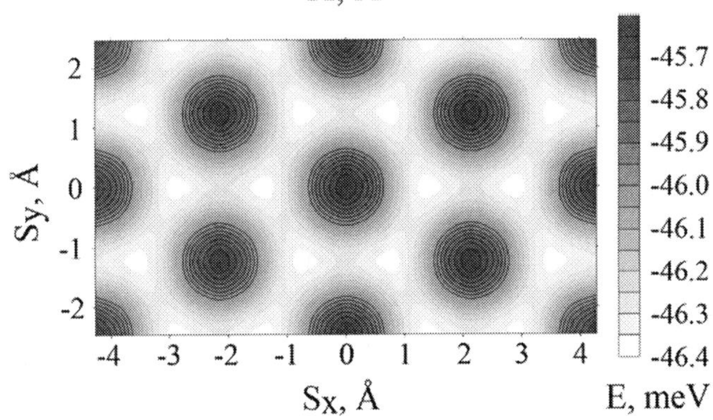

Figure 2. Dependence of the van der Waals energy (E) on the graphene layer shifts along the X and Y axes.

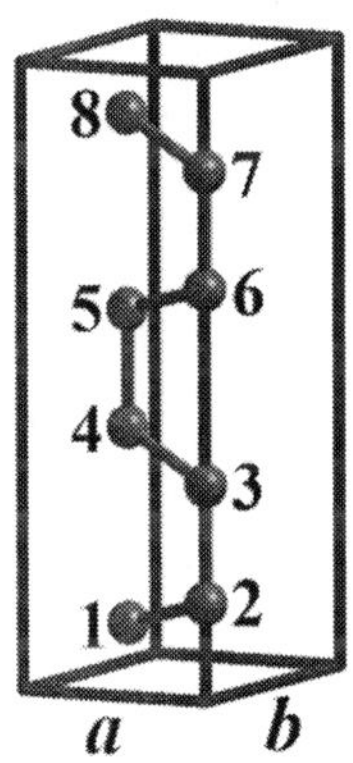

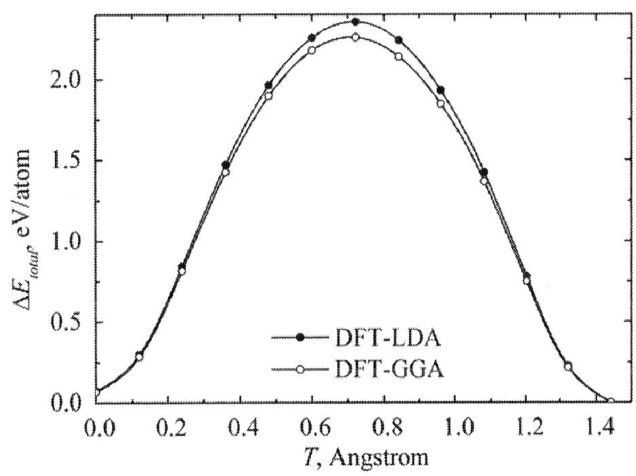

Figure 3. Hexagonal diamond unit cell.

Figure 4. Dependences of the difference total energy on the shear vector modulus for the shifted molecular layer in 2H diamond.

References

[1] Pierson H O 1993 *Handbook of Carbon, Graphite, Diamond, and Fullerenes: Properties, Processing and Applications* (Park Ridge: Noyes) p 43
[2] Lisovenko D S, Baimova Yu A, Rysaeva L Kh, Gorodtsov V A and Dmitriev S V 2017 *Phys. Solid State* **59** 820
[3] Belenkov E A and Greshnyakov V A 2013 *Phys. Solid State* **55** 1754
[4] Belenkov E A and Greshnyakov V A 2017 *Phys. Solid State* **59** 1926
[5] Enyashin A N and Ivanovskii A L 2011 *Phys. Status Solidi B* **248** 1879
[6] Greshnyakov V A and Belenkov E A 2017 *J. Exp. Theor. Phys.* **124** 265

IOP Conf. Series: Materials Science and Engineering **447** (2018) 012017 doi:10.1088/1757-899X/447/1/012017

Influence of ultrasound on the structure and properties of nickel processed by equal-channel angular pressing

A A Samigullina[1], A A Mukhametgalina[1,2], A A Nazarov[1], N Yu Parkhimovich[1], A P Zhilyaev[1], Yu V Tsarenko[3], V V Rubanik[3]

[1]Institute for Metals Superplasticity Problems, Russian Academy of Sciences, 39 Khalturin st., 450001 Ufa, Russia
[2]Bashkir State University, 32 Zaki Validi st., 450074 Ufa, Russia
[3]Institute of Technical Acoustics, National Academy of Sciences of Belarus, 13 Lyudnikova pr., 210023 Vitebsk, Belarus

E-mail: asiya_nazarova@mail.ru

Abstract. Commercially pure nickel was processed by equal-channel angular pressing (ECAP) and subjected to ultrasonic treatment (UST) with a stress amplitude of 100 MPa. The microstructure, microhardness and mechanical properties of the deformed and sonicated samples were studied. It was shown that the UST affected the microstructure of ECAP nickel. A refinement of grains in the longitudinal section, changes in the texture, and an increase in the dislocation density and microstrain were observed. Tensile tests have shown an increase in the elongation to failure without any noticeable changes in the tensile strength. The possible mechanisms of the action of ultrasound on the structure and properties of the deformed nickel are discussed.

1. Introduction

Technologies for the production and processing of metallic materials based on the ultrasonic effects are now widely used. Depending on the parameters of the ultrasonic action and the material used, it is possible to obtain its certain structural states and properties. For example, the use of ultrasound for the crystallization of melts leads to a more uniform and fine-grained structure of cast metals and alloys [1]. The impact of high intensity ultrasound is used for surface hardening of materials [2]. At present, ultrasonic welding of various materials finds its application and provides the basis for the development of a new technology of ultrasonic additive manufacturing [3].

At the same time, much attention is now paid to the study and application of materials subjected to severe plastic deformation, which demonstrate high strength, fatigue, and other improved properties but suffer from structural features that result in a low ductility, impact toughness and thermal stability of the microstructure [4]. Annealing used for the structure relaxation of deformed materials often leads to an undesirable grain growth that decreases their strength properties.

Earlier it was shown that the moderate-intensity ultrasonic treatment (UST) reduced the microstrain and dislocation density and increased the thermal stability of ultrafine-grained pure nickel processed by high-pressure torsion [5,6]. UST of commercially pure nickel subjected to ECAP can lead both to relaxation and hardening of the structure, depending on the stress amplitude that affects the mechanical properties [7-9], and can increase the impact toughness of the material [10].

In this paper, the effect of UST with a stress amplitude of 100 MPa on the structure and properties of commercially pure nickel processed by ECAP is investigated. This amplitude was chosen on the basis of the results of our previous studies [7-9].

2. Experimental materials and procedure

The rods of commercially pure nickel (grade NP2 according to the Russian classification) with a length of 100 mm and a diameter of 20 mm were subjected to 8 passes of ECAP following route Bc at a temperature of 350 °C on a tool with an intersection angle of channels 120°. Then, the resulting samples were extruded at a temperature of 300 °C to a length of 115 mm and a cross-sectional diameter of 18 mm. The total deformation degree was equal approximately to 8.

Two samples after deformation were processed by UST according to the scheme described in [8]. A standing wave with a sinusoidal stress distribution was excited in the sample. The antinode of compression-tension stresses was located in the center of the rod, their amplitude was approximately equal to 100 MPa.

Cylinders with a length of 32 mm were cut from the central part of the samples after ECAP and after ECAP + UST. The change in the amplitude of the UST at a small distance from the center of the sample is negligible, and it can be assumed that over this length the amplitude of the UST was equal to 100 MPa. From these cylinders, flat samples for tensile tests in the longitudinal direction and specimens for studying the microstructure and measuring the microhardness were cut.

Tensile tests were carried out on an Instron testing machine at room temperature. The specimens for these tests had a gauge length of 10 mm, width of 3 mm and thickness of 1.5 mm. The Vickers microhardness of the samples was measured using an Axiovert 100A equipment with an MNT_10 attachment with a load on the indenter of 100 g and an exposure time of 10 s.

EBSD analysis was performed using a TESCAN MIRA 3 LMH FEG microscope. Scanning was carried out with steps 100 and 50 nm from square grids with areas of 400 and 100 μm^2 respectively. The surface of the samples for scanning was prepared by mechanical and subsequent electrolytic polishing.

X-ray diffraction (XRD) analysis of the investigated cylindrical samples was carried out on a DRON-4 apparatus in planes perpendicular and parallel to the axis of the cylinder. The CuK_α line was used with Bragg-Brentano focusing. A Soller collimator for primary beam and a graphite monochromator for the diffracted beam were used. The experimental data were processed by means of the software «Maud» (Materials Analysis Using Diffraction).

3. Experimental results

Figure 1 shows the microstructure of nickel after ECAP and after ECAP+UST in longitudinal and cross sections, obtained by EBSD analysis. It is noticeable that after the UST the grains are visually finer. In the cross-section, twins are observed in some grains (shown by arrows in figures 1c and d).

The quantitative results of X-Ray diffraction and EBSD analyses for longitudinal and cross sections are presented in table 1. It can be seen that the grain sizes in both states are rather large (2-4 microns). UST facilitated a reduction in the average grain size and a slight increase in the microstrain in the longitudinal direction, but the microstrain in the cross direction practically did not change. The dislocation density also increased after the UST.

The UST affected the distribution of grain boundaries by the misorientation angle. Table 1 presents the fractions of three types of grain boundaries: low-angle boundaries (LABs, misorientation angle from 2° to 15°), high-angle boundaries (HABs, >15°), and boundaries close to special ones (CSL). The volume fraction of LABs did not change, but the fraction of CSL significantly decreased (especially in the longitudinal section), a part of them apparently were transformed into the HABs.

Figure 2 shows the inverse pole figures of nickel after ECAP and after ECAP+UST in longitudinal and cross sections, composed according to X-Ray diffraction data. In the initial deformed state, one can see a multicomponent texture with preferential orientations <100> and <110>, in addition, there are components <311> and <210>. After the UST, the intensity of some of them decreased, and a tendency to form a one-component texture with a preferential orientation <100> is observed.

Table 1. XRD and EBSD data for longitudinal (l.s.) and cross sections (c.s.) of samples after ECAP and after ECAP+UST with the amplitude of 100 MPa.

State	X-Ray		EBSD					
	Domain size, nm	Microstrain, $\times 10^{-4}$	Grain size d, μm Major axis	Grain size d, μm Minor axis	Dislocation density, m^{-2}	LAB, %	HAB, %	CSL, %
ECAP, l.s.	332.6±5.3	4.78±0.14	4.4	1.8	13.21×10^{13}	52	29	19
ECAP+UST, l.s.	354.6±7.1	6.02±0.32	2.4	1.2	18.80×10^{13}	54	38	8
ECAP, c.s.	240.1±3.8	6.49±0.21	3.3		13.58×10^{13}	43	21	36
ECAP+UST, c.s.	311.1±4.1	6.37±0.57	2.9		16.55×10^{13}	50	17	33

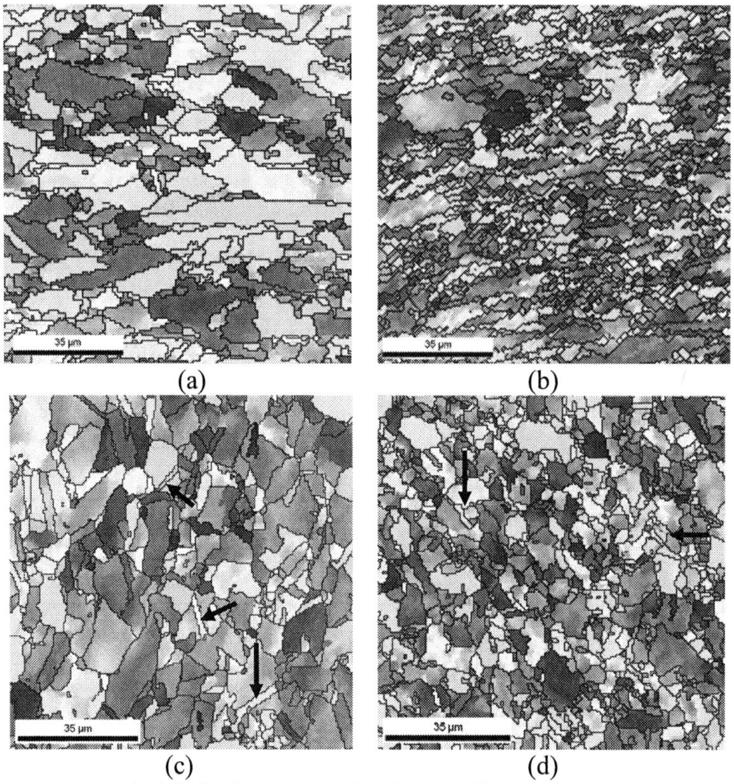

Figure 1. Microstructure of nickel after ECAP (a, c) and after ECAP+UST with the amplitude of 100 MPa (b, d). a, b - longitudinal section (the deformation axis is horizontal), c, d – cross-section. Arrows show the twins in the structure.

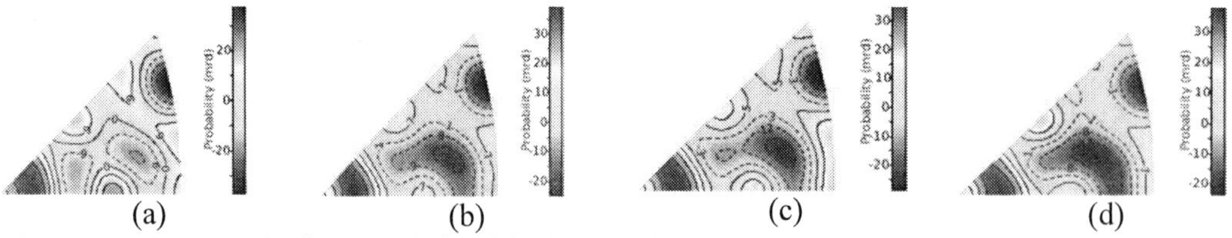

Figure 2. Inverse pole figures of nickel in longitudinal (a, b) and cross (c, d) sections: a, b - after ECAP, b, d - after ECAP+UST with amplitude of 100 MPa.

The results of tensile tests and measurements of the microhardness of nickel after the treatments are given in table 2. It can be seen that the elongation to failure increases by 8-9%, which is more than a quarter increase in the elongation as compared to the initial ECAP-deformed state. At the same time, there are no noticeable changes in the tensile strength and the microhardness.

Table 2. Mechanical properties and microhardness of nickel after ECAP and ECAP+UST with the amplitude of 100 MPa.

State	Ultimate strength, MPa	Elongation to failure, %	Microhardness Hv, МПа
ECAP	439±18	32,5±2,5	1574±27
ECAP+UST	440±21	41,3±2,0	1554±59

In our earlier paper [7], it was also shown that the UST resulted in texture changes of nickel subjected to ECAP. Moreover, after the UST a slight increase in the ductility of nickel was obtained during the tensile tests without noticeable changes in the tensile strength. However, in the present study, the high-amplitude UST obviously led not to a relaxation of the internal stresses, as it was in [7], but, on the contrary, to their increase. This is probably due to additional grain refinement and an increase in the density of dislocations. But the UST process is different from the deformation by its mechanisms. Both the generation and the accumulation of dislocations, as well as their reorganization, annihilation inside the grains take place during the action of oscillating stresses. The contribution of each of these mechanisms, as was shown in [5-8], depends on the amplitude of the UST. Therefore, a decrease of the grain size does not lead to a significant increase in the internal stresses and other characteristics, and even contributes to some increase in the ductility of the material.

Thus, our recent and present studies show that the effect of UST on the structure and mechanical properties of UFG materials can be very diverse. This can lead to an increase in the ductility and a further increase in the ultimate strength, or to an increase in the ductility without a changing the strength. The effect obviously depends on the amplitude of the ultrasound and on the initial state of the UFG material. In order to establish all regularities of the structure and property changes in UFG metals under ultrasound, further systematic studies are needed.

4. Conclusions

The results of microstructure studies, tensile tests and microhardness measurements show that the UST with an amplitude of oscillating stresses equal to 100 MPa of nickel processed by ECAP with a moderate deformation degree results in texture changes (the tendency to form one-component texture), a reduction of the average grain size, a slight increase of dislocation density and a noticeable increase in ductility without changes in tensile strength and microhardness. This effect is related to specific mechanisms of changes in the microstructure operating during the UST process.

Acknowledgments

The present work was supported by the Russian Science Foundation (grant No. 16-19-10126). Electron microscopic studies were carried out on the facilities of shared services center of the Institute for Metals Superplasticity Problems of Russian Academy of Sciences «Structural and Physical-Mechanical Studies of Materials».

References

[1] Abramov O V 1999 *High-Intensity Ultrasonics: Theory and Industrial Applications* (CRC Press, Boca Raton) pp 515-522

[2] Mordyuk B N and Prokopenko G I 2007 *J. Sound Vibr.* **308** 855

[3] Gallego-Juarez J A and Graff K 2014 *Power Ultrasonics. Applications of High-Intensity Ultrasound* (Woodhead Publishing) pp 313-332

[4] Valiev R Z, Zhilyaev A P and Langdon T G 2014 *Bulk Nanostructured Materials:*

Fundamentals and Applications (New Jersey: Wiley) p 338

[5] Samigullina A A, Mukhametgalina A A, Sergeyev S N, Zhilyaev A P, Nazarov A A, Zagidullina Yu R, Parkhimovich N Yu, Rubanik V V and Tsarenko Yu V 2018 *Ultrasonics* **82** 313

[6] Nazarova A A, Mulyukov R R, Rubanik V V, Tsarenko Yu V and Nazarov A A 2010 *Phys. Metals Metallogr.* **110** 574

[7] Samigullina A A, Tsarenko Yu V, Rubanik V V, Popov V A, Danilenko V N and Mulyukov R R 2012 *Letters on Materials* **2** 214

[8] Samigullina A A, Nazarov A A, Mulyukov R R, Tsarenko Yu V and Rubanik V V 2014 *Rev Adv. Mater. Sci.* **39** 48

[9] Zhilyaev A P, Samigullina A A, Medvedeva A E, Sergeev S N, Cabrera J M and Nazarov A A 2017 *Mater. Sci. Eng. A* **698** 136

[10] Samigullina A A, Mulyukov R R, Nazarov A A, Mukhametgalina A A, Tsarenko Yu V and Rubanik V V 2014 *Letters on Materials* **4** 52

IOP Conf. Series: Materials Science and Engineering **447** (2018) 012018 doi:10.1088/1757-899X/447/1/012018

Diamond-like phase formed of carbon C_{24} clusters

V A Greshnyakov and E A Belenkov

Physics Department, Chelyabinsk State University, Chelyabinsk 454001, Russia

E-mail: greshnyakov@csu.ru

Abstract. The crystalline structure and some properties of the cubic diamond-like CA6 phase have been calculated by the density-functional theory method in the generalized gradient approximation. As a result of calculations, it was found that the density, cohesive energy, bulk modulus, hardness and band gap width of this phase are 2.824 g/cm^3, 7.59 eV/atom, 351 GPa, and 71 GPa, respectively. The phase structure of CA6 should be stable at room temperature.

1. Introduction

Compounds with diamond-like structures are widely used as anticorrosive coatings, abrasive and structural materials. A number of new diamond-like phases that can be formed on the basis of fullerene-like molecules, carbon nanotubes, graphene layers, and 3D covalently bonded graphite-like compounds are theoretically predicted [1-4]. These phases can be close to cubic diamond by mechanical characteristics [1-10]. The diamond-like CA6 phase formed on the basis of fullerene-like C_{24} molecules may be the most stable among other diamond-like carbon C-phases [4]. Therefore, this paper is devoted to theoretical calculations of the structure, properties, and stability of cubic diamond-like CA6 phase.

2. Methods

Calculations of the structural and energy characteristics of the CA6 phase were performed in the software package Quantum ESPRESSO [11] using the density functional theory method in the generalized gradient approximation. The Perdew-Burke-Ernzerhof exchange-correlation energy functional was used for the calculations [12]. The effect of ion cores was taken into account using norm-conserving pseudopotentials. We used a $10 \times 10 \times 10$ grid of k-points in our calculations. The wave functions were decomposed in terms of a truncated basis set of plane waves with the cut-off energy of 60 Ry. The unit cell of the CA6 phase was optimized until the magnitudes of the stresses and forces acting on the atom became less than 1 mRy/Å and 0.50 GPa, respectively. The hardness and the bulk modulus were calculated using the methods described in [13,14].

3. Results

The crystalline structure of the diamond-like CA6 phase was modelled as a result of the linking of C_{24} carbon clusters according to the procedure described in [1,4]. The geometrical optimized structure of the new phase is shown in figure 1a. The CA6 phase unit cell is body-centred cubic ($Im\bar{3}m$ – № 229) with the parameter a = 4.3927 Å (figure 1b). The unit cell contains 16 atoms, the coordinates of which are given in table 1. All atoms in the CA6 phase are in symmetrically equivalent positions. As in the cubic diamond, all covalent bond lengths in the structure of CA6 are the same and equal to 1.5531 Å. However, there are two non-equivalent bond angles, the values of which are 90 and 120°.

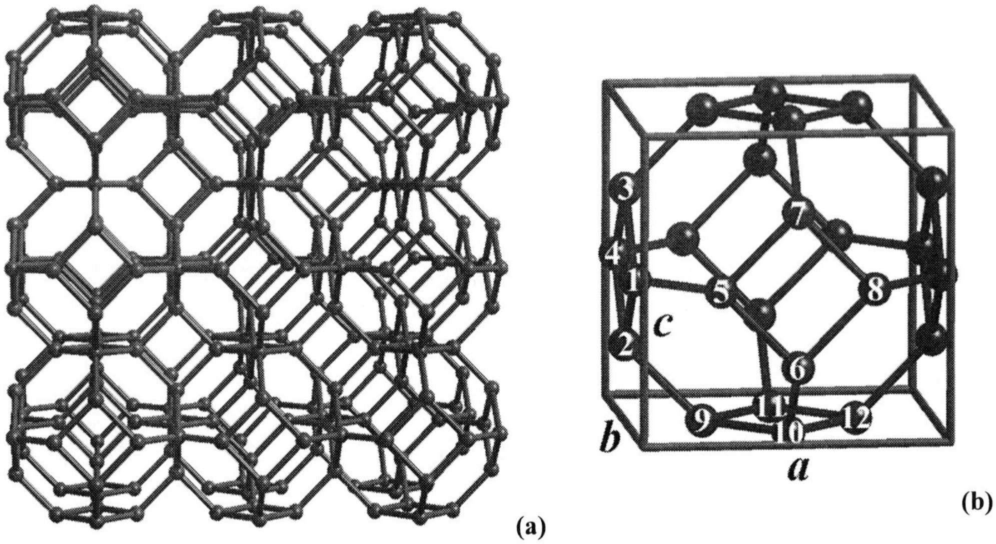

Figure 1. Fragment of the structure (a) and unit cell (b) of the diamond-like CA6 phase.

Table 1. Atomic coordinates in the CA6 phase unit cell.

Atomic number	*X* (Å)	*Y* (Å)	*Z* (Å)	Atomic number	*X* (Å)	*Y* (Å)	*Z* (Å)
1	0.0000	1.0982	2.1963	7	2.1963	0.0000	3.2945
2	0.0000	2.1963	1.0982	8	3.2945	0.0000	2.1963
3	0.0000	2.1963	3.2945	9	1.0982	2.1963	0.0000
4	0.0000	3.2945	2.1963	10	2.1963	1.0982	0.0000
5	1.0982	0.0000	2.1963	11	2.1963	3.2945	0.0000
6	2.1963	0.0000	1.0982	12	3.2945	2.1963	0.0000

Further, the following properties of the cubic diamond-like CA6 phase were determined: density, cohesive energy, bulk modulus and Vickers hardness. The cohesive energy, density, hardness and bulk modulus of a new structural modification of diamond are lower than these properties of 3C diamond polytype by 5.3, 17.5, 18.1 and 21.1%, respectively.

The electronic properties of cubic diamond-like CA6 phase were studied using the band structure and the density of electronic states (DOS) calculations. Figure 2a shows the band structure of this phase. The electron energy was calculated with six intervals between four points of high symmetry in the Brillouin zone: ΓN, NH, HΓ, ΓP, PN, and PH. The minimum difference in the electron energies between the bottom of the conduction band and the top values of the valence band is 4.72 eV and is observed at 1/3 of the length of the $\overrightarrow{H\Gamma}$ vector. As a result of the DOS calculation (figure 2b), the magnitude of the indirect band gap of the diamond-like CA6 phase is determined, and its value is 3.96 eV. Therefore, the studied structural variety of diamond is a wide-gap semiconductor with an indirect band gap, which is 27% less than the corresponding value for 3C diamond polytype.

The thermal stability of the cubic diamond-like CA6 phase was studied by the molecular dynamics method. The structure was annealed for 8 ps at a constant temperature of 400 K. The graph of the total energy change from the annealing time is shown in figure 3. As a result of the calculation it was established that the CA6 phase crystal lattice is stable when the temperature is above room temperature.

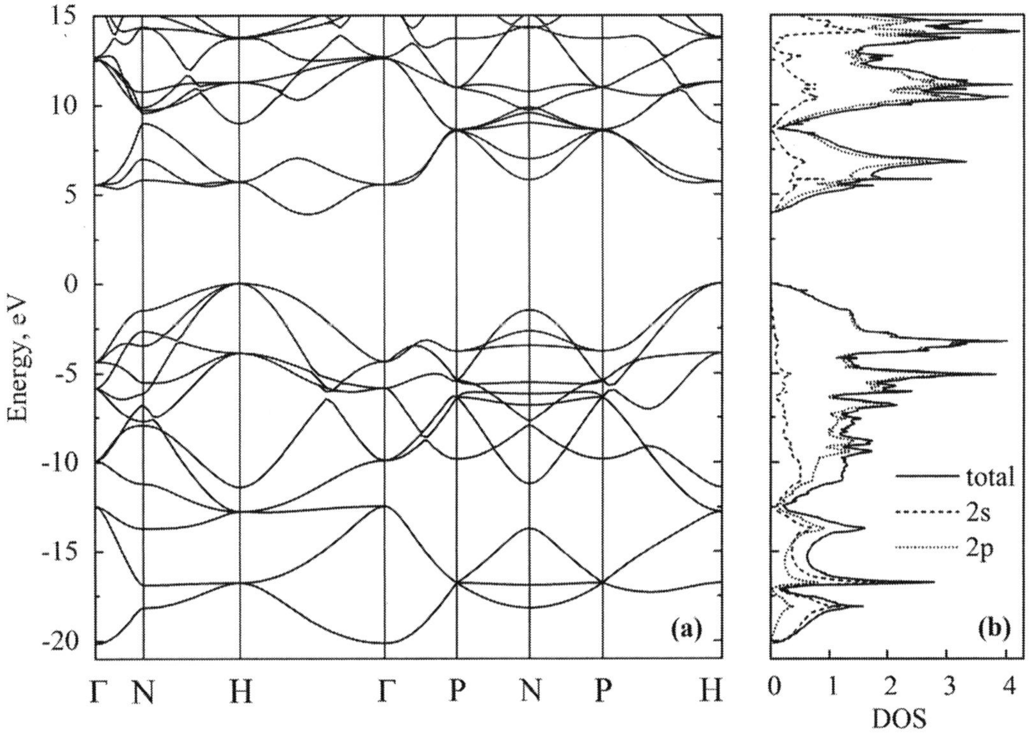

Figure 2. Band structure (a) and DOS (b) of diamond-like CA6 phase.

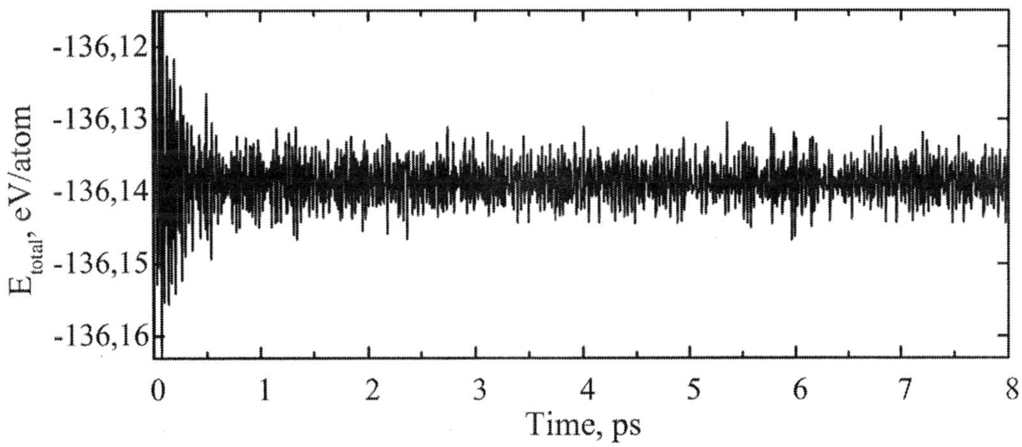

Figure 3. Dependence of the total energy on annealing time for diamond-like CA6 phase.

To enable experimental identification of the theoretically predicted phase, the calculation of its X-ray diffraction pattern was performed. The calculated powder X-ray diffraction patterns of the cubic diamond and the diamond-like CA6 phase are shown in figure 4. The positions of the most intense diffraction lines of the CA6 phase differs significantly from the positions of the diffraction maxima of the cubic diamond, which should allow its uncomplicated identification in the synthesized carbon materials.

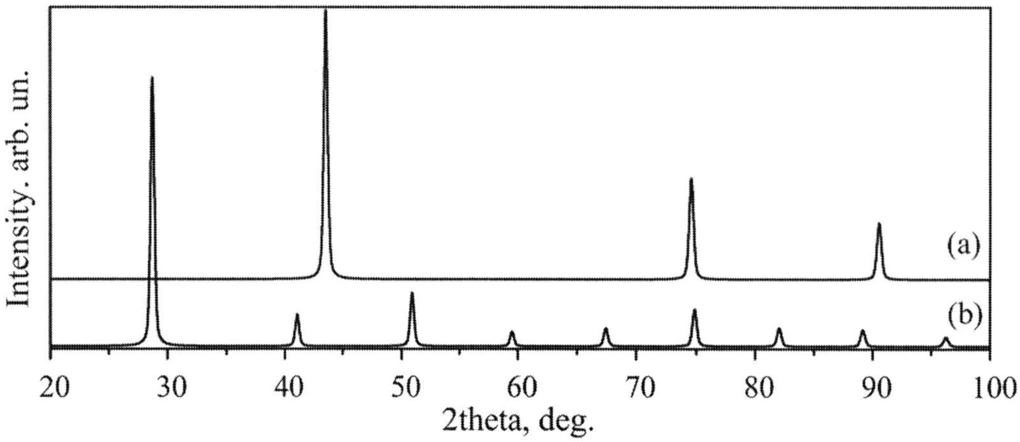

Figure 4. Powder X-ray patterns of the cubic diamond (a) and the CA6 phase (b).

4. Conclusions

As a result of the calculations performed by the density functional theory method, a geometrically optimized structure and some properties of the cubic diamond-like CA6 phase are determined. The structure of this polymorph can be formed by linking carbon fullerene-like C_{24} clusters in the form of truncated octahedra. The calculated properties of the CA6 phase are close in values to the properties of the 3C diamond polytype. Thus, the density, cohesion energy, Vickers hardness, and bulk modulus of this phase are 5-20 % lower than the corresponding properties of the cubic diamond. The CA6 phase should be a wide-gap semiconductor with a direct band width of ~ 4.7 eV. The energy and mechanical characteristics of the cubic CA6 phase, studied in detail in this paper, exceed the corresponding properties of most diamond-like phases with equivalent atomic positions [1-6,9]. It was also found that the structure of this phase is stable at a temperature of 400 K which indicates a high probability of its experimental production.

References

[1] Belenkov E A and Greshnyakov V A 2014 *J. Struct. Chem.* **55** 409
[2] Belenkov E A, Brzhezinskaya M M and Greshnyakov V A 2014 *Diam. Relat. Mater.* **50** 9
[3] Belenkov E A and Greshnyakov V A 2015 *J. Mater. Sci.* **50** 7627
[4] Belenkov E A and Greshnyakov V A 2015 *Phys. Solid State* **57** 2331
[5] Lisovenko D S, Baimova J A, Rysaeva L Kh, Gorodtsov V A, Rudskoy A I and Dmitriev S V 2016 *Phys. Status Solidi B* **253** 1295
[6] Lisovenko D S, Baimova Yu A, Rysaeva L Kh, Gorodtsov V A and Dmitriev S V 2017 *Phys. Solid State* **59** 820
[7] Greshnyakov V A and Belenkov E A 2017 *J. Exp. Theor. Phys.* **124** 265
[8] Grishakov K S, Katin K P and Maslov M M 2018 *Diam. Relat. Mater.* **84** 112
[9] Baimova J A, Rysaeva L Kh, Gorodtsov V A and Rudskoy A I 2018 *Diam. Relat. Mater.* **81** 154
[10] Greshnyakov V A and Belenkov E A 2018 *Inorg. Mater.* **54** 111
[11] Giannozzi P, Baroni S, Bonini N et al. 2009 *J. Phys.: Condens. Matter.* **21** 395502
[12] Perdew J P, Burke K and Ernzerhof M 1996 *Phys. Rev. Lett.* **77** 3865
[13] Gao F, He J, Wu E, Liu S, Yu D, Li D, Zhang S and Tian Y 2003 *Phys. Rev. Lett.* **91** 015502
[14] Greshnyakov V A and Belenkov E A 2014 *Russ. Phys. J.* **57** 731

IOP Conf. Series: Materials Science and Engineering **447** (2018) 012019 doi:10.1088/1757-899X/447/1/012019

Analysis of microstructure of additively manufactured stainless steel

G F Korznikova[1], T N Tzibizova[2], S N Sergeyev[1], V V Smirnov[2], S P Pavlinich[2], D V Gunderov[3,4], G R Khalikova[1], R R Mulyukov[1]

[1] Institute for Metals Superplasticity Problems, Russian Academy of Sciences, 39 Khalturin st., Ufa 450001, Russia
[2] Ufa State Aviation Technical University, 12 K. Marx st., Ufa 450008, Russia
[3] Saint Petersburg State University, 28 Universitetskiy pr., Saint Petersburg 198504, Russia
[4]Institute of Molecule and Crystal Physics - Subdivision of the Ufa Federal Research Centre of the Russian Academy of Sciences, 151 Prospekt Oktyabrya, Ufa 450075, Russia

E-mail: korznikova@anrb.ru

Abstract. Selective laser sintering was used for additive manufacturing of stainless steel parts. The microstructure of the initial powder and the as-processed part was investigated. It was found that high temperature gradients involved in selective laser sintering evolve non-equilibrium microstructures in the as fabricated state, e.g. retained austenite in martensite, significant dislocation density and residual stresses. The additively manufactured samples have an ultrafine grained structure with predominantly high angle grain boundaries and are free of a crystallographic texture and residual porosity.

1. Introduction

Selective laser sintering (SLS) is a promising method of modern additive production. Interest in additive technologies is associated with the possibility to produce relatively quickly and inexpensively complex three-dimensional parts without intermediate operations of mechanical or deformation-heat treatment. By now, it has been established that the complex of performance characteristics is determined by the micro- and macrostructure formed during SLS, which, in turn, depends on the conditions and modes of fusion, as well as on the quality of the initial powder [1,2]. Often the heterogeneity of the structure, impurities and porosity do not allow achieving the expected properties. Since steel is the most demanded material in various industries - first of all in engineering - studies of the structure and properties of steels obtained by additive methods were reported in a large number of works [1-3]. In addition, the allotropy of iron-based alloys in combination with large temperature gradients that occur with all additive production methods opens up opportunities for creating unique microstructures in steels. In particular, in martensitic-aging steels at high cooling rates, the presence of retained austenite is possible, which can have a significant effect on both the structure and the physical and mechanical properties of steel. In this paper, the microstructure including the grain boundary spectrum and the phase composition of the EOS PH1 martensitic alloy steel obtained by the SLS is studied. It should be noted that the study of steels produced by SLS is not only of scientific, but also of practical value both for the development of new ways of producing already existing parts, and for the manufacturing of new products.

2. Experimental

Commercial EOSINT system equipment was used for the SLS processing of EOS Stainless Steel PH1 powder. The chemical composition of the powder conforms to the compositions of DIN 1.4540 and is presented below..

Table 1. Chemical composition of EOS Stainless Steel PH1 (wt. %).

Fe	Cr	Ni	Cu	Mn	Si	Mo	N
balance	14-15.5	3.5-5.5	2.5-4.5	≤ 1	≤ 1	≤ 1	0.15-0.45

Microstructural characterization of the Stainless Steel PH1 powder and the sample processed by SLS was performed using a Tescan Mira 3LMH scanning electron microscope (SEM) with the help of secondary electrons (SE) and back-scattered electrons (BSE) detectors. The fine structure of the SLS sample was examined by a JEOL-2000 EX transmission electron microscope (TEM). The chemical composition was investigated using a Tescan VEGA 3 SBH SEM with an energy dispersive X-ray spectroscopy (EDS) attachment. Electron backscattered diffraction (EBSD) analysis was performed by a Tescan Mira 3 LMH SEM. To improve the reliability of the EBSD data, small grains comprising three or fewer pixels were automatically removed from the maps. Furthermore, to eliminate spurious boundaries caused by orientation noise, a lower limit boundary misorientation cut off of 2° was used. A 15° criterion was employed to differentiate low-angle and high-angle boundaries. The grain size was quantified by the determination of the area of each grain and the calculation of its circle-equivalent diameter, i.e., the so-called grain reconstruction [4].

3. Results and discussion

Since the powder characteristics impact the bulk material properties of the fabricated component, such as density and porosity of the part, the structure of the initial powder was analyzed by SEM. The initial powder had a form of spherical or near spherical granules with an average size of about 25 μm, as shown in figure 1 a. Some individual granules reached a size of 50 μm. With a larger magnification, separate grains were found inside the granules, the average size of which was less than 2 μm (figure 1 b). X-ray diffraction analysis showed that the powder consisted of the bcc martensite phase (90%) and fcc austenite phase (10%), and the average size of the coherent-scattering region was 200 nm, which corresponded to the SEM observation. Such a phase composition of the steel indicates a high quenching rate during the powder preparation [5].

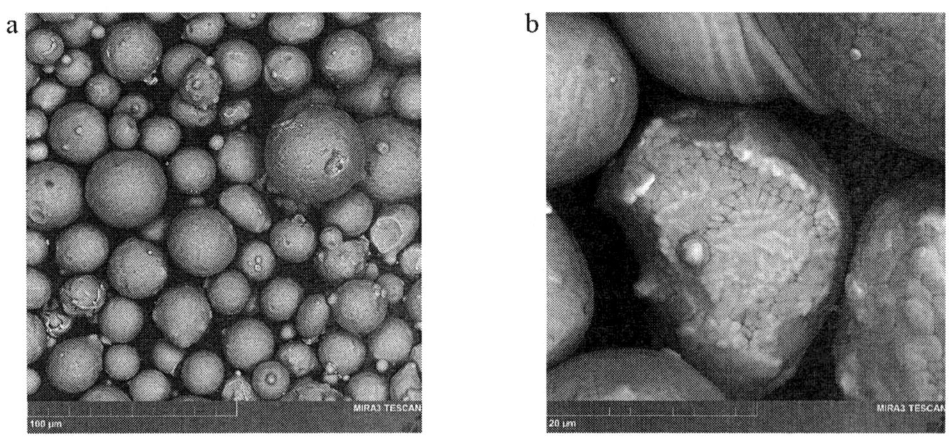

Figure 1. BSE image of the Stainless Steel PH1 powder at low (a) and high (b) magnifications.

The microstructure of the Stainless Steel PH1 sample, processed by SLS, in the longitudinal and transverse sections is shown in figure 2. The samples were found to be monolithic without visible pores and microcracks. The structure was generally uniform both in grain size and in the chemical composition of the samples. Most grains had a fairly elongated shape, but it is impossible to distinguish the preferred direction of elongation (figures 2 a and 2 b).

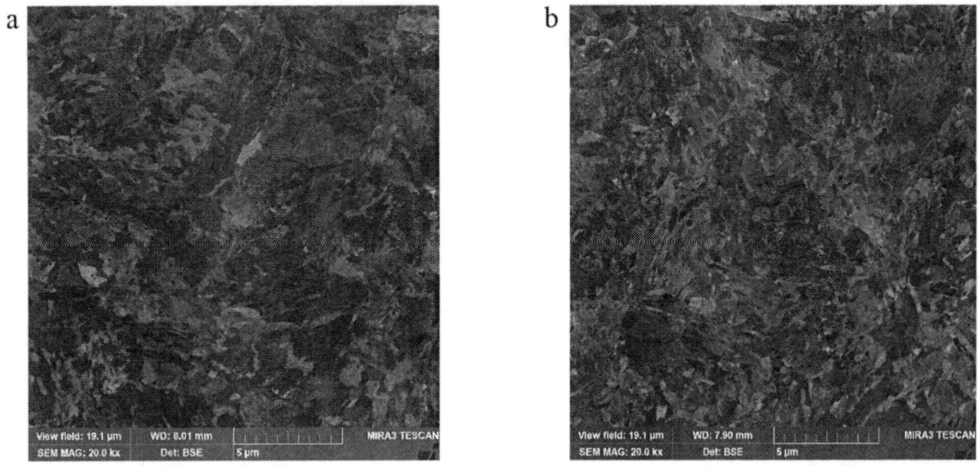

Figure 2. BSE image of Stainless Steel PH1 sample processed by SLS in the longitudinal (a) and transverse (b) sections.

X-ray diffraction analysis (figure 3) revealed no crystallographic texture, showed the same phase composition i.e. *bcc* (90%) and *fcc* (10%), high internal stresses and increased values of lattice parameters in the additively manufactured sample, obviously caused by high cooling rates during the SLS process.

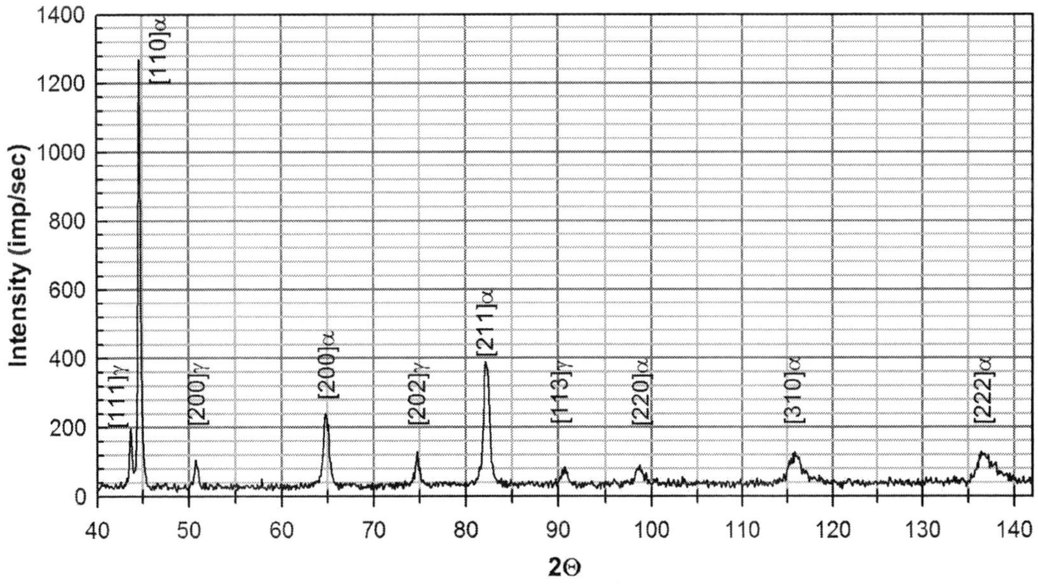

Figure 3. X-ray diffraction pattern of Stainless Steel PH1 sample processed by SLS.

Additionally, transmission electron microscopy has been carried out to obtain a qualitative view. It showed a significant dislocation density and a specific contrast of the matrix caused by small coherent particles (figure 4a), which were invisible on SEM micrographs. The grain boundaries were mainly blurred that pointed out to their nonequilibrium state and a high level of internal stresses of the additively processed steel, which coincided with the X-Ray data.

EBSD mapping revealed *bcc* and *fcc* grains in the transverse and longitudinal sections of the sample. The average size of both *bcc* and *fcc* grains was found about 0.4 μm (figure 4b). The fraction of the *bcc* phase high angle grain boundaries was about 75%, therewith 67% of them represented the boundaries between the variants corresponding to the Kurdy/umov-Sachs orientation relationship. This mechanism of martensitic transformation is a consequence of the phase instability of a crystal in high local stress fields.

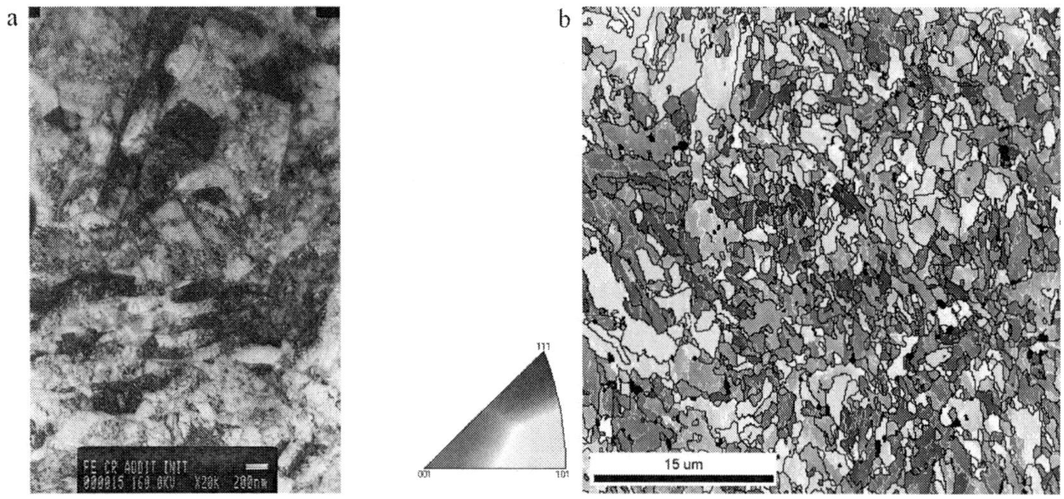

Figure 4. TEM micrograph (a) and EBSD map (b) of additively manufactured Stainless Steel PH1.

4. Summary

The results presented in this paper provided strong evidence that the additive manufacturing of Stainless Steel PH1 by SLS-processing allowed one to synthesize an ultrafine grained material with predominantly high angle grain boundaries free of crystallographic texture and residual porosity. The homogeneous spherical shape of the initial powder meets the requirements of the SLS process to achieve homogeneous spreading of the powder, as well as good packing characteristics for generation of a high relative density. High temperature gradients involved in SLS evolve non-equilibrium microstructures in the as-fabricated state, e.g. retained austenite in martensite, significant dislocation density and residual stresses.

Acknowledgments

The work was supported by the program of fundamental scientific researches of Governement Academy of Sciences No. AAAA-A17-117041310213-0. Authors wish to acknowledge assistance of their colleague Dr. S. Mironov in EBSD studies.

References

[1] Herzog D, Seyda V, Wycisk E and Emmelmann C 2016 *Acta Mater.* **117** 371
[2] Guo P, Zou B, Huang Ch and Gao H 2017 *J. Mater Process. Technol.* **240** 12
[3] Salzbrenner B C, Rodelas J M, Madison J D, Jared B H, Swiler L P, Shen Yu-L and Boyce B L 2017 *J. Mater Process. Technol.* **241** 1
[4] Humphreys F J 1999 *J. Microscopy* **195** 170
[5] Bernshtein M L and Rahshtadt A G 1983 *Metallurgy and Heat Treatment of Steel* (Moscow: Metallurgy) (in Russian)

IOP Conf. Series: Materials Science and Engineering **447** (2018) 012020 doi:10.1088/1757-899X/447/1/012020

Mixed $Ca^{2+}/Na^{+}(Mg^{2+})$ polyphosphates for polymer matrix filling and their solubility

D M Zuev[1], E S Klimashina[1,2], M Kikuchi[3], Sh Oshima[3], J Labuta[4], V I Putlyaev[1,2], P V Evdokimov[1], Ya Yu Filippov[2] and TV Safronova[2]

[1] Department of Materials Science, Lomonosov Moscow State University, Moscow, 119991, Russia
[2]Department of Chemistry, Lomonosov Moscow State University, Moscow, 119-991, Russia
[3]Bioceramics Group, Research Center for Functional Materials, National Institute for Materials Science, 1-1 Namiki, Tsukuba, 305-0044, Japan
[4] International Center for Materials Nanoarchitectonics (WPI-MANA), National Institute for Materials Science, 1-1 Namiki, Tsukuba, 305-0044, Japan

E-mail: zuev.dmitri@gmail.com

Abstract. Mixed cationic $Ca^{2+}/Na^{+}(Mg^{2+})$ polyphosphates, regarded as potential biocomposite fillers, were fabricated by solid-state reactions. Most of them demonstrated low solubility (assessed also by the Glasser and Jenkins approach) from 1 ± 0.2 g/L for $Ca(PO_3)_2$ to 5 ± 0.4 g/L for $NaCa(PO_3)_3$ and 50 ± 2 g/L for soluble phase of $NaPO_3$. Liquid-state NMR of the soluble part of polyphosphates gave the degree of polymerization of their chains ranging from 10 to 30 for $NaCa(PO_3)_3$ and ca. 1000 monomeric units for $NaPO_3$.

1. Introduction

Mixed polyphosphates of Ca/Na are beneficial for bone grafting since polyphosphates (PP_i) are rather soluble (resorbable) and reveal a cumulative effect due to the long-time splitting of their backbone with the required long-chain chelate features for biomineralization by alkaline phosphatase (ALP) enzyme as compared to orthophosphates (P_i)[1]. They comply with most basic requirements: lack of an immune response and inflammation processes, undesirable chemical reactions shifting pH level from physiological value, the closeness of mechanical characteristics to that of native bone tissue necessary for prevention of bone restoration by stress shielding [2]. Likewise, the macroporous structure of grafts usually described as their osteoconductivity is important for providing favorable conditions for bone ingrowth and neovascularization. The issue of implants with a special porous architecture can be addressed to 3D-printing, e.g. fused deposition modelling technique (FDM), which demands a thermoplastic polymer and an inorganic filler [3].

Coming from the bone mineral, hydroxyapatite (HA), we proposed to replace it as a possible filler with calcium phosphates containing different-length condensed anions such as $P_2O_7^{4-}$, $P_3O_9^{3-}$, $P_3O_{10}^{5-}$ bearing in mind a higher dissolution rate of the corresponding PP_i [4]. Being subjected to hydrolysis by ALP *in vivo,* these anions produce P_i anions that are the building material for biogenic HA. Regulation of the rate of hydrolysis is a crucial factor to maintain the optimal rate of dissolution atca.10 mm per year [1]. As for the organic polymer component, it can be represented, for example by poly(ε-caprolactone), poly(D,L-lactide) or bacterial poly(hydroxybutirate). The polymers have suitable melting points, elastic modulus, and time of biodegradation ranging from 1 to 2 years [5].Thus, this work is aiming at the synthesis and theoretical assessment of the solubility of the PP_i under study through adoption of the thermodynamic model proposed for salts with monomeric anions [6], and its verification. The lengths of PP_i chains in solid-state and in solution were also evaluated.

2. Materials and methods

Mixed cationic $Ca(PO_3)_2$, $CaNa(PO_3)_3$, $NaPO_3$ polyphosphates, occurring at the $CaO\text{-}P_2O_5\text{-}Na_2O$ system, were synthesized via solid-state reactions (1-3). The prepared material is subjected to measurement of the solubility and the determination of the length of PP_i chains in solution and in the solid state. Note that the prepared compounds have a polymeric structure and the presented formulas (1-3) are stoichiometric.

2.1. Chemical reactions

$$Ca(H_2PO_4)_2 \cdot H_2O \rightarrow Ca(PO_3)_2 + H_2O \qquad (400\text{-}700\ °C,\ 2\ h) \qquad (1)$$
$$NaH_2PO_4 \rightarrow NaPO_3 + H_2O \qquad (400\text{-}600\ °C,\ 2\ h) \qquad (2)$$
$$NaH_2PO_4 + Ca(H_2PO_4)_2 \cdot H_2O \rightarrow NaCa(PO_3)_3 + 4H_2O \qquad (400\text{-}700\ °C,\ 2\text{-}12\ h) \qquad (3)$$

2.2. Used equipment

Thermogravimetric and differential thermal analysis (TG/DTA) were provided by DTA Thermo (Rigaku, Japan) in air at a heating rate of 5 °C/min from r.t. to 1000 °C. X-ray powder diffraction analysis (XRD) was made by RINT-Ultima III (Rigaku, Japan) at Cu Kα irradiation within the range of 2Θ angles 5° - 80°. Phase identification was made by epy 'Match!' and 'WinXPOW' programs with the ICDD PDF-2 database.

The solubility of the powders was measured by ICP-OES with SPS3520UV-DD (Hitachi, Japan). The optical emission bands were as follows: Na – 589.0 nm, Ca – 396.8 nm, P – 177.5 nm. ^{31}P NMR spectra were recorded using a JEOL AL300BX NMR spectrometer operating at 300 MHz (^1H), the number of scans is in the range from 64 to 7620. Chemical shifts were referenced to 0 ppm using an external phosphoric acid standard. Each spectrum showed a peak corresponding to outerphosphate groups (α peak at ~ -9±1 ppm) and a second peak corresponding to internal phosphate groups (i at -23±1 ppm); for some samples, a third peak was also observed, corresponding to phosphates neighboring to the external residues (β peak at -21±1 ppm) and a forth peak (~ -4±1 ppm) corresponding to P_i. Quantitative analysis was performed using peak area integration. Each PP_i has 2 α residues and i internal phosphorus atoms (including the β peak when present). The total polymer lengths n equals to 2 + i[7].

3. Results and discussion

The phase diagrams for systems $CaO\text{-}P_2O_5\text{-}Na_2O(MgO)$ were revised. Potentially prospective as well as non-studied or doubtful areas of the PP_i polymer existence were outlined. Temperatures and duration for completion of the reactions (1-3) were determined using DTA and XRD (figure 1 a, b). All samples containing Na or $Ca(PO_3)_2$ met the requirements of the 400-800 °C interval to complete all required phase transformations, otherwise depolymerized P_is were formed (figure 1 a). Phase purity and possible polymorphs were checked by XRD (figure 2 b). Na containing phases appeared more crystalline (narrow peaks) due to the lower lattice energy coming from less charged Na^+. All samples of $NaCa(PO_3)_3$ annealed within the interval specified above demonstrated the same XRD pattern (figure 1 b) identical to that from PDF-file # 00-023-0669.

The solubility of different $Ca^{2+}/Na^+(Mg^{2+})PP_i$ compounds were assessed according to the volume-based thermodynamic model proposed by Glasser and Jenkins [6]. Within this approach, the absolute standard entropies of individual polyphosphate phases and the enthalpies of their lattices depend explicitly on the molar volumes of the phases derived from XRD- and/or pycnometry data. The lattice enthalpy is related to the molar volume through generalized Kapustinskii equation. The change in the standard Gibbs free energy of solubility in water can be found as $\Delta G_{\text{sol}}^{298} = \Delta H_{\text{sol}}^{298} - 298 \times \Delta S_{\text{sol}}^{298}$.

Enthalpy and entropy contributions were defined the sum of two consequent processes: the conversion of the crystal lattice into constituent gaseous ions and the following hydration of them.

The results of calculation are presented in table 1. Most Ca containing P_i phases have limited solubility, however, PP_i phases are theoretically soluble making them rather promising as a filler in the composite bone implants (bearing in mind their morphogenetic activity [8]). The main contribution into the Gibbs energy of dissolution for PP_i came from the enthalpy term. These mix-cationic polyphosphates possess comparable solubility to other promising phosphates, e.g.$CaKPO_4$ and $CaNaPO_4$[9].

Table 1. Theoretical and experimental assessment of solubility properties of mix-cationic polyphosphates.

Phases	$\log_{10}K_s$(theoretical data)	Solubility (experimental data)
$Ca(PO_3)_2$	-47.9 ± 3	1 ± 0.2 g/L
$Ca_3(PO_4)_2$	27.7 ± 2	Not measured
$Mg(PO_3)_2$	-75.8 ± 4	Not measured
$NaPO_3$	-34.3 ± 3	50 ± 2 g/L
$NaCa(PO_3)_3$	-45.2 ± 3	5 ± 0.4 g/L
$Na_4Ca(PO_3)_6$	-41.5 ± 3	Not measured
$CaMg(PO_3)_4$	-64.0 ± 4	Not measured

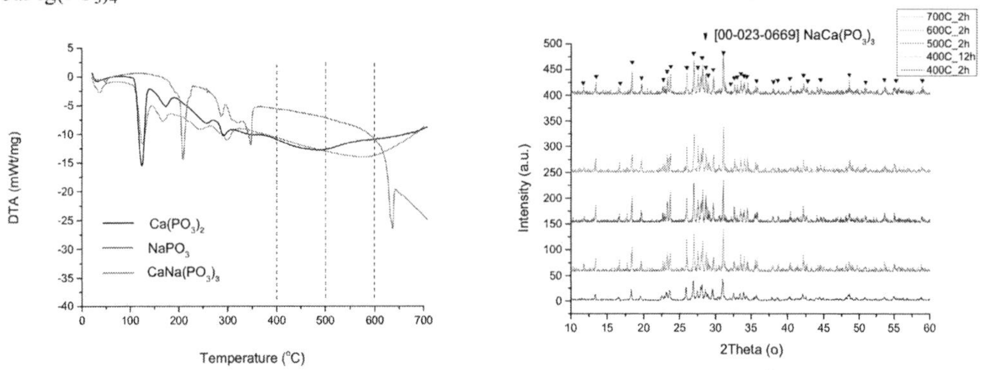

(a) (b)

Figure 1.(a) DTA for mix-cationic polyphosphates, $NaCa(PO_3)_3$, $NaPO_3$, $Ca(PO_3)_2$. (b) XRD spectra of $NaCa(PO_3)_3$ obtained at 400-700 °C for 2 h.

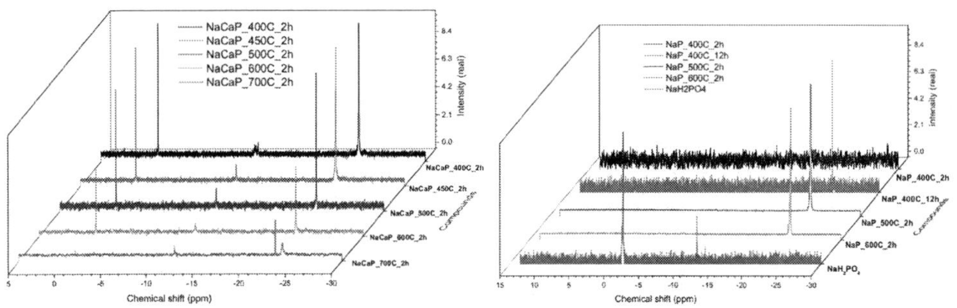

(a) (b)

Figure 2. (a) Liquid-state NMR spectra for $NaCa(PO_3)_3$ samples obtained at different temperatures. (b) $NaPO_3$ obtained at different temperatures and time of treatment (used precursor NaH_2PO_4).

Estimation of the chain length of the dissolved part of PP_is using liquid-state NMR was performed (figure 2 a). Upon increasing temperature during synthesis, the concentration of monomer P_i (peak around 0 ppm) was increasing compared to $PP_i\alpha$-peak (around -10 ppm, terminal polyphosphate groups) and PP_i internal polyphosphate peak (around -23 ppm). An unidentified wide resonance within the range from -20 to -25 ppm might be assigned to polyphosphates with a partially cross-linked structure (and therefore less soluble). NMR spectra for $NaPO_3$ (figure 2 b) demonstrate a strong difference between $NaPO_3$ (400°C, 2h), $NaPO_3$ (400 °C, 12 h), which are forming poorly soluble very long chains of PP_i, and those synthesized at higher temperatures $NaPO_3$ (500 °C-600 °C, 2 h), which are completely soluble with relatively shorter chains of PP_is.

4. Conclusions

Mixed cationic $Ca^{2+}/Na^{+}(Mg^{2+})PP_i$ which are promising for regenerative medicine were synthesized. Most PP_i materials demonstrated low solubility from 1 ± 0.2 g/L for $Ca(PO_3)_2$ to 5 ± 0.4 g/L for $NaCa(PO_3)_3$ and 50 ± 2 g/L for the soluble phase of $NaPO_3$. The discrepancy between experimental data and calculations might be caused by the limited applicability of the volume-based model to polyanions. The liquid-state NMR of the soluble part of PP_i gave the degree of polymerization of PP_i chains ranging from 10 to 30 for $NaCa(PO_3)_3$ and ca.1000 for $NaPO_3$.

Acknowledgments

The work was funded by RFBR grants №18-08-01473, 16-38-60203,18-33-00789 and 18-53-00034.D.M.Z. was supported by MSU - NIMS Cooperative Graduate School Program OH-1619-2014-5. Equipment purchased through the Program for Development of MSU was used.

References

[1] Grover L et al. 2013 *Biomaterials* **34** 6631
[2] Gibon E, Lu L Y, Nathan K and Goodman S B 2017 *J. Orthop. Transl.* **10** 28
[3] Vashisth P and Bellare J 2018 *Nanomed.: Nanotechnol., Biol. Med.* **14** 1325
[4] Neufurth M, Wang X, Wang S, Steffen R, Ackermann M, Haep N D, Schröder H C and Müller W E G 2017 *Acta Biomater.* **64** 377
[5] Pattanashetti N, Heggannavar G and Kariduraganavar M 2017 *Procedia Manufacturing* **12** 263
[6] Glasser L and Jenkins H 2011 *J. Chem. Eng. Data* **56** 874
[7] Smith S A, Choi S H, Davis-Harrison R, Huyck J, Boettcher J, Rienstra C M and Morrissey J H 2010 *Blood* **116** 4353
[8] Müller W E G, Tolba E, Schröder H C, Wang S, Glaßer G, Muñoz-Espí R, Link T, Wang X 2015 *Mater. Lett.* **148** 163
[9] Putlyaev V I, EvdokimovP V, Safronova T V, Klimashina E S and Orlov N K 2017 *Inorg. Mater.* **53** 529

IOP Conf. Series: Materials Science and Engineering **447** (2018) 012021 doi:10.1088/1757-899X/447/1/012021

Effect of annealing on the structure and phase composition of Al-Cu laminated metal-matrix composites produced by shear deformation under pressure

V N Danilenko, G F Korznikova, A P Zhilyaev, S N Sergeev, G R Khalikova, R Kh Khisamov, K S Nazarov, L U Kiekkuzhina, R R Mulyukov

Institute for Metals Superplasticity Problems, Russian Academy of Sciences, 39 Khalturin st., Ufa 450001, Russia

E-mail: vdan@anrb.ru

Abstract. This paper presents the data from the study on the effect of annealing of Al-Cu metal-matrix composites, produced by severe plastic deformation via shear under pressure. Three-layer compositions with different alternations of layers (Al-Cu-Al and Cu-Al-Cu) were subjected to deformation by high-pressure torsion, and then annealed at a temperature of 450 °C for 15 minutes. Using X-ray phase analysis and energy-dispersive analysis, we revealed the formation of the Al_2Cu, AlCu and Al_4Cu_9 intermetallic compounds, surrounded by solid solutions of Al in Cu and Cu in Al. Their fraction and distribution in the volume depend on the way of stacking of the layer components. The produced microstructure is an "in situ" composite formed in the process of deformation and annealing.

1. Introduction

The need for new materials with a predetermined level of properties promotes the development of novel composite materials. The use of solid-phase processes to produce new materials is currently an intensively developing area of physical materials science. In this area, a great attention is drawn to the possibility of using deformation techniques to produce (synthesize) new materials. Among various deformation techniques for solid-phase fabrication of compounds from dissimilar materials, high-pressure torsion is of special interest is. This technique enables producing materials with a nanocrystalline structure and a large fraction of grain boundaries [1], and also stimulates phase transformations due to a high pressure and/or a high strain rate [2,3]. At present, there are several papers showing solid-phase transformations of dissimilar materials under high-pressure torsion, for example: Al-Nb [4], Al-Cu [5,6], Al-Mg [6]. The solid-phase transformations in the above-mentioned Al-based binary systems occurred in the process of deformation and annealing under various conditions.

We demonstrated earlier [7] that, after high-pressure torsion (HPT) processing at room temperature, Cu-Al-Cu and Al-Cu-Al non-porous three-layer composites were produced. Different alternations of layers resulted in different degrees of mixing of the initial components. The present study is focused on the analysis of the structure and phase formation processes during the annealing of composites with different layer alternations of Cu-Al-Cu and Al-Cu-Al, produced by HPT.

2. Experimental

In the experiment, rods of commercially pure Al (99.5%) and Cu (99.9%) were used. To produce a coarse-grained structure, Al and Cu were subjected to preliminary annealing at 400 and 900 °C, respectively. The disks with a diameter of 12.0 mm and a thickness of 0.5 mm were stacked with layer alternations of Al-Cu-Al and Cu-Al-Cu, and were subjected to deformation via high-pressure torsion under a pressure of up to 5.2 GPa by rotating a stack of disks for 10 revolutions with a rotation rate of 1 revolution per minute at room temperature. For deformation processing, anvils having a groove with a diameter of 12 mm and a total depth of 0.5 mm were used [1]. Then, halves of the produced Al-Cu-Al and Cu-Al-Cu disks were annealed at a temperature of 450 °C for 15 minutes. The temperature and duration of annealing were selected from preliminary X-ray data.

The samples were prepared using conventional metallographic techniques, followed by polishing with a colloidal silica suspension. Microstructure analysis was performed using a Tescan Mira 3LMH scanning electron microscope (SEM) with a back scattered electron (BSE) detector. The chemical composition of the Al-Cu-Al and Cu-Al-Cu samples was investigated using a Tescan VEGA 3 SBH SEM with an energy-dispersive X-ray spectroscopy (EDS) attachment.

The structure of the composites was studied in the cross section of the samples. XRD phase analysis from the surface of the studied samples was conducted on a DRON-4-07 diffractometer.

3. Results and discussion

The vertical cross sections of the Cu-Al-Cu and Al-Cu-Al composites in the as-deformed state are presented in figure 1. The first insight into the microstructure reveals that both composites are monolithic and do not contain any pores, as shown in the SEM images at low magnification in figure 1a and e. The detailed microstructure at higher magnification revealed a complex lamellar microstructure developed in each of the composites. In the central part, highlighted with square 3, both samples have three clear layers (figure 1 d and h) and look very similar to the initial layered structure, but the layers are much thinner than in the original samples before HPT. In the mid-radius area highlighted with square 2, some fine lengthened lamellas of Al in the Cu matrix and Cu lamellas in the Al matrix appear (figure 1 c and h). The latter is much more pronounced in Al-Cu-Al sample (figure 1 h). Close to the edge of both samples, highlighted with square 1, lamellas became much thinner and longer (figure 1b and f), which is conditioned by an increase in strain from the center towards the periphery. XRD revealed the presence of α-Al and α-Cu and a negligibly small amount of the Al_2Cu phase in the Al-Cu-Al sample, whereas in the Cu-Al-Cu sample only α-Cu peaks were detected.

The microstructure and the data from the EDS analysis of the annealed samples are shown in figure 2. BSE images of the annealed samples at low magnification are not significantly different from those not annealed. The use of a higher magnification made it possible to detect the emergence of individual layers of other phases at the boundaries of the initial components (areas 2, 3 and 4, 5, 6 in figure 2). For chemical analysis, we selected the most typical areas in the center, at mid-radius and at the edge of the samples. Their magnified images and the data from the EDS analysis are shown in figure 2 b, c, d and f, j, h. In the center of the Cu-Al-Cu sample, no mixing of the Al and Cu layers is observed (figure 2 b). The EDS analysis reveals the presence of solid solutions of α-Cu and α-Al, and also in the layers between the basic metals there is a compound close in composition to the Al_4Cu_9 phase. At the mid-radius (figure 2 c) and at the edge (figure 2 d) of the Cu-Al-Cu sample, the interaction of the layers was more intensive. The EDS analysis reveals that during the transition from the Cu layer to the Al layer, there formed successively the regions of the α-Cu solid solution, of the Al_2Cu, AlCu and Al_4Cu_9 intermetallic compounds, and of the α-Al solid solution, which are different in contrast.

Since in the Al-Cu-Al sample HPT processing resulted in a more pronounced mixing and the formation of longer interphase Al/Cu boundaries than in the Cu-Al-Cu composite, after annealing phase transformations occurred almost in the whole volume of the sample (figure 2e). The EDS analysis revealed the presence of solid solutions of α-Al and α-Cu and the Al_2Cu, AlCu and Al_4Cu_9 intermetallic compounds in all the studied areas during the transition from the Al layer to the Cu layer (figure 2 f-h).

Cu-Al-Cu

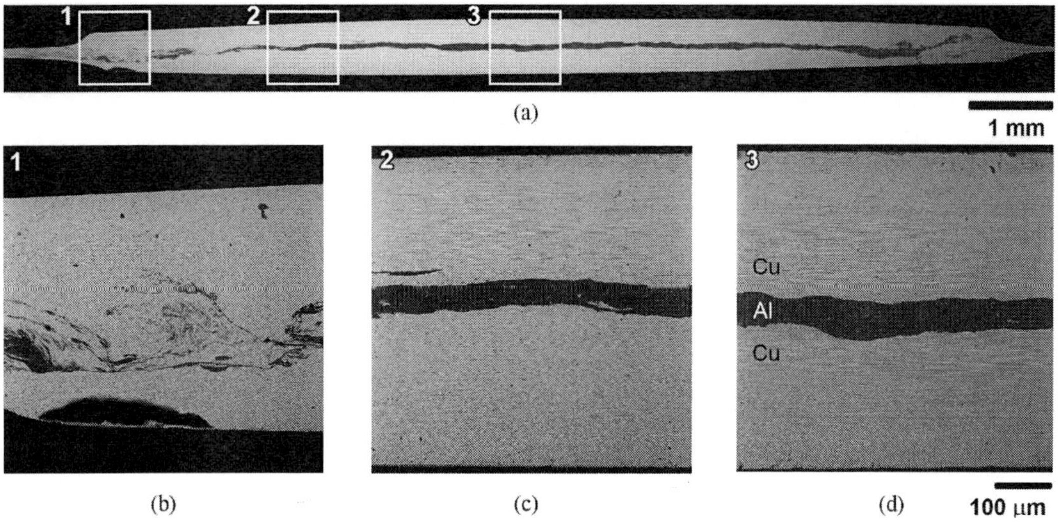

Al-Cu-Al

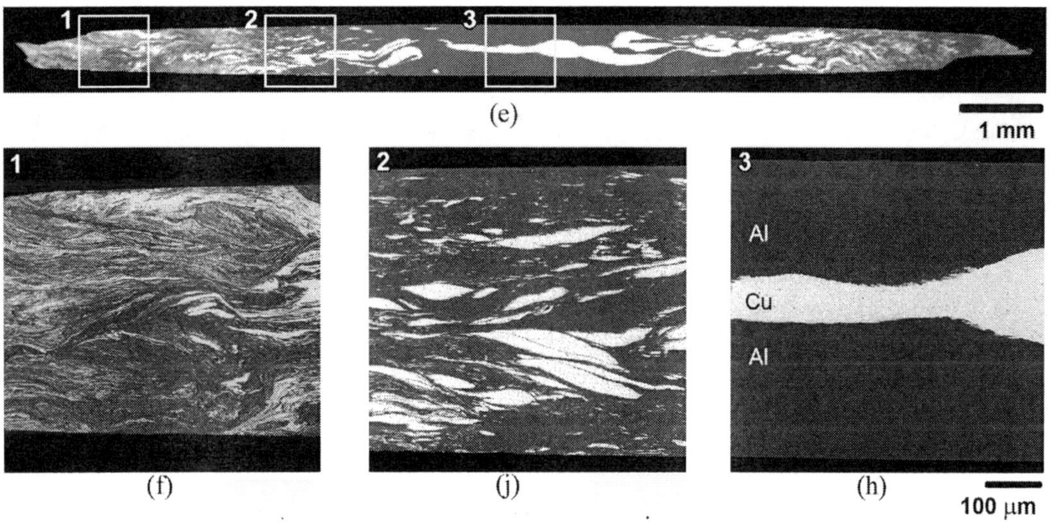

Figure 1. BSE image of the Cu-Al-Cu (a-d) and Al-Cu-Al (e-h) samples at a small magnification (a, e), microstructure details in the center (d, h), at mid-radius (c, j), at the peripheral area (b, f).

The data from the EDS analysis of the Cu-Al-Cu and Al-Cu-Al samples are in good agreement with the data from the X-ray phase analysis (figure 3). Alongside with the lines from α-Cu, diffraction peaks belonging to Al_2Cu, AlCu and Al_4Cu_9 also appear in the Cu-Al-Cu sample. In the Al-Cu-Al composite, the process of structural and phase transformations is more pronounced, which follows from a significant increase in the quantity and intensity of the diffraction peaks belonging to Al_2Cu, AlCu and Al_4Cu_9. As noted above, the energy-dispersive analysis shows that Cu and Al are present only in the form of a solid solution with different contents of the solute element. This is confirmed by a displacement of the diffraction peaks from Cu and Al, related to a change in the lattice parameter in solid solutions of α-Cu and α-Al.

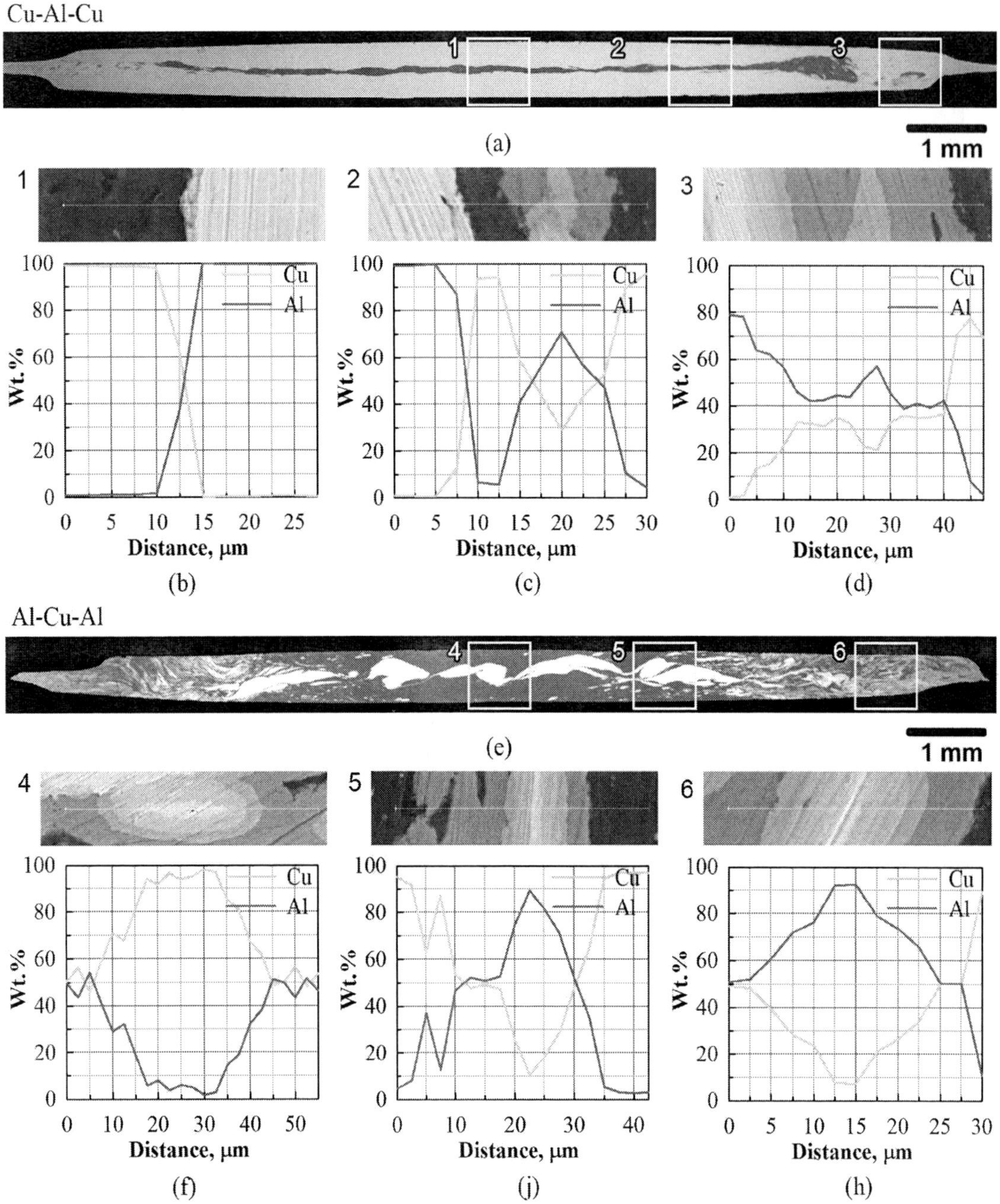

Figure 2. BSE image of the Cu-Al-Cu (a-d) and Al-Cu-Al (e-h) samples at a small magnification after annealing (a, e), microstructure details in the center (1, 4), at mid-radius (2, 5), at the peripheral area (3, 6). Elemental distribution along the line depicted at the top of the picture in the form of a graph of relative intensities with relation to the scanning distance for Cu and Al, marked in green color and in red color, respectively, in the center (b, f), at mid-radius (c, j) and in the peripheral area (d, h).

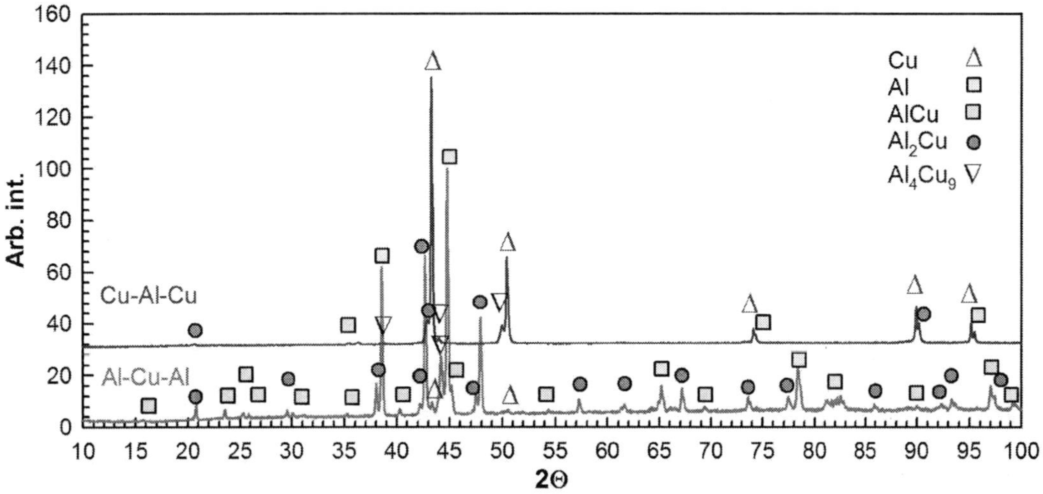

Figure 3. XRD of the Cu-Al-Cu and Al-Cu-Al samples after annealing at 450 °C for 15 minutes.

As can be seen from figure 2, the Al_2Cu, AlCu and Al_4Cu_9 intermetallic compounds are surrounded by solid solutions of Al in Cu and Cu in Al. If we take into account that the microhardness of intermetallic compounds is higher than that in the solid solution formed during annealing [4], we see a microstructure where the "hard" intermetallic phases are surrounded by a "soft" solid solution. In other words, during the annealing of the deformed three-layer composites, a structure is formed, which can be called an "in situ" composite.

4. Summary

Thus, the X-ray phase analysis and the study by scanning electron microscopy with energy-dispersive analysis showed that the three-layer stacking of the samples for deformation has an effect on the process of solid-phase mixing and phase formation during annealing. In the Cu-Al-Cu and Al-Cu-Al samples, during annealing there occurs the formation of the Al_2Cu, AlCu and Al_4Cu_9 intermetallic compounds, which are surrounded by solid solutions of Al in Cu and Cu in Al. In the Al-Cu-Al sample, the phases that have emerged are distributed throughout the sample, and their proportion is greater. The produced structure can be called an "in situ" composite.

Acknowledgements

The present work was accomplished according to the state assignment of IMSP RAS and supported by the Russian Science Foundation (Grant No. 18-12-00440).

References

[1] Zhilyaev A P and Langdon T G 2008 *Prog. Mat. Sci.* **53** 893
[2] Perez-Prado M T, Gimazov A A, Ruano O A, Kassner M E and Zhilyaev A P 2008 *Scripta Mater* **58** 219
[3] Straumal B B, Mazilkin A A, Baretzky B, Schutz G, Rabkin E and Valiev R Z. 2012 *Scripta Mater* **53** 63
[4] Danilenko V N, Galushev S N and Mulyukov R R 2009 *Perspektivnuye Materialy* **7** 94 (in Russian)
[5] Oh-ishi K, Edalati K, Kim H S, Hono K, Horita Z 2013 *Acta Mater.* **61** 3482
[6] Kawasaki M, Han D K Han J K, Jang J and Langdon T G 2017 *IOP Conf. Series: Mater. Sci. Eng.* **194** 012002
[7] Korznikova G, Mulyukov R, Zhilyaev et al., 2018 *AIP Conf. Proc.* (in press)

Deformation behaviour of zirconia with a different pore structure

I N Sevostyanova[1], T Yu Sablina[1], N L Savchenko[1], S N Kulkov[1,2]

[1] Institute of Strength Physics and Materials Science SB RAS, ¾ Akademichesky ave., Tomsk 634055, Russia
[2]National Research Tomsk State University, 36 Lenin ave., Tomsk 634050, Russia

E-mail: kulkov@ms.tsc.ru

Abstract. The structure and mechanical properties of zirconia ceramics obtained by slip casting with unimodal and bimodal size distributions and a wide range of porosity (up to 65%) were studied. Samples sintered using powder ZrO_2-3 mol.% Y_2O_3 with an average particle size of 0.2 μm with an unimodal pore size distribution were obtained by slip casting, followed by removal of the binder. Samples with a bimodal pore size distribution were sintered using spherical ultra-high molecular weight polyethylene particles with an average particle size of 100 μm. Axial and diametral (the Brazilian test) compression tests were performed for the obtained materials. The samples showed a brittle fracture, irrespective of the pore structure in the case of diametral compression tests. In the case of axial compression tests, a typical brittle failure took place in samples with porosity up to 10 vol.% and a "pseudo-plastic" failure occurred in samples with higher porosity. The latter can be ascribed to the movements of the meso-volumes to the pores and the accumulation of multiple damages, which were visually observed during the compression test.

1. Introduction

Mechanical characteristics are one of the key parameters of porous ceramic materials. The deformation behaviour and mechanical properties of porous ceramics are now thoroughly studied mainly for materials with unimodal porosity and a relatively narrow pore size distribution [1, 2]. At present, there is a need to develop materials with a multilevel pore structure, for example, for biomedical applications, when a bimodal pore structure with differing pore sizes is required [3-5]. Nevertheless, studies in which the strength properties and the deformation behaviour of similar materials are simultaneously investigated in a wide range of porosity under different loading schemes are not available in the literature. The aim of this work was to study the mechanical properties of porous zirconia with different pore structures.

2. Materials and research methods

Powder ZrO_2-3mol% Y_2O_3 with an average particle size of 0.2 μm was used for preparing cylindrical samples with diameter 5 mm and height 7 mm with different porosity types. Samples with an unimodal pore size distribution were obtained by slip casting, followed by removal of the binder. To obtain samples with a bimodal pore size distribution, spherical ultra-high molecular weight polyethylene (UHMWPE) particles with an average particle size of 100 μm in the amount of 25% by volume were added to the slip as a burning pore-forming agent. The samples were sintered in air in the temperature range 1100 – 1600 °C with a sintering time 1 hour. The rate of heating and cooling of the

IOP Conf. Series: Materials Science and Engineering **447** (2018) 012022 doi:10.1088/1757-899X/447/1/012022

samples during sintering was 240 °C / h. A metallographic analysis of the polished surfaces of the samples was carried out using an "Altami MET 1M" optical microscope. The pore size distribution was determined from images of polished surfaces of the samples using the ImageJ program. The density and porosity of the samples were determined geometrically. In doing so, the relative standard deviation was less than 5%. The mechanical properties of the sintered samples were measured using the Devotrans universal testing machine. Two different measurements were performed: axial (Ϭc) and diametral compression (Ϭt) (the Brazilian test). In both cases, the loading rate was 0.2 mm/min, and load-displacement curves were recorded.

3. Results and discussion

Typical images of the structure of the samples and the pore size distributions are shown in figure 1. One can see that the size distributions for small pores were very similar for both types of samples. Large pores were exactly stipulated by the burning pore-forming agent.

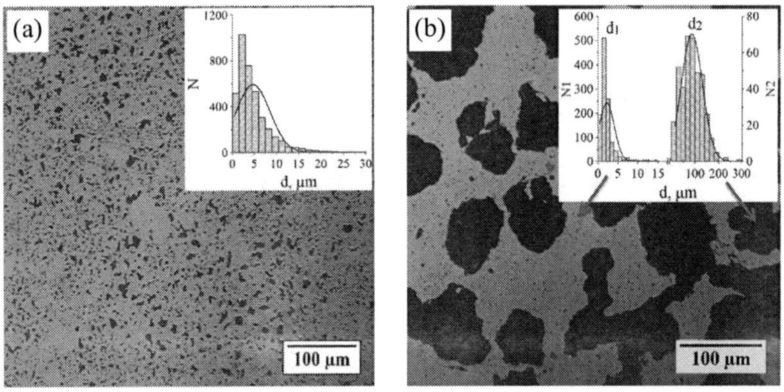

Figure 1. Typical microstructure images of ceramic samples with unimodal (a) and bimodal (b) pore size distributions.

Figure 2 shows the influence of the sintering temperature on the porosity of samples with unimodal and bimodal pore size distributions.

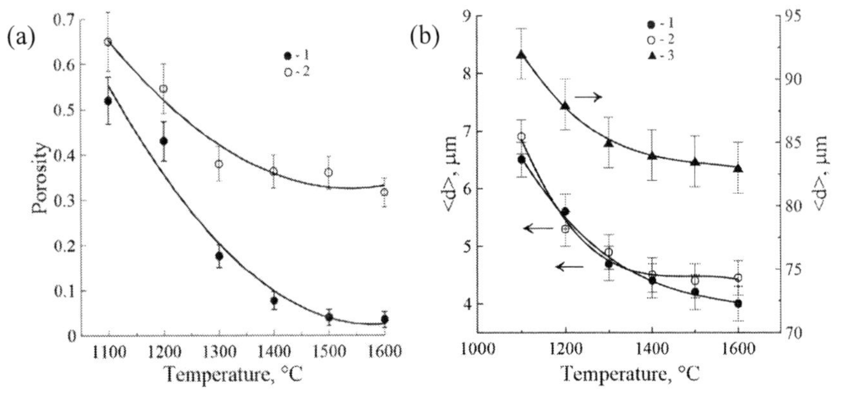

Figure 2. Porosity (a) and mean pore size (b) dependencies on the sintering temperature: 1 - for samples with an unimodal pore size distribution; 2, 3 - for samples with a bimodal pore size distribution.

One can see that with increasing the sintering temperature, the porosity of the samples monotonously decreased, and the porosity values with different types of porosity after sintering at 1400-1600 °C differed by 25-30%. That corresponds to the volume fraction of the introduced pore-forming additive. In addition, with increasing the sintering temperature, the average pore size decreased (figure 2 b). The average grain size in the sintered samples was varied from d=0.2 μm (T=1100 °C) to d=0.9 μm (T=1600 °C).

Figure 3 a represents typical "stress-displacement" curves of samples after diametral compression tests with an unimodal pore size distribution and a porosity of 3.5% and a bimodal pore size distribution and a porosity of 38%. Comparison of the "stress-displacement" curves after diametral compression tests shows that the fracture type of the samples did not depend on the type of porosity and the pore volume. All samples were characterized by elastic behaviour. On the active loading stage, sharp changes of the stress value were observed, which can be ascribed to the formation of microcracks. This did not lead to the macrofracture of samples, followed by a brittle fracture.

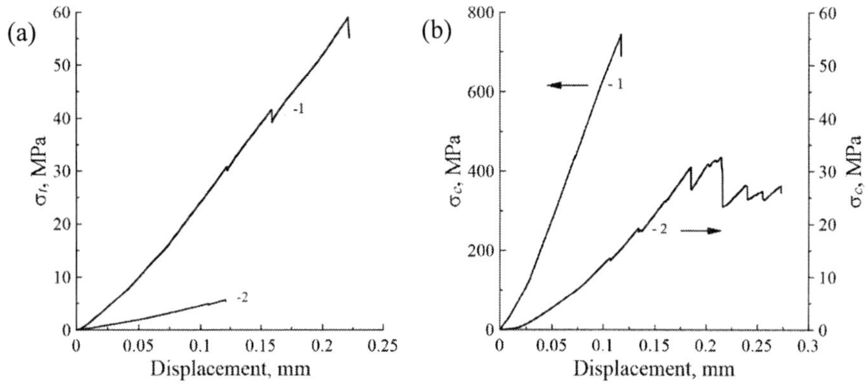

Figure 3. Typical "stress-displacement" curves of zirconia ceramic samples obtained: a - after diametral compression tests with a unimodal pore size distribution and a porosity of 3.5% (curve 1) and a bimodal pore size distribution and a porosity of 38% (curve 2); b - after axial compression tests with an unimodal pore size distribution and a porosity of 3.5% (curve 1) and a bimodal pore size distribution and a porosity of 53% (curve 2).

Figure 3 b represents typical "stress-displacement" curves of samples after axial compression tests with an unimodal pore size distribution and a porosity of 3.5% and with a bimodal pore size distribution and a porosity of 53%. Comparison of "stress-displacement" curves after axial compression tests shows that the fracture type of the samples depended on the type of porosity and the pore volume. The samples with an unimodal pore size distribution and a porosity up to 10% were characterized by elastic behaviour and brittle fracture (curve 1), whereas the samples with a bimodal pore size distribution and high porosity demonstrated saw-tooth oscillations of the stress value after reaching the ultimate compressive strength indicating accumulation of damages. Thus, under the axial compression test, a transition from typical brittle fracture in samples with porosity up to 10 vol.% to "pseudo-plastic" fracture in samples with a higher porosity took place regardless of the type of pore structure. This behaviour is typical of highly porous ceramic materials and is associated with the appearance and accumulation of multiple damages [6-9], which can be visually observed during the compression test.

Figure 4 shows the dependences of ultimate compressive and ultimate strength for diametral and axial compression samples with unimodal and bimodal pore size distributions vs. porosity. One can see that with increasing porosity, the ultimate strength decreased, and these dependences are well described by an exponential equation of the type:

$$\sigma = \sigma_0 \exp(-bP) \qquad (1),$$

where σ is the strength of a porous material, σ_0 is the strength of a material free of pores, P is the volume fraction of pores and b is a coefficient, which depends on the structure and composition of the material [10]. The analysis of the obtained dependences showed that the coefficient b was 6-7 for all dependences that is consistent with the literature data [10].

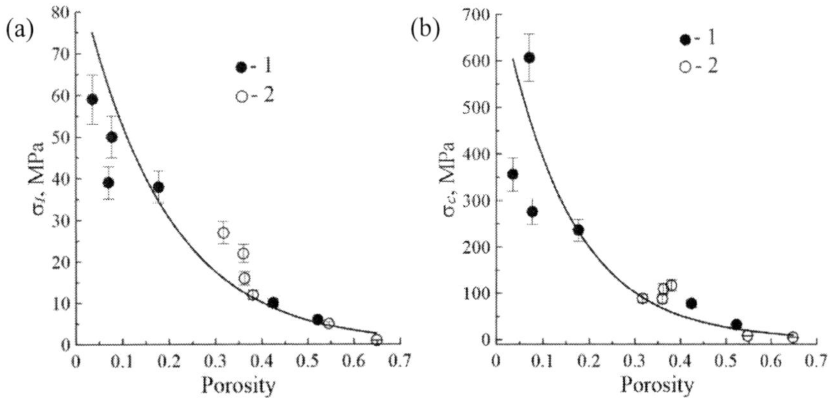

Figure 4. Dependences of the ultimate strength for diametral compression (a) and axial compression (b) on porosity obtained for zirconia samples with unimodal (1) and bimodal (2) porosity.

4. Conclusions

Zirconia ceramics samples with different types of the pore structure demonstrated a brittle fracture, and only single microcracks were formed on the active stage of loading. For the zirconia ceramics samples having high porosity and subjected to axial compression tests, the accumulation of microdamages occurred without fracture of the entire sample. All dependences of the ultimate strength on porosity were well described by an exponential equation with exponent factors in the range of 6-7.

Acknowledgments

This work was carried out according to the Program III.23 of Fundamental Scientific Research of the State Academies of Sciences for 2013-2020 and with partial financial support of the RFBR Grant No. 18-48-700056.

References

[1] Keleş Ö, García R E and Bowman K J 2013 *Acta Mater.* **61** 2853
[2] Borges P H R, Bhutta A, Teixeira Bavuzo L and Banthia N 2017 *Építőanyag – J. Silicate Based and Composite Materials* **69** 13
[3] Chuanuwatanakul C, Tallon C, Dunstan D E and Franks G V 2013 *J Amer. Cer. Soc.* **96** 1407
[4] Sadowski T and Samborski S 2003 *Comp. Mater. Sci.* **28** 512
[5] Tallon C, Chuanuwatanakul C, Dunstan D E and Franks G V 2016 *Ceram. Int.* **42** 8478
[6] Meille S, Lombardi M, Chevalier J and Montanaro L 2012 *J. Eur. Ceram. Soc.* **32** 3959
[7] Buyakova S P, Wei H, Dunmy L, Haiyun C, Sablina T Yu, Melnikov A G and Kulkov S N 1999 *Techn. Phys. Lett.* **25** (17) 44-48
[8] Savchenko N L, Sablina T Yu, Sevostyanova I N, Buyakova S P and Kulkov S N 2016 *Russian Phys. J.* **58** (11) 15
[9] Grigoriev M V, Savchenko N L, Sablina T Yu, Kurovics E, Sevostyanova I N, Buyakova S P, Gömze, L A and Kulkov S N 2018 *Építőanyag – J. Silicate Based and Composite Materials* **70** (1) 18
[10] Li L. and Aubertin M 2003 *Can. J. Civ. Eng.* **30** 644

Migration of tilt boundaries in nickel and Ni_3Al intermetallide: a molecular dynamics study

G M Poletaev[1], I V Zorya[2], R Y Rakitin[3], M A Iliina[4], M D Starostenkov[1]

[1]Altai State Technical University, 46 Lenin St. Barnaul656038, Russia
[2]Siberian State Industrial University, 42 Kirov St., Novokuznetsk 54007, Russia
[3]Altai State University, 61 Lenin St., Barnaul656049, Russia
[4]Financial University under the Government of the Russian Federation, 54 Lenin St., Barnaul 656036, Russia

E-mail: gmpoletaev@mail.ru

Abstract. The study of migration of <100> and <111> tilt boundaries in Ni and the Ni_3Al intermetallide is performed by the molecular dynamics method. It is shown that in Ni and Ni_3Al, the low-angle <100> boundaries migrate much slower than the <111> boundaries (approximately twice at a temperature of 1700 K) due to the difference in the mechanisms of migration of the low-angle <100> and <111> boundaries. According to the obtained results, the migration rate of similar boundaries in the Ni_3Al intermetallide is much lower than in Ni (approximately three times at 1700 K). The reason for this is mainly the additional energy expenditure for the formation of a disordered area in Ni_3Al behind the migrating boundary. Due to the relatively low mobility of the boundaries in Ni_3Al, the contribution of diffusion displacements of atoms in the process of boundary migration is higher than in Ni.

1. Introduction

Grain boundary migration plays a crucial role in the process of recrystallization and in many cases of phase transformations. Despite the long-standing interest in the problem of migration of grain boundaries, there are still disagreements and unresolved issues related to the migration mechanism. It is believed that low-angle tilt boundaries migrate through the combined action of two mechanisms: slip and climb of grain boundary dislocations [1]. In [2, 3], for example, the authors come to the conclusion that the main mechanism of migration of the tilt boundaries is the climbing of grain boundary dislocations. But, on the other hand, it is known that in fcc metals the low-angle <111> tilt boundaries have the highest mobility, while the <100> tilt boundaries, for example, migrate much more slowly [1, 3], although the density of jogs on grain boundary dislocations within the <100> boundaries is higher, i.e. the climb intensity should be greater than within <111> boundaries. The reason for this difference in the mobility of the <111> and <100> boundaries, as well as the migration mechanism, is not completely clear.

The migration of grain boundaries in ordered alloys and intermetallides is scantily known. One of the interesting and promising intermetallic compounds is Ni_3Al, which has a fcc lattice and ordering of the $L1_2$ type. The intermetallic compound Ni_3Al is distinguished from a number of similar ordered alloys by

IOP Conf. Series: Materials Science and Engineering **447** (2018) 012023 doi:10.1088/1757-899X/447/1/012023

unique physical and mechanical properties, which include, first of all, a positive temperature dependence of the yield stress and high thermal stability (the order-disorder phase transition temperature for Ni_3Al is above the melting point) [4]. Therefore, this intermetallic compound is of practical significance as a heat-resistant structural material.

The present work is devoted to the study of the features of the migration of <100> and <111> tilt boundaries in Ni and the Ni_3Al intermetallide by the molecular dynamics method.

2. Model description

A clearly defined boundary was created in the form of a loop or arch, as in figure 1 (black dotted line). The tension of the boundary, which, like a surface tension, arises from the tendency of the boundary to minimize its energy, is the reason for the directed movement of the boundary, which decreases its area. The force that provokes the migration and the boundary migration velocity remained constant throughout almost the entire movement of the boundary, smoothly decreasing by the end of the computer experiment.

The three-dimensional computational block in the molecular dynamics model was created in the form of a plate with a thickness of 12 atomic layers (figure 1). This thickness is sufficient for the appearance of effects associated with the jogs of grain boundary dislocations. In the case of the <111> tilt boundaries, the nickel computational block had a height of 18.0 nm, a width of 12.0 nm and a thickness of 2.4 nm. For the <100> boundaries, the dimensions were 18.2 nm, 12.1 nm and 2.2 nm, respectively. The computational blocks of Ni_3Al were slightly larger due to the difference in the lattice parameters of Ni and Ni_3Al. The blocks contained approximately 50,000 atoms. Along the Z-axis (figure 1), an infinite repetition of the structure was simulated, i.e. periodic boundary conditions were imposed. At the edge of the computational block, the grain boundaries must be fixed, which implies preserving the orientation of the crystal lattice of two different grains at the border of the block. In this connection, along the X and Y axes, the block borders (highlighted in dark gray in figure 1) were rigidly fixed to preserve the given misorientation of the grains.

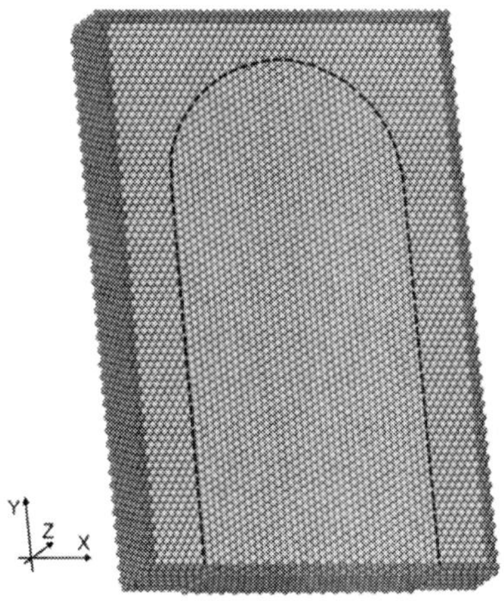

Figure 1. Computational block for modeling the migration of <111> 30° tilt boundary. The dark-gray atoms on the edge of the block remained motionless during the computer experiment (rigid boundary conditions).

To describe the interatomic interactions in Ni and in the Ni_3Al intermetallide, the Cleri–Rosato tight-binding EAM potentials were used [5]. Potentials of this type have been repeatedly used in models of molecular dynamics and have been tested for a large number of characteristics [6–10]. The experience of their application shows that their use allows describing various properties of metals and alloys. The time step in the molecular dynamics method was equal to 2 fs. The temperature in the model was set through the initial velocities of the atoms according to the Maxwell-Boltzmann distribution, wherein the thermal expansion of the calculation blocks was taken into account. To maintain a constant temperature during the simulation, the Nose-Hoover thermostat was used.

3. Results and discussion

Figure 2a shows the dependences of the migration rate of <100> and <111> tilt grain boundaries in nickel at a temperature of 1700 K on the misorientation angle. Special and symmetrical boundaries were not considered in the paper. The tension of the grain boundaries is proportional to their energy: as the misorientation angle increases, both energy and tension increase as well [11].

The boundary migration rate of during the computer experiment remained practically constant and high enough at the considered temperature of 1700 K, and at misorientation angles above 10°, which made it possible to make the required measurements in the molecular dynamics model. The angle of the grain misorientation θ varied from 10° to 45° for the <100> boundaries and to 40° for the <111> boundaries. In the case of <100> boundaries, 45° is the maximum misorientation angle. In the case of the <111> boundaries, it was taken into account that the <111> tilt boundaries with the misorientation angle of 38° have the highest mobility, according to, for example, [1, 12]. As the misorientation angle increases, the migration rate of the boundaries increases similarly to the well-known regularity [1, 11].

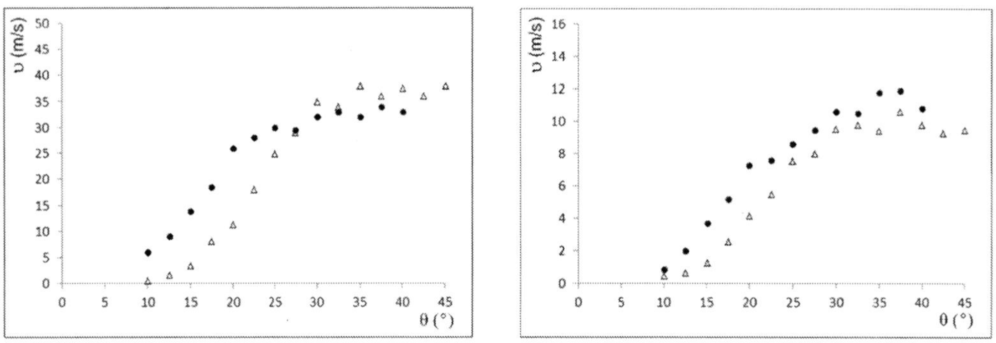

(a) (b)

Figure 2. The rate of migration of <100> (white triangulars) and <111> (black dots) tilt boundaries at 1700 K (b), depending on the misorientation angle θ in Ni (a) and the Ni_3Al intermetallide.

It should be noted that at the angles of misorientation above 25° the high-angle <100> and <111> boundaries migrate with approximately the same velocity (30–37 m/s at 1700 K), while the migration rates of the low-angle <100> and <111> tilt boundaries differ significantly – the low-angle <100> boundaries migrate about twice as slow as the <111> boundaries.

Earlier in [10], we found that the migration of the <100> boundary is carried out by splitting of the paired grain boundary dislocations with a subsequent change of the dislocations partners. The dislocations partners are replaced through the slip of split dislocations. The migration of the <111> tilt boundary is accomplished by the combined action of two mechanisms: the mechanism described above and the mechanism consisting in joint sliding of the paired grain boundary dislocations, which, in contrast to the grain boundary dislocations within <100> boundaries, have common slip planes. The joint sliding of the paired dislocations has a relatively low activation energy. As a result, the low-angle <111> tilt boundaries are more mobile than the <100> boundaries.

Figure 2b shows the dependences of the migration rate of the boundaries in Ni_3Al on the misorientation angle. Despite the relatively higher energy of the boundaries and, hence their tension, the migration rate of the boundaries in the Ni_3Al intermetallide proved to be significantly lower compared to Ni, approximately three times at the same temperature of 1700 K. This is apparently due to the additional expenditure of energy for disordering and the rupture of Ni-Al bonds during the motion of the boundaries in Ni_3Al. Figure 3 shows an example of the formation of a disordered area behind the migrating <111> 20° boundary. The order does not have enough time to recover. The ordering process in this case is much slower than the boundary migration.

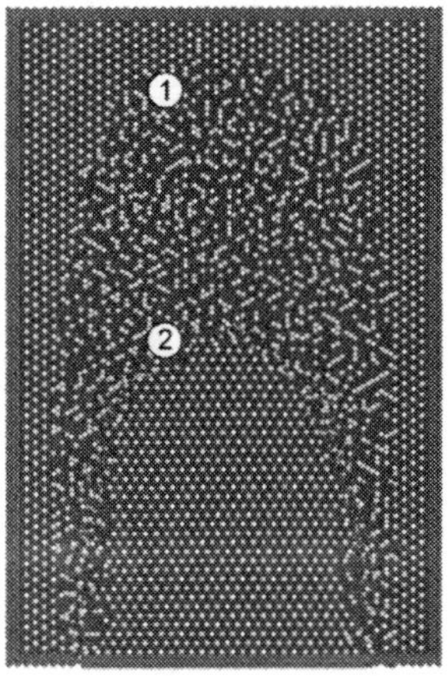

Figure 3. Formation of a disordered area behind the moving boundary (1 – initial position of the boundary, 2 – current position) during the migration of <111> 30° tilt boundary at a temperature of 1700 K in the course of 800 ps.

As in the case of boundaries in Ni, the low-angle <100> boundaries in Ni_3Al were approximately two times lower in mobility than the <111> boundaries (figure 2b). The mechanism of migration of low-angle boundaries, in general, in Ni_3Al was the same as in Ni. Due to the relatively low rate of migration of boundaries in Ni_3Al, one of the significant differences was a greater contribution of the diffusion displacements of atoms in the migration process.

4. Conclusions

In the present work, the study of the features and mechanism of migration of <100> and <111> tilt boundaries in Ni and the Ni_3Al intermetallide was performed the molecular dynamics method. The dependences of the migration rate of the considered boundaries at a temperature of 1700 K on the misorientation angle are obtained. It is shown that the high-angle <100> and <111> boundaries migrate at approximately the same rate at a given temperature, while the low-angle <100> boundaries migrate about twice as slower as the <111> boundaries. According to our previous work [10], this is due to the difference in the mechanisms of migration of low-angle <100> and <111> boundaries. As the molecular dynamics studies demonstrated, the migration rate of similar boundaries in Ni_3Al intermetallide is much lower than in Ni (approximately three times at 1700 K). In particular, the reason for this is the formation of a disordered area behind the migrating boundary in Ni_3Al. In general, the mechanism of the migration of low-angle boundaries in Ni_3Al was the same as in Ni. Due to the relatively low rate of boundary migration in Ni_3Al, one of the significant differences was a greater contribution of the diffusion displacements of atoms in the migration process.

References

[1] Gottstein G and Shvindlerman L S 2009 *Grain Boundary Migration in Metals: Thermodynamics, Kinetics, Applications. Second Edition* (Boca Raton, CRC Press)
[2] Balluffi R W and Cahn J W 1981 *Acta Metall.* **29** 493
[3] Winning M, Rollett A D, Gottstein G, Srolovitz D J, Lim A and Shvindlerman L S 2010 *Philos. Mag.* **90** 3107
[4] Ramesh R, Pathiraj B and Kolster B H 1996 *J. Mater. Process. Technol.* **56** 78
[5] Cleri F and Rosato V 1993 *Phys. Rev. B* **48** 22
[6] Poletaev G M, Zorya I V, Novoselova D V and Starostenkov M D 2017 *Int. J. Mater. Res.* **108** 785
[7] Poletaev G M, Novoselova D V, Zorya I V et al. 2018 *Phys. Solid State* **60** 847
[8] Kulabukhova N A, Poletaev G M, Starostenkov M D, Kulagina V V and Potekaev A I 2012 *Russ. Phys. J.* **54** 1394
[9] Poletaev G M and Starostenkov M D 2009 *Tech. Phys. Lett.* **35** 1
[10] Poletaev G, Zorya I and Rakitin R 2018 *Comp. Mater. Sci.* **148** 184
[11] Kaibyshev O A and Valiev R Z 1987 *Grain Boundaries and Properties of Metals* (Moscow, Metallurgy) (In Russian)
[12] Huang Y and Humphreys F J 2012 *Mater. Chem. Phys.* **132** 166

The influence of forging and extrusion on the microstructure and martensitic transformation in Ni-Mn-Ga alloys

I I Musabirov[1], I M Safarov[1], R M Galeyev[1], D R Abdullina[1], V V Koledov[2] and R R Mulyukov[1]

[1] Institute for Metals Superplasticity Problems, Russian Academy of Sciences, Khalturina 39, Ufa, 450001 Russia
[2] Kotelnikov Institute of Radio Engineering and Electronics, Russian Academy of Sciences, Mokhovaya 11-7, Moscow, 125009 Russia

E-mail: irekmusabirov@mail.ru

Abstract. The work presents the comparative analysis of the microstructure and character of martensitic transformation in Heusler Ni-Mn-Ga alloys subjected to plastic deformation by forging and complex treatment by forging with subsequent extrusion. It is shown that the treatment by forging leads to the formation of a bimodal structure with large grains of several micrometres surrounded by a fine-grained structure. After a complex treatment, a bimodal structure is also observed. But the volume fraction of the fine-grained structure is larger than in the previous case. Analysis of the martensite transformation temperature shows that thermomechanical treatment by both methods leads to a slight shift in the martensite transformation temperature to the low temperature region.

1. Introduction

Ni_2Mn-X (X=Ga, In, Sn, Co, Fe, etc.) alloys have a combination of simple and unique physical properties. In particular, the ferromagnetic shape memory effect [1-3] and the magnetocaloric effect [4-7] are observed in the region of martensitic and magnetic phase transformations. As a result, these alloys belong to the class of promising functional materials. One of the main directions of the studies is the search for ways to improve the mechanical properties of these alloys, because the multiple occurrence of martensitic transformation (MT) leads to fracture along the boundaries of crystallites. The most promising way to solve this problem is the thermomechanical treatment (TMT) of as-cast polycrystalline alloys [8-11]. The authors of the work carry out research on TMT of Heusler alloys using multiple isothermal forging (MIF) [12, 13] and extrusion (EXT). The advantage of MIF is the possibility of obtaining a massive workpiece with the required structure and a high level of internal stresses. In order to enhance the anisotropy of the alloy in the MT region, it is possible to carry out complex treatment by forging with subsequent extrusion (MIF+EXT).

2. Experimental

Two Ni-Mn-Ga alloys for the investigation were prepared by the method of arc-melting in an argon atmosphere. The elemental composition was analyzed by energy dispersive X-ray spectroscopy on a TESCAN Vega 3 SBH scanning electron microscope. The first alloy, marked as Ga16/6 has the composition $Ni_{54.1}Mn_{19.6}Ga_{24.6}Si_{1.7}$, the second alloy marked as Ga17/3 - $Ni_{52.9}Mn_{21.1}Ga_{24.6}Si_{1.5}$. The presence of Si in the alloy is explained by its diffusion from quartz glass during vacuum remelting.

Detailed information about this procedure is described in previous works [12, 13]. The Ga16/6 alloy in the initial state was subjected to TMT by MIF, and the Ga17/3 alloy in the initial state was subjected to complex TMT by MIF+EXT. The temperatures of phase transformations (martensitic and magnetic) were determined by measuring the temperature dependences of the specific magnetization and electrical resistance. The magnetization of the alloy was studied with a vibration magnetometer. The installation allows measuring the specific magnetization of the sample in magnetic fields up to 2 T in the range from liquid nitrogen temperatures to +150 °C. The electrical resistivity of the sample was measured by the four-probe method on a sample of 1 mm x 1 mm x 7 mm with a current of 70 mA. The thermocouple and the voltage drop across the sample were measured by two voltmeters V7-78/1. The microstructure of the alloys was investigated by a TESCAN Mira 3 LMH scanning electron microscope in the backscattered mode. MIF was carried out on the machine of complex loading Schenck Trebel RMC 100. Forged alloy was also extruded on a special tool, in which the output circular section has a transition 10 mm→8 mm with an extrusion ratio of 1.6.

3. Results and discussion

3.1. The analysis of phase transitions temperatures in various structural states

The curves of the temperature dependence of the specific magnetization of the Ga16/6 alloy in the initial state and after MIF are presented in figure 1 a. The curves were obtained by heating the sample in the temperature range from –100 °C to +60 °C in an external magnetic field of 80 kA/m.

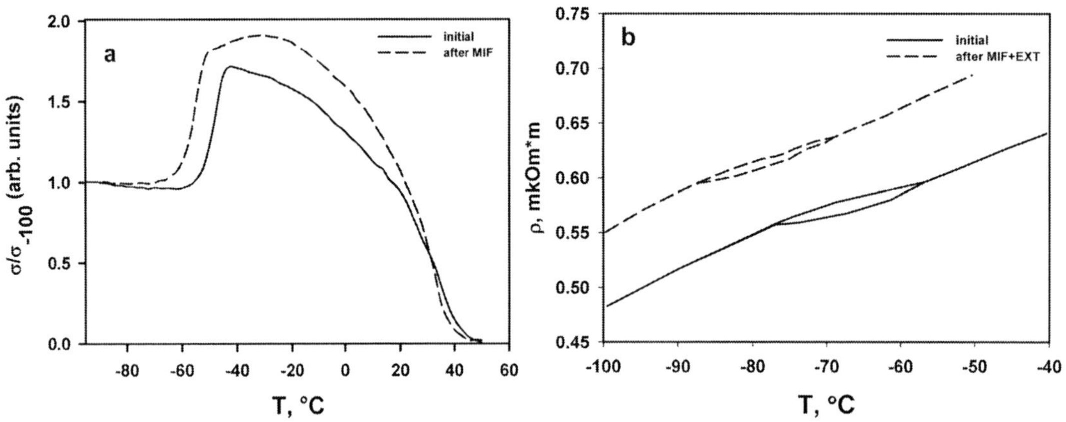

Figure 1. Temperature dependences of the specific magnetization of the Ga16/6 alloy in different states (a) and the electrical resistivity of the alloy in the initial state and after MIF+EXT (b); the specific magnetization is normalized to its value at -100°C

On the curve for the initial state in the interval of the reverse MT, a jump-like change in the magnetization characteristic of Heusler alloys is observed. The temperatures of the phase transformation have the following values: A_S= –54 °C, A_F= –44 °C. Transformation to the paramagnetic phase occurs at 43 °C. This temperature can be roughly called the Curie point, since its precise determination must be carried out by non-magnetic methods. Analysis of *σ(T)* after forging shows that the characteristic temperatures are shifted to low temperatures by ~7 °C.

The temperature dependence curves of the electrical resistivity of the Ga17/3 alloy in the initial state and after MIF+EXT are presented in figure 1 b. The *ρ(T)* curves are recorded when the sample is heated and cooled from –100 °C to –40 °C. It can be seen that in the initial state there is a deviation from a monotone increase in the electrical resistivity of the alloy in the martensitic transformation region. In the process of reverse martensitic transformation, this increase slows down slightly. There is a bend on the curve. It is known that the high temperature phase is more symmetric, and its specific electrical resistance is lower. The temperatures of the beginning and end of the transformation in the initial state are $A_S = -72$ °C and $A_F = -57$ °C. After TMT by MIF+EXT the transformation on the *ρ(T)* curve is weaker. The electrical resistance in the treatment state is higher than in the initial state. This indicates an increased density of defects in the treatment state. Subsequently, the electrical resistance of the martensitic and austenitic phases differs less as compared with the initial state. As in the case of TMT by MIF of the Ga16/6 alloy, after MIF+EXT, on the curves, there is also some displacement (~11 °C) of the characteristic temperatures in the downward direction.

It is known that in Ni-Mn-Ga alloys, a decrease in the average grain size as a result of TMT leads to a gradual suppression of the martensitic transformation [14]. This is manifested in the reduction of the transformation temperature and its severity in the analysis of magnetic, electrical or dilatometric properties. With a decrease in the grain size of less than 1 μm, the phase transformation is practically not observed. In the investigated Ga16/6 and Ga17/3 alloys TMT by forging and forging with subsequent extrusion do not lead to suppression of the martensitic transformation, but only shifts it by 7-10 °C to the region of lower temperatures.

3.2. The microstructure of the alloy after treatment by multiple isothermal forging

Microstructure of the Ni-Mn-Ga alloys in the initial cast state (after vacuum melting) was described by us earlier in [12, 13] Therefore, this paper presents the results of microstructural studies of the alloys only after TMT. Figure 2 a shows the microstructure of the Ga16/6 alloy after MIF, characterized by a bimodal distribution of grains. In this structure, large grains of several hundred microns are surrounded by a fine-grained structure. In the body of large grains there is a blurred contrast indicating the presence of a substructure and/or internal stress fields as a result of the treatment.

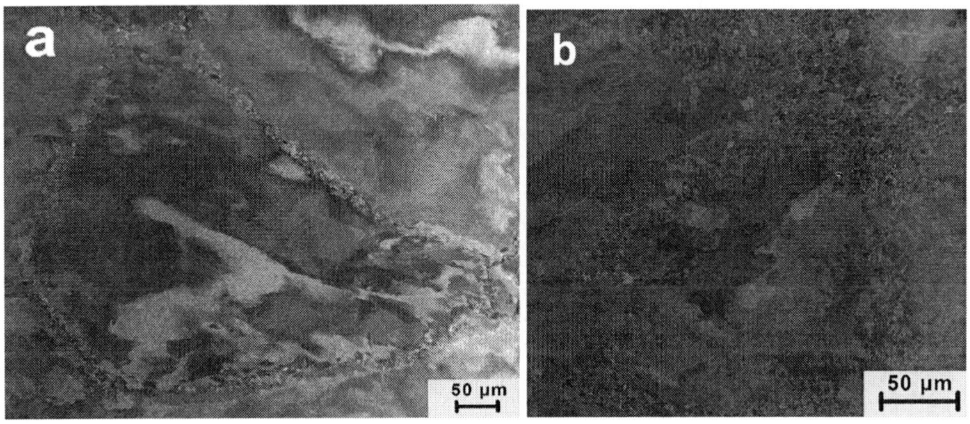

Figure 2. SEM image of the Ga16/6 alloy after MIF (a) and the Ga17/3 alloy after MIF+EXT (b).

3.3. The microstructure of the alloy after complex treatment by forging and extrusion

The microstructure of the Ga17/3 alloy after complex TMT by MIF+EXT is shown in figure 2 b. The axis of the extrusion is horizontal relative to the image. The bimodal distribution of the grains formed during the treatment by forging is not eliminated in the structure after hot extrusion. So, there are still large grains of several hundred micrometers in size, surrounded by a fine-grained layer. However, compared to the forged state, the layer of small grains has a greater width. That is, the volume fraction of small grains after extrusion deformation increases.

Thus, after forging and extrusion, the martensitic transformation is not suppressed, despite the grinding structure, and, accordingly, the manifestation of functional effects in it can be maintained. This is probably due to the fact that the layer of fine-grained structure serves as an area of relaxation of stresses arising in the process of martensitic transformation and prevents the accumulation of excessive density of defects and the development of microcracks in the material. It can be assumed that in this way it will be possible to increase the stability of the functional properties of alloys with multiple cycles of phase transformation.

4. Conclusions

Thermomechanical treatment of Ni-Mn-Ga alloys by forging and forging with subsequent extrusion leads to the transformation of a cast coarse-grained structure into a bimodal structure consisting of large grains several hundred micrometers in size, surrounded by layers of a fine-grained structure. As a result of the formation of such a structure, the martensitic transformation shifts by ~7-11 °C to the region of lower temperatures.

It is expected that due to the creation of a bimodal grain structure, it will be possible to significantly increase the thermal stability of the functional properties of Heusler alloys during cycles of multiple martensitic transformation.

Acknowledgments

The thermomechanical treatment, investigations of the microstructure and electrical properties were funded by RFBR, according to the research project No. 16-32-60159 mol_a_dk (for I.I.M.). The investigation of magnetic properties was supported by RSF 14-22-00279 (for V.V.K.). The deformation processing and microstructure studies were carried out on the facilities of shared services center of the Institute for Metals Superplasticity Problems of Russian Academy of Sciences «Structural and Physical-Mechanical Studies of Materials».

References

[1] Pagounis E, Chulist R, Szczerba M J and Laufenberg M 2014 *Appl. Phys. Lett.* **105** 052405
[2] Barmina E, Kosogor A, Khovaylo V, Gorshenkov M, Lyange M, Kuchin D, Dilmieva E, Koledov V, Shavrov V, Taskaev S, Chatterjee R and Varga L K 2017 *J. Alloys Comp.* **696** 310
[3] Li Z, Yang B, Zou N, Zhang Y, Esling C, Gan W, Zhao X and Zuo L 2017 *Materials* **10** 463
[4] Pandey S, Quetz A, Aryal A, Dubenko I, Mazumdar D, Stadler S and Ali N 2017 *J. Magn. Magn. Mater.* **444** 98
[5] Zhukova A, Rodionova V, Ilyn M, Aliev A M, Varga R, Michalik S, Aronin A, Abrosimova G, Kiselev A, Ipatov M and Zhukova V 2013 *J. Alloys Comp.* **575** 73
[6] Arumugam S, Devarajan U, Muthu S E, Singh S, Thiyagarajan R, Raja M M, Rama Rao N V and Banerjee A 2017 *J. Magn. Magn. Mater.* **442** 460
[7] Rodionova I D et al 2015 *JETP Letters* **101(6)** 38
[8] Kourov N I, Korolev A V, Pushin V G and Marchenkova E V 2012 *Phys. Solid State* **54** 2128
[9] Chulist R, Skrotzki W, Oertel C.-G, Bohm A, Lippmann T and Rybacki E 2010 *Scripta Mater.* **62** 650
[10] Chulist R, Bohm A, Rybacki E, Lippmann T, Oertel C.-G and Skrotzki W 2012 *Mater. Sci. Forum* **702–703** 169
[11] Chulist R, Potschke M, Boehm A, Brokmeier H.-G, Garbe U, Lippmann T, Oertel C G and Skrotzki W 2007 *MRS Proc.* **1050** BB09-03
[12] Musabirov I I, Safarov I M, Galeyev R M, Afonichev D D, Koledov V V, Rudskoi A I and Mulyukov R R 2017 *Mater. Phys. Mech.* **33** 124
[13] Musabirov I I, Safarov I M, Galeyev R M, Gaisin R A, Koledov V V and Mulyukov R R 2018 *Phys. Solid State* **60(6)** 1061
[14] Musabirov I I, Safarov I M, Mulyukov R R, Sharipov I Z and Koledov V V 2014 *Letters on Materials* **4** 265

Hardening mechanisms of aluminum alloys under dynamic-channel angular pressing

I G Brodova[1], **A N Petrova**[1,2]

[1]M.N. Miheev Institute of Metal Physics of Ural Branch of Russian Academy of Sciences, 18 S. Kovalevskaya St., Ekaterinburg, 620108 Russia
[2]B.N. Yeltsin Ural Federal University, 19 Mir St., Ekaterinburg, 620002 Russia

E-mail: brodova@imp.uran.ru

Abstract. An experiment on high-speed deformation of aluminum alloy A3003 by dynamic channel angular pressing is described. During dynamic pressing, the material is exposed to two deformation modes. They are compression-tension due to the interaction of shock and rarefaction waves when the sample hits the bottom of the first channel, and a simple shear localized at the joint of the channels. The features of structure formation under the action of each deformation mode are determined. It was found that a high dislocation density and the presence of localized deformation zones caused by shock-wave loading, have a significant influence on the formation of the structure upon dynamic channel-angular pressing. The mechanisms of hardening and the contribution of dislocation and grain-boundary components to the overall hardening of the alloy are determined.

1. Introduction

It was found earlier [1-3] that dynamic channel angular pressing (DCAP) is an effective technique for refining the structure and hardening of commercial aluminum alloys. The enhancement of structural and phase transformations in aluminum alloys is due to the high strain rate (10^4-10^5 s^{-1}) and two acting deformation modes. The deformation modes are compression-tension on account of the circulation of shock and rarefaction waves emerged when a sample hits the bottom of the first channel and a simple shear is localized at the joint of the channels. The aim of the present work is to study the nature of the hardening of an aluminum alloy with an ultrafine-grained (UFG) structure and to determine the contribution of each deformation mode to the strength of the material produced by DCAP.

2. Experimental

During DCAP, a cylindrical Al-Mn (A3003 alloy) billet with a diameter of 30 mm and a length of 200 mm with a coarse subgrain structure with an average size of the subgrains 1-2 μm was stopped when it was crossing the intersecting area of the channels with a velocity of 200 m s^{-1}. For the study, two cross-sectional samples were cut from the obtained DCAP L-shaped sample. Sample A was cut from the vertical part and sample B from the horizontal part.

Structural characterization of the samples was carried out using a Neophot-32 optical microscope, a Philips CM-30 Super Twin transmission electron microscope (TEM) and a Quanta-200 Pegasus scanning electron microscope (SEM) equipped with an EBSD analysis system. TEM investigations were carried out at an accelerating voltage of 200 kV. Quantitative analysis of the size of structural fragments was done from dark field TEM images by means of Siams-700 image analysis software. X-ray diffraction analysis was performed using a DRON-3 diffractometer with Cu Kα radiation. The dislocation density was calculated with respect to the mean-square lattice microdeformation and the size of coherent scattering region. Microhardness was measured using a PMT-3 device. Brinnell hardness was determined as well using standard techniques.

3. Results

Studies by SEM and TEM show that the coarse subgrain structure is preserved in sample A. The influence of high strain rate deformation results in the formation of a generous amount of defects accumulated as near the boundaries of subgrains, as well as in the subgrain body. Figure 1 shows the dislocations pile-up within subgrains. The originally plane subgrain boundaries are distorted, their width increases due to aggregation of dislocations in the boundary regions (figure 1a, b). The single-crystal type of the microdiffraction pattern obtained from three neighboring subgrains with low angle-boundaries (LAGBs) shows reflections with low azimuthal misorientation (figure 1a). According to X-ray data, the average dislocation density in this sample is ρ_d=6.0 10^{13} m^{-2} and is 1.5 times higher than that of the initial material. Moreover, localized deformation bands with the UFG structure (500-700 nm) were detected in sample A (figure 1 c). These bands appear due to the plastic flow instability during loading of the shock wave. The UFG structure of the bands consists of grains separated by high-angle boundaries (HAGBs) which is evidenced by the ring-like shape of the selected-area diffraction pattern and subgrains with LAGBs (figure 1 c). Thus, X-ray and TEM data show that the accumulation of structural defects and minor structural refinement already occur in sample A due to the shockwave induced deformation.

The structure formed due to the shock wave loading undergoes shear deformation when the billet crosses the intersection area of the channels. Features of the structure formation upon additional simple shear deformation (sample B) are shown in figure 2. One can see a fragmented severely deformed structure. The dark field TEM image shows that the shear bands are oriented at an angle of 45-50° to the longitudinal axis and consist of grains/subgraines with HAGBs and LAGBs. A ring-like diffraction pattern (figure 2 a) proves the formation of such a mixed structure. The average lateral and longitudinal size of the structural fragments calculated from the dark field images are 370 and 420 nm, correspondingly. The DCAP pressing results in an increase in the dislocation density up to ρ_d=5.0 10^{13} m^{-2}. EBSD studies were performed on the deformed alloy to obtain more detailed information about its structure. A typical EBSD map of a sample after one DCAP cycle is presented in figure 3 a, b. An analysis of the distribution of the misorientation angle (figure 3c) shows that the shock-wave loading and shear deformation cause an increase in the number of LAGBs up to 70-75%, while approximately two thirds of which are boundaries with misorientation angles from 2 to 5 degrees. It was found [4-6] that these boundaries make a significant contribution to the material hardening.

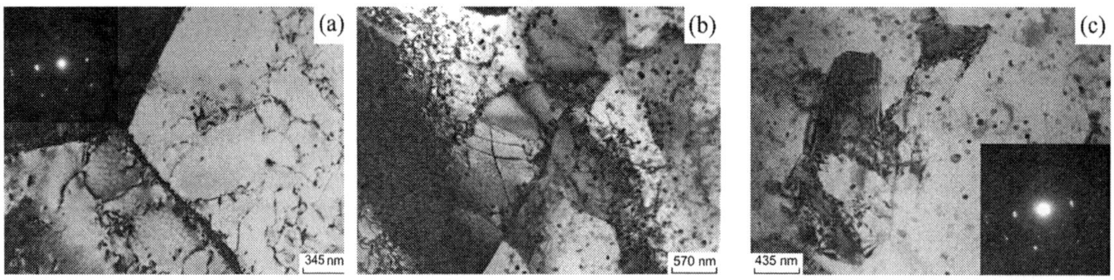

Figure 1. Bright field TEM images of the structure of the sample A (a and c combined with the selected area diffraction patterns): dislocations in the subgrain body (a); subgrain boundaries (b), localized deformation area (c).

IOP Conf. Series: Materials Science and Engineering **447** (2018) 012025 doi:10.1088/1757-899X/447/1/012025

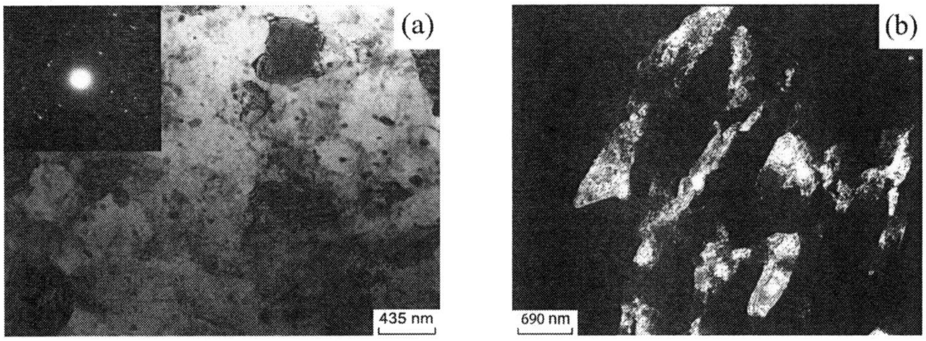

Figure 2. TEM images of the structure of sample B: bright field image combined with the selected area diffraction pattern (a), dark field image of aluminum matrix (b).

Analysis of the structure of the A3003 alloy after four cycles of DCAP showed that 60-65 % of the grain boundaries were HAGBs [1, 3].

The mechanisms of material hardening upon DCAP were determined by means of hardness measurements. One can see how the microhardness and hardness change after shock-wave loading and additional shear deformation (figure 4). According to figure 4, the first increase in the microhardness value (by 15%) is associated with a high accumulation of dislocation density and its evolution under shock-wave impact. At the next stage of simple shearing, the formation of a grain/subgrain structure resulted in further increase in microharndess values by 20%. Thus, the total increase in microhardness values is 35% or 190 MPa relative to the initial state. Brinell hardness is increased by 46% during DCAP. However, the primary increase in Brinell hardness by 35% occurs only after shear deformation, i.e. due to the formation of the UFG structure (figure 4). Considering these data and the results of TEM, we can conclude that the main contribution to the hardening of the alloy under shock-wave loading is due to the dislocation evolution. Hardening after shear deformation increases by means of the additional accumulation of defects and refinement of structural fragments, i. e. due to the hardening of grain boundaries. Therefore, a different contribution to the formation of the structure and hardening was revealed in two deformation modes of DCAP.

4. Discussion

The experiment, when the sample was stopped in a DCAP die, showed that the shock-wave loading in the initial state of dynamic pressing has a great influence on the structural formation of aluminum

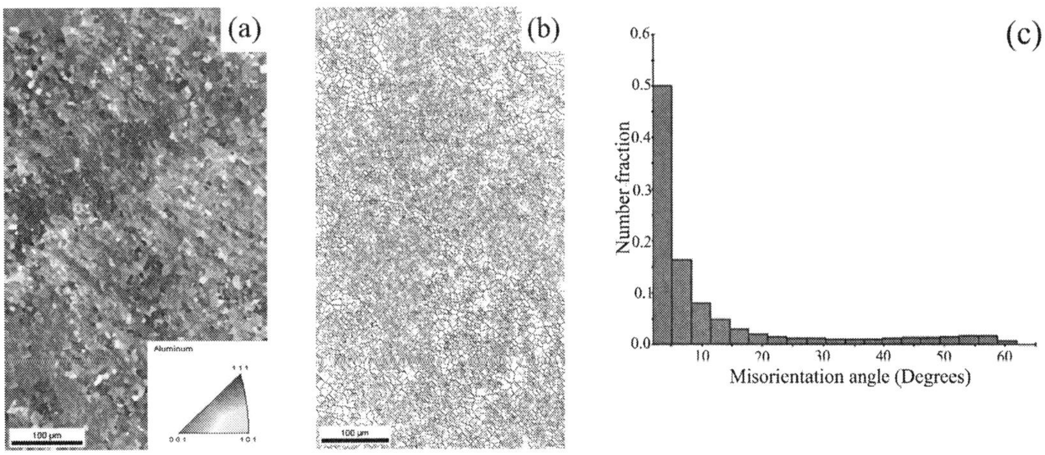

Figure 3. EBSD analysis data of sample B: a – EBSD map, b – boundaries pattern, red lines - 2-5°, green lines- 5-15°, blue lines - > 15°, c - misorientation angle distribution.

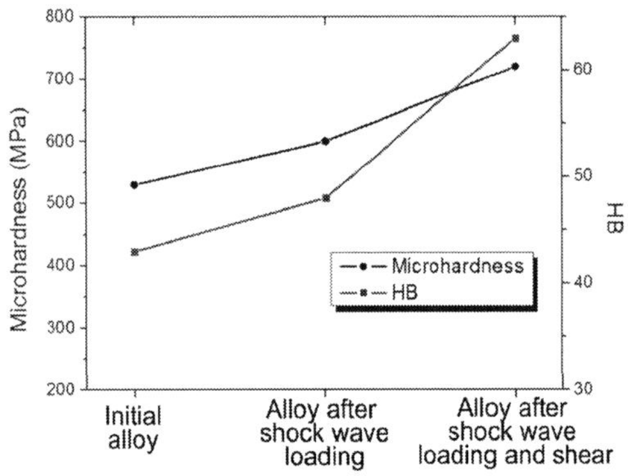

Figure 4. Changes of microhardness and hardness of the Al-Mn alloy with the deformation mode.

alloys and strength properties. The increase in the number of structural defects occurs due to the circulation and interaction of compression and rarefaction waves under shock-wave loading. At this stage, the value of the dislocation density increases in 1.5 times. High dislocation density leads to the material hardening and intensifies the structural refinement at the stage of shear deformation. Moreover, instabilities of the plastic flow under shock wave loading cause the formation of localized deformation bands. Their development and intersection during the shock wave propagation result in the dislocation assembly evolution and in the cell structure formation, which transforms in to the UFG structure with HAGBs and LAGBs under the shear deformation. Upon DCAP, both of these factors accelerate the transformation of the structure of aluminum alloys from coarse grained to UFG states. These aspects determine a higher efficiency of DCAP compared to static channel-angular pressing.

Considering X-ray and EBSD data, two hardening mechanisms of the alloy were distinguished. Firstly, these are dislocations hardening due to the accumulated high dislocation density and LAGBs with misorientation angles from 2 to 5 degrees. Secondly, it is the grain boundaries hardening due to the formation of LAGBs with a misorientation angle of 5-15 degrees and the UFG structure with HAG Bs.

The experimental results of the present study are in good agreement with the numerical simulation of the DCAP process based on the plastic flow dislocation model [7].

5. Acknowledgments

This work was performed in the framework of the «Structura» State Task № AAAA-A18-118020190116-6). The electron microscopy investigations performed at the Center of Collaborative Access "Testing Center of Nanotechnologies and Advanced Materials" of the M.N. Miheev Institute of Metal Physics of Ural Branch of Russian Academy of Sciences.

References

[1] Petrova A N, Shirinkina I G, Shorokhov E V, Minaev I V, Zhgilev I N and Abramov A V 2010 *Rev. Adv. Mater. Sci.* **25**(2) 128-135
[2] Petrova A N, Brodova I G, Plekhov O A, Naimark O B, Shorokhov E V 2014 *Techn. Phys.* **59**(7) 989-996
[3] Brodova I G, Petrova A N, Razorenov S V and Shorokhov E V 2015 *Phys. Metals Metallogr.* **116**(5) 519-526
[4] Brodova I G, Petrova A N, Naimark O B, Plekhov O A, Razorenov S V and Shorokhov E V 2017 *J. Phys.: Conf. Series* **894** 012016
[5] Kamikawa N, Huang X, Tsuji N and Hansen N 2009 *Acta Mater* **57** 4198-4208
[6] Estrin Y and Vinogradov A 2013 Acta Mater **61** 782-817
[7] Kato M 2014 *Mater.Trans.* **55** 19-24
[8] Borodin E N and Mayer A E 2013 *Techn. Phys.* **58**(8) 1159-1163

IOP Conf. Series: Materials Science and Engineering **447** (2018) 012026 doi:10.1088/1757-899X/447/1/012026

Structure, strength and superplasticity of the bulk 1570C alloy subjected to high-temperature multidirectional isothermal forging

O Sh Sitdikov, E V Avtokratova, R R Ilyuasov, O E Latypova and M V Markushev

Institute for Metals Superplasticity Problems, Russian Academy of Sciences, 39 Khalturin st., Ufa 450001, Russia

E-mail: sitdikov.oleg@imsp.ru

Abstract. The structure and mechanical properties of the cast non-age-hardenable complex-alloyed 1570C alloy (Al-5Mg-0.18Mn-0.2Sc-0.08Zr, wt%) were studied after multi-directional isothermal forging at 325 °C up to accumulative strain e ≈ 12 and subsequent annealing in the temperature interval from 325 to 500 °C. It is shown that even amid the absence of substantial alloy hardening, high-temperature multi-directional forging is an effective method for processing bulk billets, resulting in a stable and homogeneous (ultra)fine-grained structure with a grain size of about 2 μm, which provided improved ductility and extraordinary high superplastic properties.

1. Introduction

The non-age-hardenable complex-alloyed aluminum alloy 1570C is one of the advanced structural materials for aerospace application [1,2]. It can be easily hot-worked and demonstrates superior superplasticity in the ultrafine-grained (UFG) state (grain size less than 1 μm) [2], while the cold deformation of coarse-grained counterparts is limited due to the high yield stress and low ductility [1]. One of the crucial issues for the industrial use of the alloy is the lack of an efficient technology to process bulk semi-products with the UFG structure.

The UFG structures have been produced in different materials, including complex-alloyed aluminum alloys, via the so-called severe plastic deformation (SPD) imparting large effective strains e>1 onto the workpiece [2-6]. Among many other SPD techniques, multi-directional isothermal forging (MIF) has been proven to be quite efficient to produce an industrially scalable amount of UFG bulk billets [3,6]. However, the effect of MIF on the microstructure and properties of aluminum alloys has been just scarcely studied to date.

The aim of the work is to investigate the structure and mechanical behavior of the 1570-type alloy ingot subjected to MIF.

2. Material and procedure

Samples with a diameter of 80 mm and a length of 150 mm were cut from an ingot of the 1570C alloy (Al-5Mg-0.18Mn-0.2Sc-0.08Zr, wt %), homogenized at 360 °C for 6 hours. Then they were subjected to MIF to effective strain of about 12 at 325 °C ($T \approx 0.6T_m$) using a hydraulic press with an isothermal die set. Each MIF cycle consisted of several deformation steps, which involved changing the deformation axis, as shown in figure 1a. Specifically, the first step was compression in the X-direction for a billet preheated in an air furnace, followed by sequential compressions from the sides of the ribs. As a result of a series of steps, the billet elongated in the Y-direction and then in the Z-direction, so that its shape and dimensions were maintained roughly unchanged in each MIF cycle (figure 1b).

The structure of the deformed billets was investigated in the central parts of the sections parallel to the last compression axis by standard methods of optical microscopy (OM) (using a Nikon L 150 optical microscope), scanning electron microscopy with the electron backscatter diffraction (SEM-EBSD) analysis (using a field emission TESCAN MIRA 3 LMH scanning electron microscope equipped with the Oxford Instruments HKL Channel 5 system), as well as transmission electron microscopy (TEM) (using a JEOL-2000EX transmission electron microscope).

The parameters of the structure, including the average misorientation angle of the intercrystallite boundaries Θ_{ave}, the high angle boundary fraction f_{HAB}, as well as the average size of the fine grains, were derived from the SEM-EBSD analysis [2,5]. The grain size was evaluated from the EBSD data by converting the grain area measurements into the "equivalent circle diameter" (with an average equivalent diameter of each of the individual grains). The Vickers microhardness (HV) was measured at room temperature using a standard procedure at a load of 0.5N. Tensile tests were carried out at ambient and elevated temperatures on an Instron 1185 screw-driven testing machine using dog-bone shaped specimens with the gage part of 1.5 × 3 × 6 mm^3. The thermal stability of the developed structure was evaluated after one-hour annealing in the temperature interval 325-500 °C.

3. Results and discussions

Structural changes. In the initial cast and homogenized state, the alloy possessed an equiaxed grain structure with a grain size of 25 μm (figure 2 a, b) and uniformly distributed coherent precipitates $Al_3(Sc, Zr)$ of about 5-10 nm in diameter (figure 2 c). MIF to e=12 resulted in the development of a rather homogeneous UFG structure with the size and the volume fraction of new equiaxed grains of about 2 μm and 80-85 %, respectively (figure 2 d-f). Moreover, the fraction of high-angle boundaries and the average misorientation angle of intercrystallite boundaries in such a structure were quite high: f_{HAB} = 78% and Θ_{ave} = 31.4°. The formation of this structure was conditioned by the multistep character of the deformation carried out to high strains, occurring in various directions; that provided the uniformity of grain refinement throughout the material volume.

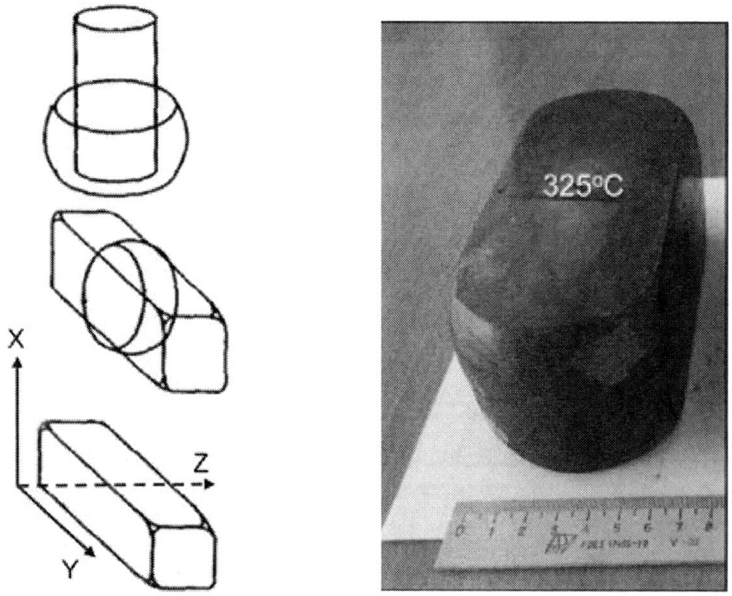

(a) (b)

Figure 1. Scheme of MIF (a) and photograph of the processed billet (b).

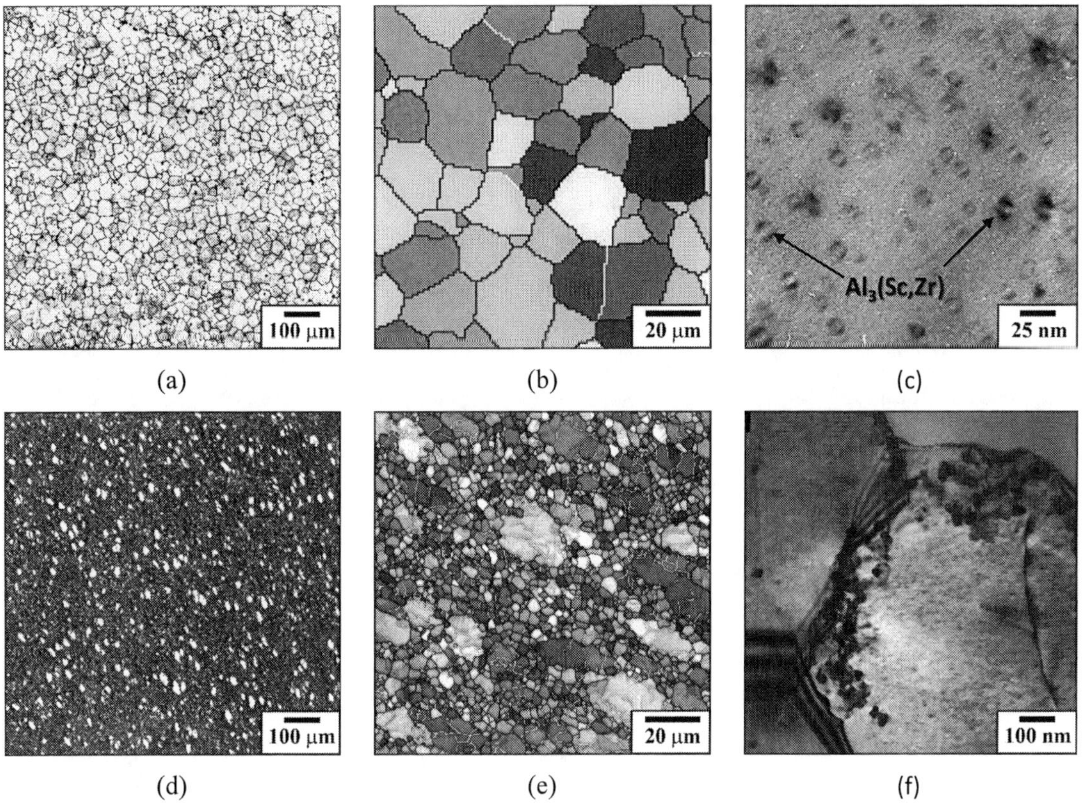

Figure 2. The microstructure of the alloy 1570C before (a-c) and after (d-f) MIF: (a, d) OM; (b, e) SEM-EBSD and (c, f) TEM.

Room-temperature hardness and tensile strength parameters of the alloy before and after MIF are represented in table 1. Besides, the data for the same alloy, which was processed by equal channel angular pressing (ECAP) to e=8 at 325 °C [4], are shown in table 1 for comparison. It is seen that, in contrast to ECAP, the processing of the alloy by MIF at the same temperature did not result in its strengthening, while the ductility after MIF increased to a greater extent. This behavior was conditioned by several reasons. The absence of a significant difference in the strength of the alloy before and after MIF was caused by the fact that its structural (Hall–Petch) strengthening due to the grain refinement (figures 2 a, d) was almost completely compensated by a nearly equal strength drop owing to the coarsening of $Al_3(Sc, Zr)$ precipitates under processing at the high temperature (figure 2 c, f) [5]. Meanwhile, a much more prominent alloy strengthening after ECAP than after MIF was reasoned by a higher (on about two orders) strain rate during ECAP [6] and, hence, by a higher defect density, as well as twice as smaller ultra-fine grain size (about 1 μm), developed in ECAP [5]. Therewith, the higher ductility was predictably exhibited by the material with a more uniform and equilibrium UFG structure processed by MIF.

Thermal stability. It was found that the MIF processed UFG structure was entirely stable up to temperatures 425-450 °C. At higher temperatures, an abnormal grain growth was noticed. As a result, the UFG structure was transformed into the bimodal one, composed by areas of ultrafine and coarse grains.

Superplasticity. Tensile tests at elevated temperatures have shown that the alloy after MIF demonstrated enhanced superplastic properties with elongations-to-failure of more than 500% in the temperature interval 400-500 °C and at strain rates up to 10^{-1} s^{-1} (figure 3 a). Therewith the maximum elongation-to-failure values of about 2800% were achieved at 450 °C and a strain rate of 5.6x10^{-3} s^{-1} and at 500 °C and a strain rate of 1.4x10^{-2} s^{-1} (figure 3 b). The data obtained testify to the achievement of superplasticity with ultra-high elongations in a wide range of temperatures - strain rates, including high-strain rate superplasticity, which significantly increases the efficiency and potential of commercial application of the alloy.

Table 1. Mechanical properties of the alloy 1570C at room temperature after MIF and ECAP

State	HV	YS, MPa	UTS, MPa	El , %
Initial	105±10	240±3	355±6	28±1
MIF at 325°C	105±10	235±5	360±5	38±3
ECAP Bcz at 325 °C [4]	-	300±4	380±6	31±2

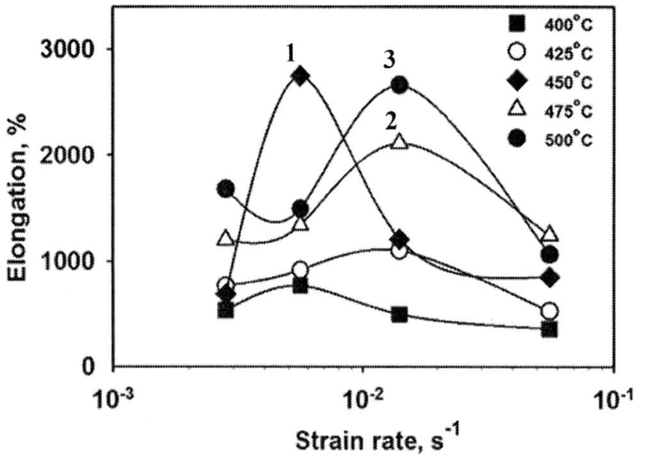

(a)

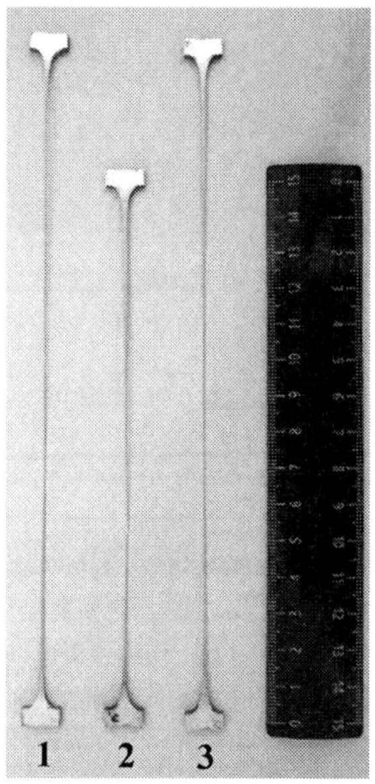

(b)

Figure 3. Elongations-to-failure of the alloy 1570C after MIF (a); photographs of the samples after tensile tests at 450 °C and 5.6x10^{-3} s^{-1} (1), 475 °C and 1.4x10^{-2} s^{-1} (2), 500 °C and 1.4x10^{-2} s^{-1} (3) showing maximum elongations (b).

4. Conclusions

Bulk billets with a fairly stable and uniform structure with a grain size of about 2 μm were processed from the complex-alloyed 1570C alloy by MIF at 325 °C (at about 0.6T_m). The data obtained testify to a high potential of commercial use of high-temperature MIF for obtaining bulk billets with a stable and homogeneous UFG structure from 1570-type aluminum alloys.

In the absence of substantial hardening, the UFG alloy 1570C processed by high-temperature MIF exhibited significantly increased room temperature ductility. Besides, the structural superplasticity with extraordinary high parameters of elongation-to-failure was achieved at elevated temperatures in a wide range of strain rates, including high-strain rate superplasticity.

Acknowledgments
This study was supported by the Russian Foundation for Basic Research (RFBR) under grant No 16-08-01189 A. The experimental work was carried out using the facilities of the Shared Service Center of IMSP RAS «Structural and Physical-Mechanical Studies of Materials».

References

[1] Filatov Y, Elagin V and Zakharov V 2000 *Mater. Sci. Eng. A* **280** 97
[2] Avtokratova E, Sitdikov O, Markushev M and Mulyukov R 2012 *Mater. Sci. Eng. A* **538** 386
[3] Imayev R, Salishchev G, Senkov O, Imayev V, Shagiev M, Gabdullin N, Kuznetsov A and Froes F 2001 *Mater. Sci. Eng. A* **300** 263
[4] Kaibyshev R, Avtokratova E and Sitdikov O 2010 *Mater. Sci. Forum* **638-642** 1952
[5] Avtokratova E, Sitdikov O, Mukhametdinova O, Markushev M, Murty S V S N and Prasad M J N V, Kashyap B P 2016 *J. Alloys Comp.* **673** 182
[6] Sitdikov O 2016 *Inorg. Mater.: Appl. Res.* **7** 149

Influence of additional severe plastic deformation at elevated temperatures on microstructure and physical-mechanical properties of ultrafine grained Al-0.4Zr alloy

T A Latynina[1], M Yu Murashkin[2,3], R Z Valiev[2,3] and T S Orlova[1,4]

[1]ITMO University, 49 Kronverkskiy pr., St. Petersburg 197101, Russia
[2]St. Petersburg State University, 7/9 Universitetskaya nab., St. Petersburg 199034, Russia
[3]Institute of Physics of Advanced Materials, Ufa State Aviation Technical University, 12 K. Marx str., Ufa 450000, Russia
[4]Ioffe Institute, 26 Politekhnicheskaya str., St. Petersburg 194021, Russia

E-mail: t.latynina13@yandex.ru

Abstract. The effect of high pressure torsion (HPT) at room temperature and subsequent HPT at elevated temperatures on the microstructure, mechanical and electrical properties of the Al-0.4Zr alloy has been investigated. It has been shown that the HPT treatment of the alloy with an ultrafine grained structure at elevated temperatures led to a drastic increase in strength, which is accompanied by an increase in conductivity. Experimental values and theoretical estimations of the yield stress and electrical resistivity at 77 K are compared. Possible reasons for the observed strengthening effect are discussed.

1. Introduction

For electrotechnical alloys, which are used for power transmission lines, it is especially important to combine high values of electrical conductivity and strength for operation at elevated temperatures of 150-200 °C [1]. Nowadays, Al-based alloys with 0.1-0.4 wt.% Zr are considered promising materials to meet these requirements [1]. However, Al-Zr alloys have low strength, hence increasing their strength while keeping a high level of heat resistance and conductivity is a challenge. Methods of severe plastic deformation (SPD) are efficient to drastically increase mechanical properties of different metals and alloys due to the formation of an ultrafine grained (UFG) structure and the introduction of an enhanced defect density in the crystal lattice. By using a two-stage processing by SPD, first at room temperature (RT) and then at an elevated temperature, a good combination of strength (365 MPa) and conductivity (58.4% IACS – International Annealed Copper Standard) has been achieved for Al-Mg-Si alloys due to the grain refinement and the nanoscale precipitate formation [2]. But secondary phase nanoprecipitates in this alloy did not provide any reasonable thermal stability of the strength. In this paper, the influence of two-stage processing by high pressure torsion (HPT) on microstructure and physical-mechanical properties of the Al-0.4 wt.% Zr alloy has been investigated.

2. Materials and experimental procedures

The initial Al-0.4 wt.% Zr alloy was processed by the method of combined casting and rolling [3], and then subjected to HPT under a hydrostatic pressure of 6 GPa at room temperature (RT) to 10 revolutions (n=10) (hereinafter referred to as HPT_RT state). As a result of such treatment, samples were obtained in the shape of discs with a diameter of 20 mm and a thickness of approximately 1.2 mm. The true strain at a distance of 5 mm from the center of the disc was $\varepsilon\approx6.6$. Then, some HPT_RT samples were subjected to additional HPT processing at elevated temperatures, T_{HPT}=230 and 280 °C, to n=1, 5, 10 and 20. The microstructure of the alloy in all the studied states was examined by X-ray diffraction (XRD), electron backscattered diffraction (EBSD) and transmission electron microscopy (TEM), as described in [4]. EBSD studies were performed using a Zeiss Merlin scanning electron microscope with a scan step of 0.2 μm on an area containing at least 2000 grains. To perform mechanical tests, blade-shaped samples with a gauge width of 2 mm and a gauge length of 6 mm were cut from HPT-processed disks at a distance of 5 mm from the disk center [5]. Uniaxial tensile tests were performed on a Shimadzu AG-XD Plus machine with a constant strain rate of $5\cdot10^{-4}$ s^{-1} [5]. Vickers microhardness (H_V) was measured using a Shimadzu HMV-G tester with a load of 1 N during 15 s, the indentations being performed along the gauge [6]. Electrical resistivity (ρ) was measured by a standard four probe technique at 77 K for comparison with theoretical estimates and at 293 K for calculation of conductivity ω at RT ($\omega=1/\rho_{293}$), more detailed data of these measurements are provided in [6]..

3. Results and discussion

The effect of a substantial increase in the strength of UFG Al-0.4 wt.% Zr (preliminary treated by HPT at RT) by its short-term annealing in the temperature range of 90-280 °C was earlier revealed [7]. Therefore, in the present work, temperatures of 230 and 280 °C were chosen for the second HPT treatment. Figure 1 shows the dependences of microhardness and resistivity at 77 K for the UFG Al-0.4Zr alloy on the number of HPT revolutions at 230 and 280 °C. For further investigation, the states after HPT at 230 °C with n=10 and at 280 °C with n=5 (indicated in figure 1 by a dashed circle and hereinafter referred to as HPT_RT-HPT_230_10 and HPT_RT-HPT_280_5, respectively) were chosen, which demonstrated a significant increase in microhardness and a decrease in electrical resistivity compared to the HPT_RT state.

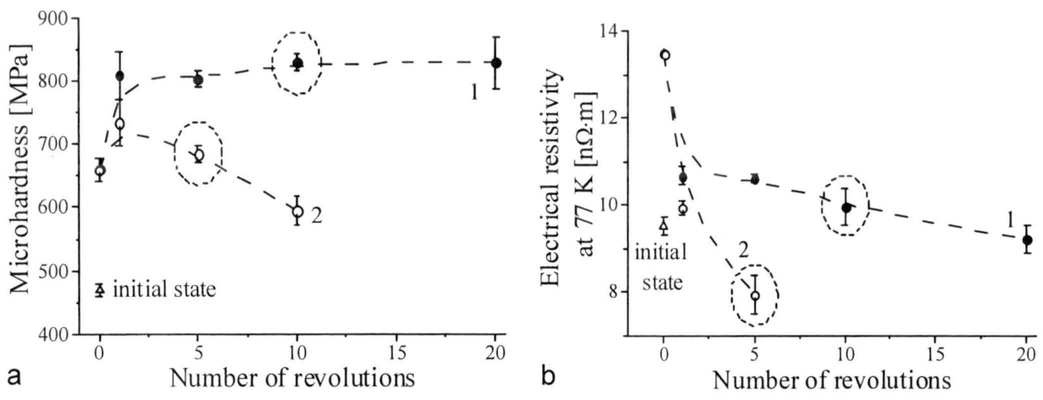

Figure 1. Dependences of microhardness H_V (a) and electrical resistivity $\rho_{77}^{\exp}$ at 77 K (b) for the Al-0.4Zr alloy preliminary treated by HPT at RT on the number of revolutions of additional HPT at 230 °C (1) and 280 °C (2). The values of H_V and $\rho_{77}^{\exp}$ for the initial state are also presented.

Quantitative changes in mechanical properties (microhardness H_V, conventional yield stress $\sigma_{0.2}^{\exp}$, ultimate tensile strength σ_{UTS} and relative elongation to failure δ), electrical resistivity at 77 K (ρ_{77})

and electrical conductivity at RT ($\omega=1/\rho_{293}$) for the alloy due to HPT at RT, and then at elevated temperatures are given in table 1. HPT processing at RT leads to a significant increase in strength, however, the electrical conductivity significantly decreases. As a result of the additional HPT processing at 230 and 280 °C, the strength properties further drastically increased ($\sigma_{0.2}$ by 89 and 65%, σ_{UTS} by 43 and 21%, respectively), and ω at RT also significantly increased (by 3.5 and 8% IACS).

Table 1. Mechanical and electrical properties of the Al-0.4Zr alloy.

State	$\sigma_{0.2}^{exp}$ (MPa)	$\sigma_{0.2}^{th}$ (MPa)	σ_{UTS} (MPa)	δ (%)	ω (% IACS)	ρ_{77}^{exp} (nΩ·m)	ρ_{77}^{th} (nΩ·m)
Initial	120±2	108	130±1	25.9±0.2	50.7	9.5±0.1	9.5
HPT_RT	140±1	137	192±1	23.2±0.6	43.9	13.5±0.1	9.9
HPT_RT-HPT_230_10	264±4	110*	276±5	7.7±0.6	47.3	10.0±0.4	-
HPT_RT-HPT_280_5	229±2	104*	232±2	15.7±0.9	52.0	7.9±0.5	-

* Estimations were made without contribution to strengthening from the precipitates

To reveal the microstructural changes, which are responsible for the obtained changes in properties, the microstructural characterization was performed. The results of EBSD (average grain size d_{av}, fraction of high angle grain boundaries (HAGBs) $f_{\geq 15}$, average misorientation angle θ_{av}) and XRD (the lattice parameter a, the size of coherent-scattering domains D_{XRD}, the elastic microdistortion level $<\varepsilon^2>^{1/2}$ and the dislocation density L_{dis}) analyses are shown in table 2. The dislocation density strongly increases after HPT at RT. The subsequent HPT at 230 and 280 °C causes its decrease by approximately 4 and 7.5 times respectively. The EBSD investigation of the microstructure shows that HPT at RT results in a refinement of the grain structure, an equiaxed grain structure is formed with an average grain size d_{av}=840 nm (figure 2 a and table 2). Most grain boundaries (GBs) are HAGBs with misorientations ≥15°.

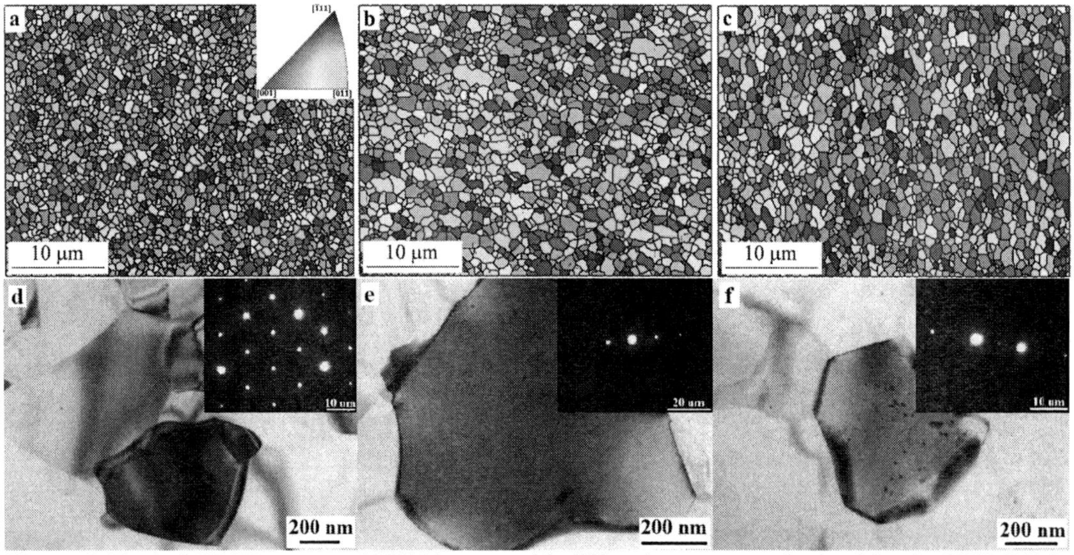

Figure 2. Typical EBSD maps and TEM images with diffraction patterns obtained for the HPT_RT (a, d), HPT_RT-HPT_230_10 (b, e) and HPT_RT-HPT_280_5 states (c, f).

TEM studies did not reveal the formation of secondary phases (figure 2 d). Additional deformation at elevated temperatures only slightly increases d_{av}, $f_{\geq 15}$ and θ_{av} (table 2) and mainly leads to dynamic aging – the formation of nanosized Al_3Zr precipitates (figure 2 e, f), thereby reducing the concentration of solute Zr atoms. The amount of precipitates is not high, it increases with HPT temperature growing (figure 2 e,f). The precipitates are non-uniformly distributed in the samples. A rough estimation of the average size of the precipitate from the TEM images gave the values of 18-19 nm for both states after HPT at elevated temperatures. TEM studies demonstrate that the grain interior is mostly free of dislocations in all the three states: HPT_RT and HPT_RT-HPT_230_10 and HPT_RT-HPT_280_5. Hence, a decrease in the L_{dis} after the additional HPT at elevated temperatures is mainly related to the regions at and around GBs, which points to GB relaxation during HPT at elevated temperatures.

Table 2. Microstructure parameters obtained by EBSD and XRD analysis for the Al-0.4Zr alloy

State	EBSD			XRD			
	d_{av} (nm)	$f_{\geq 15}$ (%)	θ_{av} (°)	a (Å)	D_{XRD} (nm)	$<\varepsilon^2>^{1/2}$ (%)	L_{dis} (m^{-2})
Initial	1420±20	25	11.5	4.0514±0.00002	252±6	0.003±0.0001	$1.4\cdot10^{12}$
HPT_RT	840±15	82	33.8	4.0515±0.00006	212±11	0.063±0.0006	$3.6\cdot10^{13}$
HPT_RT-HPT_230_10	920±15	88	36.2	4.0512±0.00003	483±6	0.036±0.0007	$9.0\cdot10^{12}$
HPT_RT-HPT_280_5	930±15	89	36.5	4.0514±0.00007	559±25	0.022±0.0002	$4.8\cdot10^{12}$

The contributions from different electron scattering mechanisms to electrical resistivity at 77 K were estimated according to the Matthiessen's rule [8]:

$$\rho_{77}^{th} = \rho^{pure} + L_{dis}\Delta\rho^{dis} + S_{GB}\Delta\rho^{GB} + C_{Zr}^{sol}\Delta\rho_{Zr}^{sol} + \Delta\rho^{pt}, \quad (1)$$

where ρ^{pure}=2.7 nΩ·m [9] is the electric resistivity of single-crystalline defect-free aluminum, $\Delta\rho^{dis}$=2.7·10-25 Ω·m^3 [10], $\Delta\rho^{GB}$=2.6·10-16 Ω·m^2 [10] are the contributions from unit densities of dislocations and grain boundaries in Al, respectively, $\Delta\rho_{Zr}^{sol}$=15.8 nΩ·m/wt.% [11] is the contribution from a unit concentration of Zr in the solid solution, L_{dis} (m^{-2}) is the dislocation density, S_{GB} (m^{-1}) is the bulk density of GBs, C_{Zr}^{sol} (wt.%) is the concentration of solute Zr; $\Delta\rho^{pt}$ – the contribution originating from the secondary phase precipitates. The estimated (ρ_{77}^{th}) and experimental ($\rho_{77}^{\exp}$) values are compared in table 1. HPT at RT led to a substantial increase in $\rho_{77}^{\exp}$ caused by changes in the microstructure. The estimation showed that the contribution from the L_{dis} change was negligibly small (~9.4·10^{-3} nΩ·m), the Zr concentration in the Al matrix did not change, the increase in the GB density gives the resistivity increase ~0.38 nΩ·m, which is an order of magnitude smaller than the change in $\rho_{77}^{\exp}$ (table 1). It is reasonable to suggest that the specific resistivity of GBs ($\Delta\rho^{GB}$) in the Al-0.4Zr alloy subjected to HPT at RT is higher than that in coarse-grained Al. This could be caused by two reasons: the formation of mainly HAGBs in non-equilibrium states similarly to the case of commercially pure Al [4] and the possible segregation of impurities at GBs. Additional HPT at 230 and 280 °C results in a substantial decrease in $\Delta\rho^{GB}$ by ~3.5 and 5.5 nΩ·m, respectively. This is mainly due to the reduction of the Zr content in the Al matrix. An additional decrease in electrical resistivity can be caused by the GB relaxation to a more equilibrium state, however, if the grain-boundary segregation process occurs during additional HPT, this will contribute to the electrical resistivity.

Contributions from different mechanisms to the total strengthening, which is their superposition, were also estimated in the same way as was done in [6]:

$$\sigma_{0.2}^{th} = \sigma_0 + \sigma_{GB} + \sigma_{dis} + \sigma_{Or} + \sigma_{sol}, \quad (2)$$

where σ_0=10 MPa is the Peierls-Nabarro stress of the Al crystal lattice, σ_{GB} is grain boundary strengthening, σ_{dis} is strain hardening due to dislocations, σ_{Or} is precipitate strengthening by the Orowan mechanism and σ_{sol} is solid-solution hardening.

The experimental ($\sigma_{0.2}^{\exp}$) and theoretical ($\sigma_{0.2}^{th}$) values of $\sigma_{0.2}$ are in good agreement for the states before and after HPT at RT (table 1). However, for the states after additional HPT at elevated temperatures the values of $\sigma_{0.2}^{\exp}$ are dramatically higher than the estimated values of $\sigma_{0.2}^{th}$ (table 1), which points to the operation of the additional strengthening mechanism caused by microstructural changes during the second stage of HPT. The amount of nano-sized precipitates is small, and their distributions throughout the samples are non-uniform, hence, the Orowan strengthening could not be efficient. According to a simple estimation, if even the Orowan strengthening takes place, the maximum possible contribution of the Orowan strengthening (if the whole amount of Zr was in the precipitates of 18 nm in size) will not exceed 80 MPa, whereas the obtained difference between $\sigma_{0.2}^{\exp}$ and $\sigma_{0.2}^{th}$ equals to 154 and 114 MPa for HPT_RT-HPT_230_10 and HPT_RT-HPT_280_5 states, respectively. Then one may conclude that similar to the case of commercially pure Al [5], the achieved dramatic strengthening of the Al-0.4Zr alloy after additional HPT at an elevated temperature is most probably caused by the rearrangement of the HAGBs structure, which is accompanied by a decrease in the dislocation density in them. In addition, in the Al-0.4Zr alloy, the influence of possible GB segregation cannot be excluded [12,13] and needs to be checked in further investigations.

4. Conclusions

A drastic increase in strength was found after additional HPT at elevated temperatures for the Al-0.4Zr alloy, which was preliminary subjected to HPT at RT. The strength enhancement was accompanied by an increase in electrical conductivity. The analysis of different hardening mechanisms to the total strengthening indicates that HPT at elevated temperatures caused an additional strengthening mechanism, which is most likely associated with the rearrangement of the HAGBs structure.

Acknowledgments

The present work was supported by the Russian Science Foundation (grant No. 17-19-01311).

References

[1] Belyi D I 2012 *Cables and Wires* **1** 8 (in Russian)
[2] Valiev R Z, Murashkin M Yu and Sabirov I 2014 *Scr. Mater.* **76** 13
[3] Machine for the continuous casting of metal rods 1953 US Patent 2659948 A
[4] Orlova T S, Mavlyutov A M, Bondarenko A S, Kasatkin I A, Murashkin M Yu and Valiev R Z 2016 *Phil. Mag.* **96** 2429
[5] Mavlyutov A M, Latynina T A, Murashkin M Yu, Valiev R Z and Orlova T S 2017 *Phys. Sol. State* **59** 1970
[6] Orlova T S, Mavlyutov A M, Latynina T A, Ubyivovk E V, Murashkin M Yu, Schneider R, Gerthsen D and Valiev R Z 2018 *Rev. Adv. Mater. Sci.* **55** 92
[7] Latynina T A, Mavlyutov A M, Murashkin M Yu, Valiev R Z and Orlova T S 2018 *Collection of Materials of XXIII Petersburg Readings on the Problems of Strength* (St. Petersburg: VVM Publishing) p 130 (in Russian)
[8] Rossiter P L 1991 *The Electrical Resistivity of Metals and Alloys* (Cambridge, Cambridge University Press)
[9] ASM Handbook 1990 *Properties and Selection: Nonferrous Alloys and Special-Purpose Materials* (ASM International)

[10] Karolik A S and Luchvich A A 1994 *J. Phys. Cond. Matter* **6** 873
[11] Kutner F and Lang G 1976 *Aluminium* **52** 322
[12] Murashkin M Yu, Sabirov I, Medvedev A E, Enikeev N A, Lefebvre W, Valiev R Z and Sauvage X 2016 *Materials & Design* **90** 433
[13] Sauvage X, Enikeev N, Valiev R, Nasedkina Y and Murashkin M 2014 *Acta Mater.* **72** 125

IOP Conf. Series: Materials Science and Engineering **447** (2018) 012028 doi:10.1088/1757-899X/447/1/012028

The effect of hot deformation by torsion under pressure and post-deformation annealing on superconducting properties of Bi(Pb)2223 ceramics

N Yu Parkhimovich[1], M F Imayev[1,2], A A Eliseev[3], P E Kazin[3]

[1] Institute for Metals Superplasticity Problems, Russian Academy of Sciences, 39 Khalturin st., Ufa 450001, Russia
[2] Bashkir State University, 32 Zaki Validi st., Ufa 450076, Russia
[3] Moscow State University, 1 Leninskie Gory, Moscow 119991, Russia

E-mail: pn@imsp.ru

Abstract. The influence of conditions of hot deformation by torsion under pressure and subsequent annealing on the parameters of superconducting transition and critical current density J_c(77 K, 100 Oe) of $(Bi,Pb)_2Sr_2Ca_2Cu_3O_{10+d}$ {Bi(Pb)2223} ceramics was studied. The optimal conditions of post-deformation annealing for superconducting transition recovery were determined to be T= 835 °C, t= 14 h. Post-deformation annealing led to a change in the size of grain colonies and a weakening of the texture of the Bi(Pb)2223 matrix phase. The position of the peak on the temperature dependence of the imaginary component of the AC magnetic susceptibility, $T_{\chi''}$, and the critical current density J_c were determined. The samples twisted with a rate ω= 1.85×10^{-4} rpm at T= 855 °C to α=25 ° and at T= 875 °C to α=25° possessed the best properties: $T_{\chi''}$=104.4 K, J_c= 9295 A/cm^2 and $T_{\chi''}$=104.1 K, J_c= 8098 A/cm^2, respectively. Post-deformation annealing resulted in a minimal weakening of the texture in these samples. The weakening of the basal texture in the the post-deformation annealing process was explained by the growth of grain colonies of Bi(Pb)2223 with a non-basal orientation occurring during dynamic recrystallization.

1. Introduction

High values of the critical current density J_c in high-T_c superconductors can be obtained by strenghtening their texture, increasing the density of magnetic flux pinning centers, and achieving the optimal oxygen content [1]. In bulk superconducting ceramics, a strong texture is obtained by the melt-textured growth [2] and deformation techniques, including plastic deformation [3]. Ceramics $(Bi, Pb)_2Sr_2Ca_2Cu_3O_{10+d}$ {Bi(Pb)2223} can be textured by deformation, because the melt texturing growth techniques are inapplicable due to the narrow range of existence of this phase on the phase diagram [4]. High-temperature torsion under quasi-hydrostatic pressure (HPT) is a promising technique of texturing, since it allows deforming the material to high strains [5]. Immediately after deformation, Bi(Pb)2223 ceramics usually does not exhibit high superconducting properties due to the non-optimal oxygen content and the presence of residual stresses. In order to recover the superconducting properties in deformed ceramics, it is necessary to carry out annealing, at which the optimal combination of oxygen content and structural properties of the superconducting phase will be achieved. Usually, Bi(Pb)2223 ceramics is annealed above 800 °C, while the temperature T and the duration of annealing t are chosen depending on the gas environment composition and the deformation technique [3]. The goal of the present work was to study the effect of hot deformation conditions and post-deformation annealing on the parameters of the superconducting transition and the critical current density in Bi(Pb)2223 ceramics.

2. Experimental

The initial sintered pellets of Bi(Pb)2223 (Ø≈ 8 mm and $h\approx$ 1.5 mm) were deformed by HPT under the following conditions: deformation temperature T_d= 845 – 885 °C, pressure P= 10 MPa, twist rate ω_1= 1.85×10^{-4} rpm, twist angle α=5 – 30°; T_d= 865°C, P=10 МПа, ω_2= 5.14×10^{-5} rpm, α= 1 – 10° [6]. Post-deformation annealing was carried out in air. To determine the optimal annealing temperature, the sintered undeformed Bi(Pb)2223 pellets were annealed in the range of T = 825-845 °C for 5 hours. After determining the optimal annealing temperature, the annealing time was optimized. The annealing time was optimized using samples deformed under the following conditions: ω_1/865 °C/15° and ω_2/865 °C/6° (twist rate/temperature of deformation/twist angle). The deformed samples were annealed for t= 6-18 hours. The degree of the basal plane texture was determined by the Lotgering method using the peaks (00.10) and (109) of the X-ray diffraction pattern [7]. The mean length of the colonies of grains of the Bi(Pb)2223 phase before (L_0) and after (L) annealing was calculated by measuring the length of individual colonies with statistics of at least 400 colonies per sample. The temperature of the intergranular superconducting transition $T_{\chi''}$ was determined by the peak on the temperature dependence of the imaginary component of the AC magnetic susceptibility. The frequency and amplitude of the field of the modulating coil were 88 Hz and 2 Oe, respectively. The determination of J_c (77 K, 100 Oe) value was described elsewhere [8].

3. Results and discussion

Figure 1 a shows the results of the AC susceptibility measurements for the undeformed Bi(Pb)2223 samples annealed in the temperature range T= 825 – 845 °C. The highest value $T_{\chi''}$= 91.2 K corresponds to the annealing temperature T= 835 °C. The dependence of $T_{\chi''}$ on the time of annealing of the deformed samples is shown in figure 1 b. The $T_{\chi''}$ value achieves saturation at t> 10 h in both samples. Therefore, the time of annealing of the deformed samples t= 14 h was chosen as optimal. A study of the microstructure has shown that post-deformation annealing causes a change in the mean length of the Bi(Pb)2223 phase colonies: in most samples the value of L increased, while in the sample ω_1/865 °C/15° the mean colony length decreased from $L_0\approx$3 μm to $L\approx$2.4 μm. Figure 2 shows the degree of basal texture before and after annealing of the deformed Bi(Pb)2223 samples. Post-deformation annealing causes a weakening of the texture, which is the result of recrystallization of the Bi(Pb)2223matrix phase. In the samples deformed in the conditions ω_1/855 °C/25° and ω_1/875 °C/25°, the texture almost did not weaken. The obtained results can be explained in the following way. It was previously established that during HPT of Bi(Pb)2223 ceramics after achieving the maximum of the texture, the process of dynamic recrystallization (DRx) intensifies, as a result of which an increase in the twist angle causes the refinement of colonies of the matrix phase and weakening of the texture [9].

In the process of post-deformation annealing the growth of colonies, which were formed during DRx and have a predominantly non-basal orientation, occurs that results in a weakening of the texture. The degree of weakening of the texture depends on the number of such colonies. The twist angle of the samples ω_1/855 °C/25° and ω_1/875 °C/25° corresponds to the texture maximum. In these samples, the number of small colonies with a non-basal orientation seems to be low, so the weakening of the texture during post-deformation annealing is minimal or practically absent. The refinement of Bi(Pb)2223 colonies in the sample ω_1/ 865 °C/15° can be explained by the fact that for this temperature the texture maximum corresponds to the angle α = 10° and during deformation to α = 15° many small grains with a non-basal orientation were formed due to DRx. For this reason, annealing leads to weakening of the texture alongside with a reduction of the mean colony size.

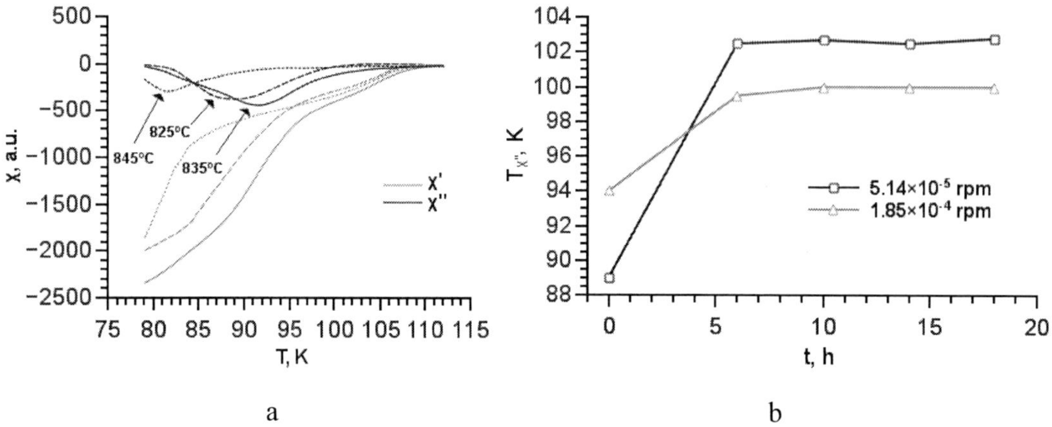

Figure 1. Optimization of annealing: (a) the dependence of AC susceptibility of the undeformed Bi(Pb)2223 samples on the annealing temperature and (b) the dependence of the value $T_{\chi''}$ of the deformed Bi(Pb)2223 samples on the annealing time at T= 835 °C.

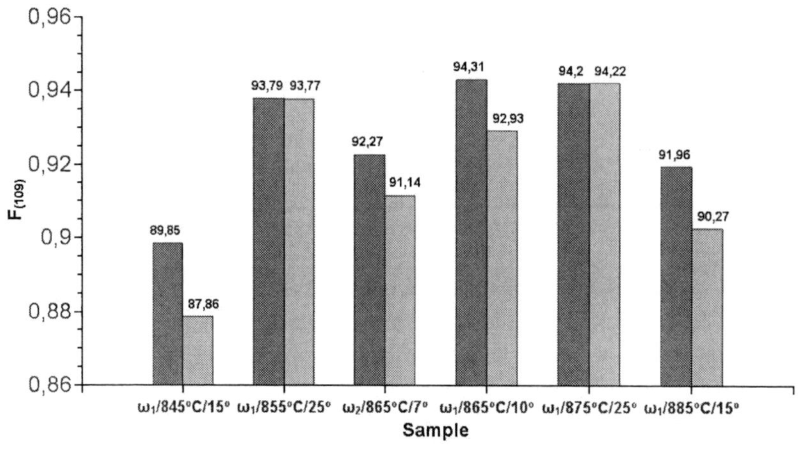

Figure 2. The effect of post-deformation annealing on the degree of basal texture for the deformed Bi(Pb)2223 samples.

The superconducting properties of Bi(Pb)2223 samples, deformed in different conditions and subjected to post-deformation annealing, are shown in table 1. For T_d= 865 °C and the twist rate ω_1 of the sample ω_1/865 °C/10° has the highest superconducting properties: $T_{\chi''}$= 90.9 K, J_c= 1127 A/cm^2. At T_d= 865 °C the decrease in the twist rate from ω_1 to ω_2 significantly increases the superconducting properties. The sample ω_2/865 °C/7° has highest properties at a given temperature and twist rate: $T_{\chi''}$= 103.2 K, J_c= 5647 A/cm^2. Thus, a decrease in the twist rate by 3.5 times at T_d= 865 °C increases J_c by 5 times in the best samples. Among the studied samples, the highest superconducting properties are observed after HPT at ω_1/875 °C/25° and ω_1/855°C/25°: $T_{\chi''}$= 104.4 K, J_c= 9295 A/cm^2 and $T_{\chi''}$= 104.1 K, J_c= 8098 A/cm^2, respectively. The highest superconducting properties of these samples are apparently caused by a small weakening of the deformation texture during post-deformation annealing in comparison with other samples.

Table 1 – Superconducting properties of the deformed Bi(Pb)2223 samples after annealing.

HPT conditions ($\omega/T_d/\alpha$)	ω_1/845°C/15°	ω_1/855°C/25°	ω_2/865°C/7°	ω_1/865°C/10°	ω_1/875°C/25°	ω_1/885°C/15°
$T_{\chi''}$, K	98.3	104.1	103.2	90.9	104.4	104.4
J_C, A/cm^2	1500	8098	5647	1127	9295	5588

Acknowledgments
The present work was accomplished according to the state assignment of IMSP RAS No AAAA-A17-117041310216-1.

References

[1] Parinov I A 2013 *Microstructure and Properties of High-Temperature Superconductors* (Springer Science & Business Media) p 779
[2] Jin S, Tiefel T, Sherwood R, Van Dover R, Davis M, Kammlott G and Fastnacht R 1988 *Phys. Rev. B* **37** 7850
[3] Rouessac V, Poullain G, Provost J, Gomina M and Desgardin G 1998 *Eur. Phys. J. -Appl. Phys.* **2** 145
[4] Majewski P 1997 *Supercond. Sci. Technol.* **10** 453
[5] Imayev M, Daminov R, Reissner M, Steiner W, Makarova M and Kazin P 2007 *Physica C: Superconductivity and Its Applications* **467** 14
[6] Imayev MF and Parkhimovich N Yu 2013 *Letters on Materials* **3** 188 (in Russian)
[7] Yoo J and Mukherjee K 1994 *Physica C: Superconductivity and Its Applications* **222** 241
[8] Kazin P, Os' kina T and Tretyakov Y D 1993 *Appl. Supercond.* **1** 1007
[9] Imayev MF and Parkhimovich N Yu 2016 *Letters on Materials* **6** 343 (in Russian)

IOP Conf. Series: Materials Science and Engineering **447** (2018) 012029 doi:10.1088/1757-899X/447/1/012029

Formation of the $L1_2$-type superstructure in Cu-5.9 at.%Pd and Cu-8 at%Pd alloys

E G Volkova, O S Novikova and A Yu Volkov

M.N. Mikheev Institute of Metal Physics, Ural Branch, Russian Academy of Sciences, 18 S. Kovalevskaya Str., Ekaterinburg, 620108 Russia.

E-mail: volkova@imp.uran.ru

Abstract. There was performed a resistometric study of the Cu-8 at.%Pd alloy during heating and cooling with a constant rate. There was made a comparison with the change in the electrical properties of the Cu-5.9 at.%Pd alloy. In the investigated Cu-Pd alloys, the anomalous behavior of electrical properties was revealed: long low-temperature annealing increased the electrical resistivity. It was concluded that nuclei of the ordered $L1_2$ phase were formed in the Cu-5.9 at.% Pd alloy. The obtained result is in contradiction with the currently known Cu-Pd phase diagram, where the Cu-5.9Pd alloy is the fcc-solid solution. The critical temperatures of the order→disorder phase transformation in the alloys were determined.

1. Introduction

Dilute Cu-Pd alloys have a great potential as corrosion-resistant conductors of electric current. However, their structure and properties are far from being sufficiently investigated. For example, abnormal annealing hardening was detected in the Cu-7 at.%Pd alloy [1]. The enhancement of the conductor strength is an important practical task. Therefore, it is of interest to understand the nature of the revealed phenomenon.

According to the phase diagram of Cu-Pd system [2], the content of palladium in the alloy should exceed 7.5 at% for the formation of the $L1_2$ superstructure. This means that a single-phase solid solution is formed in the alloy Cu-7 at.%Pd. Therefore, it is difficult to understand the reason for the increase in microhardness during annealing of the alloy. Nevertheless, several theoretical studies [3, 4] have recently shown that an ordered $L1_2$-phase can be formed in the Cu-5 at.% Pd alloy. The results of theoretical calculations [3, 4] and experimental data [1] are matching each other well. Indeed, the formation of nuclei of a new phase will lead to a significant increase in strength of the material in the course of annealing.

The purpose of this work was to clarify the temperature boundary of the order↔disorder ($L1_2$↔A1) phase transformation in copper-palladium alloys with low palladium content.

2. Materials and methods

Cu-5.9 at.%Pd and Cu-8.0 at.%Pd alloys were prepared by melting Cu and Pd of 99.99% purity. For the sake of simplicity, the alloys were designated as Cu-5.9Pt and Cu-8Pd, respectively. Pure Cu was also melted and used as a reference material.

The selection of alloys was due to the following considerations. According to [2], the Cu-8Pd alloy is ordered with type $L1_2$ but the temperature of order-disorder transformation was not established. Before conducting our study, it was not known whether the ordered $L1_2$ phase was formed in the Cu-5.9Pd alloy and at what temperature this occurred.

Although the electrical resistivity of many superlattice alloys decreases with increasing degree of atomic order [5], in the Cu_3Pd alloy, the value of the electrical resistivity just below the order-disorder transition temperature (T_c) becomes higher than just above T_c [6]. This is ascribed to the fact that the temperature coefficient of electrical resistivity in the order state is larger than that in the disorder state. Therefore, the resistometric method of investigation was chosen as the main one in this work. We used resistometric measurements of the Cu-8Pd alloy to compare the results with the electrical properties of the Cu-5.9Pd alloy.

Ingots with a diameter of 5 mm were homogenized at 800 °C for 3 hours and then drawn to obtain thin wires with a diameter of 0.22 mm. Thus, drawing played the role of severe plastic deformation (SPD). Deformed wires were annealed at 700 °C for 1 hour followed by quenching in water. The following long low-temperature heat treatments were applied to form the nuclei of the ordered phase in the alloys. For the Cu-8Pd alloy, the heat treatment included multiple anneals in the temperature range of 200-300 °C followed by slow cooling to room temperature. The total heat treatment time was 3.5 months. Long-term low-temperature heat treatment of the Cu-5.9Pd alloy included annealing at 200 °C for 2 months.

Measurements of electrical resistivity were used as the principal method of investigation. This method was selected because it is known that the electrical resistivity of dilute Cu-Pd alloys changes noticeably during the disorder-order phase transformation [1, 6]. The electrical resistivity of thin wires was measured by the four-point method during heating and cooling in the temperature range of 20-550 °C under a low pressure of about 1×10^{-3} Pa. The rate of heating and cooling during the resistometric measurements was 120 °C/h. The electric current passing through the specimen was 20 mA.

3. Results and discussion

As has been established earlier [5,7], the rate of the ordering in the Cu-Pd system after SPD is higher than after SPD, followed by annealing and quenching. Therefore, at first we investigated the samples in two structural states: (1) after SPD and (2) after SPD, followed by annealing at 700 °C and subsequent water quenching. However, we had great difficulties in separating the effects of recrystallization and disordering on the electrical resistivity. Therefore, this paper deals only with the results obtained for the quenched samples.

The temperature dependence of the electrical resistance of the quenched Cu-8Pd alloy is shown in figure 1a.

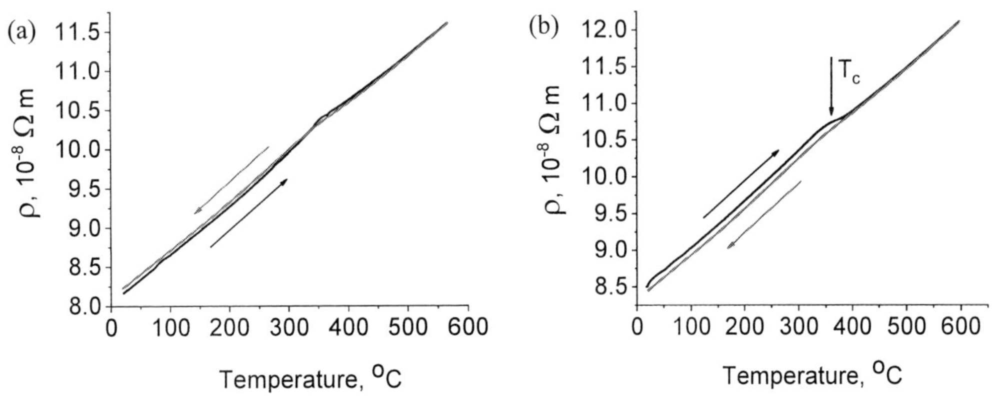

Figure 1. Temperature dependences of the electrical resistivity of the Cu-8Pd alloy: the initial quenched state (a) and the state after long-term low-temperature heat treatment (b).

The temperature dependence of the electrical resistivity of the Cu-8Pd alloy within the range from 300 to 400 °C has a nonlinear form. It is seen that the rate of the electrical resistivity change increases in temperature range from 300 to 350 °C, but further it decreases between 350 and 400 °C. With further heating, the temperature dependence of the electrical resistivity becomes linear. It should be noted that after the heating-cooling cycle, the electrical resistivity of the alloy under investigation slightly increased at room temperature. This is in full compliance with the results of work [6]. Indeed, in the course of slow cooling from a temperature above T_c, the $L1_2$-superstructure is formed in this alloy, which increases the electrical resistivity. As a result, after cooling to room temperature, the electrical resistivity of the alloy becomes higher.

The quenched samples of the Cu-8Pd alloy were subjected to the long-term low-temperature heat treatment to obtain a two-phase state with a maximum volume fraction of the ordered $L1_2$-phase. We were guided by the following considerations when selecting the heat treatment conditions. As is known, diagrams of temperature-time-transformation of diffusion controlled reactions have a *C*-shaped form [8]. Thus, there is some optimal temperature at which the rate of formation of a new phase is maximal. Above this temperature, the rate of nuclei formation of the ordered phase will be slower. Below the optimal temperature, the rate of the nuclei formation of the ordered phase will be reduced due to lower diffusivity.

Based on the phase diagram of the Cu-Pd system [2], we assumed that the temperature of the phase order-disorder transformation in the Cu-8Pd alloy lies in the range from 350 to 400 °C. Therefore, we used the following heat treatment conditions. First, the quenched samples of the alloy were annealed at the temperature of 300 °C for 1 month, then cooled to the temperature of 250 °C and annealed for 1 month. Then, the temperature was lowered to 200 °C and the samples were annealed again for 1 month. After that, the alloy was cooled to room temperature at a rate of 50 °C/week. The total heat treatment time was 3.5 months. As shown by a TEM study [7], after such a long treatment, fine grains with an ordered $L1_2$ phase were found in the disordered matrix of the Cu-8Pd alloy.

Figure 1b shows the temperature dependence of the electrical resistivity of the Cu-8Pd alloy after long-term low-temperature annealing. It is seen that the electrical resistivity of the alloy has increased as compared with the initial quenched state (figure 1 a, b). This is explained by the formation of some volume of the ordered $L1_2$ phase in the alloy. The greatest interest in figure 1 b is the presence of a kind of "step" on the graph during heating. The end point of this "step" and the transition of the graph to a linear dependence is sufficiently accurate to indicate the temperature of the order-disorder phase transition (T_c). It was reliably established earlier by various methods in the study of Cu-Pd alloys containing from 12 to 18 at.% of palladium [6]. Thus, the critical temperature of the order-disorder phase transition in the Cu-8Pd alloy is $T_c \approx 380$ °C (figure 1b).

We also attempted to elucidate whether the ordered $L1_2$ phase can be formed in the Cu-5.9Pd alloy. Note that most investigators believe that this alloy is a single-phase fcc-solid solution [2, 6, 9]. The possibility of atomic ordering with the formation of the $L1_2$ phase is discussed only in [3, 4] by the results of computer simulation. In addition, in our previous study [10] there were noted anomalies of electrical properties, which were explained by the formation of nanosize nuclei of the new phase.

Figure 2 a shows the temperature dependence of the electrical resistance of the quenched Cu-5.9Pd alloy. It is seen that the graph in figure 2 a differs from the temperature dependence of the quenched alloy Cu-8Pd (figure 1a). Moreover, based on this result, it is very difficult to make an unambiguous conclusion whether the ordered phase can be formed in the Cu-5.9Pd alloy.

To verify the possibility of ordering in the Cu-5.9Pd alloy, we also used a long low-temperature annealing. We assumed that the temperature of the phase order-disorder transformation of this alloy is about 300 °C. Therefore, the annealing of the alloy was carried out at a temperature of 200 °C for two months. Figure 2 b shows the temperature dependence of the electrical resistivity of this alloy in the obtained structural state. In general, the temperature dependences obtained for the Cu-8Pd and Cu-5.9Pd alloys after long-term low-temperature heat treatment are very similar (figures 1 b and 2 b). Thus, the critical temperature of the order-disorder phase transformation in the Cu-5.9Pd alloy is $T_c \approx 350$ °C.

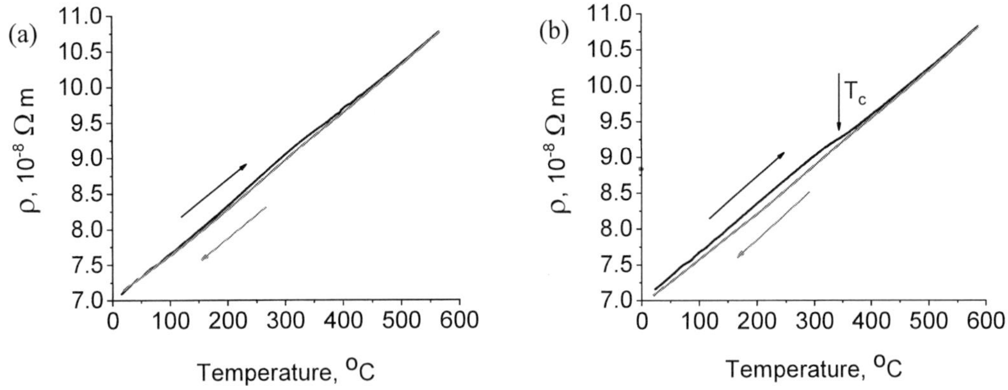

Figure 2. Temperature dependences of the electrical resistivity of the Cu-5.9Pd alloy: the initial quenched state (a) and the state after long-term low-temperature annealing (b).

4. Conclusions

1) The temperatures of the order→disorder phase transformation (T_c) in the Cu-8Pd and Cu-5.9Pd alloys were found to be T_c≈380 °C and T_c≈350 °C, respectively.
2) The results of first-principle calculations of the $L1_2$ phase stability in binary Cu-Pd alloys obtained in refs. [3, 4] were experimentally confirmed.
3) Addressing to the abnormal annealing hardening in the quenched Cu-7Pd alloy [1], one can conclude that this was caused by the formation of nuclei of the ordered $L1_2$-phase.

Acknowledgments
The work was carried out within the framework of the State task of the Federal Agency for Scientific Organizations of Russia (theme: Pressure № AAAA-A18-118020190104-3) with partial support of the Urals Branch of the RAS (project № 18-10-2-24) and the Russian Foundation for Basic Research (project № 18-03-00532).

References

[1] Marcovic I, Ivanov S, Stamenkovic U, Todorovic R and Kostov A 2018 *J. Alloys Comp.* **768** 944
[2] Subramanian P R and Laughlin D E 1991 *J. Phase Equil.* **12** 231
[3] Li M, Guo C and Li C 2008 *Comp. Coupl. Phase Diagr. Thermochem.* **32** 439
[4] Freudenberger J, Kauffmann A, Klaub H, Marr T, Nenkov K, Subramanya Sarma V and Schultz L 2010 *Acta Mater.* **58** 2324
[5] Volkov A Yu 2004 *Platinum Metals Rev.* **48** 3
[6] Mitsui K 2001 *Phil. Mag. B* **81** 433
[7] Volkov A Yu, Kostina A E, Volkova E G, Novikova O S and Antonov B D 2017 *Phys. Metals Metallogr.* **118** 1236
[8] Christian J W 2002 *The Theory of Transformations in Metals and Alloys* parts I and II 3ed (Pergamon) p 1200
[9] Huang P, Menon S, de Fontaine D 1991 *J. Phase Equil.* **12** 3
[10] Volkov A Yu, Novikova O S, Kostina A E and Antonov B D 2016 *Phys. Metals Metallogr.* **117** 945

Subsonic *M,N*-crowdions in 2D Morse crystal

E A Korznikova[1], I A Shepelev[2], A P Chetverikov[2], S Yu Fomin[3], S V Dmitriev[1]

[1] Institute for Metals Superplasticity Problems of RAS, 39 Khalturin St., Ufa 450001, Russia

[2] Saratov Chernyshevsky State University, 83Astrakhanskaya St., Saratov 410012, Russia

[3] Ufa State Aviation Technical University, 12 K. Marx St., Ufa 450008, Russia

E-mail: elena.a.korznikova@gmail.com

Abstract. Crowdion is a zero-dimensional defect of the crystal lattice in the form of an extra atom in a close-packed atomic row. Crowdions can be at rest or move with subsonic or supersonic velocity. A moving crowdion transports the mass of one atom. Recently it has been revealed that crowdion clusters can be more efficient in mass transfer as compared to single crowdions. For example, in our recent work it has been shown that a supersonic *N*-crowdion can be excited by giving a sufficiently large equal initial momentum to *N* neighbouring atoms in a close-packed row along the row. This work presents an extension of the concept of *N*-crowdion to *M,N*-crowdion, the latter one is excited in a 2D crystal by the initial kick of a block of $M\times N$ atoms along a close-packed direction. The effect of the initial velocity on the dynamics of *M,N*-crowdions is analysed for various values of *M* and *N*. It is revealed that the initial velocity affects not only the propagation length, but also the resulting defect configurations. It is suggested that *M,N*-crowdions can be initiated from the surface by bombardment with nanoparticles......

1. Introduction

Highly mobile crystal defects, such as crowdions and prismatic dislocation loops, exhibit anomalous temperature independent mobility [1]. Due to this fact, their possible role in mass transfer in various crystals has been a subject for numerous studies [1-10]. For example, in [2] it was shown that the in-plane migration of self-interstitials in α-zirconium was dominated by one-dimensional crowdion motion along <11-20> directions, from time to time interrupted by off-line or out-of-plane jumps. Another analysis of the crowdion contribution to the mass transfer in the biatomic structure of W-He was performed in [3]. During the formation of helium clusters, interstitial tungsten atoms were produced and evolved into bundles of <111> crowdions, which were reported to stay constrained around the helium clusters for a long time [3].

Crowdions can demonstrate different types of mobility. They can move with supersonic or subsonic velocity or be at rest. An increase in the crowdion velocity usually results in a growth of its degree of spatial localization. The concept of a supersonic an *N*-crowdion was introduced in [9]. In this case, not one, but *N* atoms in a close-packed atomic row move simultaneously at a high speed, and the defect propagation length increases considerably.

A subsonic crowdion can bear a localized vibrational mode [10], and, thus, it has features of both a crowdion and a discrete breather. Numerous investigations of localized excitations, such as crowdions [1-10] and discrete breathers [11-20], have shown that they carry energy and/or mass thus contributing to the evolution of the crystal structure. An analysis of the crowdion dynamics performed in [10], showed that it was strongly dependent on the initial velocity. In this work, our aim is to study the effect of the configuration and the initial velocity of the crowdion on its dynamics, the mass propagation length, and the resulting defect structure.

2. Simulation details

We consider a 2D close-packed triangular lattice with an interatomic distance equal to a. Interatomic interactions are described by the Morse pair potential

$$V(r) = D(e^{-2\alpha(r-r_m)} - 2e^{-\alpha(r-r_m)}), \qquad (1)$$

where r is the distance between two atoms, D, α, r_m are the parameters. The function $V(r)$ has a minimum at $r=r_m$. D is the depth of the potential (bond breaking energy) and α is the parameter defining the bond stiffness. By a proper choice of units of energy, distance, and time, one can set D, r_m, and atom mass m to be equal to 1. For the bond stiffness, a typical value of α=4 is taken. We choose the cut-off radius $r_c=5r_m$ and find the equilibrium interatomic distance to be equal to a=0.9656. The computational cell includes 300×100 atoms. Periodic boundary conditions are used in the simulation.

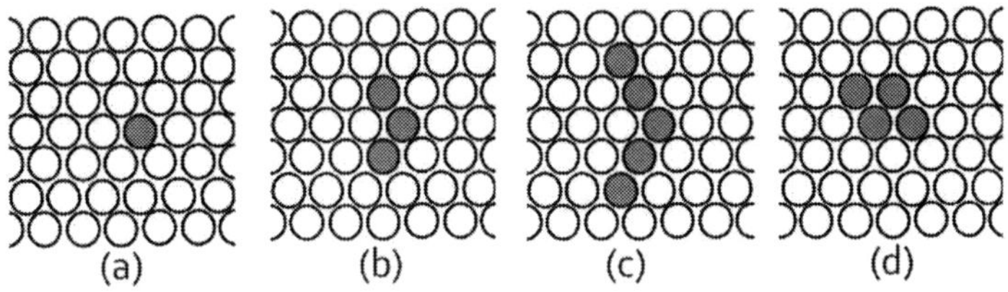

Figure 1. Representative examples of initial conditions for an initiation of M,N-crowdions: (a) M=1, N=1, (b) M=3, N=1, (c) M=5, N=1, and (d) M=2, N=2. The initial velocity v_0 is given to the colored atoms along the horizontal close-packed atomic rows (along x-axis). The initial velocities of other atoms and the initial displacements of all atoms are equal to zero.

Initial conditions are set as follows (see figure 1). The initial velocity v_0 is given to a block of $M\times N$ atoms (colored in figure 1) along the close-packed direction, where N is the number of neighboring atoms in one close-packed row, and M is the number of neighboring rows. Thermal fluctuations are not taken into account, which means that initially the crystal has zero temperature, with only $N\times M$ atoms having an initial momentum. Cases for M from 1 to 5 and N from 1 to 4 are investigated. For every M and N we have found the minimal velocity required to initiate a moving subsonic M,N-crowdion. The structure evolution induced by kicking a group of atoms was visualized with the OVITO software package [21].

3. Results and discussion

Figure 2 demonstrates the evolution of the energy distribution for the cases M=1, N=1 (figure 2, a-c) and M=3, N=1 (figure 2, d-e). Here blue (red) atoms have minimal (maximal) potential energies. The launch of one atom with the initial velocity v_0 = 10 results in the formation of a vacancy (a) and a moving breathing crowdion (b), which loses stability when its velocity becomes below the threshold value and transforms into an immobile interstitial atom (c). For the case of M=3, N=1 one can observe the formation of a three-vacancy complex (d) and a crowdion complex, which includes three extra atoms, one extra atom in each of the three neighboring rows (e). Figure 3 represents the evolution of the defect structure after launching of (a, b) M=5, N=1 and (c, d) M=2, N=4 crowdions. Detailed features are given in the figure caption. It should be noted that (i) in case of M>4, the vacancy loop and the crowdion cluster transform into prismatic dislocation loops of the opposite topological charge (a, b); (ii) for M=2, N=4 a 2×2 vacancy complex and two 1,2-crowdions are formed, one of them moving in the close-packed direction different from the direction of the initial kick (c) and another one along this direction (d).

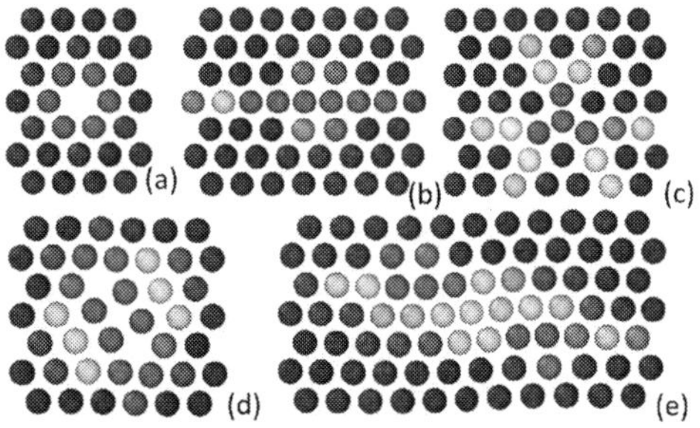

Figure 2. The potential energy distribution in a a 2D Morse crystal after the launch of (a-c) 1,1- crowdion with v_0 = 9.05 and (d,e) 3,1- crowdion with v_0 = 10. The launch of one atom leads to the formation of (a) a vacancy, simulation time t=4.5, (b) a moving breathing crowdion, t=4.5, and (c) an immobile interstitial atom, t=10. The launch of three atoms results in (d) the formation of a three-vacancy complex, t = 4 and (e) the propagation of a stable crowdion with three extra atoms, t = 4.

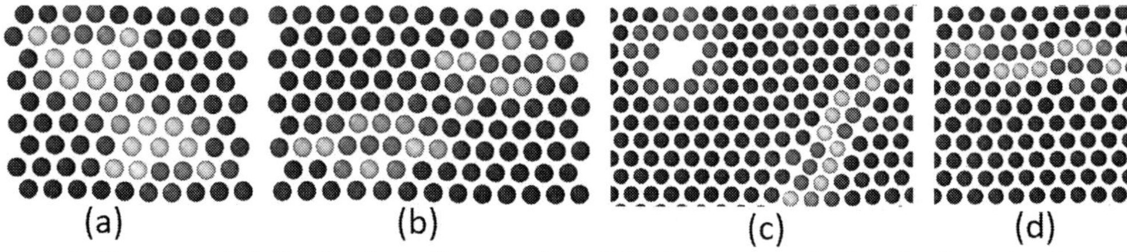

Figure 3. The potential distribution of energy in a 2D Morse crystal after the launch of (a, b) 5,1-crowdion and (c, d) 2,4-crowdion, both with v_0 = 9.0. The launch of five atoms results in the formation of prismatic dislocation loops of the opposite topological sign, namely (a) a five-vacancy agglomerate and (b) a crowdion cluster with five extra atoms shown at the simulation time t=17. Initiation of the 2,4-crowdion results in the formation of a 2×2 vacancy complex and two 2,1-crowdions, one moving along the close-packed rows tilted by 60° with respect to the initial kick direction (c) and the other one along the direction of the initial kick (d).

The propagation of a 3,1-crowdion is shown in figure 4 a. The atomic displacement as a function of time, $X_n(t) = x_n(t) - x_n(0)$, is plotted for the atoms of the middle atomic row among the three rows along which the crowdion propagates. One can notice high frequency vibrations on the trajectories of atoms. The vibration frequency is above the phonon spectrum of the considered crystal [10,12] that allows us to claim that in this case we are dealing with a breathing subsonic crowdion similar to that analyzed in [10]. A detailed analysis of atomic displacement plots allows one to estimate the propagation length S of crowdions. The results are shown in figure 4 b as a functions of M for N=1, 2, and 4, where S is normalized to the interatomic distance a. One can see that an increase in N from 1 to 2 results in an increase in S by three (for M=5) to four (for M=1) times. For N=1 and N=2 one can observe a monotonous increase in S with growing M. However, for N=4 a non-monotonous $S(M)$ dependence can be seen. This is explained by a qualitative change in the formation of defects for $N\geq4$. In particular, crowdions propagating along directions different from the initial launch direction are observed for $N\geq4$.

4. Conclusions

The dynamics of M,N-crowdions has been studied for the first time for various $M\geq1$ and $N\geq1$. It was found that (i) an increase in N from 1 to 2 drastically increases the propagation length of crowdions (by three to four times, see figure 4 b; (ii) a further increase in N weakly affects this parameter due to the formation of crowdion complexes propagating along crystallographic directions different from the initial kick direction (see figure 3 c); (iii) M,1-crowdions carry an internal mode oscillating with a frequency above the upper edge of the phonon spectrum (see figure 4 a). We note that M,N-crowdions can be initiated from the crystal surface by its bombardment with nanoparticles [22]. This process will be analyzed in a forthcoming publication.

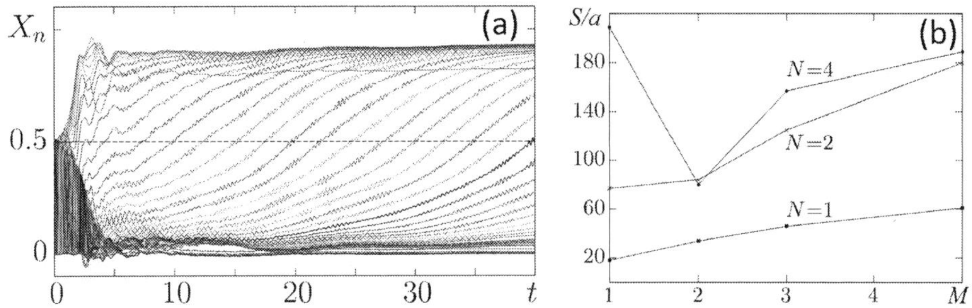

Figure 4. (a) Propagation of the M=3, N=1 crowdion launched with v_0=9. Shown are the atomic displacements as functions of time, $X_n(t) = x_n(t) - x_n(0)$, for atoms in the middle row of three rows along which the crowdion complex moves. (b) Distance of mass transfer as a function of M for N=1, 2, and 4, for the initial velocity v_0=9. The distance is normalized to the interatomic distance a.

Acknowledgements
The work of A.P.Ch., I.A.Sh. and S.Yu.F. was supported by the Russian Science Foundation, grant No. 16-12-10175 (obtaining numerical results). E.A.K. and S.V.D. were supported by the Russian Foundation for Basic Research, grant no 17-02-00984-a (statement of the problem and writing the paper). This work was partly supported by the IMSP RAS State assignment.

References
[1] Dudarev S and Sutton A 2014 *Phys. Rev. Lett.* **113** 215501
[2] Woo C, Huang H and Zhu W 2003 *Appl.Phys. A* **76** 101
[3] Li M, Cui J, Wang J and Hou Q 2014 *Nucl. Instr. Meth. Phys. Res., Sect. B* **337** 45
[4] March N and Pushkarov D 1996 *J. Phys. Chem. Solids* **57** 139.
[5] Saralidze Z, Galustashvili M and Driaev D 2006 *Phys. Solid State* **48** 1298
[6] Volosyuk M, Volosyuk A and Rokhmanov N 2015 *Funct. Mater.* **22** 51
[7] Natsik V and Nazarenko Y 2002 *Eur. Phys. J. B* **29** 285
[8] Dmitriev S, Korznikova E and Chetverikov A 2018 *J. Exp. Theor. Phys.* **126** 347
[9] Dmitriev S, Medvedev N, Chetverikov A and Zhou K 2017 *Phys. Status Solidi RRL* **11** 1700298
[10] Chetverikov A, Shepelev I, Korznikova E, Kistanov A, Dmitriev S and Velarde M 2017 *Comput. Cond.. Matt.* **13** 59
[11] Dmitriev S, Korznikova E, Baimova Y and Velarde M 2016 *Physics-Uspekhi* **59** 446
[12] Korznikova E, Fomin S, Soboleva E and Dmitriev S 2016 *JETP Lett.* **103** 277
[13] Barani E, Lobzenko I, Korznikova E, Soboleva E, Dmitriev S, Zhou K and Marjaneh A 2017 *Phys. Lett. A* **381** 3553
[14] Murzaev R, Bachurin D, Korznikova E and Dmitriev S 2017 *Phys. Lett. A* **381** 1003
[15] Evazzade I, Lobzenko I, Korznikova E, Ovid'Ko I, Roknabadi M and Dmitriev S 2017 *Phys. Rev. B* **95** 035423
[16] Barani E, Lobzenko I, Korznikova E, Soboleva E, Dmitriev S, Zhou K and Marjaneh A 2017 *Eur. Phys. J. B* **90** 38
[17] Murzaev R, Babicheva R, Zhou K, Korznikova E, Fomin S, Dubinko V and Dmitriev S 2016 *Eur. Phys. J. B* **892** 168
[18] Barani E, Korznikova E, Chetverikov A, Zhou K and Dmitriev S 2017 *Phys. Lett. A* **381** 3553
[19] Sergeev K, Dmitriev S, Korznikova E and Chetverikov A 2018 *Nelineinaya Dinamika* **14** 195
[20] Zakharov P, Korznikova E, Dmitriev S, Ekomasov E and Zhou K 2019 *Surf. Sci.* **679** 1
[21] Stukowski A 2010 *Modelling Simul. Mater. Sci. Eng.* **18** 015012
[22] Marjaneh A M, Saadatmand D, Evazzade I, Babicheva R, Soboleva E, Srikanth N, Zhou K, Korznikova E and Dmitriev S 2018 *Phys. Rev. E* **98**, 023003

IOP Conf. Series: Materials Science and Engineering **447** (2018) 012031 doi:10.1088/1757-899X/447/1/012031

Electrophysical and thermal properties of $Na_xCu_{2-x}S$ (x = 0.05, 0.075, 0.10) and $Na_{0.125}Cu_{1.75}S$ semiconductor alloys

M M Kubenova[1], K A Kuterbekov[1], E T Abseitov[1], A M Kabyshev[1] , A. Kozlovskiy[1] , S N Nurakov[1] , R Kh Ishembetov[2], and M Kh Balapanov[2]

[1]L.N. Gumilyov Eurasian National University, 2 Satbayev st., Astana 010008, Kazakhstan
[2]Bashkir State University, 32 Z. Validi st., Ufa 450001, Russian Federation

E-mail: kubenova.m@yandex.kz

Abstract. The results of measurements of electronic conductivity and the Seebeck coefficient, as well as X-ray phase analysis, electron microscopy, differential scanning calorimetry (DSC) and studies of the heat capacity of nanocrystalline $Na_xCu_{2-x}S$ (x = 0.05, 0.075, 0.10) and $Na_{0.125}Cu_{1.75}S$ semiconducting alloys in the temperature interval of 300 - 600 K are presented. At ambient temperature, the alloys under study consist of three phases: monoclinic $Na_2Cu_4S_3$, hexagonal chalcocite Cu_2S and rhombohedral digenite Cu_9S_5. Measurements show that the Seebeck coefficient increases, and the conductivity decreases with increasing sodium concentration in the alloys. Thermal effects near the temperatures of 360 K, 450 K and 640 K have been observed by the DSC method.

1. Introduction

In nature, copper and sulfur form chemical compounds of a wide variety of compositions, ranging from chalcocite (Cu_2S) to villamaninite (CuS_2) with various intermediates: covellite (CuS), jurleite ($Cu_{1.96}S$) and anilite ($Cu_{1.75}S$) [1]. Copper sulfides exhibit mixed electronic-ionic conductivity and possess high potential for practical applications as p-branches of solar cells, thermoelectric devices, superionic materials, cathode materials with high energy density for lithium batteries, nanosized switches, catalysts, and nonlinear optical materials [2-5].

We have previously studied the effect of lithium substitution on the electronic and ionic transport in $Cu_{2-\delta}S$ and $Cu_{2-\delta}Se$ [6-8]. The substitution resulted in a decrease in both the ionic and electronic conductivities of the compounds. The $Li_xCu_{2-x-\delta}S$ and $Li_xCu_{2-x-\delta}Se$ alloys are mixtures of two or three phases at ambient temperature and exhibit transition to a single phase state at elevated temperatures forming solid solutions on the base of nonstoichiometric copper sulfide (selenide) or an individual chemical substance as, for example, $Li_{0.25}Cu_{1.75}Se$ [7]. As a rule, copper sulfide and selenide are chemically unstable in air. Insertion of lithium decreases the temperatures of phase transformations and increases the chemical stability of copper sulfide and copper selenide. For example, the high temperature f.c.c. cubic phase of copper sulfide becomes stable at ambient temperature for $Li_{0.15}Cu_{1.85}S$ [6]. Shannon effective ionic radii of Li^+ and Cu^+ ions are near 0.59 and 0.60 Å in tetrahedral coordination correspondingly [9]) and this fact favors the formation of a solid solution in $Li_xCu_{2-x-\delta}S$. The ionic radii of Na^+ ions are significantly larger (0.99 Å) than the ionic radii of Li^+ ions, so, in our opinion, ordering effects are probable in sodium doped copper sulfides. Na-containing copper sulfides are investigated rather poorly. As an example, the existence of a quasi-one-dimensional $Na_2Cu_4S_3$ compound is known [10]. Along with this fact, the crystal structure of copper sulfide does not change after light sodium doping. The authors of [11] obtained homogeneous $Na_xCu_9S_5$ samples with x = 0.025, 0.05, 0.15 and achieved the maximal thermoelectric figure of merit ZT = 1.1 at 773 K for the composition x = 0.05 (in our terminology it corresponds to the $Na_{0.01}Cu_{1.80}S$ composition).

Nanocrystalline alloys of the $Na_{0.15}Cu_{1.85}S$, $Na_{0.17}Cu_{1.80}S$ and $Na_{0.20}Cu_{1.77}S$ compositions were synthesized in recent work [12]. A quasi-one-dimensional monoclinic phase of $Na_2Cu_4S_3$ and two ordinary sulfide phases, the hexagonal Cu_2S and the cubic Cu_9S_5 ones, were observed at ambient temperature. A very low heat conductivity in the interval of 0.1-0.7 $Wm^{-1}K^{-1}$ was found. The achieved thermoelectric figure of merit ZT ≈0.28 for the $Na_{0.15}Cu_{1.85}S$ composition at 590 K is comparable with the result [11] for the $Na_{0.01}Cu_{1.80}S$ composition at the same temperature. In other recent paper [13], high ionic conductivity values of 1.5-2.3 S/cm in the temperature interval of 620 K – 680 K and low heat conductivity about 0.3-0.4 $Wm^{-1}K^{-1}$ in the temperature interval of 620 K – 680 K were observed for $Na_{0.2}Cu_{1.8}S$ and good semiconducting electron properties for the $Na_{0.1}Cu_{1.9}S$ and $Na_{0.2}Cu_{1.8}S$ compositions were found. In this paper we present the results of more detailed investigations of the $Na_xCu_{2-x}S$ (x = 0.05, 0.075, 0.10) and $Na_{0.125}Cu_{1.75}S$ compositions including phase analysis, electron microscopy, differential scanning calorimetry and heat capacity studies with in order to obtain a better understanding of the transport phenomena in these multiphase alloys.

2. Experimental methods

Materials for investigations were synthesized at temperatures of about 165 °C. The reaction was performed in a mixed medium of NaOH and KOH hydroxides. All reagents (CuCl, NaCl, Na_2S*9H_2O) were simultaneously placed in a heated Teflon vessel. A nanocrystalline powder was formed inside the vessel in 8 hours. The resulting product was washed with distilled hot water and then with pure ethanol. The particle sizes of the obtained powder were in the range from 20 to 600 nm. X-ray phase analysis was performed with a D8 ADVANCE ECO diffractometer (Bruker, Germany) using the CuK_α radiation. The international database ICDD PDF-2 was used to identify the phases. Images of the surfaces of the samples were taken using scanning electron microscopes (SEM) MIRA3 (TESCAN) and JSM 7500F (JEOL). For differential scanning calorimetry of the samples and measurements of the heat capacity in the temperature range from 300 to 700 K, the DSC404 F1 Pegasus device (NETZSCH, Germany) was used. The heating rate was 10 K/min, a slow argon flow was applied (30 ml/min). Electronic conductivities and the Seebeck coefficients were measured simultaneously in an argon atmosphere with ZEM-3 equipment (ULVAC Inc., Japan).

3. Results of X-ray phase analysis, electron microscopy and differential scanning calorimetry

Figure 1 shows the X-ray diffraction patterns of the samples obtained at room temperature. Analysis of diffraction lines using the Rietveld method shows that the samples consist of the following phases: $Na_2Cu_4S_3$ (PDF – 01-082-6340), the lattice type is monoclinic, space group is C2/m (12); chalcocite Cu_2S (PDF – 00-026-1116), the lattice type is hexagonal, space group is $P6_3/mmc$ (194); and digenite Cu_9S_5 (PDF - 00-047-1748), the lattice type is rhombohedral, space group is R-3m (166).

It is known [1] that chalcocite Cu_2S has a hexagonal structure with the space group $P6_3/mmc$ with a_h =3.95 Å, c_h = 6.75 Å between 377 K and 708 K only, so the presence of this structure at ambient temperature is very unusual and probably could be caused by the influence of introducing Na into the crystal lattice. The hexagonal modification of the chalcocite exhibits superionic properties and, therefore, high ionic conductivity can be expected in the alloys under study even at room temperature.

Figure 2 a shows a SEM image of the etched surface of a $Na_{0.05}Cu_{1.95}S$ sample. On this image, the porous surface structure can be seen. Individual grains predominate in the form of melted cubes or parallelepipeds with dimensions of 100-200 nm, and larger blocks with an unresolved substructure are also visible. Figure 2 also presents SEM images of powder particles of $Na_{0.075}Cu_{1.925}S$ (b), $Na_{0.10}Cu_{1.90}S$ (c) and $Na_{0.125}Cu_{1.75}S$ (d) samples. The synthesized powder consists of irregularly shaped plates, having a granular substructure with sizes from 20 nm to several hundred nanometers.

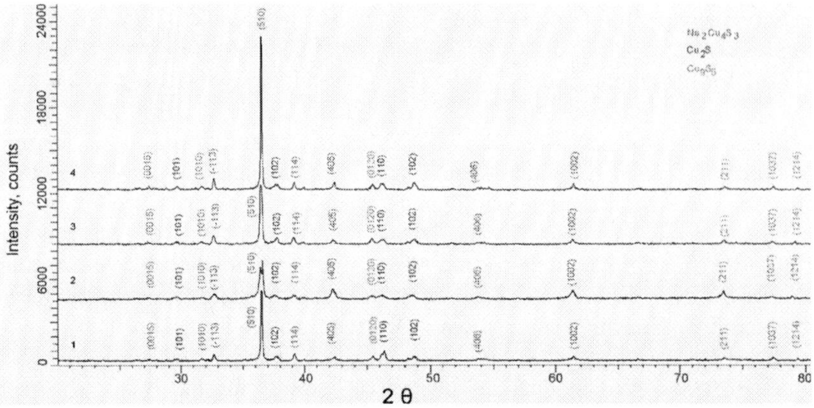

Figure 1. X-ray diffraction patterns of the samples at ambient temperature (1- $Na_{0.05}Cu_{1.95}S$, 2- $Na_{0.075}Cu_{1.925}S$, 3- $Na_{0.10}Cu_{1.90}S$, 4- $Na_{0.125}Cu_{1.75}S$).

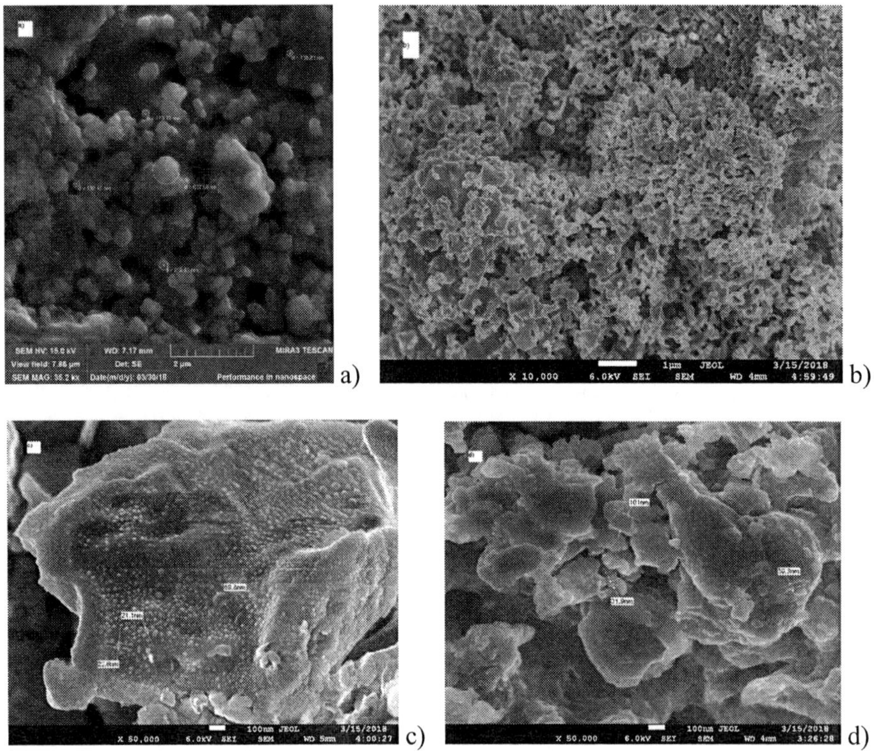

Figure 2. Image of the etched surface of $Na_{0.05}Cu_{1.95}S$ **(a)**, obtained on a scanning electron microscope and SEM images of powder particles of $Na_{0.075}Cu_{1.925}S$ **(b)**, $Na_{0.10}Cu_{1.90}S$ **(c)**,$Na_{0.125}Cu_{1.75}S$ **(d)** samples.

Differential scanning calorimetry curves and temperature dependences of specific heat are presented on figure 3. According to G. Will et al. [14], two modifications of digenite exist, low-temperature (below 364 K) and high-temperature (above 364 K). Sharp endothermic peaks about 370 K recorded on all DSC and heat capacity curves in figure 3 confirm the X-ray phase analysis data on the presence of the digenite phase in the studied samples. As is known, the high-temperature cubic phase of digenite exhibits superionic conductivity, carried by copper Cu^+ ions [5, 6].

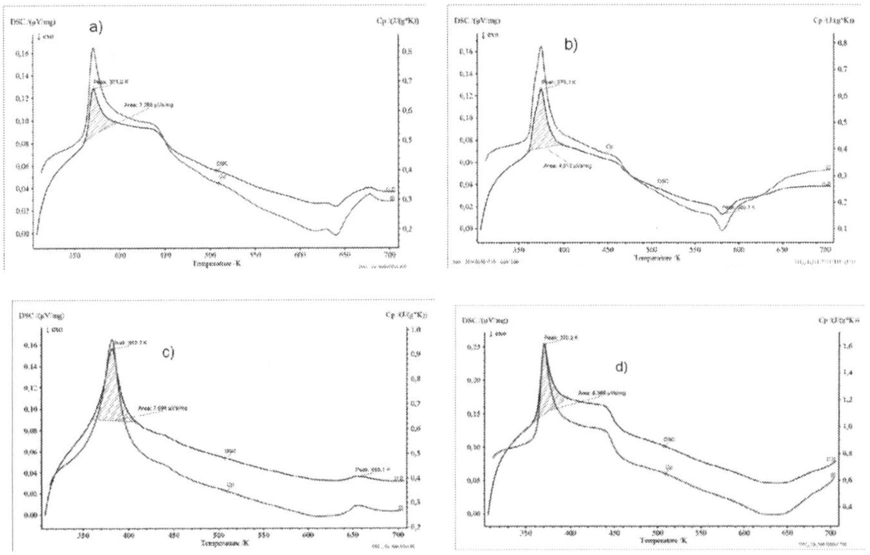

Figure 3. DSC curves and temperature dependencies of specific heat of $Na_{0.05}Cu_{1.95}S$ (a), $Na_{0.075}Cu_{1.925}S$ (b), $Na_{0.10}Cu_{1.90}S$(c), $Na_{0.125}Cu_{1.75}S$ (d) samples.

The temperature of the peak tip increases with increasing sodium content from 371 K to 382 K in the $Na_{0.05}Cu_{1.95}S$, $Na_{0.075}Cu_{1.925}S$, $Na_{0.10}Cu_{1.90}S$ row. The $Na_{0.125}Cu_{1.75}S$ alloy deviates from the general trend exhibiting the peak tip temperature of 370 K. The value of the enthalpy of phase transformation, measured by the peak area in figure 3, increases with increasing sodium content for all samples from 3398 J/kg*K for $Na_{0.05}Cu_{1.95}S$ up to 8366 J/kg*K for $Na_{0.125}Cu_{1.75}S$. It is possible that sodium partially dissolves in the structure of the digenite or accumulates at the grain boundaries, and its presence leads to an increase of thermal costs during structural transformation in the digenite.

In the range of 440-460 K, the curves in figure 3 exhibit an anomaly in the form of a stepped fall, the origin of which is unclear. There is no data on phase transitions in $Na_2Cu_4S_3$ above room temperature in the literature. Since digenite has the phase transition temperature of 364 K, at 440 K it is already in the high-temperature cubic phase [14]. The hexagonal chalcocite of Cu_2S composition is transformed into the f.c.c. modification above 713 K [1].

It is possible that the substitution by sodium in the Cu_2S lattice reduces the temperature of the phase transition to the cubic phase to 440-460 K or the phase transformation takes place in $Na_2Cu_4S_3$ phase.

For all samples, the exothermic effect is exhibited around 630-640 K. For the $Na_{0.075}Cu_{1.925}S$ sample, this effect is shifted to 580 K. Also a weak endothermic peak is observed for the $Na_{0.05}Cu_{1.95}S$ and $Na_{0.10}Cu_{1.90}S$ samples at 677 K and 653 K, respectively. High temperature X-ray structural studies are needed to shed light on the nature of all these thermal effects.

4. Electronic conductivity and the Seebeck coefficient

Figure 4 shows the temperature dependences of the electronic conductivity and the Seebeck coefficients of nanocrystalline $Na_{0.05}Cu_{1.95}S$, $Na_{0.075}Cu_{1.925}S$, $Na_{0.1}Cu_{1.9}S$, $Na_{0.125}Cu_{1.75}S$ samples in the range from 300 to 600 K. All the presented curves show several extrema at the points of phase transitions assumed on the basis of figure 3. The conductivity of all samples is small. The conductivities of $Na_{0.05}Cu_{1.95}S$ and $Na_{0.125}Cu_{1.75}S$ samples is significantly higher than that of others.

The $Na_{0.05}Cu_{1.95}S$ sample contains less impurity (Na), scattering on which usually reduces the mobility of carriers. The $Na_{0.125}Cu_{1.75}S$ sample contains more sodium than the $Na_{0.075}Cu_{1.925}S$ and $Na_{0.1}Cu_{1.9}S$ samples, and this should lead to a decrease in charge carrier mobility, but the chemical composition of the $Na_{0.125}Cu_{1.75}S$ sample also has a large metal deficiency, which, according to the electroneutrality rule, should greatly increase the concentration of electron holes, and, consequently, increase the conductivity,. We believe that as a result of the compromise of these two factors, the conductivity of the non-stoichiometric $Na_{0.125}Cu_{1.75}S$ sample is higher than that of the stoichiometric $Na_{0.075}Cu_{1.925}S$ and $Na_{0.1}Cu_{1.9}S$ compositions.

In the low-temperature phase of the samples, the semiconductor character of conductivity is observed with the same impurity activation energy of 0.057 ± 0.001 eV, as shown in figure 4, only for the $Na_{0.075}Cu_{1.925}S$ alloy, the impurity activation energy is slightly higher and equal to 0.067 ± 0.001 eV. The Seebeck coefficient of the samples is positive over the entire temperature range studied. This fact is characteristic for copper sulfides, which usually exist with a deficiency of copper in the crystal lattice. Figure 4 shows that the Seebeck coefficients increase with increasing sodium content. An exception is the $Na_{0.125}Cu_{1.75}S$ composition, which exhibits the lowest Seebeck coefficient, which can be explained by a strong deficiency of cations and, consequently, a high concentration of holes, at which carrier degeneracy most likely occurs.

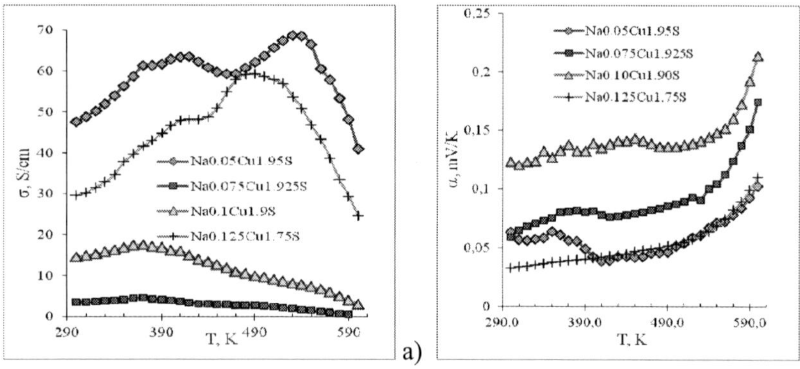

b)

Figure 4. Temperature dependencies of electronic conductivities (a)

5. Conclusions

At ambient temperature, the studied alloys consist of three phases: monoclinic $Na_2Cu_4S_3$, hexagonal chalcocite Cu_2S and rhombohedral digenite Cu_9S_5. The complex phase composition of the mixture is confirmed by differential scanning calorimetry, which recorded a number of thermal effects at temperatures of 370–382 K, 440–460 K, 630–640 K, 653 K, and 677 K. The first heat effect occurring at 370–382 K, in our opinion, is associated with the superionic phase transition in digenite. The enthalpy of phase transformation for the studied samples increases with increasing sodium content in the samples. Since the superionic transition is associated with the disordering of the structure, it can be assumed that the presence of sodium in the lattice hampers the disordering and retards the rearrangement of the lattice. This assumption is confirmed by an increase in the temperature of phase transitions with increasing sodium content. The chalcocite Cu_2S phase in the studied alloys has a hexagonal structure at room temperature and can be a superionic conductor already at room temperature.

The electronic conductivity of the alloys is small, it has a tendency to decrease with increasing sodium content in the alloy, while the Seebeck coefficient increases. The analysis of a three-phase system is comprehensive and cannot give unambiguous conclusions. It is clear that sodium atoms play the role of a compensating impurity when doping non-stoichiometric copper sulfide, reducing the concentration of holes. At the same time, the carrier mobility should also decrease due to scattering by impurity sodium atoms. Nano-dimension of the grains also usually reduces conductivity due to the increasing role of scattering on the grain boundaries. X-ray structural studies at high temperatures are necessary to verify the possibility of obtaining a single-phase state of the alloys under study.

Acknowledgments
The authors are grateful to the Ministry of Education and Science of the Kazakhstan Republic for financial support the Program "Development of Hydrogen Energy Technologies in the Republic of Kazakhstan" (2018–2020).

References
[1] Chakrabarti D J and Laughlin D E *Phase Equil.* 1983 **4** 254
[2] Shuai X, Shen W, Hou Zh, Ke S, Xu Ch and Jiang Ch 2014 *Nanoscale Res. Lett.* **9** 513
[3] Jache B, Mogwitz B, Klein F and Adelhelm P 2014 *Power Sources* **247** 703
[4] Zhao Y, Pan H, Lou Y, Qiu X, Zhu J and Burda C 2009 *J. Am. Chem. Soc.* **131** 4253
[5] Cassaignon S, Pauport Th, Guillemoles J F and Vedel J 1998 *Ionics* **4** 364
[6] Balapanov M Kh, Gafurov I G, Mukhamedyanov U Kh, Yakshibaev R A and Ishembetov R Kh 2004 *Phys. Stat. Sol. (b)* **241** 114
[7] Balapanov M Kh, Bikkulova N N, Mukhamedyanov U Kh, Asilguschina, Musalimov R Sh and Zeleev M Kh 2004 *Phys. Stat. Sol. (b)* **241** 3517 G N
[8] Balapanov M Kh, Ishembetov R Kh, Kuterbekov K A, Nurakhmetov T N, Urazaeva E K and Yakshibaev R A 2014 *Inorg. Mater.* **50** 930
[9] Shannon R D 1976 *Acta Cryst. A* **32** 751
[10] Savelsberg G and Schafer H 1981 *Mat. Res. Bull.* **16** 1291
[11] Ge Z H, Liu X Y, Feng D, Lin J Y and He J Q 2016 *Adv. Energy Mater.* **6** 1600607
[12] Balapanov M Kh, Kubenova M M, Kuterbekov K A, Kozlovskiy A, Nurakov S N, R Kh Ishembetov and Yakshibaev R A 2018 *Eurasian J. Phys. Func. Mat.* **2** 231
[13] Balapanov M Kh, Ishembetov R Kh, Kuterbekov K A, Kubenova M M, Almukhametov R F and Yakshibaev R A 2018 *Ionics* **24** 1349
[14] Will G, Hinze E and Abdelrahman A R 2002 *Eur. J. Miner.* **14** 591

IOP Conf. Series: Materials Science and Engineering **447** (2018) 012032 doi:10.1088/1757-899X/447/1/012032

Optical properties of Al nanoparticles prepared by laser ablation method

D A Afanasyev[1,2], **N Kh Ibrayev**[1], **A–F E Serik**[3]

[1]Institute of Molecular Nanophotonics, Buketov Karaganda State University, Karaganda 100028, Kazakhstan
[2]Institute of Applied Mathematics, Karaganda 100028, Kazakhstan
[3]Karaganda State Technical University, Karaganda 100027, Kazakhstan

E-mail: a_d_afanasyev@mail.ru

Abstract. The results of the synthesis and study of the optical properties of aluminum nanoparticles (NPs) processed by the laser ablation method are presented. Water, ethanol and chlorobenzene are used as solvents for the synthesis of nanoparticles. Measurements of the size and shape of the obtained NPs are carried out. The absorption and luminescence spectra of NPs are measured. The extinction spectra of aqueous solutions of Al nanoparticles and Al-Al_2O_3 nanostructures were simulated. Studies demonstrate the formation of Al-Al_2O_3 nanostructures during laser ablation of aluminum in water and ethanol liquids.

1. Introduction

The development of methods for the synthesis of metal nanoparticles is one of important directions of the modern materials science. The main attention is given to the synthesis of such nanoparticles (NPs) as silver and gold [1, 2]. Less attention is paid to the development of methods for the synthesis of other types of metal NPs. Aluminum NPs are promising materials for the use in various optoelectronic devices [3, 4]. The search for methods for the synthesis of aluminum NPs with known characteristics is an important task.

The method of laser ablation in a liquid is one of simple methods for the preparation of NPs. However, both the aluminum NPs and complex nanostructures (NSs) containing Al and aluminum oxide are formed during this process depending on the solvents used. For example, the laser ablation of Al in water [5] results in a formation of aluminum oxide (Al_2O_3) along with Al NPs. The Al-Al_2O_3 NSs formed can have a core/shell structure, where the core is an Al nanoparticle and the shell is an oxide layer Al_2O_3. It is necessary to use various methods of studies including optical methods to obtain more complete information about the synthesized nanostructures.

The results of the synthesis of Al nanoparticles by laser ablation and studies of their optical properties are presented in this paper.

2. Experiment

Laser ablation was used as a method for the synthesis of Al NPs in a liquid. The second harmonic of a Nd:YAG laser SOLAR LQ 215 is used (wave length λ_{gen} = 532 nm, pulse energy E_{imp} = 90 mJ, pulse duration τ = 7 ns, pulse frequency ν = 20 Hz). The laser radiation was directed by a mirror and focused by a lens onto a horizontally placed target - an aluminum plate in a cell with liquid.

Various solvents were used for ablation. Data on the ablation time are given in table 1. The average size of the obtained Al nanoparticles was determined by dynamic light scattering using a Zetasizer Nano ZS particle size analyzer. The morphology and energy dispersive spectra of Al NPs were obtained using a Tescan Mira 3 scanning electron microscope. The absorption spectra were recorded on a Cary 300 (Agilent) spectrophotometer. The luminescence spectra were measured by Cary Eclipse (Agilent).

The program Mie plot V.4.4.11 was used for simulating the extinction, absorption and scattering spectra of Al NPs and nanostructures of Al-Al_2O_3 in water. The numerical calculation algorithm used is described in detail in [6, 7]. The dielectric constants of Al and Al_2O_3 solid materials are given in [8, 9].

3. Results and discussion

The sizes of NPs formed in various solvents are shown in table 1. As shown in [10], the size of NPs depends on the boiling point of the solvent. Indeed, NPs with the smallest size are formed for the liquid with the highest boiling point (chlorobenzene). On the contrary, NPs with the largest size are obtained for the solution with the lowest boiling point (ethanol). The concentration of NPs in water was determined. The mass of the target was weighed before and after the ablation process. The concentration of Al NPs in water was equal to 1.38×10^{-9} mol/L.

Table 1. Characteristics of the laser ablation process and optical properties of aluminum NPs

Medium	Ablation time (min)	Boiling point, ^{0}C	Average particle size r (nm)	λ_{abs}^{max} (nm)	$\lambda_{1/2}$ (nm)
Ethanol	10	78	2.5	199	114
Water	4	100	19	1) 196; 2) 209; 3) 263	1) 34; 2) 17; 3) 40
Chlorobenzene	2	131	7.5	1) 229; 2) 265	1) 13; 2) 41

Main characteristics of the absorption spectra are given in table 1. A three-fold dilution was used to measure the absorption spectrum of Al NPs in water. The absorption spectra for NPs in water and ethanol have a main maximum in the range 195-200 nm (figure 1, curves 1 and 2). These spectra are in a good agreement with the spectral data obtained in other works [5, 11]. In water, the absorption spectrum contains additional bands with maxima at 209 and 263 nm. The absorption spectrum has peaks at 229, 264 nm for Al NPs in chlorobenzene. The solution becomes yellow during ablation of the aluminum target in chlorobenzene. The spectrum of Al_2O_3 NPs with a size of 125 nm in water also is shown in figure 1, curve 4.

The formation of aluminum oxide (Al_2O_3) instead of aluminum is possible during laser ablation of an aluminum target in water. However, the measured spectrum of Al_2O_3 NPs with a radius of r = 62.5 nm did not show the presence of the absorption maxima in the range of 200 -600 nm (figure 1, curve 4). The absorption spectrum of Al_2O_3 NPs differs significantly from the absorption spectrum of an aqueous solution of Al NPs obtained by laser ablation. The presence of Al_2O_3 luminescence may be an additional factor indicating the formation of aluminum oxide as a result of laser ablation of an Al target. The absorption spectra of Al are located in the ultraviolet region. The glow of solvents is observed after ultraviolet irradiation of chlorobenzene and ethanol. Therefore, only aqueous solutions were used to study the luminescence of NSs based on aluminum. The luminescence spectra of Al_2O_3 and Al NPs in water are shown in figure 2.

A comparison of spectra in figure 2 shows the presence of Al_2O_3 particles in solutions. There are luminescence bands with peaks at 380, 403, 426, and 486 nm in the luminescence spectrum of Al NPs solutions.

Luminescence bands with maxima at 425 and 486 nm are observed for the luminescence spectrum of Al_2O_3 NPs (figure 3, curve 2). Similar results were obtained in [5, 12]. These bands can be attributed to the luminescence of various F centers of the crystalline Al_2O_3 [12].

Electron microscopic investigation indicates the spherical shape of Al NPs. Energy dispersive analysis (EDX) shows the presence of Al_2O_3 oxide in the composition of NSs. The results of the analysis are shown in figure 3. The oxygen constitutes 60 mass % and the aluminum the remaining 40 mass % in NSs prepared in water. The concentration of oxygen in NSs prepared in ethanol is lower than in NSs prepared in water. This indicates a lower thickness of the oxide layer for Al processed in ethanol.

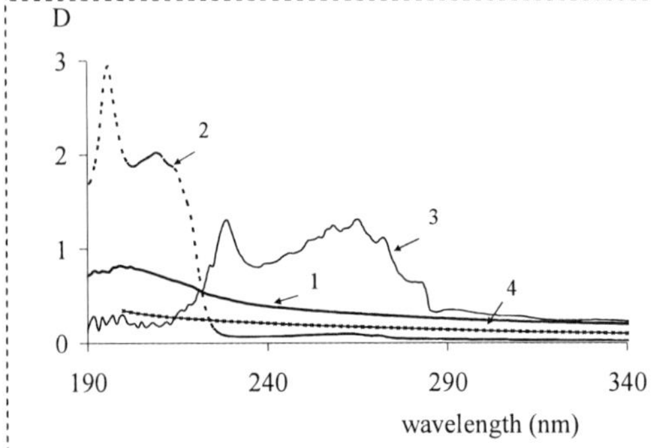

Figure 1. Absorption spectra of Al NPs (1-3) and Al_2O_3 NPs (4) in different solutions: 1 – ethanol; 2,4 – water; 3 – chlorobenzene.

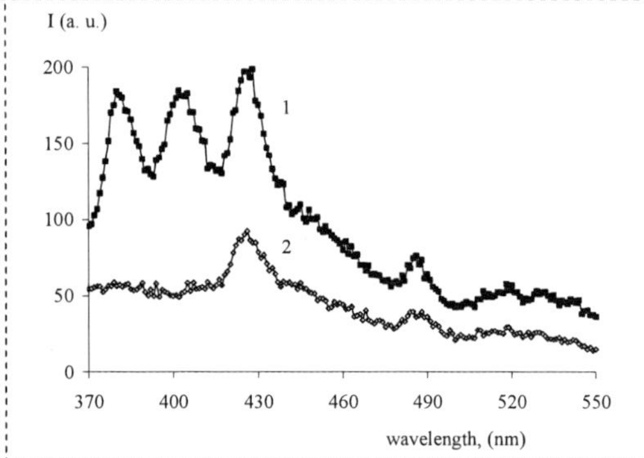

Figure 2. . Luminescence spectra of Al NPs (1) and Al_2O_3 NPs (2) in water.

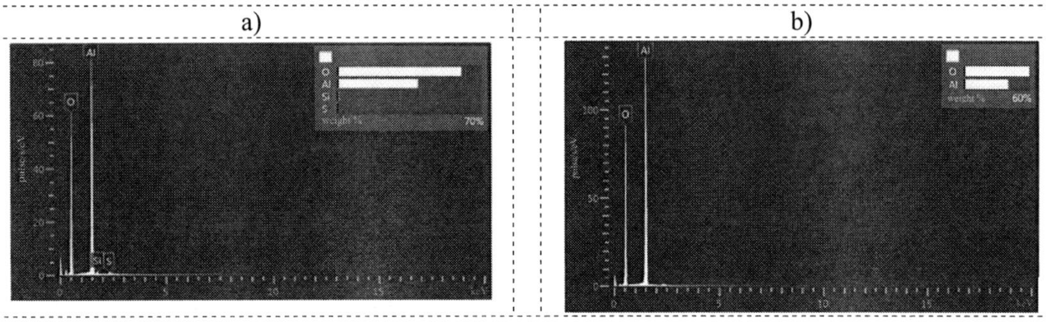

Figure 3. EDX–spectra of Al NPs obtained in the water (a) and in the ethanol (b).

The extinction spectra for Al nanoparticles in water obtained by simulations are shown in figure 4. The simulation results show that the experimental absorption spectrum of aluminium NPs in water (figure 1, curve 2) may be due to the NPs with radius r in the range from 9 to 11 nm. However, the measurement of sizes of NPs by the dynamic light scattering method shows the presence of NPs with a radius of r = 19.5 nm in water. Also, the distribution of the NPs is monodisperse. Thus, the results of direct measurements of the NPs size in water by dynamic light scattering do not correspond to the results of analysis of the absorption spectra and the simulation of extinction spectra of NPs in water. The appearance of the absorption band with a maximum at 263 nm can be explained by the presence of NPs with a size of 26-27 nm.

The extinction spectra of core/shell NSs were simulated with Al as the core and Al_2O_3 as the shell. Sizes of the core and shell were varied to obtain the extinction spectrum of Al NPs in water. The extinction spectra were simulated starting from λ = 200 nm due to the capabilities of the code and availability of dielectric constants of Al and Al_2O_3. Therefore, the spectra were compared to the maximum at 209 nm. Two bands (figure 6, curve 2) are present in the extinction spectrum of NPs with a large core radius r_{core} = 24 nm and Al_2O_3 shell thickness h_{shell}=3 nm. A shift to the shortwave region of the absorption maxima occurs with a decrease in the particle size.

The short-wave maximum of extinction spectra becomes inaccessible for size of NSs r_{NSs} = 20 nm (r_{NSs}= r_{core}+ h_{shell}). The extinction band with a maximum at 210 nm is observed for a NS with r_{core} = 12.5 nm and h_{shell} = 0.5 nm (figure 5, curve 2). This size exceeds the size of NPs obtained by dynamic light scattering. However, the size of the simulated NSs is closer to the data of table 1 than the size from simulations of Al NPs.

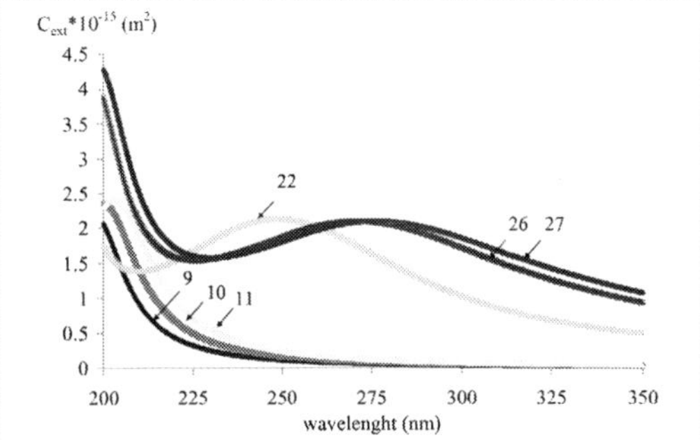

Figure 4. Extinction spectra of Al NPs with various sizes (r) in the water.

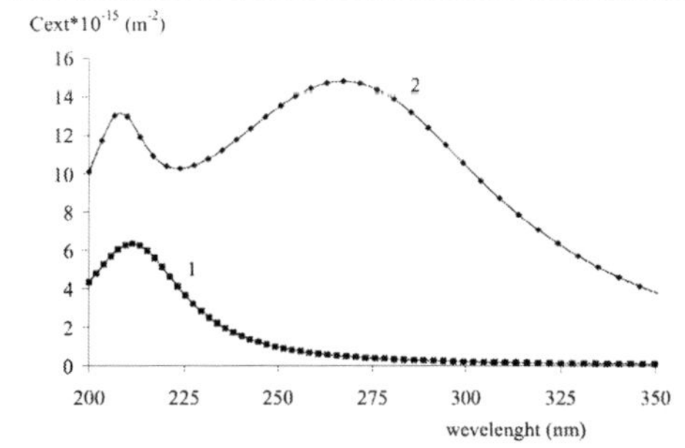

Figure 5. Extinction spectra Al–Al_2O_3 NSs in the water 1 - r_{core} = 12.5 nm, h_{shell} = 0.5 nm; 2 - r_{core} = 24 nm, h_{shell} = 3 nm.

Thus, the results of direct measurements of the size of Al NPs in water do not correspond to the results of the analysis of the absorption spectra and the simulation of the extinction spectra of Al NPs in water. The results of the measuring of the luminescence spectra and EDX-analysis of aluminum NPs indicate the presence of Al_2O_3 in the solution. A comparison of the obtained spectral data for Al NPs with the data on the absorption of Al_2O_3 and simulation of spectra of Al_2O_3 NPs shows that the spectra of synthesized NPs can't be explained by the formation of only the Al_2O_3 NPs. The obtained data indicate the formation of nanostructures based on Al and Al_2O_3.

References

[1] Xia Y, Xiong Y, Lim B and Skrabalak S E 2009 *Angew. Chem. Int. Ed. Engl.* **48(1)** 60
[2] Ghosh R, Paria Ch and Paria S 2012 *Chem. Rev.* **112** 2373
[3] Ikhsanov R Sh, Protsenko I E and Uskov A V 2013 *Techn. Phys. Lett.* **39(5)** 450
[4] Zhou L, Tan Y, Wang J, Xu W, Yuan Y, Cai W, Zhu Sh and Zhu J 2016 *Nature Photon.* **10** 393
[5] Kumar Bh and Thareja R K 2010 *J. Appl. Phys.* **108** 064906-1.
[6] Grainger R G, Lucas J, Thomas G E and Ewen G B L 2004 *Appl. Optics* **43(28)** 5386
[7] Šileikaite A, Prosycevas I, Puiso J, Juraitis A and Guobiene A 2006 *Mater. Sci. (Medžiagotyra)* **12(4)** 287
[8] Rakic A D 1995 *Appl. Optics* **34(22)** 4755
[9] Malitson I H, Dodge M J J 1972 *Opt. Soc. Am.* **62** 1405.
[10] Golightly J S 2007 *Formation and Characterization of Nanopartiles via Laser Ablation in Solution* (Thesis in Chemistry, Pennsylvania State University)
[11] Baladi A and Mamoory R S 2009 *Fifth International Conference on MEMS, NANO, and Smart Systems* 218
[12] Abdolvahab A, Farjami Sh S and Taheri Gh R 2017 *Chin. Phys. B* **26(3)** 036101.

IOP Conf. Series: Materials Science and Engineering **447** (2018) 012033 doi:10.1088/1757-899X/447/1/012033

Two-dimensional discrete breathers in hcp titanium

O V Bachurina[1], R T Murzaev[2], M N Semenova[3], A S Semenov[3], D S Ryabov[4], G M Chechin[4], E A Korznikova[2] and S V Dmitriev[2,5]

[1] Ufa State Petroleum Technological University, Ufa 450062, Russia
[2] Institute for Metals Superplasticity Problems, Ufa 450001, Russia
[3] North-Eastern Federal University n.a. M.K. Ammosov, Polytechnic Institute (branch) in Mirny, Mirny 678170, Sakha (Yakutia), Russia
[4] Southern Federal University, Rostov-on-Don 344006, Russia
[5] National Research Tomsk State University, Tomsk 634050, Russia

E-mail: dmitriev.sergey.v@gmail.com

Abstract. Recently delocalized nonlinear vibrational modes (DNVMs) have been derived for a two-dimensional (2D) triangular lattice and their frequencies as the functions of amplitude have been calculated for a 2D Morse crystal. A 3D hcp lattice is formed by certain stacking of 2D triangular lattices. In the present work, we attempt to excite eight DNVMs with in-plane atomic displacements in a single plane of hcp titanium, modelled using many-body interatomic potentials. It is found that three of eight DNVMs produce long-lived 2D discrete breathers (DBs) in a titanium crystal. A 2D DB is delocalized in two spatial dimensions and localized in the third one, so that atoms of only one atomic plane have large vibrational amplitudes. The fFrequencies of the newly found DBs lie above the small-amplitude phonon spectrum of titanium and grow with increasing amplitude. Thus, we report on three novel 2D DBs in hcp metal. Our results open the route to finding new types of DBs in pure metals.

1. Introduction

Discrete breathers (DBs) are spatially-localized, large-amplitude vibrational modes in discrete nonlinear lattices [1-3]. DBs have frequencies outside the spectrum of small-amplitude lattice vibrations. Their frequency, for sufficiently large amplitudes, can leave the spectrum due to the nonlinearity of the lattice, since the frequency of the nonlinear vibrational mode depends on its amplitude. For the soft (hard) type of nonlinearity, DB frequency decreases (increases) with its amplitude. The existence of DBs in crystal lattices has been proved by a number of molecular dynamics studies [2-18], and recently in several works using the *ab initio* approach in frames of the density functional theory [19,20]. Experimental observation of DBs in crystals is a challenge due to their atomic-scale size and the lifetime from 10 to 1000 atomic vibrations. Nevertheless, their successful identification has been summarized in the recent review [2]. Until 2011, it was believed that pure metals could not support DBs because interatomic interactions were soft, and, in this situation, only crystals having a gap in the phonon spectrum could support DBs [16]. However, Haas et al. demonstrated the existence of DBs in fcc Ni and bcc Nb [11]. Later it was shown that interatomic interactions were actually soft at large distances and hard at small distances [18]. When the hard core of the interatomic potential makes a greater contribution to the dynamics of nonlinear excitation than

IOP Conf. Series: Materials Science and Engineering **447** (2018) 012033 doi:10.1088/1757-899X/447/1/012033

its soft tail, the frequency of excitation can grow with amplitude and cross the upper edge of the phonon spectrum. DBs in pure metals were analyzed in the works [12-14].

Recently it was realized that 2D crystals could support 1D DBs (delocalized in one dimension and localized in another), while in 3D crystals both 1D and 2D DBs were possible. In particular, a 1D DB has been analyzed in fcc aluminum with large-amplitude oscillations of atoms in a single close-packed atomic row [15]. In the present study, we discuss 2D DBs in hcp titanium, simulated using molecular dynamics. The considered nonlinear, large amplitude oscillations are localized in a single basal atomic plane of the crystal. The amplitudes of atomic vibrations decay exponentially with a deviation from the plane, where a 2D DB is excited.

The main problem in excitation of a DB in molecular dynamics simulations is to find the proper initial conditions. It has been proved that delocalized nonlinear vibrational modes (DNVMs), also called one-dimensional bushes of nonlinear normal modes, can be very helpful in solving this problem [5,18]. DNVMs are the exact solutions to the nonlinear equations of atomic motion, regardless the interatomic potentials used in simulations, because DNVMs are dictated by the lattice symmetry and are found with the use of the group theory [21]. Note that the basal plane of the hcp crystal is a 2D triangular lattice, for which Chechin and Ryabov derived eight DNVMs with in-plane atomic displacements [22]. We excite these modes in a single basal plane of titanium and analyze the lifetime of such excitations. If the lifetime is long, the obtained excitation is called a 2D DB.

2. Simulation details

We consider hcp titanium modeled with the use of many-body interatomic potentials obtained with the help of the embedded-atom method (EAM) [23]. The computational cell includes 20 basal planes with 12×12 atoms in each plane. The lattice parameters of titanium are a=2.951 Å,c=4.684 Å. Periodic boundary conditions are used. Thermal fluctuations are not introduced in the system.

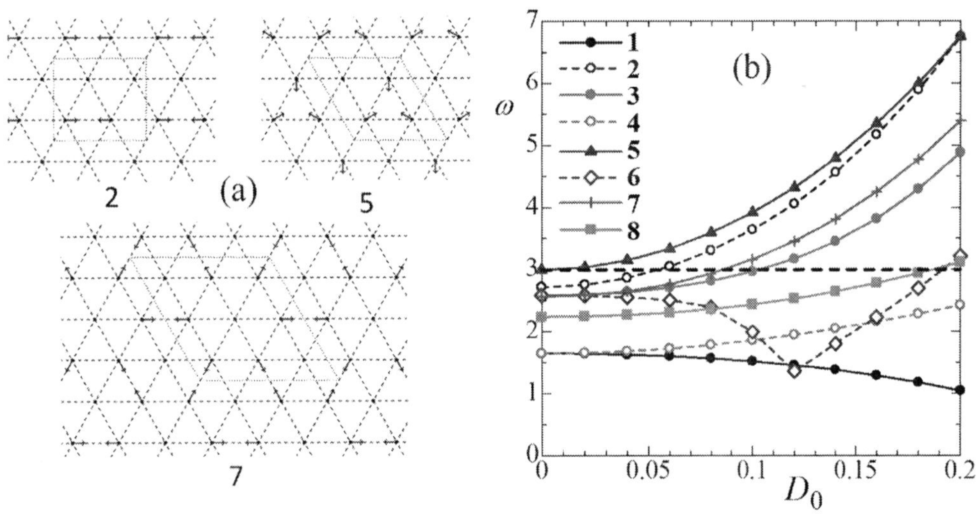

Figure 1 (a) Three of the eight studied DNVMs with in-plane atomic displacements supported by a triangular lattice, as found by Chechin and Ryabov [22]. (b) Results for 2D triangular Morse lattice [22]. Shown is the frequency as a function of initial displacements of atoms for all eight DNVMs. Only DNVMs 2, 3, 5, and 7, in the studied range of D_0, have frequencies above the phonon band of the Morse crystal, which is shown by the horizontal dashed line.

The initial conditions are set by applying an initial shift to the atoms in single basal plane according to the pattern of one of the eight DNVMs with in-plane atomic displacements derived by Chechin and Ryabov [22]. Three of the eight DNVMs are shown in figure 1 a, the other modes can be found in [22]. The length of the arrows is equal to D_0, which defines the initial amplitude of the DNVM. The initial atomic displacements in all other basal planes are equal to zero, and the initial velocities of all atoms in the computational cell are equal to zero. In the work [22], the frequency as a function of amplitude was calculated for all eight DNVMs excited in a 2D Morse crystal. This result is shown in figure 1 b. It can be seen that in the studied range of D_0, only DNVMs 2, 3, 5 and 7 have frequencies above the upper edge of the phonon spectrum, which is indicated by the horizontal dashed line.

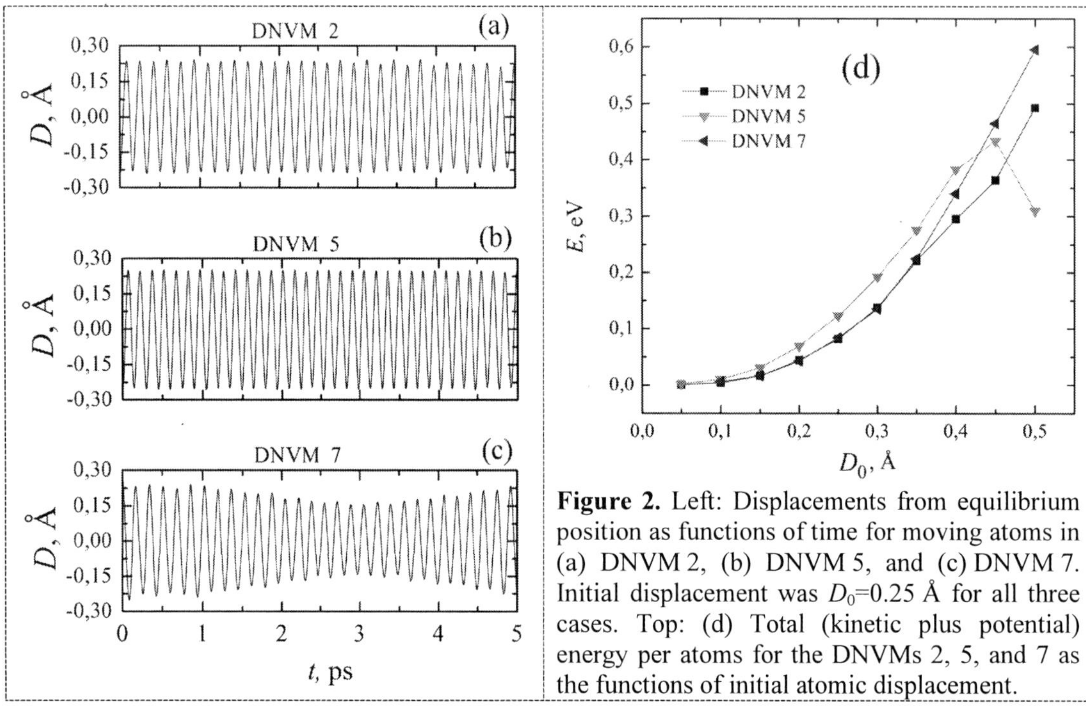

Figure 2. Left: Displacements from equilibrium position as functions of time for moving atoms in (a) DNVM 2, (b) DNVM 5, and (c) DNVM 7. Initial displacement was D_0=0.25 Å for all three cases. Top: (d) Total (kinetic plus potential) energy per atoms for the DNVMs 2, 5, and 7 as the functions of initial atomic displacement.

3. Numerical results and discussion

In our simulations, all eight DNVMs with in-plane displacements derived for a triangular lattice have been analysed. Only three of them have produced long-lived 2D DBs. These are DNVMs 2, 5 and 7 shown in figure 1 a. It is not surprising that DNVMs 1, 4, 6, and 8 did not produce a long-lived vibrational mode, because in a 2D Morse lattice they have frequencies within the phonon band. DNVM 3 has a frequency above the phonon band in a 2D Morse lattice only for sufficiently large vibrational amplitude. It should be pointed out that when a DNVM is excited in a 2D triangular lattice, it does not interact with other vibrational modes and it keeps its energy constantly in time. On the other hand, when a DNVM is excited in a single basal plane of the hcp lattice, a part of its energy is given to the neighbouring atomic planes, and this is the reason for the decay of DNVM 3, which has a frequency above the phonon spectrum of the 2D Morse lattice only for sufficiently large amplitude.

In figure 2 (d), the total (kinetic plus potential) energy per atom in a single basal plane is plotted for DNVMs 2, 5 and 7 as a function of the initial displacements of atoms, D_0. It can be seen that the maximal energy per atom is approximately equal to 0.5 eV and the maximal initial displacement is about 0.5 Å. For larger initial displacements, strong energy radiation from the excited basal plane is observed.

4. Conclusions

For the first time, three new 2D DBs were obtained and analysed in an hcp crystal using titanium as an example. Large amplitude vibrations in the described 2D DBs take place in a single basal plane. The patterns of atomic vibrations in this plane correspond to DNVMs 2, 5, and 7 described in [22]. It would be interesting to analyze the possibility of the existence of similar 2D DBs in fcc pure metals.

Acknowledgements
The work of A.S.S.(discussion of the numerical results) was supported by the Russian Foundation for Basic Research, grant no. 18-32-00171_mol_a. D.S.R and G.M.Ch. have worked under state assignment no. 3.5710.2017/8.9 (derivation of DNVM and discussion of the results). E.A.K. thanks the Russian Foundation for Basic Research, grant no. 17-02-00984_a (writing the paper). S.V.D. thanks the Russian Science Foundation, grant no. 16-12-10175, for their financial support (design of the research). This work was partly supported by the state assignment of IMSP RAS.

References
[1] Flach S and Gorbach A V 2008 *Phys. Rep.* **467** 1
[2] Dmitriev S V, Korznikova E A, Baimova Yu Aand Velarde M G 2016 *Physics-Uspekhi* **59** 446
[3] Baimova J A, Korznikova E A, Lobzenko I P and Dmitriev S V 2015 *Rev. Adv. Mater. Sci.* **42** 68
[4] Korznikova E A, Fomin S, Soboleva E and Dmitriev S V 2016 *JETP Lett.* **103** 277
[5] Barani E, Lobzenko I P, Korznikova E A, Soboleva E, Dmitriev S V A 2017 *Phys. Lett. A* **381** 3553
[6] Murzaev R T, Bachurin D V, Korznikova E A and Dmitriev S V 2017 *Phys. Lett. A* **381** 1003
[7] Evazzade I, Lobzenko I P, Korznikova E A, Ovid'ko I, Roknabadi M and Dmitriev S V 2017 *Phys. Rev. B* **95** 035423
[8] Barani E, Lobzenko I P, Korznikova E A, Soboleva E, Dmitriev S V, Zhou K and Marjaneh A 2017 *Eur. Phys. J. B* **90** 38
[9] Barani E, Korznikova E A, Chetverikov A P, Zhou K and Dmitriev S V 2017 *Phys. Lett. A* **381** 3553
[10] Dmitriev S V, Medvedev N N, Mulyukov R R, Pozhidaeva O V, Potekaev A I and Starostenkov M D 2008 *Russ. Phys. J.* **51** 858
[11] Haas M, Hizhnyakov V, Shelkan A, Klopov M and Sievers A J 2011 *Phys. Rev. B* **84** 144303
[12] Murzaev R, Babicheva R I, Zhou K, Korznikova E A, Fomin S, Dubinko V I and Dmitriev S V 2016 *Eur. Phys. J. B* **892** 168
[13] Murzaev R T, Kistanov A A, Dubinko V I, Terentyev D A and Dmitriev S V 2015 *Comp. Mater. Sci.* **98** 88
[14] Terentyev D A, Dubinko A V, Dubinko V I, Dmitriev S V, Zhurkin E E and Sorokin M V 2015 *Model. Simul. Mater. Sci.* **23** 085007
[15] Bachurina O V, Murzaev R T, Korznikova E A and Dmitriev S V 2017 *Mater. Phys. Mech.* **33** 49
[16] Kiselev S A, Bickham S R and Sievers A J 1993 *Phys. Rev. B* **48** 13508
[17] Dmitriev S V 2016 *Letters on Materials* **6** 86
[18] Korznikova E A, Kistanov A A, Sergeev K S, Shepelev I A, Davletshin A R, Bokij D I and Dmitriev S V 2016 *Letters on materials* **6** 221 (in Russian)
[19] Lobzenko I P, Chechin G M, Bezuglova G S, Baimova Y A, Korznikova E A and Dmitriev S V 2016 *Phys. Solid State* **58** 633
[20] Chechin G M, Dmitriev S V, Lobzenko I P and Ryabov D S 2014 *Phys. Rev. B* **90** 045432
[21] Chechin G M and Sakhnenko V P 1998 *Physica D* **117** 43
[22] Semenova M N, Semenov A S, Bebikhov Yu V, Ryabov D S, Chechin G M, Rakhmatullina Zh G, Korznikova E A and Dmitriev S V 2018 *Fundamental'nye Problemy Sovremennogo Materialovedenia* **15** 257 (in Russian)
[23] Hennig R G, Lenosky T J, Trinkle D R, Rudin S P and Wilkins J W 2008 *Phys. Rev. B* **78** 054121

Functions of interlayers based on ultradisperse metal powders at diffusion bonding of heterogeneous materials

A V Liushinskii

JSK "Ramenskoye Design Company" 2, Gurieva Str., Ramenskoye, Moscow region 140103 Russia

E-mail: nilsvarka@yandex.ru

Abstract. The features of the formation of compounds produced by means of diffusion bonding using intermediate layers based on ultradisperse nickel powder and binary mixtures made of Ni + Cu, Ni + Co, Cu + Co ultradisperse powders, have been considered. It has been demonstrated that application of such layers provides obtaining of welded joints with guaranteed quality within the temperature range of 300–350 °C and at pressures 10-20 MPa lower than those applied in the conditions of bonding without inter layers.

1. Introduction

The simplest and most convenient method of intensifying the process of diffusion bonding and ensuring the efficient operation of mechanical and thermal activation channels is the use of interlayers, which may play different functions during welding.

Interlayers are made of gold, silver, copper, nickel, aluminum and other metals in form of wire, foil, powders and films deposited on the surfaces to be bonded by electroplating or in the form of vacuum condensates. The type of interlayer is determined by the possibilities of technological production and depends on the grade of welded materials, bonding conditions, service conditions of resultant joints, etc.

A method has been developed of producing ultradisperse powders (UDPs) of metals, characterized by high activity in compacting and sintering as a result of the developed free surfaces and the presence of a large number of crystal structure defects (dislocations, vacancies, etc.) [1,2].

The effect of dispersion and the method of producing different powders on the quality of formation of the joint was examined. It showed that the diffusion bonding parameters ensuring high-quality bonds decrease with an increase in the specific surface of powder particles and an increase in the distance of their state from thermodynamic equilibrium.

2. Model description and simulation procedure

The study of diffusion bonding (temperature T=550 °C, welding pressure P=20 MPa, holding time t=30 minutes) using intermediate layers of various nickel powders such as electrolytic (PNE-1), carbonyl (PNCOT-1) and ultradisperse, thermally decomposed nickel formate Ni $(COOH)_2 \cdot 2H_2O$, with different dispersities (d=39.75, 7.63 and less than 0.1 μm) and the specific surface area (S_{yd}=0,12; 0,48 and 17,32 m^2/g, respectively) showed that the maximum strength ($\sigma_u \geq 20$ kgF/mm^2) is achieved when bonding through UDP.

IOP Conf. Series: Materials Science and Engineering **447** (2018) 012034 doi:10.1088/1757-899X/447/1/012034

The operation of applying UDP on welded surfaces is not technological, since it is impossible to ensure uniformity of the deposited layer in terms of thickness and mass. These features of the intermediate layer can cause instability of the structure and properties of the bonding joints, as well as lead to their poor quality. The most acceptable form of such a layer, which excludes the above mentioned drawbacks, is a porous tape obtained by rolling of the UDP of nickel. In this case, the total value of the specific surface area of the powder particles constituting the tape decreases by no more than 10%. This occurs due to the formation of interparticle contacts.

3. Simulation results and discussion

Figure 1 shows the porosity θ of the rolled strip from its thickness τ at different rolling speeds υ. As the rolling speed increases, the thickness decreases and the porosity of the belt increases.

To determine the optimal values of the parameters of the rolling pattern, we investigated the effect of θ, τ, υ tapes on the properties of 10890 steel compounds obtained by diffusion bonding through these layers. Parameters of the bonding mode: T=500 °C, P=2 kgF/mm^2, t=30 minutes. The welded joints were tested for static stretching and the tensile strength σ_u was determined. Analysis of the extreme dependences $\sigma_u(\tau)$ (figure 2) and $\sigma_u(\theta)$ (figure 3) shows that the maximum strength values are achieved with the following initial characteristics of the rolled strip: θ=50-65%, τ=60-80 μm. With increasing υ, the values of σ_u decrease.

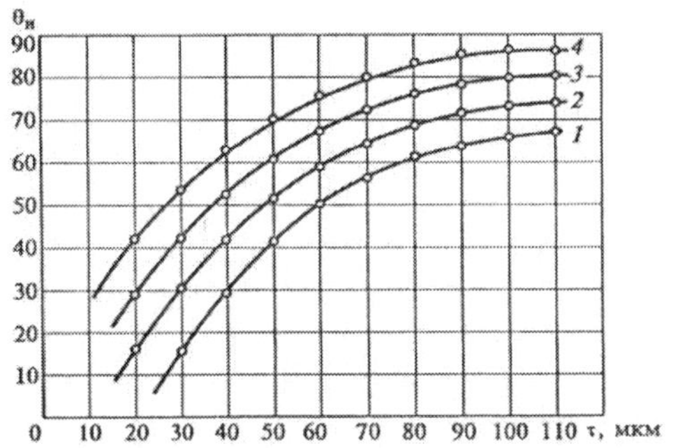

Figure1. Dependences of $\theta(\tau)$ for various υ: 1 - 190; 2 - 470; 3 - 790; and 4 - 1570 mm/min

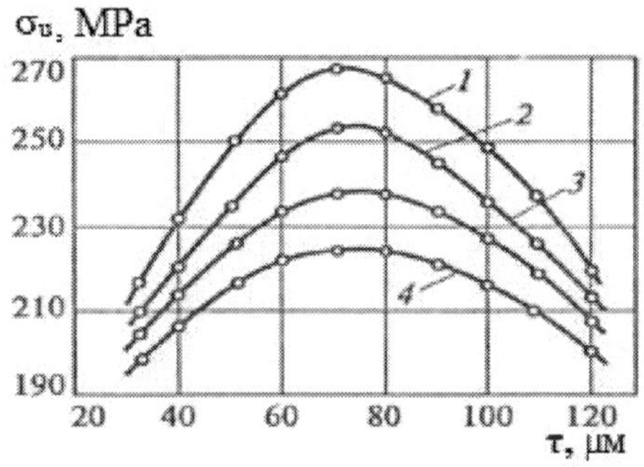

Figure 2. Dependences of $\sigma_u(\tau)$ for various υ: 1 - 190; 2 - 470; 3- 790 and 4 - 1570 mm/min

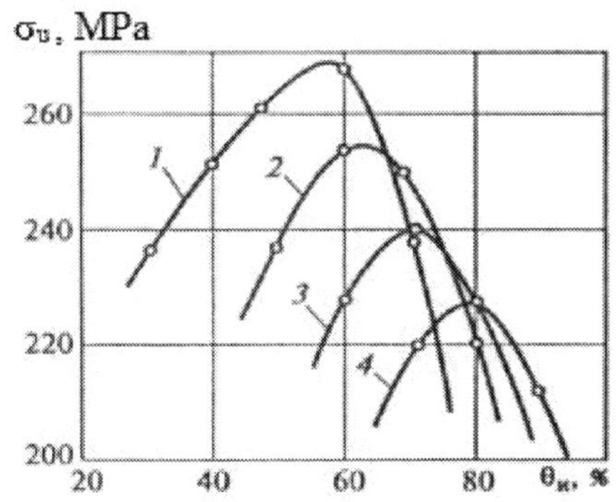

Figure 3. Dependences of $\sigma_u(\theta)$ for various υ: 1 - 190; 2 - 470; 3 - 790 and 4 - 1570 mm/min

Figure 4 shows the dependencies $\overline{\sigma}_u(t)$, reflecting the change in the relative tensile strength of the bonded joints of steel 10890 + steel 10890. These results show that with diffusion bonding parameters T=500-600 °C, P=1.5-2.0 kgf/mm^2 and t=20-30 min σ_u approaches 0.8. When bonding without a powder intermediate layer, this strength can be obtained at T=900 °C, P=2-2.5 kgF/ mm^2 and t≥30 min. Such an increase in temperature and such bonding pressure is especially relevant when connecting special materials, for example, permanent magnets with magnetic circuits and fittings, in which the loss of the initial magnetic properties occurs at temperatures above 550 °C.

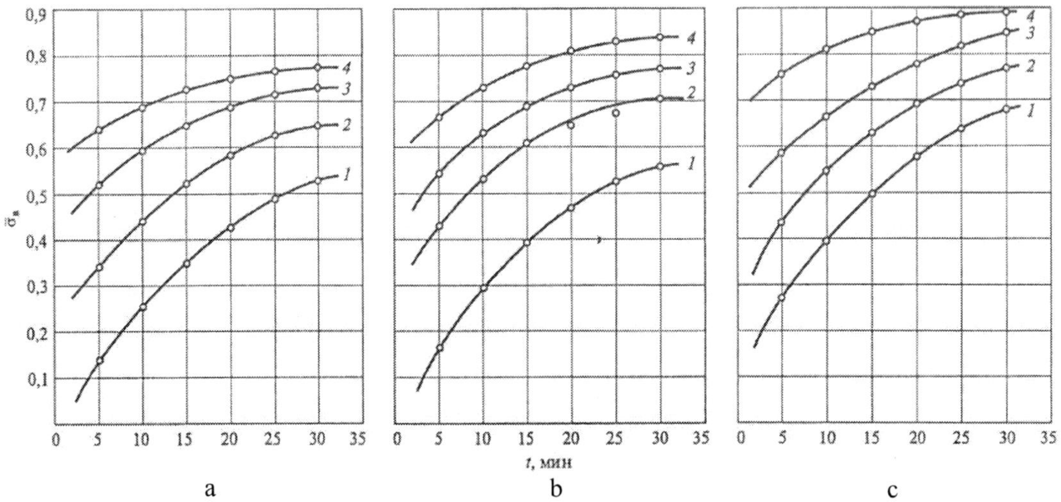

Figure 4. Dependences of $\overline{\sigma}_u(t)$ at T=400 (a), 500 (b) and 600 °C (c), P=0.5 (1), 1 (2), 1.5 (3) and 2 kgF/mm^2 (4)

The sintering of a powder layer consisting of one component is accompanied by a decrease in free energy due to a decrease in the total surface area of the interparticle pores. An intermediate layer containing several components, together with excess free energy due to the presence of defects in the structure and the developed surface of its constituent dispersed particles, also has excess energy associated with the possibility of forming an alloy. Another important feature of the powder intermediate layers is that, by varying their composition, it is possible to ensure its compliance with the chemical composition of the welded materials. Consequently, it is possible to obtain welded joints, which have not only high strength characteristics, but also the same electrophysical properties as the main materials.

The investigations were carried out using UDP mixtures obtained from mechanical mixtures of nickel + copper, nickel + cobalt and copper + cobalt formats with a different ratio of components.

The strength of the joints produced through the interlayer of mechanical mixtures of nickel + copper UDP at 400 °C varies in the range of 20-50 MPa, depending on the ratio of the components (figure 5 a). In mechanical tests of these specimens, the contact surface contains individual areas of bonding. With increasing temperature, the strength naturally increases, and the dependence becomes "wavy" with the maximum strength corresponding to the composition 25%Ni+75% Cu and the minimum strength to 75%Ni+25% Cu.

The strength of the joints bonded through an interlayer of mechanical mixtures of nickel + cobalt increases with increasing the nickel content in the mixture (figure 5 b). With an increase in the cobalt content in mechanical mixtures of the cobalt + copper UDP, the strength of the welded joints rapidly decreases (figure 5 c).

4. Conclusion

Optimum results in welding dissimilar materials (formation of full-strength joints, reduction of temperature and pressure of the process, formation of special properties, etc.) can be obtained using UDPs and mixtures of UDPs produced from formats or mechanical mixtures of formats.

IOP Conf. Series: Materials Science and Engineering **447** (2018) 012034 doi:10.1088/1757-899X/447/1/012034

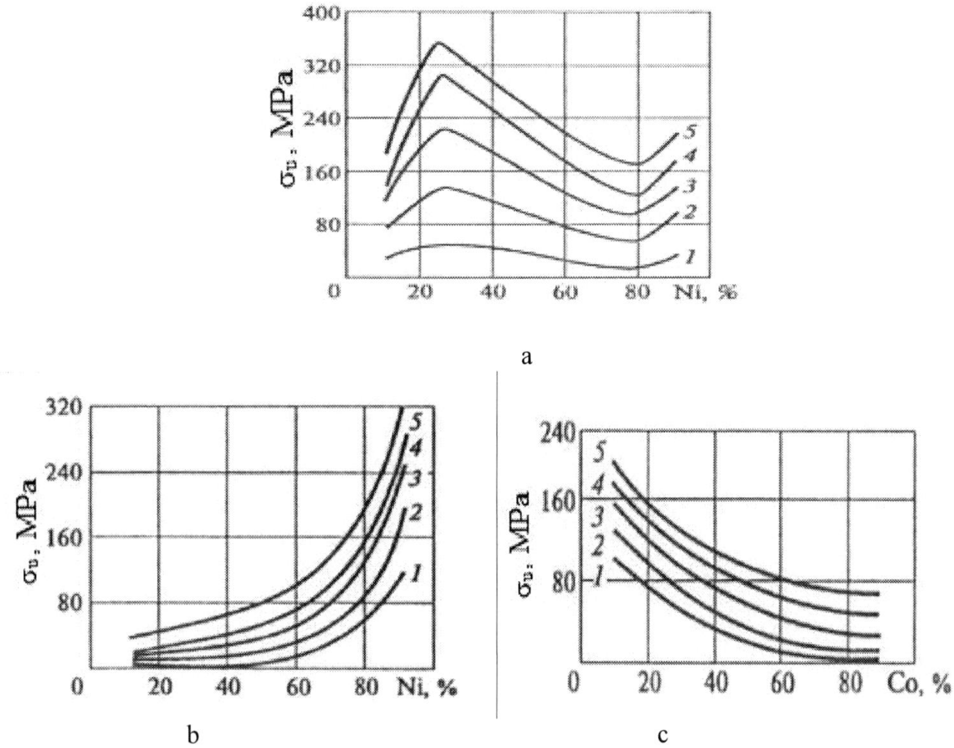

Figure 5. Strength of bonding joints produced through interlayers of UDPs of mechanical mixtures: Ni + Cu (a); Ni + Co (b); Co + Cu (c); 1-5 – T=400, 500, 600, 700 and 800 °C, P=2 кgF/мм^2, t=30 min

References

[1] Liushinskii A V *Diffusion Bonding of Heterogeneous Materials* 2006 (Moscow, IC "Academy") (In Russian)
[2] Liushinskii A V 2011 *The Paton Welding Journal* **5** 31

Deformation behaviour of re-entrant carbon honeycomb structures

L Kh Rysaeva[1], A I Klyavlina[2], F Kh Galiachmetov[1], J A Baimova[1,2], D S Lisovenko[3] and V A Gorodtsov[3]
[1]Institute for Metals Superplasticity Problems of the Russian Academy of Sciences, Khalturina 39, Ufa 450001, Russia
[2]Bashkir State University, Validy Str. 32, Ufa 450076, Russia
[3]Ishlinsky Institute for Problems in Mechanics of the Russian Academy of Sciences, Prospekt Vernadskogo 101-1, Moscow 119526 Russia

E-mail: lesya813rys@gmail.com

Abstract. Recent successes of researchers in the fabrication of new carbon structures of different morphologies and dimensions open up broad prospects for the development of carbon materials with qualitatively new properties. Such promising materials are cellular structures (or carbon aerogels) that have high deformability, good adsorption and conductivity characteristics. In this work, we consider a carbon structure consisting of re-entrant honeycomb cells, which allow the structure to be auxetic. The stability and deformation behaviour of re-entrant carbon honeycomb structures is studied by molecular dynamics simulation. The effect of structural parameters on the stability of carbon aerogel is discussed. It is shown that structural changes significantly depend on the strain rate and the loading direction. At a high strain rate in tension along the *z*-axis, all cells open slightly at the same time. On the contrary, at small strain rates, certain cells open, while others almost completely collapse, which is expected to result in changes in physical properties, for example, in conductivity. During tension along the *x*-axis, a transformation from a re-entrant honeycomb to a traditional honeycomb took place.

1. Introduction

Various carbon nanostructures with complex geometries are of great interest because of novel physical, mechanical and electronic properties. Very promising results have been obtained for three-dimensional carbon structures of different morphologies, such as crumpled graphene [1, 2, 3], carbon aerogels [4], peapods [5], diamond-like phases [6, 7] etc. Graphene aerogels or cellular honeycomb structures comprise a promising new class of materials, since they exhibit nanoscale mechanical properties of graphene, while also achieve new properties that can be tuned by designing their morphology. Among such structures, cellular materials based on graphene flakes are of interest [4, 8, 9, 10]. The idea of creating such morphology is quite simple and inspired by the morphology of graphene itself - carbon atoms stacked in a hexagonal lattice. As it was shown experimentally [8, 11], these structures are stable and can be obtained on the basis of graphite. One of the important advantages of aerogel structures is a high rate of gas adsorption, which makes it possible to predict their usage in the storage and transportation of hydrogen [8, 11, 12]. In addition to chemical properties, their mechanical properties play an important role in the use of cellular structures in various nanodevices. For example, it was shown that cellular structures exhibit unique strength and ultra-high extensibility [10, 13, 14].

These structures can be presented, for instance, as a system of stacked graphene nanoribbons. Among the previously studied carbon aerogels are structures with square [4] and honeycomb [8, 11, 12] cells, while many other opportunities to combine graphene nanoribbons to a complex architecture are not considered.

In the present work, the effect of structural features on the stability of the re-entrant honeycomb cellular structure is investigated by molecular dynamics (MD) simulation. The simulation is carried out to study the deformation behaviour of this structure under one-dimensional tensile loading.

2. Simulation details

The initial structure of re-entrant honeycomb cells is created using a specially developed code that allows combining nanoribbons into a structure with the main required parameters such as the angle of inclination θ, the size of the cell sides l, h and the width of the nanoribbon a.

Figure 1 a shows a part of the initial structure with the main geometrical parameters. To create a cellular structure, a carbon nanoribbon of the "armchair" type is chosen. When constructing the simulation cell, the following parameters were used: $\theta = 60°$ and $l/h = 0.5$. These parameters were proposed in [15], where auxetic materials (i.e. materials with a negative Poisson's ratio [16, 17, 18]) of the same structural configuration were investigated. Values of the nanoribbon width in the y direction from 12 Å to 60 Å are considered. It is found that no significant changes in the relaxation behavior of the structures occur. Thus, $a = 12$ Å is considered below.

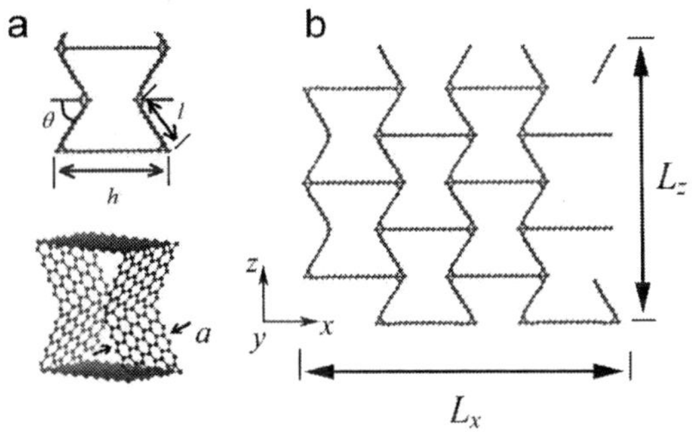

Figure 1. An example of a cellular structure (a) is the elementary cell in the projection onto the xz plane and in perspective (θ is the slope angle, h and l are the sides length, and a is the nanoribbon width), (b) the periodicity cell used in the projection on to the xz plane.

All the structures and their properties have been studied using the freely distributed program package LAMMPS with the interaction between carbon atoms defined by the well known interatomic interaction potential AIREBO [19]. Periodic boundary conditions are applied to the x, y and z directions. The periodicity cell in projection to the xz plane is shown in figure 1 b. At first, the structure is minimized for 20 ps to an equilibrium state with a minimal potential energy. A Nose-Hoover thermostat is used with a target temperature of 1 K with a constant number of atoms and standard velocity-Verlet time integration with a timestep of 1 fs.

To study the deformation behaviour, the uniaxial tensile stress $\sigma_x(\sigma_z)$ is applied to the structure along the x or z axes and the corresponding stress components $\sigma_z(\sigma_x)$ are calculated. The range of elongation rates $\dot{\varepsilon}$ from 2 Å/ps to 0.1 Å/ps is considered, but only two typical elongation rates are presented: $\dot{\varepsilon} = 1$ Å/ps and $\dot{\varepsilon} = 0.1$ Å/ps. The length of the simulation cell L_z (L_x) along the $z(x)$-axis allows one to find the final strain $\varepsilon = (L_{z(\text{final})}-L_{z(0)})/L_{z(0)}$ and track the change in the cell size. The stress component σ_y does not change in the process of deformation, remaining close to zero. The stress component σ_z for tension along z - axis and σ_x for tension along the x - axis changes linearly during the deformation process. Therefore, $\sigma_z(\sigma_x)$ and σ_y are not presented.

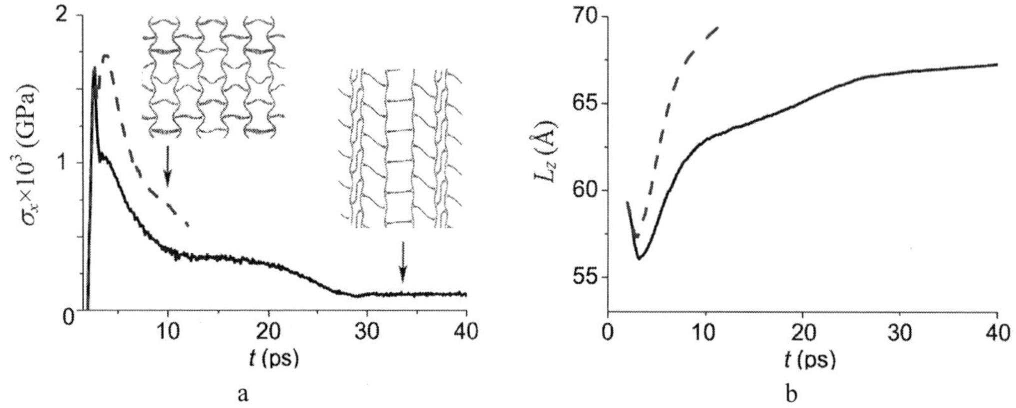

Figure 2. The stress σ_x (a) and the length of the simulation cell along the z axis (b) as a function of time for two strain rates: $\dot{\varepsilon} = 0.1$ Å/ps (black solid line) and $\dot{\varepsilon} = 1$ Å/ps (blue dashed line).

3. Results and discussion

3.1. Tension along z direction

In figures 2 a, b, the stress σ_x and the length of the simulation cell along the z axis are presented as a function of time for $\dot{\varepsilon} = 1$ Å/ps and $\dot{\varepsilon} = 0.1$ Å/ps. As can be seen from figure 2 a, a sharp increase in the stresses σ_x occurs during the first 2 ps, which is typical for all deformation rates. The growth of stresses along the x axis until 2 ps is accompanied by a slight decrease in the cell size in the z-axis which means the relaxation process. After 2 ps, a gradual relaxation of σ_x takes place for at least 20 ps. The structure of re-entrant honeycomb in the projection to the xz plane after deformation at two strain rates, shown by arrows on the graph, is presented in figure 2 a. At high strain rates (1 Å/ps), the structure is stretched uniformly, while at small strain rates ($\dot{\varepsilon} = 0.1$ Å/ps), the deformation occurs non-uniformly in each layer, which leads to a transformation of the structure - some cells open and others collapse. Reducing the strain rate leads to the relaxation of stresses during the stretching process. For $\dot{\varepsilon} = 0.1$ Å/ps, both covalent and Van-der-Waals interactions are taken into account, since atoms in a slower regime can easily interact with the neighbors that is close to the structural relaxation. It can be concluded that at too high strain rates such relaxation does not occur, which has an effect on the final result. As a result, it is possible to obtain a structure with the required parameters (uniformly or non-uniformly stretched cells).

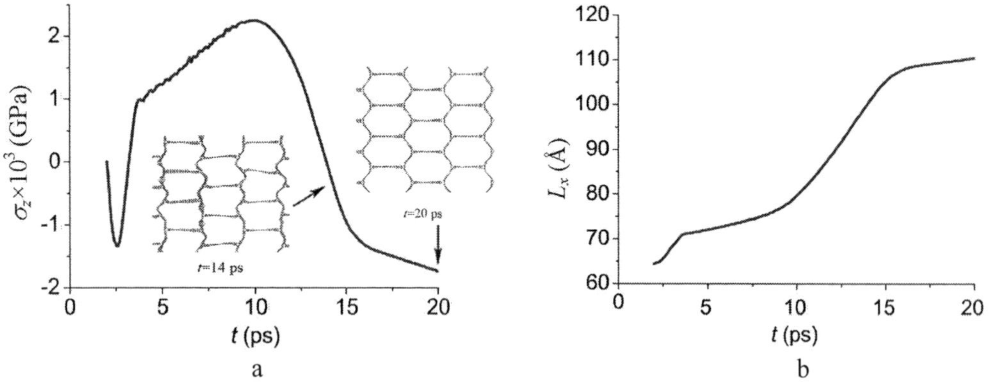

Figure 3. The stress σ_z (a) and the length of the simulation cell along the x axis (b) as a function of time for $\dot{\varepsilon} = 2.3$ Å/ps.

3.2. Tension along x direction

Here, the behavior of the structure under uniaxial tension along the *x*-axis is presented for $\dot{\varepsilon}$ =2.3 Å/ps. In figures 3a, b, σ_z and the size of the simulation cell along *x*-axis as a function of time is shown. The corresponding structural states during deformation are also shown in figure 3 a. As it can be seen, re-entrant honeycombs transform to squares in about 13-14 ps, and then to traditional honeycomb cells after tension in 16-17 ps. In the initial state, before 2 ps, the relaxation of the structure takes place, while after that an abrupt increase of the stress can be seen, which is connected with the transformation of are-entrant honeycomb to its traditional form as described above. After transformation to almost square cells (at about 10 ps), the stress relaxation can be seen (for 11-15 ps). Then, all structural changes are completed, and further deformation will lead to a fraction of the carbon aerogel.

4. Conclusions

To summarize, MD simulations are performed to investigate the stability and deformation behavior of the class of graphene superstructure aerogels with an emphasis on structural parameters. The most favourable angle of inclination, equal to $\theta = 60°$ with a favourable ratio *l/h,* equal to 0.5, is chosen. It is shown that the width of the nanoribbon at periodic boundary conditions has no effect on the stability and properties of the honeycomb structure at periodic boundary conditions.

It is noted that the deformability of the structure depends on the strain rate. At too small strain rates, uniformly stretched cells appear, while tension along the *z*-axis with a high strain rate leads to a redistribution of stresses in the structure and the formation of a banded morphology, where two types of cells are obtained. The tension along the *x*-axis leads to the transformation of the re-entrant honeycombs to the stretched honeycomb cells.

Acknowledgments

Initial simulation setup was conducted by D.S.L. and V.A.G. and supported by RFBR, grant 16-01-00325 and Government Assignment AAAA-A17-117021310373-3. Stability analysis was conducted by J.A.B. and supported by grant of the President of the Russian Federation for state support of young Russian scientists MD-1651.2018.2. Calculations were conducted by R.L.Kh. and supported by the RSF, grant 14-13-00982. Work is conducted in the frame of the program of fundamental researches of Government Academy of Sciences of IMSP RAS.

References

[1] Korznikova E A, Baimova J A, Dmitriev S V, Korznikov A V and Mulyukov R R 2014 *Rev. Adv. Mater. Sci.* **39** 92
[2] Baimova J A, Korznikova E A, Dmitriev S V, Liu B and Zhou K 2014 *Rev. Adv. Mater. Sci.* **39** 69
[3] Baimova J A, Liu B, Dmitriev S V, Srikanth N and Zhou K 2014 *Phys. Chem. Chem. Phys.* **16** 19505
[4] Morris B, Becton M and Wang X 2018 *Carbon* **137** 196
[5] Katin K P, Prudkovskiy V S and Maslov M M 2016 *Physica E* **81** 1
[6] Belenkov E A and Greshnyakov V A 2017 *Letters on Materials* **7** 318
[7] Belenkov E A, Brzhezinskaya M M and Greshnyakov V A 2014 *Diam. Rel. Mater.* **50** 9
[8] Krainyukova N V 2016 *J. Low Temper. Phys.* **187** 90
[9] Wu M, Wu X, Pei Y, Wang Y and Zeng X C 2011 *Chem. Commun.* **47** 4406
[10] Zhang Z, Kutana A, Yang Y, Krainyukova N V, Penev E S and Yakobson B I 2017 *Carbon* **113** 26
[11] Krainyukova N V and Zubarev E N 2016 *Phys. Rev. Lett.* **116** 055501
[12] Antipina L Y, Avramov P V, Sakai S, Naramoto H, Ohtomo M, Entani S, Matsumoto Y and Sorokin P B 2012 *Phys. Rev. B* 86 085435
[13] Pang Z, Gu X, Wei Y, Yang R and Dresselhaus M S 2016 *Nano Lett.* **17** 179

IOP Conf. Series: Materials Science and Engineering **447** (2018) 012035 doi:10.1088/1757-899X/447/1/012035

[14] Gu X, Pang Z, Wei Y and Yang R 2017 *Carbon* **119** 278
[15] Grima J N, Oliveri L, Attard D, Ellul B, Gatt R, Cicala G and Recca G 2010 *Adv. Eng. Mater.* **12** 855
[16] Goldstein R V, Lisovenko D S, Chentsov A V and Lavrentyev S Y 2017 *Letters on Materials* **7** 81
[17] Goldstein R V, Gorodtsov V A and Lisovenko D S 2016 *Phys. Status Solidi B* **253** 1261
[18] Goldstein R V, Gorodtsov V A, Lisovenko D S and Volkov M A 2015 *Letters on Materials* **5** 409
[19] Stuart S J, Tutein A B and Harrison J A 2000 *J. Chem. Phys.* **112** 6472

IOP Conf. Series: Materials Science and Engineering **447** (2018) 012036 doi:10.1088/1757-899X/447/1/012036

Mechanical properties of cryogenically-rolled type 321 metastable austenitic steel

A F Aletdinov[1], **G F Korznikova**[1], **S Yu Mironov**[2], **R G Zaripova**[3], **M M Myshlyaev**[4,5]

[1] Institute for Metals Superplasticity Problems, Russian Academy of Science, Ufa 450001, Russia
[2] Department of Materials Processing, Graduate School of Engineering, Tohoku University, Sendai 980-8579, Japan
[3] Ufa State Aviation Technical University, Ufa 450000, Russia
[4] Baikov Institute of Metallurgy and Material Science, Russian Academy of Science, Moscow 119334, Russia
[5] Institute of Solid State Physics, Russian Academy of Sciences, Chernogolovka 142432, Russia

E-mail: korznikova@anrb.ru

Abstract. In this work, the mechanical behaviour of cryogenically-rolled type 321 metastable austenitic steel was studied. For this purpose, the material was rolled at a temperature of liquid-nitrogen to degrees of rolling reduction, the evolved grain structure was characterized by the electron-backscatter diffraction (EBSD) technique and mechanical properties were evaluated by tensile tests. It has been shown that rolling to 10 pct. of thickness reduction resulted in the extensive strain-induced martensitic transformation thus giving rise to abrupt material strengthening as well as essential degradation of ductility. However, with a further increase in the rolling strain, the volume fraction of martensite tended to saturate thereby leading to stabilization of the mechanical characteristics.

1. Introduction

The possibility of a significant improvement of mechanical properties provides considerable interest in the development of ultrafine-grained (UFG) materials. Typically, this is achieved by applying severe plastic deformation methods [1], but these methods are laborious and difficult to use for manufacturing of commercial-sizable products. However, in metastable austenitic steels, the UFG structure can be obtained by a combination of conventional cold rolling followed by annealing. This approach is based on the effect of the strain-induced martensitic transformation occurring during cold rolling and the subsequent austenite reversion occurring during appropriate heat treatment [2]. It is believed that the martensitic transformation essentially enhances the grain-refinement effect and an UFG structure can be produced at relatively low strain achievable during conventional rolling [3]. Moreover, lowering the deformation temperature down to the cryogenic range may additionally improve the grain-refinement efficiency. Specifically, it should (a) enlarge the driving force for both the strain-induced martensitic transformation and the thermal-assisted austenite reversion, and (b) stimulate mechanical twinning. In light of these advantages, the cryogenic processing of metastable austenitic steels has recently attracted considerable interest [4,5].

In the previous work, the effect of cryogenic rolling on the microstructure of type 321 metastable austenitic steel was studied in detail [5]. As expected, the imposed strain contributed to martensitic transformation which preferentially nucleated in the most-heavily strained areas, i.e. in the deformation bands. This, however, stagnated the normal development of the dislocation boundaries in *austenite* and thus retarded the grain refinement in this phase. The martensitic transformation was found to follow the well-known Kurdjumov-Sachs (K-S) orientation relationship, viz. $\{111\}\gamma//\{110\}\alpha$ and $<110>\gamma//<111>\alpha$, and was characterized by a noticeable variant selection. The present work was undertaken to evaluate the mechanical properties of cryogenically-rolled material and elucidate the relationship between the microstructure and properties.

2. Material and Experimental Procedures

The program material comprised type 321 metastable austenitic stainless steel with chemical composition shown in table 1. The obtained material was rolled to 85% of hickness reduction at 950°C and then annealed at 1200 °C for 1 hour. This produced a fully recrystallized austenitic grain structure with a mean grain size of ~100 μm and a high fraction of annealing twins. This material condition is denoted throughout as the base material.

Table 1. Chemical composition of type 321 metastable austenitic stainless steel, wt%

Fe	C	Cr	Ni	Ti	Si
balance	0.12	18.6	10.2	0.7	0.76

The material was cryogenically rolled to 5, 10, 15, 23, 30 и 45 pct. of total thickness reduction. To ensure the conditions of cryogenic deformation, the rolling preform was soaked in liquid nitrogen and held for 15 minutes prior to rolling. Rolling was performed in a single pass, using a rolling speed of 160 mm/s in a cluster mill with working rolls with a diameter of 65-mm. The rolling stain to 45 pct. of reduction led to the significant adiabatic heating and concomitant loss of the cryogenic condition. Microstructural characterization of the cryogenically-rolled material was performed by electron backscatter diffraction (EBSD) and focused on the mid-thickness rolling plane. To provide reasonable statistics for measurements, a relatively large scan step size of 1 μm was used. The α′-martensite was indexed as a body-centered-cubic phase. To eliminate spurious boundaries caused by orientation noise, the lower limit boundary-misorientation cut-off of 2° was employed. A 15° criterion was applied to differentiate low-angle boundaries (LABs) from high-angle boundaries (HABs). Mechanical behavior was evaluated using tensile tests. The tests were conducted at ambient temperature and a nominal strain rate of 10^{-3} s^{-1} using an Instron 5982 testing machine.

3. Results and discussion

The effect of cryogenic deformation on the grain structure is shown in figure 1, on the EBSD maps HABs are depicted as black solid lines. The key microstructural characteristics derived from the maps is summarized in table 2. Taking into account that a relatively coarse scan step size was used in the present study (1 μm), the measured martensite volume fraction, as well as the LAB fraction were most likely overestimated. Nevertheless, the obtained results are believed to *qualitatively* reflect real trends in the microstructure evolution.

The cryogenic rolling to 5 pct. of reduction provided only a minor effect on the grain structure, which was still dominated by relatively coarse austenite grains containing annealing twins (figure 1a). However, the increase in rolling reduction to 10 pct. resulted in principal changes. Specifically, the LAB fraction abruptly increased (table 2) thus evidencing a significant acceleration of the deformation-induced processes. Moreover, the evolved microstructure was characterized by a significant proportion of α' and ε martensite (figure 1b, table 2) thus indicating the strain-induced martensitic transformation. Two characteristic morphologies of α' martensite can be distinguished, i.e.,

(a) regular arrays of nearly parallel bands within austenite grains and (b) relatively coarse regions resembling the original grains. The first morphological type of martensite suggested the preferential development of phase transformation in the deformation bands (or mechanical twins), whereas the second morphological type may be interpreted as a plane-view of the $\alpha^{/}$-transformed deformation bands.

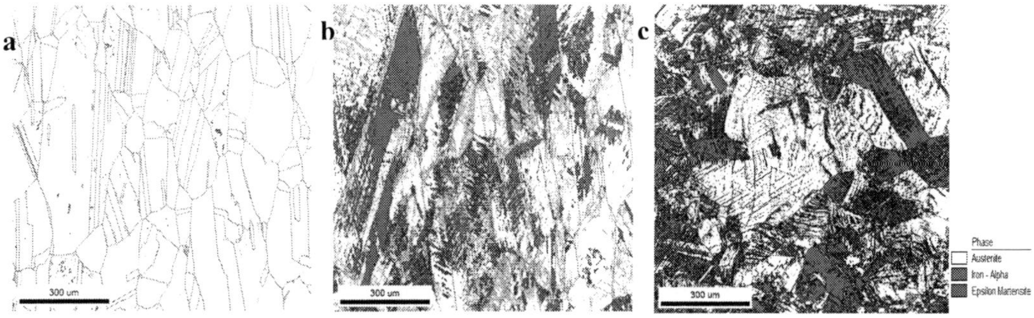

Figure 1. Selected portions of EBSD grain-boundary maps showing grain structure and phase composition after cryo-rolling to (a) 5%, (b) 15%, (c) 30%. The phase color code is shown in the bottom right corner of the figure.

Table 2. Phase composition, area fractions of HABs LABs, after cryo-rolling.

Roll reduction	5%	10%	15%	23%	30%
Austenit	0998	0.69	0.739	0.723	0.619
α'-martensite	0.001	0.232	0.205	0.205	0.354
ε-martensite	0.001	0.078	0.056	0.072	0.027
HABs	0.86	0.477	0.462	0.327	0.309
LABs	0.14	0.518	0.536	0.669	0.691

A further increase in strain led to only minor changes in the microstructure morphology (figure 1c), the LAB fraction and the proportions of the martensitic phases (table 2), thus suggesting a stagnation in the microstructure evolution. Assuming the preferential nucleation of martensite in the deformation bands, as suggested above, the revealed saturation of its volume fraction may be explained in terms of depletion of the volume fraction of the deformation bands. After 30 pct. of accumulated strain, a subtle increase in the content of the $\alpha^{/}$ martensite at due to the ε martensite is only worthy of remark (table 2). This may indicate a $\varepsilon \rightarrow \alpha^{/}$ transformation.

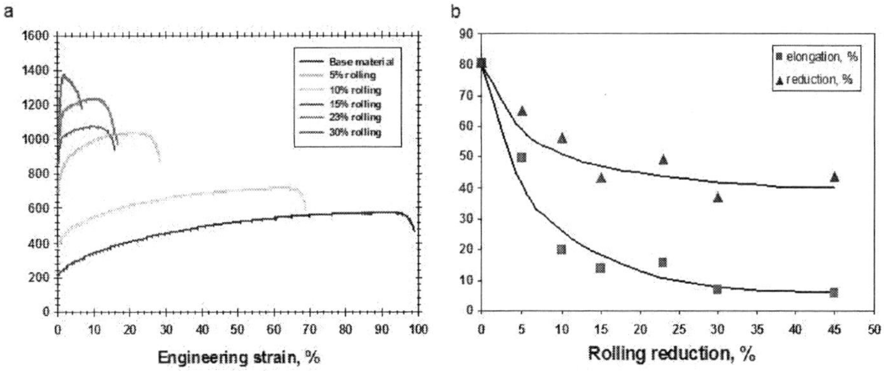

Figure 2. Stress-strain curves of type 321 austenitic stainless steel with different cryo-rolling strains (a) and variation of reduction and elongation with cryo-rolling strain (b).

Tensile diagrams of the cryo-rolled material are summarized in figure 2 a. The effect of the rolling strain on the ductility and yield stress are shown in figures 2 b and 3 a, respectively. To quantify the effect of the rolling strain on the hardening efficiency, the strengthening rate was calculated as $h = \Delta\sigma/\Delta e$, where $\Delta\sigma$ is the increment in yield stress, and Δe is the increase in rolling strain, the obtained results are shown in figure 3b.

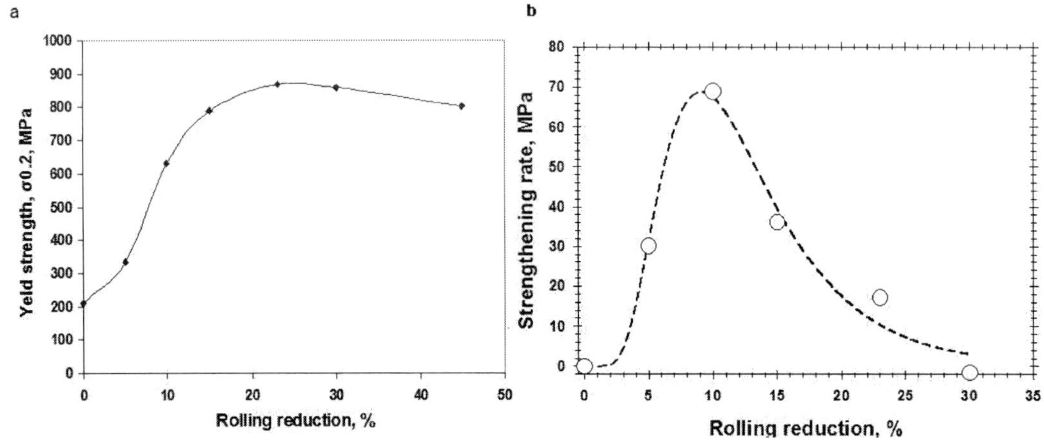

Figure 3. Variation of yield stress (a) and strengthening rate (b) with cryo-rolling strain of type 321 austenitic stainless steel.

As expected, the rolling strain resulted not only in an essential degradation of ductility (figure 2), but also in a significant strengthening of the material (figures 2 a and 3 a). Remarkably, both effects tended to saturate after the rolling strain of 10 pct. (figures 2 b and 3 a). The saturation of the strengthening effect is also clearly seen in figure 3 b. It is evident that rolling to 10 pct. of strain provided the maximal strengthening effect, whereas a further increase in the rolling strain resulted in abrupt reduction of the hardening rate. All these observations are in excellent agreement with the evolution of the total martensite volume fraction (table 2) thus evidencing the dominant role of martensite in the microstructure-property relationship.

4. Summary

The results presented in this paper provide a strong evidence that rolling at cryogenic temperature promotes the strain-induced martensitic transformation, which preferentially nucleated in the deformation bands and resulted not only in a drastic (fourfold!) strengthening effect, but also in a dramatic degradation of ductility. It was found that the martensitic transformation began after the rolling strain to 10 pct. of thickness reduction, but the martensite volume fraction rapidly achieved a nearly saturation level and a further microstructure evolution was only characterized by the transformation of ε martensite to α' martensite. The mechanical behavior broadly mirrored the evolution of the martensite volume fraction thus evidencing the dominant role of this phase in the microstructure-properties relationship.

Acknowledgements

The present work was partially supported by the Russian Fund for Fundamental Research (project No.17-42-020426). The authors would also like to thank P. Klassman for technical assistance during cryogenic rolling.

References

[1] Hirsch J, Lucke K and Hatherly M 1988 *Acta Metall.* **36** 2905
[2] Song R, Ponge D, Raabe D, Speer J G and Matlock D K 2006 *Mater. Sci. Eng. A* **441** 1

[3] Nakada N, Ito H, Matsuoka Y, Tsuchiyama T and Takaki S 2010 *Acta Mater.* **58** 895
[4] Wang T S, Peng J P, Gao Y W, Zhang F C and Jing T F 2005 *Mater. Sci. Eng. A* **407** 84
[5] Korznikova G F, Mironov S, Konkova T, Aletdinov A F, Zaripova R G, Myshlyaev M M and Semiatin S L 2018 *Mater. Character.* (to be published)

IOP Conf. Series: Materials Science and Engineering **447** (2018) 012037 doi:10.1088/1757-899X/447/1/012037

Characteristics of cathode materials for solid oxide fuel cells with mixed conductivity

K A Kuterbekov [1], K Zh Bekmyrza [1], A V Nikonov [2], N K Aydarbekov [1]

[1]L.N. Gumilyov Eurasian National University, 2 Satpaev st., Astana 010008, Kazakhstan

[2]Institute of Electrophysics, Ural Branch of the Russian Academy of Sciences, 106 Amundsen st., Yekaterinburg 620216, Russia

E-mail: kbekmyrza@yandex.kz

Abstract. In this article, we define promising materials for creating the structure of a non-porous cathode-electrolyte. The synthesis of initial powders by the method of self-propagating high-temperature synthesis is studied and described. The specific surface area, morphology and phase composition of powders were investigated.

1. Introduction

When developing solid oxide fuel cells (SOFCs), an important task is to increase their power, which depends on the internal resistance consisting of ohmic resistances of the electrolyte and electrodes, as well as the polarization losses caused by the processes on the electrodes. The generally accepted structure of a SOFC consists of a dense layer of an electrolyte separating the porous layers of electrodes. In this structure, electrode reactions occur on the three-phase (electrolyte-electrode-gas) boundary, and the values of polarization resistances depend on the structure of electrodes (composition, porosity of the electrolyte-electrode boundaries, etc.).

One can avoid many difficulties related to the formation of gas-diffusion electrodes and increase the specific power of SOFCs by forming a new nonporous cathode-electrolyte structure, in which the electrode reaction zone would be located on the electrode-gas interface. The main condition for the realization of the proposed structure is a high mixed electron-ion conductivity of the cathode material. In addition, the cathode material must have stable chemical, phase, dimensional properties in an oxidizing environment, chemical compatibility with the electrolyte material and the value of thermal expansion coefficient (TEC) similar to the one of the electrolyte. Also, to enable manufacturing the cathode-electrolyte structure by means of simultaneous sintering, the shrinkage on sintering of the cathode material and the electrolyte should be similar.

Finding a material that meets all the above requirements is not an easy task. Especially, considering that all characteristics of materials are determined by their chemical composition and structure and, therefore, are interrelated, it is not possible to change one parameter without affecting the others. A possible way to solve this contradiction is to create a composite cathode material. Thus, it is possible to increase the ionic conductivity and make the TEC of the cathode close to the one of the electrolyte.

2. Selection of cathode materials for research

Basing on the literature review, we have determined the following promising materials for developing the structure of a nonporous cathode-electrolyte: $SmSrCo_2O_{5+\delta}$ and $NbBa_{0.5}Sr_{0.5}Mn_{0.5}O_{5+\delta}$. However, additional studies have shown that these materials interact with ZrO_2 electrolyte stabilized with Y_2O_3 (YSZ). Therefore, basing on the whole set of parameters, such as electronic and ionic conductivity, TEC, surface exchange coefficient (*k*), and interaction with electrolyte YSZ, the following compositions were chosen for further investigation: materials with the perovskite structure $Pr_{1-x}Sr_xFe_{1-y}Co_yO_3$, (x = 0, 0.2, 0.3, y = 0.2, 0.5) and materials with the Ruddlesden-Popper structure $La_2Ni_{1-x}Co_xO_4$ (x = 0, 0.1, 0.2, 0.3) and Pr_2NiO_4. Characteristics of some of the selected compositions are presented in table 1. It should be noted that the determination of the ionic conductivity is a rather difficult task, therefore literature data on this parameter are rare. In this regard, the table shows the oxygen diffusion coefficient *D**, which is related to the ionic conductivity σ_i through the Nernst-Einstein equation.

Table 1. Characteristics of cathode materials (at 800 °C) [1-3].

Material	σ_e, S/cm	σ_i, S/cm	D*, cm^2/s	k, cm/s	TEC ($\times10^{-6}$), K^{-1}
$Pr_{0,8}Sr_{0,2}Fe_{0,8}Co_{0,2}O_3$	76	0.0015			12.8
$Pr_{0,7}Sr_{0,3}Fe_{0,8}Co_{0,2}O_3$	171				12.4
La_2NiO_4	47		1.2×10^{-7}	1.7×10^{-6}	13.0
Pr_2NiO_4	94		1.2×10^{-7}	3.1×10^{-6}	13.6
$La_2Ni_{0.8}Co_{0.2}O_4$			9.8×10^{-8}	2.2×10^{-6}	11.3

In addition, a nonporous cathode-electrolyte structure from a perovskite-based composite $La_{0.7}Sr_{0.3}MnO_3$ (LSM) also was fabricated. This material does not have mixed ion-electron conductivity, is a well-studied and widely used cathode for high-temperature SOFCs. Thus, the characteristics of this structure will be the reference point for all studies.

3. Experimental

The ZrO_2 electrolyte powder, stabilized with 8.5mole%Y_2O_3 (8.5YSZ), was obtained by the laser evaporation method at the Institute of Electrophysics, Urals Branch of the Russian Academy of Sciences [4]. Powders of cathode materials were obtained by self-propagating high-temperature synthesis. The starting reagents were hydrogen carbonates and nitrates of the corresponding elements.

The specific surface area of the powders was determined by the Brunauer-Emmett-Teller (BET) method on a TriStar 3000 automatic analyzer. The morphology of the powders was examined using a LEO 982 electron microscope, and the phase composition was studied on a D8 DISCOVER diffractometer.

The sintering kinetics of the electrolyte and cathode materials was studied in an air atmosphere in the temperature range of 20-1500 °C using a Dil 402C dilatometer at a heating rate of 5 °C/min. Using the same instrument, the TEC of the compositions under study was determined in the temperature range of 20-1000 °C. To measure the electrical conductivity, the powders were pressed to bars of a rectangular cross section, sintered to a density close to the theoretical one (the sintering regime was determined from the dilatometric data), and platinum wire probes were deposited on the samples. The electrical conductivity measurements were carried out by a 4-probe method at a direct current in the temperature range of 20-800 °C in the air atmosphere using a Solartron Sl-1260/1267 impedancemeter.

4. Results and discussion

The initial nano-sized powder of electrolyte 8.5YSZ was characterized by a specific surface S_{BET} = 40.1 m^2/g, which corresponds to an average particle size <d> = 21 nm. In the following, we will denote this powder YSZ_{20}. To change the sintering kinetics, YSZ_{20} was heated at 1000 °C for 4 hours. The resulting powder YSZ_{20} was characterized by S_{BET} = 22.8 m^2/g (<d> = 45 nm). The results of dilatometric studies of samples made of YSZ_{20} powders and their mixtures are shown in figure 1.

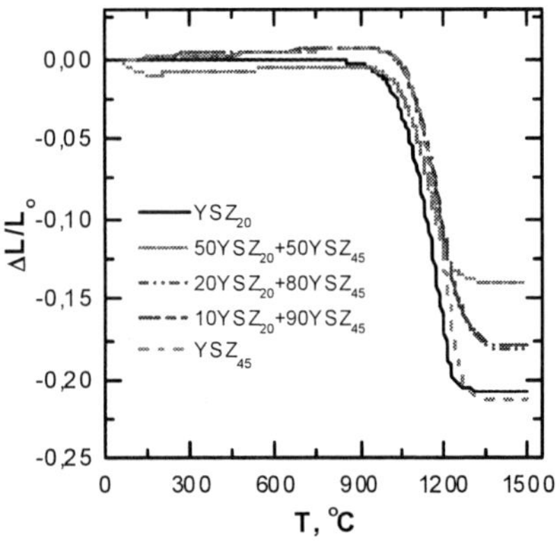

Figure 1. Curves of linear shrinkage of samples made of YSZ_{20} powders and their mixtures (figures in the notations of the mixtures correspond to weight percents.)

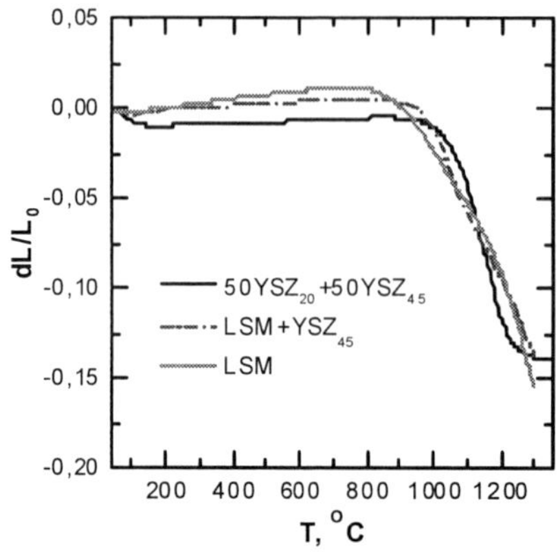

Figure 2. Curves of linear shrinkage of electrolyte material, composite LSM + YSZ45 and cathode material LSM/

From a comparison of the shrinkage curves for the cathode material based on LSM and electrolyte materials with different dispersions, it was determined that the composition $50YSZ_{20}$ + $50YSZ_{45}$ is the most suitable for realizing simultaneous sintering (figure 2).

5. Conclusions

On the basis of the literature review, promising cathode materials were determined for the development of a nonporous cathode-electrolyte structure: perovskites based on praseodymium ferrite and lanthanum and praseodymium with the Ruddlesden-Popper structure. The initial powders were synthesized by the method of self-propagating high-temperature synthesis. From a comparison of the shrinkage curves for the cathode material based on LSM and electrolyte materials with different dispersions, it was determined that the composition $50YSZ_{20}$ + $50YSZ_{45}$ is most suitable for realization of simultaneous sintering.

Acknowledgments

The present work was supported by the Ministry of Education and Science of the Republic of Kazakhstan within the program of funding of research activities in 2017-2020 «Development of Hydrogen Energy Technologies in the Republic of Kazakhstan» (grant No. BR05236795).

References

[1] Ullmann H, Trofimenko N, Tietz F, Stöver D and Ahmad-Khanlou A 2000 *Solid State Ionics* **138** 79
[2] Serra J M, Vert V B, Betz M, Haanappel V A C, Meulenberg W A and Tietz F 2008 *J. Electrochem. Soc.* **155** B207
[3] Boehm E, Bassat J-M, Dordor P, Mauvy F, Grenier J-C and Stevens Ph 2005 *Solid State Ionics.* **176** 2717
[4] Ivanov M, Osipov V, Kotov Yu, Lisenkov V, Platonov V and Solomonov V 2006 *Adv. Sci. Technol.* **45** 291

IOP Conf. Series: Materials Science and Engineering **447** (2018) 012038 doi:10.1088/1757-899X/447/1/012038

Effect of precipitates on static and fatigue strength of a severely forged aluminum alloy 1570C

E V Avtokratova[1], O Sh Sitdikov[1], O E Latypova[1], M V Markushev[1], M L Linderov[2], D L Merson[2] and A Yu Vinogradov[2,3]

[1] Institute for Metals Superplasticity Problems, Russian Academy of Sciences, 39 Khalturin st., Ufa 450001, Russia
[2] Institute of Advanced Technologies, Togliatti State University, 14 Belorusskaya st., Togliatti 445020, Russia
[3] Department of Mechanical and Industrial Engineering, Norwegian University of Science and Technology—NTNU, N-7491 Trondheim, Norway

E-mail: mvmark@imsp.ru

Abstract. Bulk billets from commercial ingots of the alloy 1570C with different size of aluminides of transition metals were subjected to severe deformation up to the equivalent strain e~24 via multi-step isothermal forging (MIF) with an inter-step temperature decrease from 325 to 175 °C. The high potential of MIF to produce UFG billets with enhanced balance of static and fatigue properties has been demonstrated. The role of precipitates of aluminides of transition metals and regimes of the MIF in the structure of the alloy and for the control of properties is discussed.

1. Introduction

Non-heat-treatable Al-Mg alloys with complex additions of transition metals (TM) belong to a modern class of materials for a wide range of commercial applications due to the unique combination of service properties which they possess [1,2]. However, the range of properties, which can be controlled during industrial processing, is limited by two main tempers of semi-finished products, namely, annealed and work-hardened ones. The primary strengthening mechanism involved in the processing of these alloys is dislocation hardening which has an obviously limited capacity. An alternative popular (though admittedly less studied) approach to the improvement of properties of the alloy is based on grain boundary hardening, assuming the formation of ultrafine grain (UFG) and nanocrystalline (NC) structures [3-5]. This approach is facilitated by a fairly high content of TM, which are sufficient to form high densities of aluminides during homogenization and stabilize the resultant deformation microstructures [6]. Such precipitates also play an active role as strengthening agents, providing remarkable hardening potential to the alloys [1]. It has been recently demonstrated that the most attractive method towards extreme enhancement of properties in complex alloys is severe plastic deformation (SPD) (with $e > 1$) combining all specified strengthening mechanisms in a single framework [3-5]. Besides the control of geometrical parameters of the microstructure (grain/subgrain size and shape, volume fraction of the recrystallized grains and etc.), SPD is capable of varying the distribution of second phases and their dispersion in the matrix as on the whole as well as in individual grains. The multi-step isothermal forging (MIF) is one of the modern metal processing techniques that meets the specified requirements along with high efficiency to process UFG/NC bulk billets [3,7].

The present study is aimed at exploring the efficiency of MIF for the structuring and strengthening of the Al-Mg-TM alloys and clarifying the role of aluminides and their distribution in microstructure and mechanical behavior.

2. Material and procedure

In order to change the size of secondary TM aluminides, an ingot of the alloy 1570C of a standard chemical composition (Al-5Mg-0.18Mn-0.20Sc-0.08Zr, wt. %) was subjected to one (at 360 °C, 6h) and two-step (360 °C, 6h + 520 °C, 1h) homogenization processes. In this way, the aluminum solid solution abnormally supersaturated by TM was decomposed and coherent to the matrix. $Al_3(Sc,Zr)$ precipitates (dispersoids) were formed and grown. Further MIF of billets ∅80×150 mm was carried out in three stages with an inter-step decrease in temperature. Each step included several deformation cycles after which the shape and dimensions of the billet were maintained as much as possible. For this purpose, in each cycle, several consecutive settings and drawings were implemented along all axes of the billet [3]. In the first stage, consisting of 12 cycles at 325 °C, the cumulative equivalent strain reached e ~12. In the second and third stages, conducted at 250 and 175 °C, respectively, e was about 6 at each stage. The microstructure of the alloy was studied by means of transmission and scanning electron microscopy (TEM and SEM) and X-ray analysis. Most of the parameters of the aluminum matrix were determined using SEM-EBSD data. The size and density of dispersoids were detected by computer analysis of TEM images of at least 2000 precipitates. Samples for the studies were prepared by mechanical grinding and polishing and electropolishing at -28 °C in 20% solution of HNO_3 in CH_3OH. The microhardness of the alloy was determined at 15 seconds loading of 1 N. Tensile and fatigue tests were carried out at ambient temperature on mechanically pre-polished flat dog-bone and corset-shaped samples with a gage section of 1.5×3×6 mm^3 and a cross-section of 1×0.9 mm^2, consequently.

3. Results and discussion

In contrast to the one-stage homogenization, after which the modal size of the dispersoids and their number density were about 10 nm and 1×10^4 µm^{-3}, respectively, after two-stage annealing their size increased more than two times, with a corresponding reduction in their density to 5×10^3 µm^{-3}. However, the microstructure of the aluminum matrix was almost the same, with the mean size of equiaxed grains of 25 µm. In both billets, new fine subgrains and grains, surrounded by low and high-angle boundaries (HABs) were formed during the very first cycles of the first stage of MIF near the boundaries of initial grains along the free zones of precipitates formed during homogenization.

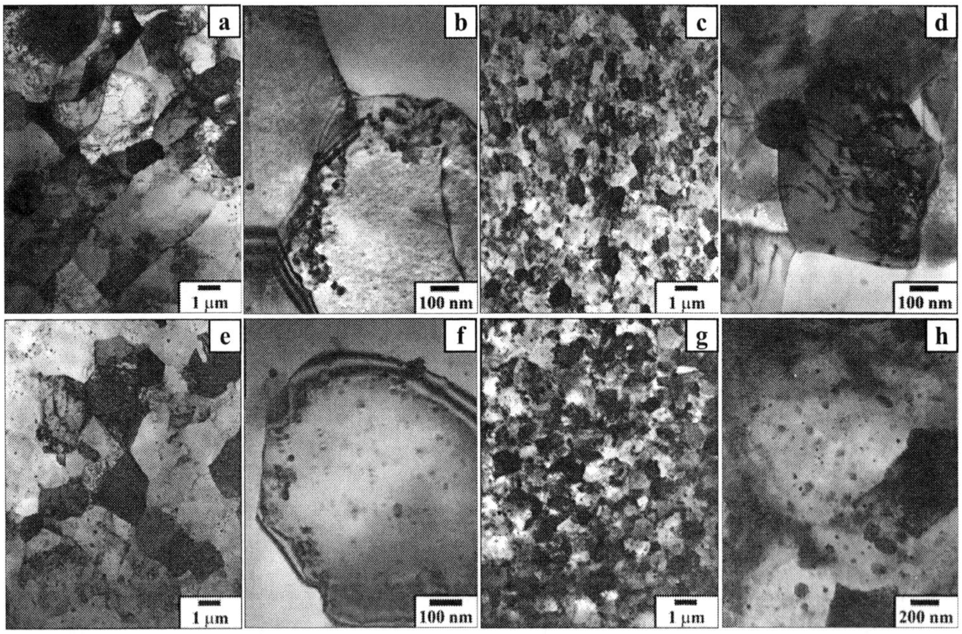

Figure 1. TEM structures of the alloy 1570C after the 1st and the 3rd stages of MIF at 325 (a,b,e,f) and 175 °C (c,d,g,h), consequently. (a-d) fine precipitates, (e-h) coarse precipitates.

With increasing strain, the fraction and misorientation of these grains increased, which led to a gradual replacement of the coarse-grained structure by the fine-grained one, regardless of the initial size of the aluminides (figure 1). The main mechanism for grain refinement was continuous dynamic recrystallization, controlled by the interaction of lattice dislocations and (sub)grain boundaries with dispersoids that prevent migration of boundaries. This process is inevitably accompanied by a rearrangement of dislocation and recovery. The effect of precipitate size on the evolution of microstructure was rather small: in both states of the alloy, the structure was partially recrystallized with almost equal subgrains (in the range of 1.8-2.0 μm) and slightly recrystallized with almost equal subgrains (in the range of 1.8-2.0 μm) and slightly recrystallized with smaller grains in the alloy with smaller precipitates (2.2 and 2.7 μm).

Analysis of TEM and SEM showed that after the third stage of MIF, the structure of both states of the alloy became more refined: the size of the subgrains was almost equal and decreased to ~1 μm, while the grain size of 1.1 and 1.3 μm was found in the alloy with fine and coarse precipitates, respectively. Simultaneously, the structure became less homogeneous and equilibrium. The majority of (sub)grains were surrounded by boundaries with the uneven TEM extinction contrast and contained increased dislocations density (X-ray data indicate an increase by two orders of magnitude). Due to the higher suppression of the dynamic recovery caused by the decrease in the deformation temperature, the structural heterogeneity of the alloy with smaller precipitates was also slightly larger. Besides, when processing at temperatures below the solvus point (~250 °C), rather coarse (50–250 nm) precipitates of the β-phase (Al-Mg) were formed in the crystallites in addition to TM aluminides (figure 1).

Measurements of alloy hardness and tensile strength parameters showed their good correlation with changes in the microstructure discussed above. Thus, coagulation of dispersoids in the second stage of homogenization of the alloy resulted in a decrease in hardness and yield stress from 110 to 90 Hv and from 240 to 180 MPa, respectively, due to a partial loss of the dispersion hardening. A similar difference in the hardness of the alloy and its tensile behavior was observed after the first stage of MIF (table 1). This was due to the fact that high-temperature forging did not lead to the strengthening of the alloy, despite the remarkable grain refinement. The latter can be explained by the compromise of strengthening the grain boundaries by partial softening of the alloy due to a decrease in the hardening of the dispersion caused by coagulation of the precipitates, which was continued during high-temperature straining and inter-cycle annealing. This difference is due to the formation of an almost equilibrium recrystallized structure with low dislocation densities in the alloy states, giving a small work hardening effect, and resulting also in higher ductility of the alloy.

After the third stage of the MIF, the precipitate size effect was accompanied by a noticeable increase in strength, especially yield stress, and an insignificant decrease in ductility (table 1). This behavior was reasonably expected from the processing of the nonequilibrium UFG structure with a smaller number of grains/subgrains, higher dislocation density and β-precipitates.

Table 1. Properties of the alloy 1570C at ambient temperature.

Condition	Vickers Hardness, HV	Yield Stress, MPa	Ultimate Tensile Strength, UTS, MPa	Elongation to failure, El, %
Ingot	110/90[a]	240/180	355/340	28/37
MIF 1st stage	105/95	230/215	355/345	38/36
MIF 3rd stage	125/110	315/285	385/370	32/32

[a]- fine/coarse dispersoids

Thus, based on these observations, it could be concluded that: (i) the grain refinement with a decrease in the grain size by more than an order of magnitude down to the UFG range, caused only a slight hardening effect on the parameters of the static strength, which was simultaneously compensated by a decrease in dispersion strengthening due to TM aluminides coarsening in the high-temperature stage of MIF processing; (ii) MIF, resulted in work-hardened mixtured/bimodal structure, consisting of ultrafine grains and subgrains with increased dislocation density (and virtually without a strong metallographic and crystallographic texture, as frequently observed in conventional hot pressing and rolling), gave a rather high impact in the strengthening of the alloy, especially in yield stress, and (iii) the higher strength of the alloy with finer precipitates is due to the formation of a less regular and less developed UFG structure under the MIF, demonstrating the indirect influence of the dispersion of the strengthening phases. Thus, if we take into account the absolute values of hardness and tensile strength parameters, their increase with decreasing temperature of the MIF is mainly conditioned by: (i) work-hardening due to an increase in dislocation density (ii) structural strengthening due to continuation of the (sub)grain refinement. As for the literature data and our estimates, the dispersion hardening due to the decomposition of the Mg-rich solid solution of aluminum with formation of β-phases is not very large and is usually able to compensate for the loss of only the aluminum solid solution strengthening [8].

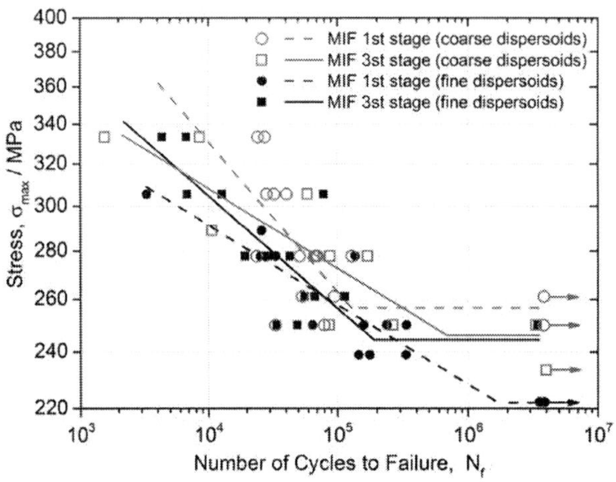

Figure 2. 1570C alloy fatigue dependences after 1st and 3rd MIF stages.

The S-N graphs of the alloy after the first stage of MIF revealed a higher fatigue resistance of the material with larger TM aluminides (figure 2) despite the lower yield stress value (table 1). Considering the conventional endurance and fatigue limit determined in 4×10^6 cycles, the behavior of the strengthened alloy with less precipitation was preferable both in low and high-cycle fatigue regions. However, after the third stage of the MIF, the effect of precipitates disappeared and the alloys in both states showed almost the same behavior.

Virtually, the fatigue behavior of the alloy has the same origin as in static tests, emphasizing the effect of dispersion of TM aluminides, which is especially noticeable in the intensity of dynamic recovery. Thus, this effect is more pronounced in fatigue testing of an alloy with an almost equilibrium fine-grained structure of a matrix, containing different densities of disperse phases. Improvement of fatigue resistance of the alloy by means of fine dispersoids after the third stage of the MIF was controlled by dislocation structures formed during low-temperature forging. However, their formation was also conditioned by the interaction of dislocations with precipitates of both the TM aluminides and the β-phase. This phenomenon should be studied in more detail in order to discuss the nature of the observed behavior of the alloy.

4. Conclusion

The role of precipitates of transition metal aluminides in the formation of UFG structures and the properties of Al-Mg alloys is rather complicated, especially when determining the alloy's resistance to fatigue. However, its origin is one and the same under different loading conditions and is controlled by the mechanisms of interaction of precipitates with moving dislocations and grain boundaries that affect the intensity and kinetics of recrystallization and matrix recovery.

Acknowledgments

The work was supported by the grant No. 16-19-10152 of Russian Science Foundation on the basis of the Shared Service Center of IMSP RAS «Structural and Physical-Mechanical Studies of Materials».

References

[1] Filatov Y, Elagin V and Zakharov V 2000 *Mater. Sci. Eng. A* **280** 97
[2] Avtokratova E, Sitdikov O, Markushev M and Mulyukov R 2012 *Mater. Sci. Eng. A* **538** 386

[3] Sitdikov O, Garipova R, Avtokratova E, Mukhametdinova O and Markushev M 2018 *J. Alloys Compd.* **746** 520
[4] Valiev R and Langdon T 2006 *Prog.Mater. Sci.* **51** 881
[5] Vinogradov A and Estrin Y 2013 *Acta Mater.* **61** 782
[6] Avtokratova E, Sitdikov O, Mukhametdinova O, Markushev M, Murty S V S N, Prasad M J N V and Kashyap B P 2016 *J. Alloys Compd.* **673** 182
[7] Markushev M 2011 *Letters on Mater.* **1** 36
[8] Markushev M and Murashkin M 2004 *Phys. Metals Metallogr.* **98** 221

IOP Conf. Series: Materials Science and Engineering **447** (2018) 012039 doi:10.1088/1757-899X/447/1/012039

Improving the accuracy of finite element modeling of superplastic hemisphere forming

O P Tulupova[1], V R Ganieva[1], A A Kruglov[1,2], F U Enikeev[1]

[1]Ufa State Petroleum Technological University, Ufa, Russia
[2]Institute for Metals Superplasticity Problems RAS, Ufa, Russia

E-mail: box_mail_2011@mail.ru

Abstract. A novel procedure to determine the values of the material constants K and m for the standard power superplastic law from the results of the bulge tests of the hemispheres is suggested. Unlike other approaches known in the literature.theinvolved procedure requires only two experimental measurements of the time of formingunder constant pressure, with the thickness at the pole being used to refine the values *K* and *m*. Experimental justificationof the developed techniques wasfulfilled on the example of aluminum alloyAA5083. The validity of the suggested approach is confirmed by comparing the experimental data with the corresponding finite element solutions of the boundary value problem in the theory of creep productionusing the ANSYS software. It is found that the maximum deviation of experimental records from the corresponding finite element solutions does not exceed 1% both for the time of formation and for the thickness at the pole.

1. Introduction

Hemispheres are widely used in various industries to produce spherical pressure vessels by welding two hemispheres. For such vessels, high demandsare placed on the reliability and the structural strength; therefore, they are made of titanium alloys and stainless steel. The most promising method for the manufacture of hemispherical products is superplastic forming (SPF)of sheet metals. Instead of labor-intensive, full-scale experiments on thestudy of the formation ofhemispheres, mathematical modeling,in particular, computer models based on the finite element method,is effectively used. The degree of adequacy of the computer model depends on the accuracy of determining the values of the constants of the material included in the determining ratio. The most commonly used in practice is the standard power law of superplasticity [1]:

$$\sigma = K\xi^m \tag{1}$$

where σ is the flow stress, ξ is the strain rate, K and m are the material constants to be determined experimentally.

In [2], we proposed an approach that enables determining the height of the dome attwo experimental points of length forming the hemispheres with elevated gas pressure values. In [3], the well-known technique was modified and made it possible to determine the height of the dome thickness at the pole domefor two experimental points at the same time depending on the duration of the hemisphere formation.

According to the method proposed in [2], the material constants K and m included in (1) are determined by the results of two technological experiments, which are a set of experimental data: $\{t_i, p_i, h_i, s_i\}$, i=1, 2, where t_i, the duration of the hemisphere forming at constant pressure p_i=*const* to the height of the dome h_i= *const* and the thickness at the pole s_i.

2. Fitting of the material constants

The method of three stages presented in [2] is a reverse analysis with a consistent refinement of the material constants K and m. To improve the estimation of K and m in the second and third stages, one experimental data set is selected from the two available (i.e. i=1 or i=2). The application of finite element modeling inthe ANSYS 10ED software with the subsequent comparison of the obtained numerical results with a set of experimental dataallowed us to confirm the reliability of the obtained calculations.

The presented method was tested in experiments on SPF hemispheres of sheet blanks of aluminum alloy AA5083 with a thickness of 1.2 mm in a cylindrical matrix with a diameter of 100 mm with an input radius of 5 mm at a molding temperature of 450 °C (table 1).

Table 1. A set of experimental data on AA5083 sheet alloy [4]

Pressure, [MPa]	Forming time of the hemisphere, [sec]	Height of the hemisphere, [mm]	Thickness at the pole, [mm]
0.29	1080	46.5	~0.48
0.56	120	43.2	~0.51

The results of the finite element modeling using the material constants K and m, calculated using the set ofthe experimental data on the hemisphere obtained at SPF with a constant gas pressure p_1=0.29 MPa (table 2) and p_2=0.56 MPa (table 3) are presented below.

Table 2. Comparison of the results of the finite element modeling with m=0.43 and$K$=216.9MPa·s^m (s_{ANS}, t_{ANS}) with experimental data (s_{Exp}, t_{Exp})

Pressure, [MPa]	Thickness of the dome at the pole, [mm]			Forming time of the hemisphere, [sec]		
	s_{Exp}	s_{ANS}	Δ, %	t_{Exp}	t_{ANS}	Δ, %
0.29	0.48	0.481	≈0.23	1080	1076.83	≈0.37
0.56	0.51	0.53	≈3.9	120	194.58	≈62

Table 3. The comparison of the results of finite element modeling with m=0.37 and K=150.18 MPa·s^m (s_{ANS}, t_{ANS}) with experimental data (s_{Exp}, t_{Exp})

Pressure, [MPa]	Thickness of the dome at the pole, [mm]			Forming time of thehemisphere, [sec]		
	s_{Exp}	s_{ANS}	Δ, %	t_{Exp}	t_{ANS}	Δ, %
0.29	0.48	0.464	≈3.3	1080	770.02	≈28.7
0.56	0.51	0.509	≈0.01	120	120.23	≈0.19

From the comparison of the numerical values with the corresponding experimental data presented in tables1 to 3, it can be seen that anerror of no more than 1% is observed with the set, which was chosen to improve the estimation ofthe material constants K and m at the second and third stage of the calculations according to the method proposed in [2].

This is due to the fact that the standard model of the SP material (1) does not allow achieving the same accuracy of the simulation, both in the duration of molding and the thickness of the dome atthe pole at different values of constant pressure with the same values of the constants K and m.

Time dependence of the dome height H, mm (figure 1) and the thickness of the pole s, mm (figure 2), calculated with the help of the ANSYS software complexiscompared with corresponding experimental data. Finite element models of the SPF of the hemisphere in ANSYS were constructed using the constant material m=0.563 and K=1128.4 MPa·s^m, obtained by selecting the first

IOP Conf. Series: Materials Science and Engineering **447** (2018) 012039 doi:10.1088/1757-899X/447/1/012039

experimental data set (i=1) for calculations at the second and third stages of the method (a) and m=0.53, K=810.9 MPa·s^m when choosing the second experimental data set (i=2) for calculations (b).

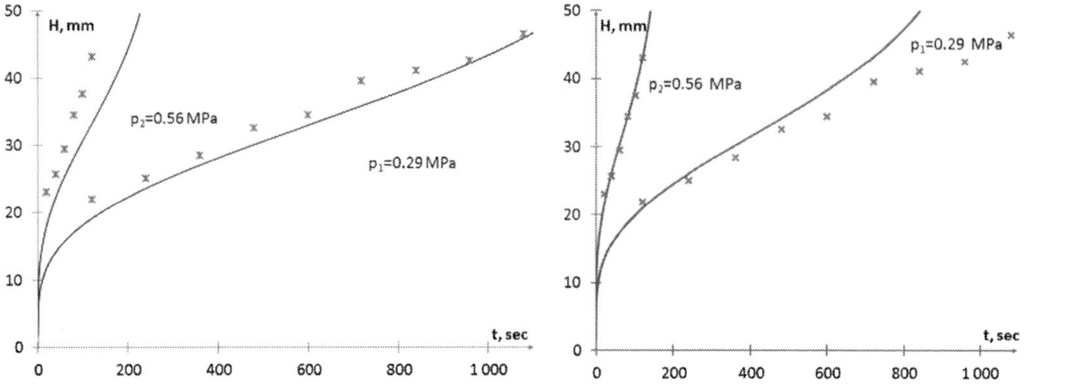

a) b)

Figure 1. Time dependencies of the dome height, H (mm), obtained as a result of the finite element modeling with ANSYS software (solid line). Markers present the experimental data.

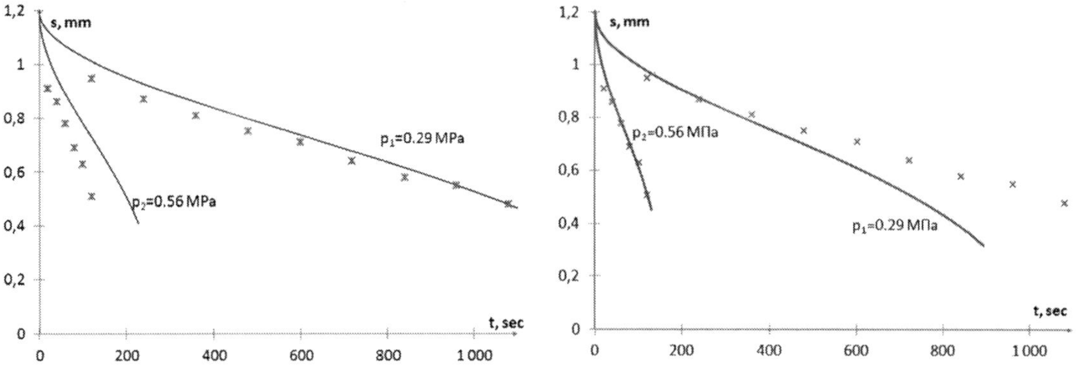

a) b)

Figure 2. Time dependencies of the thickness at the pole, s (mm), obtained as a result of the finite element modeling with ANSYS software (solid line). Markers present the experimental data.

It is known from [5] that the phenomenon of SP manifests itself in a narrow range of deformation rates, and therefore it is necessary to regulate the gas pressure supply for the process of formingwith a constant deformation rate at the pole of the dome. To ensure this mode, it is necessary to change the value of the gas pressure, according to the calculation formulas (2) and (3),

$$p(\alpha) = 2\cdot\sigma_{opt}\cdot\left[\frac{s_0}{(R_0 + r_0)}\right]\cdot\left[\frac{\sin^3\alpha}{\alpha^2}\cdot\frac{1}{1 - r_0\sin\alpha}\right] \tag{2}$$

$$t(\alpha) = \frac{1}{\xi_{opt}}\cdot 2\cdot\ln\left(\frac{\alpha}{\sin\alpha}\right) \tag{3}$$

where p is the gas pressure, t is the forming time, α is the angle between the symmetry axis and the current dome radius passing through the center of curvature of the input radius of the matrix from the work [3], s_0 is the initial thickness of epy sheet; R_0 is the die radius; $r_0$is thematrix input radius; $\sigma_{opt}$is theoptimal flow stress, $\xi_{opt}$is theoptimal strain rate.

An example of the application ofthe suggested algorithm is presented in figure 3.

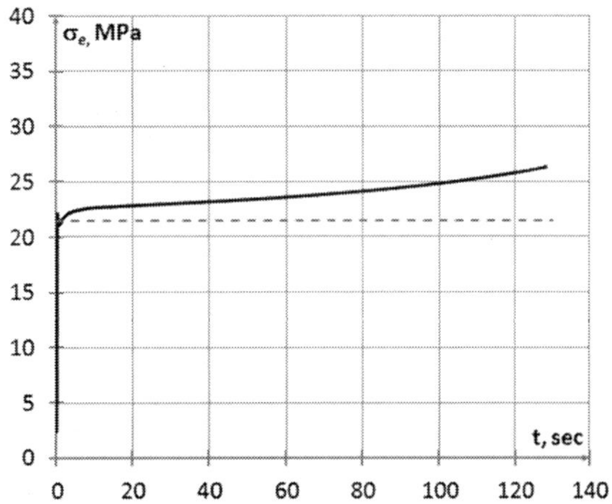

Figure 3. Time dependence of the Mises equivalent stress at the dome apex, σ_e, MPa, calculated bymeans of the finite element method with the target stress σ_{opt} =21.5 MPa

As seen in figure 3, the flow stress goes to the steady-state stage in the SPF time interval from 20 to 80 sec, after that a slight increase is observed.

3. Conclusions

The method presented in [2] allows one to calculate the values of the material constants *K* and *m* included in the standard model SP, the use of which allows building a finite element model of the SPF process of the hemisphere with engineering accuracy (the error does not exceed 1%).

The practical application of the proposed algorithm makes it possible to avoid rupture of the sheet preform and to increase the uniformity of thickness in the product.

References

[1] Backofen W A, Turner I R and Avery D H 1964 *Trans. ASM* **57** 980
[2] Tulupova O, Ganieva V, Kruglov A and Enikeev F 2017 *Letters on Materials* **7** 68
[3] Enikeev F U and Kruglov A A *Int. J. Mech. Sci.* 1995 **37** 473
[4] Jarrar F, Jafar R, Tulupova O, Enikeev F and Al-Huniti N 2016 *Mater. Sci. Forum* **838–839** 512
[5] Jovane F 1968 *Int. J. Mech.Sci.* **10** 405

IOP Conf. Series: Materials Science and Engineering **447** (2018) 012040 doi:10.1088/1757-899X/447/1/012040

Scenarios of mass transfer in fcc copper: the role of point defects

A M Bayazitov[1], E A Korznikova[2], I A Shepelev[3], A P Chetverikov[3], S Kh Khadiullin[4], E A Sharapov[4], P V Zakharov[5], S V Dmitriev[2,6]

[1] Bashkir State University, 32 Zaki Validi st., 450000 Ufa, Russia
[2] Institute for Metals Superplasticity Problems of RAS, 39 Stepan Khalturin st., 450001 Ufa, Russia
[3] Saratov Chernyshevsky State University, 83 Astrakhanskaya st., 410012 Saratov, Russia
[4]Ufa State Aviation Technical University, 12 K. Marx st., 450008 Ufa, Russia
[5]Altai State Humanitarian and Pedagogical University, 53 Korolenko st., 659333 Biysk, Russia
[6] National Research Tomsk State University, 36 Lenin Ave., 634050 Tomsk, Russia

E-mail: elena.a.korznikova@gmail.com

Abstract. It is well known that mass transfer in crystalline materials in most cases is assisted by the motion of point defects. Interstitial atoms are highly mobile, because they have relatively small migration energy as compared to vacancies. Crowdion is a type of interstitial defect being an extra atom inserted in a closely packed atomic row and having even smaller migration energy than an interstitial atom occupying a pore. In this work, we study scenarios of collisions of supersonic *N*-crowdions in Cu by means of molecular dynamics simulation. It was found that 2-crowdions collide leading to the emergence of two new 2-crowdions, which is an example of the soliton-type interaction.

1. Introduction

The mass transfer by the point defects in crystalline solids facilitates structure transformations occurring during plastic deformation, heat treatment and irradiation [1-3]. In thermal equilibrium, vacancy migration is the main mechanism of diffusion, while in nonequilibrium conditions, interstitials also contribute noticeably to the mass transfer [1]. The interstitial atoms occupying the pore are relatively immobile, while those located in a closely packed atomic row, called crowdions, are highly mobile. Interestingly, crowdions usually have a lower potential energy than immobile interstitials [4]. Crowdions can be at rest, and they can move both with subsonic and supersonic velocities. Standing or subsonic crowdions have a kink profile, the width of which is typically half a dozen of atoms. Often they can bear a vibrational mode with a frequency above the phonon spectrum [5,6]. In this case, one can speak about a combination of a discrete breather and a crowdion, which is an object of particular importance because discrete breathers can localize energy [7-18], thus promoting the generation and/or migration of defects. Supersonic crowdions are strongly localized on a single atom. Recently, the notion of supersonic *N*-crowdions has been introduced [4-6], where not one, but *N* atoms move simultaneously at high speed along a close-packed atomic row. It has been shown that N-crowdions are very mobile and much more efficient in the mass transfer as compared to classical 1-crowdions [4-6].

Interest in the study of moving crowdions in crystals is stimulated by recent experimental observation of annealing of defects deep inside a germanium crystal after surface plasma treatment [19], and also in connection with the study of tracks visible in mica crystals with the naked eye [20-22]. Crowdions can make a significant contribution to the mass transfer observed in metals and alloys during severe plastic deformation [1-3]. It has been shown that severe plastic deformation even at room temperature can stimulate phase transformations occurring in the absence of deformation at much higher temperatures [1,4]. These phase transitions occur very quickly and are accompanied by accelerated mass transfer, which cannot be explained only by conventional mechanisms, such as bulk and grain boundary diffusion, even in the presence of a high concentration of vacancies [1]. Direct experimental observation of crowdions moving at supersonic speed is a serious technical problem, so atomistic modeling can be extremely useful for studying them. The aim of this work is to study the motion and collisions of supersonic N-crowdions in a fcc Cu crystal.

2. Simulation details

In this paper, the fcc lattice of copper with the lattice parameter a = 3.615 Å and the interatomic distance $d = a/\sqrt{2}$ =2.556 Å is considered. Simulations are carried out using the LAMMPS package, which includes the EAM many-body interatomic potential for Cu. The x and y axes are oriented along the close-packed crystallographic directions {110} and $\{\bar{1}10\}$, respectively, and the z axis is oriented along the {001} direction (figure 1a). The number of translational cells along the x, y and z axes was equal to 20, 4 and 2, respectively. The total number of atoms in the simulation cell was equal to 1280. Note that supersonic crowdion produces perturbations only in a certain cone behind it. Periodic boundary conditions are used. Thermal vibrations were not introduced, which means that simulations were carried out at 0 K. N-crowdion is excited by giving equal initial velocity V_x to N neighbouring atoms in a close packed atomic row along the row. The initial velocities of all other atoms and the initial displacements of all atoms are equal to zero. Thus, the total energy of the system is equal to the kinetic energy of the atoms excited at t=0.

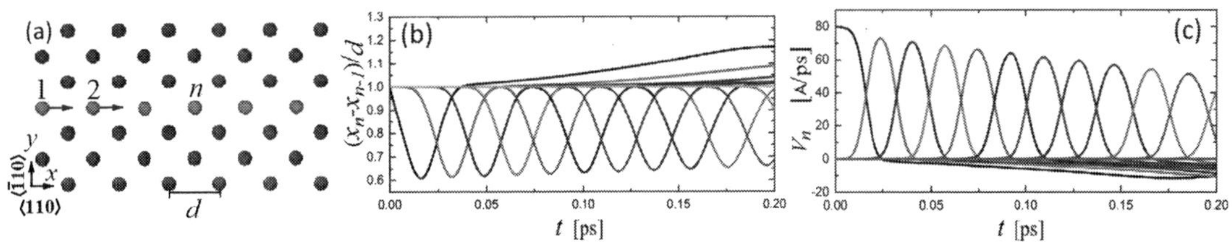

Figure 1. (a) Simulation setup. Cu atoms projected to {001} plane. Atoms of one close-packed atomic row are shown in red, they are numbered by index n. N-crowdion is excited by giving initial velocities to N neighbouring atoms along the row (the case of N=2 is shown). (b) Interatomic distances normalized to equilibrium distance d and (c) x-component of velocities for the atoms in the row where classical 1-crowdion moves, as a functions of time.

3. Simulation results and discussion

First, we show the dynamics of 1-crowdion moving along the x axis in the close-packed row (figure 1 a). The initial velocity V_x, given to a single atom, was 80 Å/ps. The distances between neighbouring atoms normalized to equilibrium distance d as a function of time are shown in figure 1 b. One can see that the minimal distance between two atoms, normalized to d, is about 0.6, so that the condition of self-focusing motion, $(x_n-x_{n-1})/d > 1/2$, is satisfied [4]. Under this condition,1-crowdion is able to propagate a relatively long distance. Figure 1 b also shows that the distance between neighboring atoms may exceed the equilibrium value $(x_n-x_{n-1})/d = 1$ due to the oscillations of atoms behind the crowdion. The velocities of atoms in the row where 1-crowdion moves are shown in figure 1 c (the x-component of the velocity is shown). It can be seen that 1-crowdion continuously radiates its energy, and the velocities of atoms gradually decrease with time, starting from the initial value of 80 Å/ps.

In figure 2 a, b we plot the same as in figure 1 b, c, but for 2-crowdion propagating along the close-packed row in fcc Cu. Here again, the condition of self-focusing motion, $(x_n-x_{n-1})/d > 1/2$, is satisfied [4] because $(x_n-x_{n-1})/d$ is not smaller than 0.6. However, the dynamics of energy radiation by 1- and 2-crowdions differs considerably. The function $V(t)$ for the case of 2-crowdion, plotted in figure 2 b demonstrates the same average level of V during the simulation time, while for 1-crowdion (figure 1 c), a noticeable decrease in velocity is observed within the same simulation time. Thus, 2-crowdion demonstrates a more robust propagation as compared to 1-crowdion.

The next step of our study is to analyze the collision scenarios between 1- and 2-crowdion, moving toward each other from the left and right, respectively. Figure 3 a schematically shows the collision process. At the first stage, the 1-crowdion collides with the first atom of the 2-crowdion, resulting in their repulsion and momentum exchange. Then the two atoms, which initially belonged to the 2-crowdion, collide and exchange by the momentum. As a result, at the end of the collision we have 1-crowdion moving to the right and a new 2-crowdion moving to the left, as if they passed through each other.

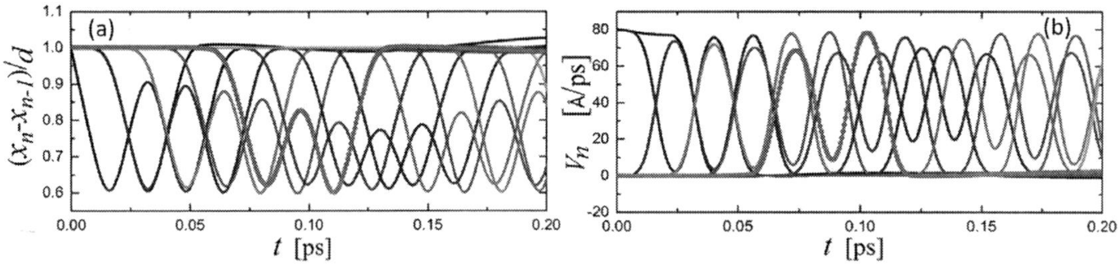

Figure 2. (a) Interatomic distances normalized to the initial distance d and (b) velocities of the atoms along x-axis as functions of time, showing propagation of 2-crowdion in a close-packed atomic row of Cu. The curves for one particular atom are highlighted by dots both in (a) and (b).

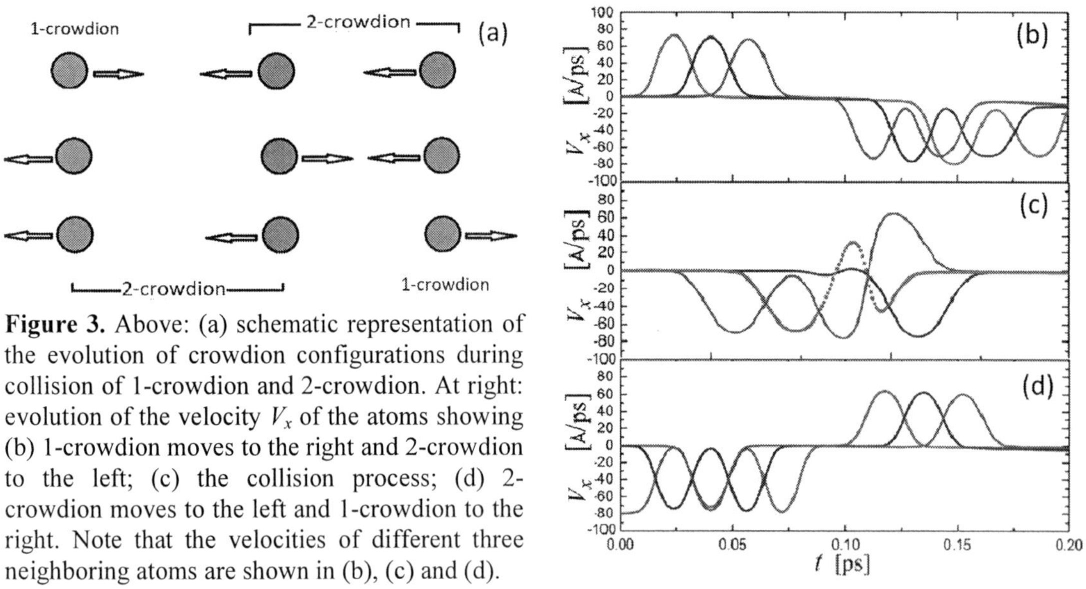

Figure 3. Above: (a) schematic representation of the evolution of crowdion configurations during collision of 1-crowdion and 2-crowdion. At right: evolution of the velocity V_x of the atoms showing (b) 1-crowdion moves to the right and 2-crowdion to the left; (c) the collision process; (d) 2-crowdion moves to the left and 1-crowdion to the right. Note that the velocities of different three neighboring atoms are shown in (b), (c) and (d).

To provide additional information on the collision between 1- and 2-crowdion, in figure 3 b-d we plot the x-component of the velocity for the atoms in the close-packed atomic row where the collision takes place. In figure 3 b the three nearest atoms which are to the left of the collision point are shown. In figure 3 c the three nearest atoms at the collision point are shown. Finally, in figure 3 d the three nearest atoms which are to the right of the collision point are shown. Thus, in figure 3b, first the 1-crowdion passes through the three nearest atoms with positive velocity and then the 2-crowdion, generated as a result of the collision, passes through these atoms with a negative velocity. The results shown in panels c and d are interpreted in a similar way.

4. Conclusions

The investigation of the motion and collision of supersonic 1- and 2-crowdion in Cu was performed by means of molecular dynamics. The simulation results can be summarizes as follows:

(i) Supersonic crowdions in Cu can propagate for long distances along a close-packed atomic row with the velocities of atoms satisfying the self-focusing condition [4].

(ii) 2-crowdion propagates a longer distance and radiates less energy than 1-crowdion.

(iii) The collision of 1-crowdion and 2-crowdion results in their passing through each other, demonstrating soliton-like interaction.

Acknowledgement

For A.M.B., I.A.Sh. and A.P.Ch. the research was supported by the Russian Science Foundation, grant no. 16-12-10175 (obtaining numerical results). E.A.K. and S.V.D. were supported by the Russian Foundation for Basic Research, grant no. 17-02-00984-a (design of the research and writing the paper).P.V.Z. gratefully acknowledges the support of Russian Foundation for Basic Research, grant no 18-42-220002 r_a (discussion of results). The work was partly supported by the State Assignment of IMSP RAS.

References

[1] Korznikova E A, Schafler E, Steiner G and Zehetbauer M 2006 *TMS Annual Meeting* **2006** 97

[2] Konkova T, Mironov S, Korznikov A, Korznikova G F, Myshlyaev M and Semiatin S 2015 *J. Alloys Compd.* **629** 140

[3] Wang L, Wang Y, Zhilyaev A P, Korznikov A V, Li S, Korznikova E A and Langdon T 2014 *Scripta Mater.* **77** 33.

[4] Dmitriev S V, Medvedev N N, Chetverikov A P and Zhou K 2017 *Phys. Status Solidi RRL* **11** 1700298

[5] Chetverikov A P, Shepelev I A, Korznikova E A, Kistanov A A, Dmitriev S V and Velarde M G 2017 *Comput. Condens. Matt.* **13** 59

[6] Dmitriev S V, Korznikova E A and Chetverikov A P 2018 *J. Exp. Theor. Phys.* **126** 347

[7] Dmitriev S V, Korznikova E A, Baimova Y A and Velarde M G 2016 *Physics-Uspekhi* **59** 446

[8] Korznikova E A, Fomin S Yu, Soboleva E and Dmitriev S V 2016 *JETP Lett.* **103** 277

[9] Barani E, Lobzenko I, Korznikova E A, Soboleva E, Dmitriev S V, Zhou K and Marjaneh A 2017 *Phys. Lett. A* **381** 3553

[10] Murzaev R T, Bachurin D V, Korznikova E A and Dmitriev S V 2017 *Phys. Lett. A* **381** 1003

[11] Evazzade I, Lobzenko I P, Korznikova E A, Ovid'ko I, Roknabadi M and Dmitriev S V 2017 *Phys. Rev. B* **95** 035423

[12] Barani E, Lobzenko I P, Korznikova E A, Soboleva E G, Dmitriev S V, Zhou K and Marjaneh A 2017 *Eur. Phys. J. B* **90** 38

[13] Murzaev R T, Babicheva R I, Zhou K, Korznikova E A, Fomin S Yu, Dubinko V I and Dmitriev S V 2016 *Eur. Phys. J. B* **892** 168

[14] Barani E, Korznikova E A, Chetverikov A P, Zhou K and Dmitriev S V 2017 *Phys. Lett. A* **381** 3553

[15] Zakharov P V, Korznikova E A, Dmitirev S V, Ekomasov E G, Zhou K 2019 *Surface Sci.* **679** 1

[16] Kistanov A A, Murzaev R T, Dmitriev S V, Dubinko V I and Khizhnyakov V V 2014 *JETP Lett.* **99** 353
[17] Baimova J A, Korznikova E A, Lobzenko I P and Dmitriev S V 2015 *Rev. Adv. Mater. Sci.* **42** 68
[18] Dmitriev S V, Chetverikov A P and Velarde M G 2015 *Phys. Status Solidi B* **252** 1682
[19] Archilla R, Coelho S, Auret, D, Dubinko V I and Hizhnyakov V 2015 *Physica D* **297** 56
[20] Russell F 1967 *Nature* **217** 51.
[21] Russell F, Archilla J, Frutos F and Medina-Carrasco S 2017 *Europhys. Lett.* **120** 46001
[22] Archilla J and Russell F 2016 *Letters on Materials* **6** 3

Effect of SPD on the structure formation and mechanical properties of the Cu-10% Zn alloy

L I Zaynullina[1,2], **V D Sitdikov**[1,3], **I V Alexandrov**[1,2], **W Wei**[2,4]

[1]Ufa State Aviation Technical University, 12 K. Marx st., Ufa 450008, Russia
[2]Sino-Russian Joint Laboratory of Functional Nanostructured Materials, Changzhou University, Changzhou, Jiangsu 213164, Peoples Republic of China
[3]Laboratory for Mechanics of Bulk Nanomaterials, Saint Petersburg State University, 28 Universitetskiy pr., St. Petersburg, Peterhof 198504, Russia
[4]School of Materials Science and Engineering, Changzhou University, Changzhou, Jiangsu 213164, Peoples Republic of China

E-mail: lele4ka.06@mail.ru

Abstract. In this paper, we report on the microstructure characteristics and strength properties of the Cu–10 wt.% Zn alloy subjected to equal-channel angular pressing (ECAP) and subsequent flat rolling (FR), in order to form an ultrafine-grained (UFG) structure and enhance the strength properties. It is shown that the formed subgrain structure increases the strength characteristics. An increase in the degree of reduction at FR resulted in an increase in the dislocation density and also in an increase in strength. The presence of high-angle grain boundaries increases the ductility of the alloy.

1. Introduction

Today, considerable scientific interest in structural bulk ultrafine-grained (UFG) materials is based on the fact that their mechanical, physical and performance properties differ considerably from the properties of coarse-grained analogues. Features of the structure of UFG materials (grain size, fraction of high-angle boundaries, etc.) are determined by methods of production and have a significant impact on their properties. The UFG structure of the materials is formed by severe plastic deformation (SPD) methods. Equal-channel angular pressing (ECAP) [1, 2], high pressure torsion [3, 4], screw extrusion [5], etc. which belong to the SPD methods are widely applied in scientific research. The SPD forms the UFG structure and increases the density of defects. As a rule, an increase in strength occurs along with a decrease of ductility. Consequently, low ductility becomes a critical limitation in the application of these metals.

The aim of the conducted research was to obtain an increased strength and sufficient ductility in the Cu-10wt.%Zn alloy by means of ECAP and subsequent rolling.

2. Experimental

The material used in this work is the Cu-Zn alloy with 10 wt.% Zn. The stacking fault energy (SFE) of this alloy is equal to 35 mJ/m^2. The rods of the alloy with a square cross section 8×8 mm^2 were cut into sections 60 mm long, annealed at 800 °C for 2 hours, slowly cooled to 500 °C and then quenched in water to obtain a homogeneous microstructure before pressing. Then the treated the so-called "initial rods" were processed by ECAP for 8 passes via route B_c (90° clockwise rotation around the axis of the billet between consecutive passes) at a temperature of 150 °C with a pressing rate of 37 mm/min using a die with the inner and outer channel intersection angles φ = 90° and ψ = 0°,correspondingly. The as-processed rods were flat rolled into thin plates with a reduction of thickness of 30%, 40%, 60%, 80%, 90% and 95% (thereinafter designated as samples 8 ECAP + 30% rolling, 8 ECAP + 40% rolling, 8 ECAP + 60% rolling, 8 ECAP + 80% rolling, 8 ECAP + 90%

rolling, 8 ECAP + 95% rolling) at a temperature of 150 °C. The rolling direction was parallel to the longitudinal axis of the rod. Besides, the initial rods of the alloy were rolled under the conditions described above (thereinafter designated as samples SS + 30% rolling, SS + 40% rolling, SS + 60% rolling, SS + 80% rolling, SS + 90% rolling, SS + 95% rolling).

Thin foils were cut from the longitudinal plane to prepare samples for transmission electron microscope (TEM) investigations. The final thinning of the foils was performed by the method of ink-jet electrolytic polishing in a special unit Tenupol-5 using the electrolyte of the following composition: 25% phosphoric acid, 25% ethanol and 50% deionized water with a current of 50-80 mA. Thin foils were examined by a JEM-2100 transmission electron microscope in dark and bright fields at an accelerating voltage of 160-200 kV. The surface density of the twins was analyzed by calculating the twin area, assuming that it was rectangular with respect to the investigated TEM image area. For statistical analysis, over 15 TEM images were analyzed.

X-ray diffractograms were obtained on a RigakuUltima IV diffractometer with the "Bragg-Brentano" scheme, using Cu *Kα* radiation generated at a voltage of 40 kV and a current of 40 mA. The lattice parameter, the X-ray crystallite size d_{XRD}, the dislocation density ρ, and the proportion m_{ixp} of edge dislocations among all dislocations were determined. At that, the diffractograms were analyzed within the framework of the Whole Powder Pattern Modeling algorithm implemented in the PM2K program.

The tensile specimens were cut by electric spark-cutting technique from the longitudinal plane and had the shape of a dog-bone with a total length of 12 mm and a gauge length of 4 mm. Tensile tests were carried out on an Instron 8801 tensile testing machine at room temperature with a ram speed of 0.24 mm/min. The values of tensile strength and yield stress were obtained from the engineering stress-strain curves.

3. Results and analysis

3.1. Microstructural evolution

Figure 1 shows TEM images of the initial coarse-grained (CG) samples with equiaxial grains with an average size of about 100 ± 40 μm after FR with 95% reduction and the sample after 8 passes of ECAP with subsequent FR to the same reduction.

Deformation twins are clearly observed in the fine structure after FR of the initial state with 95% reduction (figure 1 a). The surface density of twins is 1.2%. The average grain size, determined by the dark-field image is 170 ± 60 nm.

Grains are refined to a size of 210 ± 60 nm in the 8 ECAP + 95% rolling state (figure 1 b). Also, there are deformation twins in the structure. The surface density of twins is 0.5%.

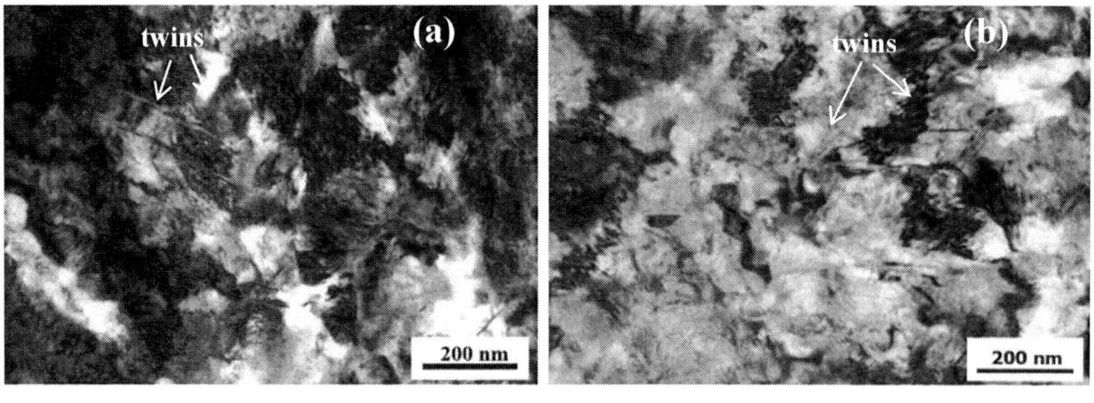

Figure1. TEM images of the samples in SS + 95% rolling (a) and 8 ECAP + 95% r rolling (b) states.

3.2. X-Ray results

Table 1 shows the estimated lattice parameter, the X-ray crystallite size d_{XRD}, dislocation density and proportion of edge dislocation obtained by X-ray diffraction (XRD). An increase in the rolling reduction of the FR of the initial CG samples leads to a regular decrease in the lattice parameter.

Table 1. Microstructure parameters obtained by the XRD analysis

	Lattice parameter, nm	d_{XRD}, nm	Dislocation density, 10^{15} m^{-2}	Proportion of edge dislocations, %
SS	0.363789(11)	452(79)	-	28
SS +30% rolling	0.363786(18)	390(54)	0.62(18)	34
SS +40% rolling	0.363782(8)	324(61)	0.89(13)	39
SS +60% rolling	0.363777(12)	226(44)	1.14(26)	58
SS +80% rolling	0.363778(10)	170(57)	2.18(17)	63
SS +90% rolling	0.363771(19)	104(23)	2.44(9)	70
SS +95% rolling	0.363769(12)	76(19)	2.51(12)	68
8 ECAP	0.363629(7)	35(3)	8.63(2)	89
8 ECAP+30% rolling	0.363635(9)	32(2)	8.84(11)	95
8 ECAP+40% rolling	0.363619(9)	31(2)	9.22(8)	90
8 ECAP+60% rolling	0.363627(9)	26(4)	9.79(12)	96
8 ECAP+80% rolling	0.363625(5)	30(3)	9.79(15)	91
8 ECAP+90% rolling	0.363623(8)	27(3)	9.72(11)	95
8 ECAP+95% rolling	0.363637(5)	29(2)	9.66(19)	89

The lattice parameter of pure Cu is 0.361491 nm. Thus, it is observed that the lattice parameter tends to the lattice parameter of pure Cu.

The decrease in d_{XRD} indicates a grain refinement, which is in good agreement with the results of TEM investigations for the SS + 95% rolling state. The density of dislocations increases as the rolling reduction rises (table 1). According to Taylor's theory [6], $\sigma \approx \sqrt{\rho}$, this process should lead to an increase in the strength characteristics, which is observed experimentally in figure 2. In the SS state the microstructure contains mainly screw dislocations, and after the rolling reduction of 95% the dislocations refer primarily to the edge ones (table 1).

FR of a sample subjected to 8 passes of ECAP does not lead to a regular change in the lattice parameter (table 1). Changing of the d_{XRD} parameter confirms the refinement of grains (table 1). The density of dislocations, which belong to the edge type, increases with the rise of the rolling reduction, which agrees with the growth of the strength characteristics of the alloy.

3.3. Tensile properties

Figure 2 shows the engineering stress-strain curves of the investigated states. The values of tensile strength calculated from the presented above engineering stress-strain curves for the initial samples after FR are equal to 410 MPa for reduction of 80%, to 421 MPa for the reduction of 90% and to 515 MPa for the reduction of 95%.

The yield stress of these samples is 380 MPa for the reduction of 80%, 400 MPa for the reduction of 90% and 496 MPa for the reduction of 95%. After 8 passes of ECAP and subsequent FR the values of tensile strength are equal to 682 MPa for the reduction of 80%, to 679 MPa for the reduction of 90% and to 724 MPa for the reduction of 95%, while the values of the yield stress are equal to 661 MPa for the reduction of 80%, to 657 MPa for the reduction of 90% and to 589 MPa for the reduction of 95%.

As can be seen from figure 2, 8 passes of ECAP +95% rolling allow achieving strength which is 1.4 times higher as compared to the SS samples after analogous FR, while the ductility also increases up to 14% from 9% for the corresponding states.

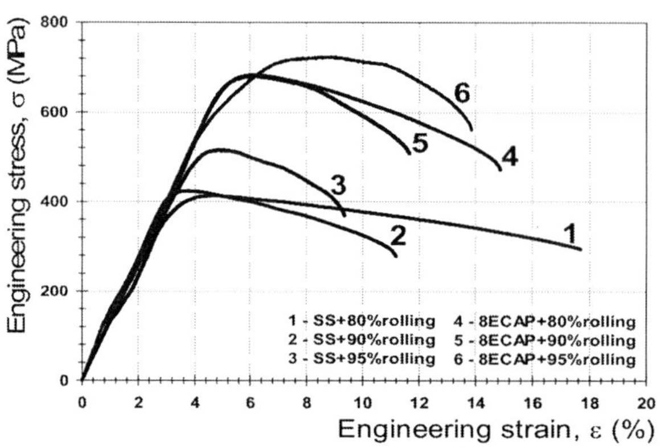

Figure 2. Engineering stress-strain curves.

4. Discussion

FR of the Cu-10 wt.% Zn alloy to different reduction values promotes the change in strength and ductility. When rolling the initial CG samples, the strength increases logically, and the ductility varies little. The lattice parameter, crystallite size and dislocation density change monotonically as the rolling reduction increases. FR of the UFG alloy allows achieving high strength and sufficient ductility. The greatest effect is achieved for the state of 8 ECAP + 95% rolling (figure 2 curve 6). Analysis of the microstructure of 8 ECAP + 95% rolling (figure 1b) shows that a slight increased grain size (according to TEM investigations) in this case cannot lead to the observed increase in strength. However, as follows from table 1, while the crystallite size decreases as the rolling reduction increases, the refined subgrain structure increases the strength characteristics. An increase in the density of dislocations as the rolling reduction increases also leads to an increase in strength. According to [7], with SPD, the edge type dislocations are mainly located at the grain boundaries. This statement corresponds to the high fraction of high-angle grain boundaries detected in this report, which corresponds to the increased ductility of the UFG alloy [8]. In the ECAP + 95% rolling state, the highest strength is 724 MPa and sufficient ductility is 14% compared to the initial CG samples after rolling.

5. Conclusions

The formation of the UFG structure in the Cu-10 wt.% Zn alloy provides an increase in both strength and ductility during subsequent FR with an increasing rolling reduction. This behavior of the UFG state is significantly different from that characteristic of the case of rolling the CG state, characterized by an increase in strength and a decrease in ductility. The increased strength and ductility of the UFG state are explained by the developed substructure and the predominance of high-angle grain boundaries.

Acknowledgments

L.I. Zaynullina, V.D. Sitdikov, I.V. Alexandrov are grateful for the financial support of the Ministry of Education and Science within the framework of the project No. 16.1969.2017/4.6.W. Wei is grateful for a support to the National Natural Science Foundation of China under Grant No. 51561001.

References

[1] Segal V M 1994 *Processes of Plastic Structuring of Metals* (Minsk: Science and Technology)
[2] Valiev R Z and Langdon T G 2006 *Prog. Mater. Sci.* **51** 881
[3] Bridgman P V 1955 *Investigation of Large Plastic Deformations and Rupture* (M: IL) 444

IOP Conf. Series: Materials Science and Engineering **447** (2018) 012041 doi:10.1088/1757-899X/447/1/012041

[4] Zhilyaev A P, Langdon T G 2008 *Prog. Mater. Sci.* **53** 893
[5] Beygelzimer Y, Orlov D and Varyukhin V 2002 *Ultrafine Grained Materials II,* ed Y T Zhu, T G Langdon, R S Mishra, S L Semiatin, M J Saran and T C Lowe (The Minerals, Metals & Materials Society) p 297
[6] Zeng X 2007 *Modeling Hardening and Softening due to High-Angle Grain Boundaries in Crystalline Solids* (Cuvillier Verlag)
[7] Zehedbauer M 1993 *Acta Mater.* **41** 589
[8] Zehetbauer M and Seumer V 1993 *Acta Metall. Mater.* **41** 577

IOP Conf. Series: Materials Science and Engineering **447** (2018) 012042 doi:10.1088/1757-899X/447/1/012042

Solid-phase joining of the cast intermetallic Ni_3Al-based alloys and wrought nickel-based superalloys via using superplastic deformation

E V Galieva and R Ya Lutfullin

Institute for Metals Superplasticity Problems, Russian Academy of Sciences, 39 Khalturin st., Ufa 450001, Russia

E-mail: galieva_elvina_v@mail.ru

Abstract. The paper summarizes the results of solid-phase joining experiments performed for intermetallic single-crystal Ni_3Al-based alloys (known as Russian VKNA-alloys) and wrought fine grained nickel-based superalloys using superplastic deformation of the wrought superalloys. For this, two single-crystal VKNA-alloys, VKNA and VKNA-25 and two Russian nickel-based superalloys, EK61 and EP975, were taken. It was shown that the formation of an ultrafine grained structure (d<1 μm) in the EK61 superalloy made it possible to obtain the solid-phase joint of EK61/VKNA under pressure welding at T=800 °C using superplastic deformation of the nickel-based superalloy. Another nickel-based superalloy, EP975, was taken in a fine grained condition (d=8 μm). High-quality solid-phase joints of EP975/VKNA and EP975/VKNA-25 were obtained under pressure welding at T=1125 °C using the superplastic deformation of the EP975 superalloy. The obtained solid-state joints EP975/VKNA and EP975/VKNA-25 were subjected to heat treatment, which led to the formation of coarse γ grains in the EP975 superalloy. This also led to an increase in thickness of the γ phase interlayers up to 500 nm in both VKNA-alloys near the solid-phase joints

1. Introduction

Nowadays one of the most important problems in the aviation industry is the development of methods for producing reliable solid-phase joints of similar and dissimilar nickel-based alloys. As a rule, nickel-based superalloys possess low weldability, with the exception of some them having lower high temperature capability, such as Inconel 718 (the Russian analogue is EK61). In spite of the fact that this problem is of great practical importance for the aircraft industry, there are only a few papers in publicly available sources devoted to the development of robust joining techniques. Particularly, there are engineering challenges related to the development of the disk combined with the blades. The disk and the blades have different operating temperatures, and therefore it is necessary to develop solid-phase joining techniques not only for similar but also for dissimilar nickel-based alloys. As has been demonstrated, pressure welding using superplastic deformation of joined materials is a promising processing route for joining both similar and dissimilar materials [1-9]. Note that this technique is more feasible in the case of fine grained structures in both joined materials. A more difficult case is joining of dissimilar materials, if only one of them has a fine grained structure and the other is made of a single-crystal. This paper presents a brief review of our latest data on the investigation of solid-phase joining of both similar and dissimilar nickel-based alloys using superplastic deformation of one of the joined materials.

2. Materials and methods

Two Russian wrought polycrystalline nickel-based superalloys having different strengthening phases, EK61 and EP975, were taken as starting materials. The EK61 and EP975 superalloys are strengthened by precipitations of the intermetallic based phases Ni_3Nb (γ'' or δ) and Ni_3Al (γ'), respectively. In the EK61 superalloy, the ultrafine grained duplex-type microstructure (with a grain size of less than 1 μm) was prepared by the multiple forging procedure [4-6]. In the EP975 superalloy, a fine grained microstructure with a mean γ grain size of d=8 μm was also prepared by forging [5]. Two intermetallic Ni_3Al-based alloys (known as Russian VKNA-alloys) containing γ' and γ phases were also used for the experiments [5, 9, 10, 11]. Single-crystals of the VKNA-alloys were manufactured by high-gradient directional crystallization. This technique is described elsewhere [10, 11]. The chemical compositions of the nickel-based alloys are represented in table 1 [4, 5, 9, 11].

Table 1. Chemical compositions of the nickel-based alloys under study.

Alloy abbreviation	Ni	Al	Cr	W	Mo	Ti	V	Nb	Fe	Co	Re	Cu	La
EK61	Base	1	16.6	-	3.9	0.8	0.5	5	15	-	-	0.55	-
EP975	Base	4.8	8.2	10.2	1.2	2.4	-	1.5	-	15.1	-	-	-
VKNA	Base	8.5	5	2.3	5	1.5	-	-	-	-	-	-	0.015
VKNA-25	Base	8.37	5.7	3	5.14	0.52	-	-	-	4.5	1.6	-	-

The pressure welding experiments were carried out using uniaxial compression and press forces, providing superplastic deformation of fine grained samples of the nickel-based superalloys, i.e. pressure welding led to the strain localization within fine grained samples. Thus, a solid-state joint was formed simultaneously with the superplastic deformation of one of the joined materials. The experiments were performed for small samples in vacuum. After pressure welding of the EP975/VKNA and EP975/VKNA-25 alloys, the samples were subjected to annealing at 1200 °C, followed by air cooling and aging at 950 °C. This was done to evaluate the effect of post welding heat treatment on the solid-phase joint.

Table 2 describes the microstructural states of the samples subjected to pressure welding and the pressure welding parameters. The compressive stresses were chosen as corresponding to the superplastic flow stresses of the fine grained superalloy samples.

Table 2. The pressure welding conditions of the nickel-based alloys under study.

Joined alloys	Microstructural states*	Solid-phase joining temperature, °C / time, min	Compressive stress, MPa	The strain of the fine grained sample, %
EK61/EK61	UFG / CG	800 / 60	100	45
EK61/VKNA	UFG / S-C	800 / 60	100	45
EP975/VKNA	FG / S-C	1125 / 42	45	24
EP975/VKNA-25	FG / S-C	1125 / 42-68	45	24-40

*UFG - ultrafine grained, FG- fine grained, CG - coarse grained, S-C - single-crystal.

The microstructure of the solid-phase joints was examined by scanning electron microscopy (SEM) in the secondary electron (SE) mode. The quality of the solid-phase joints was evaluated taking into account the pore length relative to the total length of the solid-phase joint.

3. Results and Discussion

3.1. Pressure welding of similar superalloys

The samples prepared from the wrought nickel-based EK61 superalloy had the $\gamma+\delta$ duplex-type ultrafine grained microstructure. The average grain size of the γ and δ phases was less than 1 μm. The EK61 superalloy (as mentioned, this is the Russian analogue of Inconel 718) was used as a convenient high workable material having good weldability [4-6].

As has been earlier revealed, pressure welding of coarse grained samples made of the EK61 alloy is unfeasible at a temperature of 800 °C [4]. Pressure welding of coarse grained and ultrafine grained samples made of the EK61 alloy allowed us to obtain solid-phase joints (figure 1a, b). In the course of the pressure welding, the strain was localized within the ultrafine grained sample, and the strain value was about 45%.

In the central joining zone of the welded samples, i.e. in the hard-to-deform zone, which is characteristic of the uniaxial compression scheme, there are areas where the relative pore length reached 80%. At the same time, in the peripheral joining zone there are areas located at an angle to the axis of the applied load, where the solid-phase joint of a better quality was reached. In some of the peripheral joining areas, the relative pore length reached not more than 10%. This can be ascribed to the fact that the strain was localized in the peripheral areas that is typical of the solid-phase joints obtained during uniaxial compression. This suggests that the stress-strain distribution in the joining zone during pressure welding under the uniaxial compression was rather inhomogeneous. This resulted in the different relative pore length along the solid-phase joint.

The experiment performed for the EK61/EK61 samples showed that the solid-phase joint can be obtained by pressure welding at relatively low temperatures, if one of the joined samples has the ultrafine grained structure and the pressure welding is accompanied by superplastic deformation of this sample. From a practical point of view, it is more interesting to use this effect in respect of nickel-based alloys with a higher high-temperature capability and in the case of dissimilar alloys.

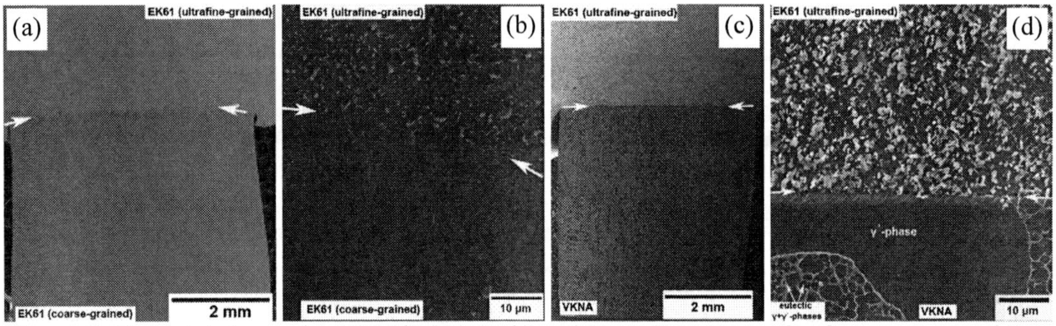

Figure 1. The jointed samples EK61/EK61 (a, b) and EK61/VKNA (c, d) after pressure welding at T=800 °C: a and c – the general view, b and d – the solid-phase joint zones; a and b – the coarse-grained samples were joined through the ultrafine grained interlayer; c and d – the VKNA-alloy sample was joined with the EK61 superalloy sample having the ultrafine grained structure. The solid-phase joints are arrowed.

3.2. Pressure welding of dissimilar alloys

The influence of a different type of surface relief on the solid-phase joining was earlier investigated by the finite element modeling [3, 7] that was well agreed with the experimental results obtained in the present work. Particularly, the computer modeling showed that the stagnant zone in the central part of the samples, joined by pressure welding using uniaxial compression, is preserved that degrades the quality of the joint. To exclude the stagnant zone and improve the quality of the solid-phase joint, special reliefs on the joined surfaces can be made. An alternative way can be associated with using the

IOP Conf. Series: Materials Science and Engineering **447** (2018) 012042 doi:10.1088/1757-899X/447/1/012042

superplastic deformation during pressure welding by uniaxial compression. This idea was tested for dissimilar nickel-based alloys with lower weldability.

Figure 1c, d shows a general view of the joined samples and a part of the solid-phase joint zone of the EK61/VKNA alloy samples. Pressure welding of the EK61/VKNA alloy samples resulted in the strain localization within the ultrafine grained EK61 alloy. This allowed us to avoid plastic deformation of the intermetallic alloy that was important for preserving its mechanical properties. One can see that the so-called diffusion zone was formed in the joint area [6]. In the general case, the width of the diffusion zone depends on the welding temperature and the chemical and phase composition of the materials being joined.

EP975 is a well-known Russian superalloy with a higher high-temperature capability and lower weldability in contrast to the EK61 alloy. The samples of the alloy were taken in a fine grained condition and joined to the VKNA-alloys. The samples of the EP975 superalloy had a $\gamma+\gamma'$ microstructure, the average size of the γ and γ' phases was 8 and 3 μm, respectively. Figure 2 represents the local areas of solid-phase joints of the EP975/VKNA and EP975/VKNA-25 samples subjected to pressure welding at T=1125 °C. It should be noted that such joined combinations can be used to produce real aircraft parts, e.g., BLISK-type parts [5, 8, 9].

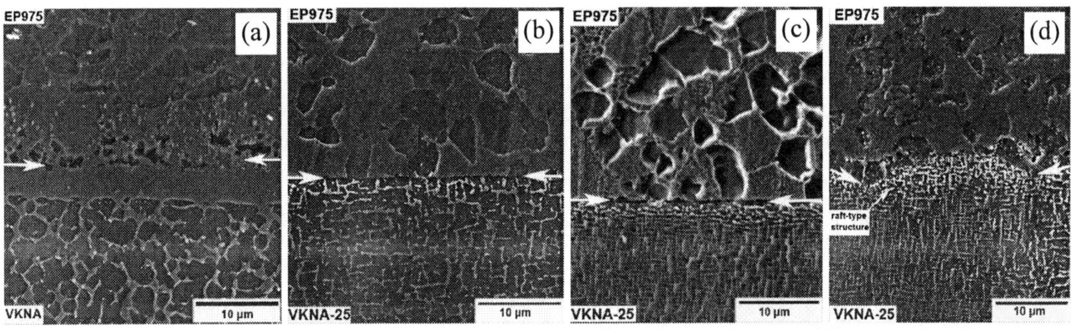

Figure 2. Local sections of the solid-phase joint zones after pressure welding at T=1125 °C: a - EP975/VKNA, b-d - EP975/VKNA-25. The strain of the EP975 sample was: b – 24%; c – 30%; d – 40%. The solid-phase joints are arrowed.

A distinct interface is visible between the alloys after the solid-phase joining. One can see that after pressure welding at T=1125 °C, the microstructure of the EP975 superalloy remains fine grained (figure 2). No microstructural changes in the VKNA and VKNA-25 alloys far from the joining zone were revealed as compared to the VKNA and VKNA-25 alloys before pressure welding (figure 2). At the same time, a diffusion zone with a width of up to 40 μm was formed. Apparently, this occurs during the superplastic deformation of the EP975 alloy. It was shown that the diffusion of chromium and cobalt developed from the high-alloying superalloy EP975 to the VKNA-alloys, while the diffusion of nickel took place in the opposite direction [5, 9].

The influence of the superplastic strain value on the solid-phase joint formation was studied for the EP975/VKNA-25 alloy samples. An increase in the strain value resulted in the formation of an uneven contact interface in the joining zone, which improved the quality of the solid-phase joint (figure 2 b-d). The microstructures of the VKNA-25 alloy differed from the joining zone and close to the joint zone, and this difference increased with increasing the superplastic strain value. Apparently, the deformation was localized in the areas adjacent to the joint at a depth of 3-5 μm, where a raft-type microstructure was formed (figure 2 d).

Heat treatment of the joined samples resulted in the formation of a coarse-grained microstructure in the EP975 superalloy. The average size of the γ phase grains reached 165 μm, the γ' phase particles of about 500 nm in size were precipitated. In the VKNA-alloys there were no significant microstructure changes after heat treatment. Nevertheless, some increase in the width of the γ phase interlayers up to

500 nm was observed near to the solid-phase joint zone. The heat treatment, including aging, contributed to the activation of diffusion of refractory alloying elements. After the heat treatment a monotonous change in the content of alloying elements in crossing the solid-phase joint was observed.

4. Conclusions

Pressure welding of the EK61 and EP975 superalloys in fine grained states with single-crystals of the Ni_3Al-based alloys (VKNA and VKNA-25) was successfully fulfilled using relatively low joining temperatures and compressive stresses providing superplastic deformation of the fine grained superalloys during pressure welding. Using this technique, sound solid-phase joints were obtained.

With decreasing the grain size of one of the joined alloys, the pressure welding temperature can be significantly reduced. This is associated with decreasing the superplasticity temperature. As was demonstrated for the EK61/VKNA alloys, the solid-phase joint was attained after pressure welding at the temperature as low as T=800 °C. The use of the EP975 superalloy with a fine grained microstructure allowed us to obtain the solid-phase joint of good quality after pressure welding at T=1125 °C. Thus, the superplastic deformation of one of the joined materials can assist in attaining high-quality solid-phase joints in the case of nickel-based superalloys and single-crystals of the Ni_3Al-based alloys.

Acknowledgments

The present work was supported by the Russian Science Foundation (Grant No. 18-19-00685) in the part related to the investigation of pressure welding of dissimilar nickel-based alloys. The investigation part devoted to the pressure welding of similar superalloys was accomplished according to the state assignment of fundamental scientific researches of Government Academy of Sciences No. AAAA-A17-117041310221-5.

5. References

[1] Lutfullin R Ya and Mukhametrakhimov M Kh 2010 *Rev. Adv. Mater. Sci.* **25** 142
[2] Hossein R. 2011 *Diffusion-fundamentals.org* **15** 2
[3] Galieva E V, Valitov V A, Lutfullin R Ya, Dmitriev S V, Akhunova A Kh and Mukhametrakhimov M Kh 2016 *Mater. Sci. Forum* **838-839** 350
[4] Valitova E V, Lutfullin R Ya, Mukhametrakhimov M Kh, Valitov V A, Akhunova A Kh and Dmitriev S V 2014 *Letters on Materials* **4** 291
[5] Povarova K B, Valitov V A, Obsepyan S V, Drozdov A A, Bazyleva O A and Valitova E V 2014 *Russian Metall. (Metally)* **9** 733
[6] Galieva E V, Valitov V A, Lutfullin R Ya and Bikmukhametova A A 2018 *Def. Diff. Forum* **385** 150
[7] Galieva E V, Lutfullin R Ya, Akhunova A Kh, Valitov V A and Dmitriev S V 2018 *Sci. Technol. Weld. Join.* **27** 612
[8] Russian Federation Patent No 2608118, 2017
[9] Drozdov A A, Valitov V A, Povarova K B, Bazyleva O A, Galieva E V and Ovsepyan S V 2015 *Letters on Materials* **5** 142
[10] Povarova K B, Bondarenko Yu A, Drozdov A A, Bazyleva O A, Antonova A V, Morozov A E and Arginbaeva E G 2015 *Russian Metall. (Metally)* **2015** 43
[11] Bazyleva O A, Povarova K B, Arginbaeva E G, Shestakov A V and Drozdov A A 2015 *Russian Metall. (Metally)* **2015** 916

IOP Conf. Series: Materials Science and Engineering **447** (2018) 012043 doi:10.1088/1757-899X/447/1/012043

Correlation between heat treatment process parameters, phase composition, texture, and mechanical properties of 12H18N10T stainless steel processed by selective laser melting

A L Kameneva[1], A A Minkova[1, 2], N N Cherkashneva[2] and V V Karmanov[1]

[1] Perm National Research Polytechnic University, 29 Komsomolski ave., Perm 614000, Russia
[2] JSC «UEC-Aviadvigatel» 93 Komsomolski ave., Perm 614000, Russia

E-mail: minkova20@gmail.com

Abstract. The phase composition, texturing degree, oriented stresses and strains of the crystal lattice, the physical and mechanical properties of austenitic steel 12H18N10T (AISI 321) manufactured by selective laser melting (SLM) before and after heat treatment with various temperatures and protective gas composition have been investigated. SLM stainless steel has better mechanical properties, given that the phase volume content is as follows: γ-Fe ~ 65%, α-Fe ~20%, NiTi ~ 5%; the texturing degree equals to T_{111}~0.70 and the dominant crystal-lattice orientation is (111). Such a phase composition is formed in the steel after annealing at 620 °C in argon. As the temperature rises during the stabilizing annealing process up to 825 °C, the phase transformation γ→α takes place in SLM steel with a change in the phase concentration γ-Fe and α-Fe in the range from 40 to 50%. In the high-temperature intermetallic phase NiTi, dissolution at 825 °C and a multiple rise of the FeO concentration lead to a change in the crystal lattice parameter of the γ-Fe and α-Fe phases and a decrease in the mechanical properties of SLM steel. This study showed that the improvement of mechanical properties was due to the separation of high-temperature intermetallic phase from austenite and an increase in the concentration of iron carbide Fe_3C in SLM steel.

1. Introduction

Selective Laser Melting allows manufacturing components of virtually any complexity, obtaining high performances of composite materials. The application of austenitic stainless steel powders in additive technologies, in particular, SLM, is of utmost interest both for research and application needs. Numerous studies have revealed a variety of features in the structure and properties of the laser melted austenitic steel powder. Despite the large number of various studies on stainless steels manufactured by SLM, not much focus is given to the influence of heat treatment on their structure and properties [2]. Sedlaka and Rican (2015), Hussein and Hao (2013), Wang and Yang (2013) determined that the above mentioned steels do not have a phase transformation γ↔α [1] over a wide range of temperatures and cooling rates. This paper aims to investigate the effect of heat treatment process parameters on the phase composition, internal stresses, texturing degree, as well as the physical and mechanical properties of the 12H18N10T item manufactured by SLM.

2. Materials and Experimental Procedure

Austenitic steel 12H18N10T (equivalent to AISI 321) was chosen as the subject of this study. Table 1 shows its chemical composition.

Table 1. Chemical composition of austenitic steel 12H18N10T, wt.%.

Fe	C	Cr	Ni	Ti	Si	Mn	P	S
Balance	0.093	17.67	9.59	0.55	0.70	0,95	0.06	0.009

SLM of powder from steel 12H18N10T was performed under a protective atmosphere of nitrogen on the Concept LaserM2 Cusing Machine (Germany). The size of SLM manufactured samples in the first batch was 10x10x15 mm. The dimensions of the second batch of test samples were in accordance with GOST 1497-84. Annealing of the samples from steel 12H18N10T manufactured by SLM was performed in a KP-110A Furnace and stabilization annealing - in a MonoThermHK.669.N.6gr Vacuum Furnace by ALD. The heat treatment (HT) process parameters are shown in table 2.

Table 2. HT process parameters of SLM manufactured samples from 12H18N10T (air-cooling).

Specimen pos.no.	HT type	Temperature, °C	Time at temperature, h.	Environment during HT
0	No heat treatment	-	-	-
1	Annealing	620	2	Air
2				Argon
3	Stabilization annealing	825	2	Argon
4				Vacuum

Phase composition, structure and strained condition of the samples from the first batch were studied by X-ray diffractometry.

The studies were conducted on a Bruker D8 Advance Diffractometer. The diffraction analysis parameters were as follows: voltage 50 kW, power 1000 μA (1mA), copper anode with wavelength of Kα1=0,154055μm. Diffraction patterns were processed in the DIFRAC.SUITE EVA software. The phase concentrations were determined by means of their volume content in accordance with the technique described in papers [3, 4]. The α-Fe and γ-Fe phase ratios was determined by the Rietveld method using Topas Software. The texturing degree T_{111} (T_{110}), the oriented stresses Δs and strains Δa of the crystal lattice, and the crystal lattice parameter change was specified in accordance with the techniques [5], [6], and [7], respectively. Static tests of the second batch samples were carried-out according to GOST 1479-84. Short-time strength tests of heat-treated samples were performed on the LFMZ-100 Machine.

3. Results

3.1 The influence of temperature values and medium during heat treatment on the phase composition, texture, stresses and strains in the crystal lattice of SLM steel

Steel 12H18N10T, manufactured layer-by-layer by SLM, before and after heat treatment, has a two-phase structure with a two-component texture. The structure of the SLM steel prior to heat treatment is γ-Fe with the crystal lattice parameter – a=3.5853 Å and α-Fe with crystal lattice parameter – a=2.8723Å (table 3). Intermetallic compound FeNi, iron carbide Fe_3C and oxide FeO are also formed in SLM steel. The texturing degree T_{111}=0.75 shows that the dominant orientation of the crystal-lattice is (111). In the process of annealing at 620 °C in air or argon, the dominant phase is also austenite (FCC), dominated by the (111) crystal orientation (table 3, 4). The crystal lattice of SLM steel also does not change, which is confirmed by stability of the crystal lattice parameter. However, an increase in the α-phase content in the steel up to 29% (air) and 26% (argon) leads to a decrease in the texturing degree T_{111} by 7%. The minimum content of Cr_3C_2 was formed in SLM steel during air annealing. The formation of a high temperature intermetallic compound NiTi and an increase in the volume content of iron carbide Fe_3C were observed in SLM steel after annealing in an argon environment (table 4).

The phase transition γ↔α was not found in SLM steel after annealing at 620 °C. However, the γ→α phase transition took place in SLM steel when rising the temperature during stabilizing annealing up to 825 °C (table 4). The concentration of the α-phase in the volume was raised up to 56%, which is two times higher than the similar parameter in SLM steel. The preferential crystallographic orientation from (111) changes to (110) (table 4). The volume content of FeO oxide is multiply increased. The high-temperature dissolution of the NiTi intermetallic alloy at 825 °C and a multiple increase in the FeO concentration lead to reduction of the crystal-lattice parameter of the γ-Fe and α-Fe phases. In this regard, we can conclude that Fe-dependent defects constitute the majority. The phase ratio of the α-Fe and γ-Fe phases is reversed when the stabilizing annealing medium changes from argon to vacuum (table 4). There is no essential difference in the T_{111} and T_{110} texturing degree of the α-Fe and γ-Fe phases (table 4). Negative and minimum oriented stresses in the crystal lattice of the γ- and α-phases indicate insignificant compression stresses.

Table 3. HT process parameters of SLM steel, crystal lattice parameter and oriented stresses of the crystal lattice.

	HT parameters			$a/\Delta a \cdot 10^{-3}$, Å		Δs	
	T, °C	Environ ment	*T*, hrs.	(111)γ-Fe	(110)α-Fe	γ-Fe	α-Fe
No heat treatment	-	-	-	3.5853/1.6	2.8723/-2.1	-1.0038	-1.0035
Annealing	620	air	2	3.5853/1.6	2.8723/-2.1	-1.0043	-1.0040
Annealing	620	argon	2	3.5853/1.6	2.8723/-2.1	-1.0045	-1.0035
Stabilization annealing	825	argon	2	3.5818/2.6	2.8708/-1.5	-1.0028	-1.0029
Stabilization annealing	825	vacuum	2	3.5836/2.1	2.8708/-1.5	-1.0036	-1.0037

Table 4. Phase composition and structure characteristics of SLM steel.

Spec.	Volume ration of all possible phases in SLM steels, %							Phase weight ratio		Texturing degree	
Pos.no.	γ-Fe	α-Fe	Fe_3C	FeNi	FeO	NiTi	Cr_3C_2	γ-Fe	α-Fe	T_{111}	T_{110}
0	77.17	18.40	1.87	0.77	1.77	-	-	76	24	0.75	0.25
1	73.31	22.84	1.03	0.93	1.86	-	0.01	71	29	0.70	0.30
2	67.00	20.45	5.52	0.51	1.59	4.92	-	74	26	0.70	0.30
3	39.44	54.86	0.08	-	5.60	-	-	44	56	0.37	0.63
4	50.60	39.17	4.51	0.33	5.38	-	-	57	43	0.51	0.49

3.2 Effect of phase composition, texture, stresses and strains in the crystal lattice of SLM steel on its mechanical properties

Table 5 shows the mechanical properties of steel 12H18N10T manufactured by SLM before and after HT. Table 4 represents the properties of the cast steel 12H18N10T for the sake of comparison. Figure 1 shows the dependence of the mechanical properties of steel on the volume content of all incoming phases.

Table 5. Mechanical properties

Specimen No	UTS, MPa	0.2%YS, MPa	δ, %
Cast workpiece (GOST 5949-75)	510	196	40
0	651	481	47
1	667	488	46
2	679	487	42
3	660	446	56
4	638	429	51

It is found that steel 12H18N10T, manufactured layer-by-layer by the SLM process, has higher mechanical properties in comparison with cast steel and steel without HT. SLM steel has maximum values of 0.2% YS (0.2 % yield strength) and UTS (ultimate tensile strength) and a minimum value of tensile strain δ, similar to cast steel giving that its phase composition is as follows: γ-Fe ~ 65%, α-Fe ~20% and NiTi ~ 5%.

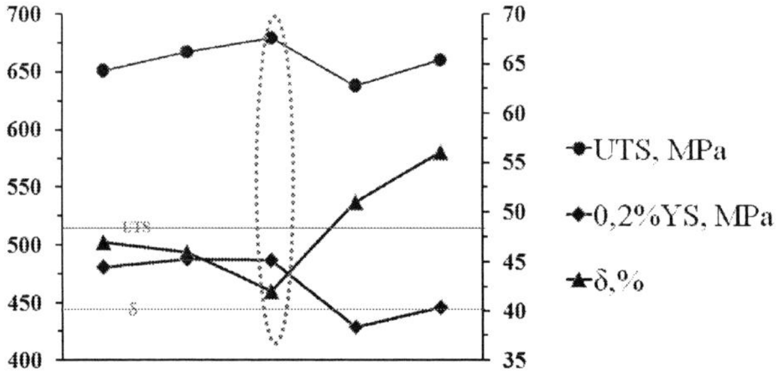

γ-Fe	77,17	73,31	67	50,6	39,4
α-Fe	18,40	22,84	20,45	39,17	54,86
Fe_3C	1,87	1,03	5,52	4,51	0,08
FeNi	0,77	0,93	0,51	0,33	-
FeO	1,77	1,86	1,59	5,38	5,60
NiTi	-	-	4,92	-	-

Figure 1. The dependence of mechanical properties of SLM steel

The texturing degree is T_{111}=0.70, and the preferential crystallographic orientation is (111). In case of an increase in the volume content of the γ-Fe phase and a decrease in the volume content of the α-Fe phase in SLM steel, its mechanical and tensile properties decline (figure 1, table 5). The limit 0.2% YS of SLM steel is 2.5 times higher than that of cast steel.

4. Conclusion

In this work, we studied the influence of the parameters of the heat treatment process on the phase composition, oriented strains of the crystal lattice, the texturing degree and the mechanical properties of steel 12H18N10T manufactured by SLM.

Steel 12H18N10T, manufactured by SLM, before and after HT, has a two-phase structure with a two-component texture. The phase transition γ↔α does not occur in a wide range of temperatures and cooling rates during the SLM process, as well as during the annealing process of SLM steel at 620 °C in the air or argon environment. The dominant phase retains the same, i.e. austenite (FCC), orientated in the (111) plane. However, the volume ratio of the Fe_3C iron carbide increases and the high temperature intermetallic compound NiTi is formed in SLM steel during annealing at 620 °C in argon.

Such a change in the phase composition results in a maximum increase in 0.2% YS up to 679 MPa, UTS up to 487 MPa and tensile strain δ up to 42%, which is close to the properties of cast steel. The volume content of incoming phases in SLM steel is: γ-Fe - 67%, α-Fe - 20.45%, NiTi - 4.92%, Fe_3C - 5.52%, FeNi - 0.51%, FeO - 1.59%.

Phase transformation γ→α takes place in SLM Steel with an increase in temperature during stabilizing annealing up to 825 °C. The dissolution of the high-temperature intermetallic alloy NiTi at 825 °C and a multiple increase in the concentration of FeO lead to a decrease in the crystal lattice parameter of the γ-Fe and α-Fe phases and mechanical properties SLM steel.

References

[1] Zhukov A, Deev A and Kuznetsov P 2017 *Phys. Procedia* **89** 172
[2] Saeidi K *Stainless Steels Fabricated by Laser Melting*. Doctoral Thesis. Department of Materials and Environmental. Chemistry Arrhenius Laboratory, Stockholm University SE-106 91 Stockholm, Sweden. **115**

[3] Kameneva A L and Karmanov V V 2013 *J. Alloys Comp.* **546** 20
[4] Kameneva A L, Guselnikova L N and Soshina T O 2011 *E-J. Surf. Sci. Nanotechnol.* **9** 34
[5] Kameneva A L , Karmanov V V and Dombrovsky I V 2014 *Res. J. Pharmac., Biol. Chem. Sci.* **5** 762
[6] Kameneva A L 2014 *Res. J. Pharmac., Biol. Chem. Sci.* **5** 965
[7] Kameneva A L 2015 *Res. J. Pharmac., Biol. Chem. Sci.* **6** 1381

IOP Conf. Series: Materials Science and Engineering **447** (2018) 012044 doi:10.1088/1757-899X/447/1/012044

Physical features of Leidenfrost effect on the surface of a graphene nanocomposite for the problems of thermal and nuclear energy

D D Babenko, A S Dmitriev, V V Mikhailov, I A Mikhailova

National Research University "Moscow Power Engineering Institute", 14 Krasnokazarmennaya st., Moscow 111250, Russia

E-mail: asdmitriev@mail.ru

Abstract. We investigate the possibility of overcoming the Leidenfrost point to achieve the stable boiling of a pool (not a film) at temperatures much higher than the Leidenfrost temperature for water on smooth surfaces. We have studied the Leidenfrost effect (droplets on a surface overheated above the film boiling temperature). Surfaces made of hybrid graphene nanocomposites are used as substrates. We investigated two types of nanocomposites based on graphene flakes (the size of the latter being about 3-5 nm in thickness, longitudinal dimensions of about 10-20 μm), sintered at a pressure of 200-300 bar, and hybrid nanocomposites consisting of graphene flakes, metal microspheres (a diameter of about 100-200 microns) and a small amount of polymer (epoxy resin).

1. Introduction

The film boiling of liquid droplets on highly superheated surfaces (the Leidenfrost effect) has been an area of extensive studies in recent years, both in connection with new problems of thermal physics, and with applications, in particular, for jet cooling of high-temperature surfaces in power engineering and electronics [1-3]. Heat dissipation from hot surfaces through cooling droplets is limited to the Leidenfrost point (LFP), in which an insulating vapor film prevents direct contact between the cooling droplet and the hot surface. Evaporation and boiling are the most important processes in a wide range of applications in various industries, such as power generation (thermal and nuclear), refrigeration and cooling, water purification, electronics etc. [1-5]. For boiling heat transfer, the latent heat of the liquid-vapor phase transformation allows one to achieve a large dissipated heat flux. At low superheat values, the direct contact of the liquid with the hot surface ensures an efficient heat transfer. However, at high superheat values, a vapor film is formed between the liquid medium and the hot surface and minimizes the rate of heat transfer. The onset temperature of this phenomenon is indicated by the Leidenfrost point (LFP). Investigation of the possibility of increasing the temperature of Leidenfrost or, in general, eliminating the film boiling (overcoming the Leidenfrost limit) is one of the most important problems of heat and mass transfer on new mesoscopic and nanostructured surfaces. Recently, a surface concept of decoupled hierarchical structures [6,7] was developed which could eliminate the LFP.

2. Material and methods

Graphene substrates consisting of pure graphene flakes or graphene flakes, metallic monodisperse pellets and a small amount of polymer as an adhesive filler were used in the work. The method of obtaining composite materials is as follows. Graphene flakes produced by the NanoGraphen Group were used to prepare a nanocolloid solution (nanofluid) for which graphene flakes of various volumes were added to pure distilled water and stirring was carried out for 20-30 minutes in a rotating centrifuge. Graphene flakes 5-10 microns in size with 3-10 layers of graphene (few-layer graphene, FLG) were obtained (figure 1a). In addition, hybrid nanocomposites were fabricated using the technology presented in [4] (figure 1b). The production of the tablets was carried out by means of an automatic machine for hot pressing the Buehler SimpliMet 1000.

IOP Conf. Series: Materials Science and Engineering **447** (2018) 012044 doi:10.1088/1757-899X/447/1/012044

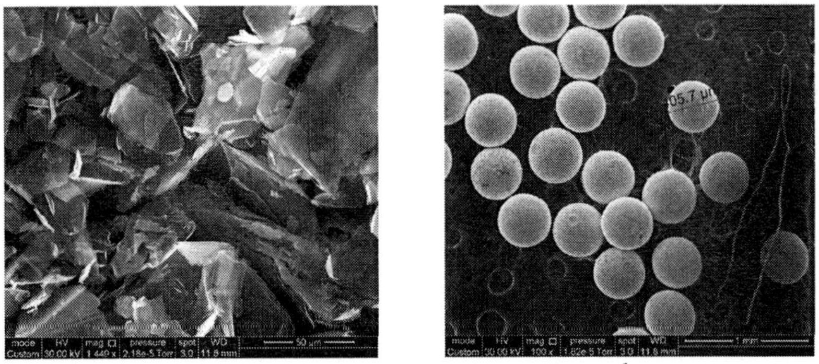

a b

Figure 1. SEM graphene nanoflakes (a) and hybrid graphene + microspheres nanocomposites (b)

Then the press-fitting parameters are selected on the control panel of the press and the process is started. At the end of the process, the cap is opened and the mold with the obtained graphene tablet is lifted using the control buttons. For this work, samples of graphene tablets were made with various press parameters and powder volumes.

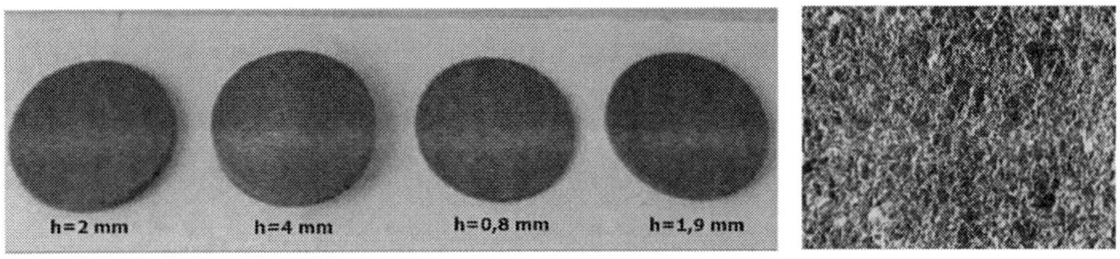

Figure 2. Graphene tablets (left) and optical image of the surface (sample h = 2 mm) (on right)

As a result, samples of different thicknesses were obtained. In production, pressures of 200 and 300 bar were used. At each pressure value, four samples of different thicknesses were made. The thickness depended on the amount of graphene powder used. The quantity was measured in volumetric units with a syringe, which allowed the measurement of a certain amount and was convenient for loading the metered powder into a mold. Since this press is highly specialized equipment, mainly intended for metallographic research, it constantly heats the material loaded in the mold. Therefore, the lowest possible temperature of 150 °C was used for manufacturing. As mentioned above, four samples with powder volumes equal to 2, 5 and 10 ml were prepared for each pressure value, respectively. Different values of pressure and volume gave different thicknesses of the obtained samples (figure 2, left). Figure 2 (on right) shows the optical image of the surface (sample $h = 2$ mm).

3. Contact angles and evaporation rate on graphene tablet surfaces

We investigated two types of nanocomposites based on graphene flakes (the size of the latter being about 3-5 nm in thickness, longitudinal dimensions of about 10-20 μm), sintered at a pressure of 200-300 bar, and hybrid nanocomposites consisting of graphene flakes, metal microspheres (a diameter of about 100-200 microns) and a small amount of polymer (epoxy resin). The Kruss EasyDrop measuring system was used with a special heater and temperature control system. First, contact angles were studied as a function of temperature for different samples of graphene substrates.

We also investigated the evaporation rates of droplets of distilled water on the surface of graphene tablets at different temperatures. To heat the substrates, the ITO (Indium Tin Oxide substrate) was used, the measurements were made by a non-contact infrared thermometer and thermocouples.

Various properties of hybrid nanocomposites based on microspheres and graphene were studied. In some cases, the microspheres coated with graphene were placed in a polymer matrix.

The temperature dependences of the contact angles of water droplets on hybrid graphene substrates over a wide temperature range are studied. We also investigated the evaporation rates of droplets of distilled water and nanocolloids of water + silver nanoparticles (AgNP - 0.5% by volume; the average nanoparticle size is about 20 nm), as well as water + graphene flakes (GNF - 1% by volume). As an example, figure 3 shows the dependencies of the evaporation rate of distilled water and nanocolloid (water + silver nanoparticles) droplets on the graphene tablet surface (pressure of 300 bar). It is important to note that water droplets with silver nanoparticles evaporate more strongly, which, apparently, is associated with an increased effective thermal conductivity of such nanofluids. According to estimates of the evaporation rates for nanocolloid droplets, such an increase can be about 11-14%, which roughly corresponds to our experimental data.

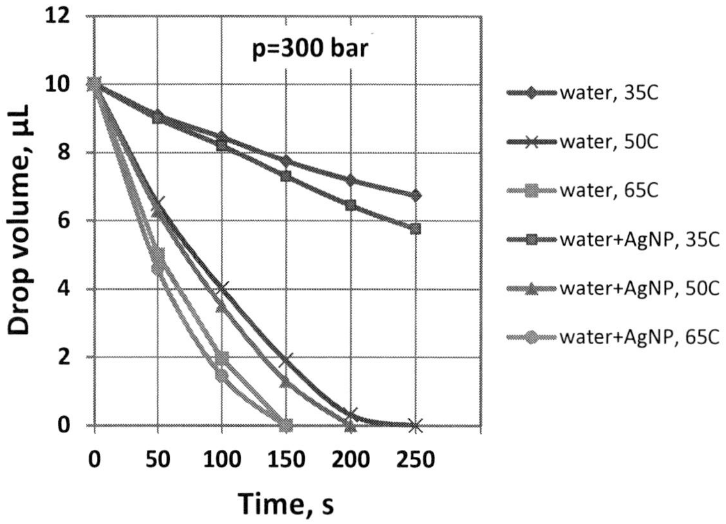

Figure 3. Evaporation rate of distilled water and nanocolloid (water + silver nanoparticles) droplets on the graphene tablet surface

4. Leidenfrost effect on graphene tablet surfaces

After experiments on the measurement of contact angles and evaporation rate on graphene tablet surfaces, studies of the boiling processes on graphene substrates at various temperatures were conducted. In this paper, we investigate the possibility of overcoming the Leidenfrost point for organizing a stable boiling of the pool (not the film) at temperatures much higher than the Leidenfrost temperature for water on smooth surfaces. We studied the Leidenfrost effect (droplets on a surface overheated above the film boiling temperature). Surfaces made of hybrid graphene nanocomposites, which were first developed and created in [4-6], are used as substrates.

A special feature of hybrid graphene nanocomposites is a noticeable increase in the Leidenfost temperature T_L. If on metal substrates for water $T_L \sim$ 175-190 °C, then the minimum value for graphene substrates is $T_L \sim$ 210-230 °C. At temperatures below $T_L \leq 120$ °C normal evaporation and pool boiling droplets on graphene substrates are observed (figure 4a). The next fact that has been clarified is the partial destruction of graphene substrates starting from a temperature $T_L \geq 120$ °C with the removal of graphene flakes from the surface (figure 4b). This, apparently, can be explained by the presence in the material of air caverns, which when heated leave the surface and destroy graphene substrates. It is interesting to note that for hybrid composites containing metallic microspheres, flocculation of smaller-sized graphene flakes was observed at temperatures above $T_L \geq 200$ °C, in the case of pure graphene samples at temperatures $T_L \geq 120$ °C.

Figure 4. Pool boiling droplets (a) and destruction of substrates (b) on graphene tablet surfaces

A study of the evaporation of droplets with a transition of the surface temperature above the film boiling point (Leidenfrost temperature) is performed. In particular, figure 5 shows a drop of water with a diameter of 2 mm on the surface of a graphene nanocomposite in the Leidenfrost regime (figure 5a) and a violation of the stability of levitation at times of about 5-6 seconds (figure 5b). When the temperature is above $T_L \sim 230$ °C, in the first seconds on graphene nanocomposites a very stable Leidenfrost effect is observed (figure 5a). At the same time, after the regime of stable film evaporation, the stability of levitation is disturbed at times about 5-6 sec (figure 5b). The figure shows that the drop is in contact with the hot surface, without a vapor film, quickly evaporates and boils. The mechanism of this phenomenon is not very clear, but perhaps the reason for this is the melting of the polymer that is part of the hybrid nanocomposite and the observation of the breakdown of the Leidenfrost regime on mesoscopic surfaces, as was first noted in [7].

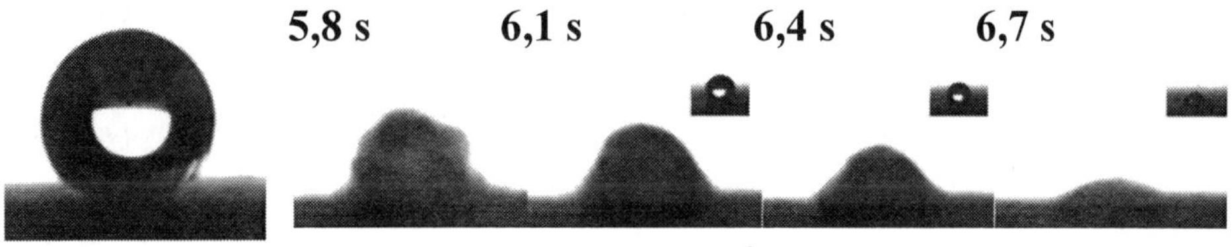

Figure 5. Water drop on the surface of a graphene nanocomposite in the Leidenfrost regime (a) and violation of the stability of levitation (b)

5. Conclusions

Due to high-efficiency the boiling heat transfer is frequently used in the heat transfer devices, such as a heat exchanger, boiler, and so on. The evaporation rate of the drops, which is important for the cooling systems of power plants in the cooling systems of jets and droplets, showed a very large amount on the graphene nanocomposite surfaces. This allows us to recommend such surfaces for efficient cooling systems for power and electronic equipment. We investigated the critical heat fluxes for the transition from the pool boiling to the film boiling (Leidenfrost effect). Unfortunately, the accuracy of such measurements is not more than 15-20%. At the same time, it was possible to establish that there is a noticeable difference in the critical heat fluxes for the four types of surfaces studied: the mesoscopic surface (carbon microspheres, 5-7 μm in diameter), the graphene tablet 200 bar and 300 bar, and the graphene nanocomposite.

We conducted studies of critical heat fluxes for various types of nanocomposites, which may be important for energy applications. Figure 6 shows a comparison of the critical heat fluxes for these surfaces. Mesoscopic surfaces based on carbon microspheres (diameter 5-7 μm), graphene tablets made at 200 bar and 300 bar, graphene nanocomposites were studied. It is clearly seen that the highest critical heat fluxes are observed in graphene nanocomposites - up to 140 W/m^2, which makes them very promising for many applications in thermal and nuclear energy. In addition, the latter materials are significantly more stable at higher temperatures than graphene tablets. This allows us to recommend hybrid graphene nanocomposites for use as materials in thermal and nuclear power engineering.

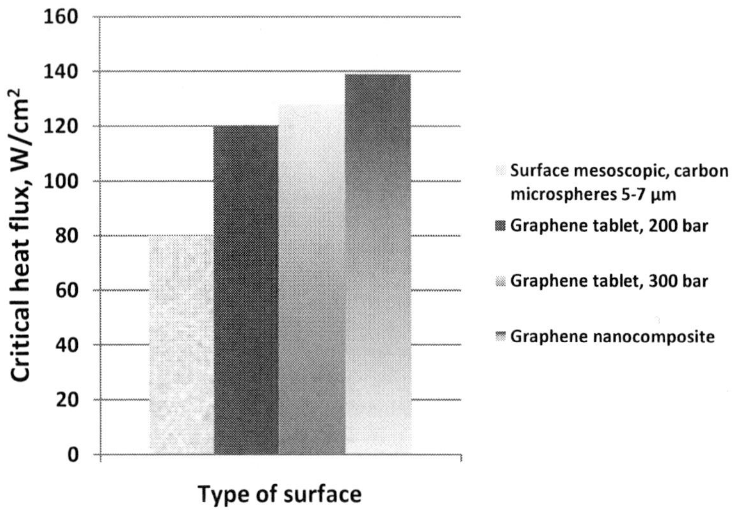

Figure 6. Critical heat fluxes for various surfaces

The features described above play an important role in the development and creation of new functional surfaces for thermal and nuclear power engineering, in the problems of thermal stabilization of electronic and optoelectronic devices [8]. New functional surfaces, as a rule, should have a high Leidenfrost temperature, a stable levitation of droplets or, conversely, a rapid transition when the droplets interact with superheated surfaces, into the pool boiling mode and efficient evaporation.

Acknowledgments
In conclusion, the authors thank S. Alekseev (Graphene Star, UK) for help in manufacturing of graphene flakes. The work is supported by the Russian Science Foundation (Grant No. 17-19-01757).

References

[1] Dmitriev A S 2015 *Introduction to Nanothermophysics* (Moscow: BINOM Press) 792 p. (In Russian)
[2] Kwon H, Bird J and Varanasi K 2013 *Appl. Phys. Lett.* **103** 201601
[3] Quéré D 2013 *Annu. Rev. Fluid Mech.* **45** 197
[4] Zhang P, Lv F 2015 *Energy* **82** 1068
[5] MacGregor-Ramiasa M and Vasilev K 2017 *Adv. Mater. Interfaces* **4** 1700381
[6] Alekseev S A, Dmitriev A S, Makarov P G and Mikhailova I A 2017 *J. Phys.: Conf. Series* **891** 012361
[7] Dmitriev A S and Romanov A S 2013 *Bulletin MPEI* **2** 14
[8] Attinger D, Frankiewicz C, Betz A, et al 2014 *MRS Energy and Sustainability Review J.* **1** 1

IOP Conf. Series: Materials Science and Engineering **447** (2018) 012045 doi:10.1088/1757-899X/447/1/012045

Microstructure and mechanical properties of a novel high Co+W polycrystalline nickel-based blade superalloy

R V Shakhov [1], A A Ganeev [1], Sh Kh Mukhtarov [1], A V Logunov [2], V M Imayev [1], R M Imayev [1]

[1] Institute for Metals Superplasticity Problems, Russian Academy of Sciences
39 Khalturin st, Ufa 450001, Russia
[2] PJSC "UEC-Saturn", 163 Lenin Ave, Rybinsk 152903, Russia

E-mail: vimayev@mail.ru

Abstract. A novel polycrystalline nickel-based superalloy SLZhS-1R, heavily alloyed with Co+W and intended as a structural material for turbine blades in gas turbine engines, has been studied. The initial cast superalloy was prepared by vacuum induction melting in the form of an ingot with a size of Ø 100×180 mm. Compression and tensile tests were performed for the cast condition. Before compression testing, the superalloy was subjected to homogenization and heterogenization annealing, which led to an improvement of chemical homogeneity and coagulation of γ' precipitates. As a result, the flow stresses during compression at elevated temperatures were reduced as compared to those obtained for the as-cast condition. Before the tensile testing, the superalloy was subjected to homogenization annealing (solution treatment) followed by air cooling and ageing. This improved the chemical homogeneity and led to the formation of fine γ' precipitates with a size in the range of 0.1-0.2 μm. The obtained mechanical properties in tension were compared with those of the typical cast superalloy ZhS6U.

1. Introduction

To increase the efficiency of the gas turbine engine (GTE), novel materials and processes are required that would meet the growing operational requirements due to the increase in temperature and loading of components [1]. To meet these requirements, novel nickel-based superalloys are designed to achieve tailored mechanical properties, such as higher creep resistance along with higher strength, fatigue strength and other service properties. Particularly, in respect to polycrystalline nickel-based superalloys, intended for blade manufacturing, alloying with inexpensive elements, which provide efficient solid-solution strengthening, is of interest. Note that the use of rare-earth alloying elements for solid-solution strengthening and the single crystal technology widely applied for blade manufacturing in modern high performance GTEs, are very expensive and therefore efficient alloying of polycrystalline superalloys with common refractory elements is of a great practical significance [1-4]. In this regard, there is a risk of formation of topologically close packed (TCP) phases during long-term high temperature exposures that can deteriorate the mechanical properties of a superalloy. Thus, the alloying additions should be precisely chosen to avoid the formation of undesirable phases.

The present work is a preliminary study of the microstructure and mechanical properties of a new polycrystalline blade superalloy highly alloyed with Co and W and prepared by casting. The obtained tensile properties were compared with those of the typical Russian blade superalloy ZhS6U.

2. Experimental

The nickel-based superalloy Ni-47(Al, Cr, Co, Ta, W, Hf)-0.2(C, B) (in wt. %) named SLZhS-1R was taken as the starting material. The superalloy ingot with a size of Ø 100×180 mm was manufactured by vacuum induction melting. The γ′ solvus temperature was determined by differential scanning calorimetry and quenching experiments and found to be equal to T_s=1185±5 °C. Before compression testing, the cast material was homogenized at temperatures below and slightly above the γ′ solvus temperature for 8 hours and then subjected to slow furnace cooling with a rate of 25 °C/h. Before tensile testing, the cast material was subjected to homogenization and solution annealing at the same regimes followed by air cooling and ageing at T=1000 °C (2 h) and T=850 °C (8 h). The as-cast and heat treated materials were used for preparation of specimens for compression and tensile tests. All specimens for mechanical testing were prepared by electrospark cutting followed by fine grinding of work surfaces. Compression tests were carried out at T=1100-1200 °C with an initial strain rate of $\dot{\varepsilon}\approx10^{-3}$ s^{-1}. Tensile tests were carried out at T=20 and 650-850 °C with an initial strain rate of $\dot{\varepsilon}\approx7\times10^{-4}$ s^{-1}. Two samples per point were tested. The microstructure examination was carried out using scanning electron microscopy (SEM) in backscattering electron (BSE) mode.

3. Results and discussion

3.1. Initial cast materials

A typical coarse grained microstructure was observed in the as-cast condition (not presented here). Nonequilibrium eutectic colonies resulted from dendritic segregation were sometimes distinguished. The γ′ phase precipitates had a size mostly in the range of 0.1-0.2 μm and predominantly near cuboidal morphology. White particles identified as carbides are observed and have a size of 0.5-5 μm. The volume fraction of the γ′ phase can be visually estimated to be about 60%.

Figure 1 represents the BSE images of SLZhS-1R in the cast and heat treated conditions. One can see that homogenization and heterogenization annealing led to the disappearance of the dendritic structure and coagulation of the γ′ phase precipitates. Their size increased up to 0.5-2 μm. Near cuboidal morphology was often retained (figure 1a, b).

After heat treatment including air cooling and ageing, fine nearly round γ′ precipitates with a size of 0.1-0.2 μm were obtained (figure 1c, d). The obtained microstructural conditions were used for preparation of samples for subsequent compression and tensile tests.

3.2. Compression properties

Figure 2 shows the true stress-strain curves obtained as a result of compression tests of the superalloy at elevated temperatures. All the deformed samples had cracks on their lateral surfaces, but the depth of the cracks was smaller in the samples, subjected to heat treatment. This suggests that homogenization and heterogenization annealing improved the hot workability of the superalloy. As the test temperature increased, the flow stresses decreased that is associated with the dissolution of γ′ precipitates. One can see that the samples in the annealed condition showed lower flow stresses during compression at T=1100 and 1150°C than the samples in the as-cast condition. This is explained by the fact that coarse coagulated γ′ precipitates exerted a smaller strengthening impact than fine γ′ precipitates in the as-cast condition. The compression at T=1200 °C showed the same flow stresses in the as-cast and annealed conditions. That is explained by the complete dissolution of the γ′ precipitates and indirectly confirms that the γ′ solvus temperature was below 1200 °C.

3.3. Tensile properties

Table 1 represents the tensile properties obtained for the superalloy at room and elevated temperatures. The strength and ductility decreased with increasing test temperature. Note that the obtained strength characteristics and rather low ductility are typical of cast nickel-based superalloys. For comparison, the tensile properties of the typical Russian blade superalloy ZhS6U in the cast condition are also

IOP Conf. Series: Materials Science and Engineering **447** (2018) 012045 doi:10.1088/1757-899X/447/1/012045

represented. The SLZhS-1R superalloy showed higher strength and similar ductility at room temperature as compared to the ZhS6U alloy. At elevated temperatures, the rather low strength and ductility of the SLZhS-1R superalloy can be associated with the formation of undefined plate-like phase(s), which arose after heat treatment.

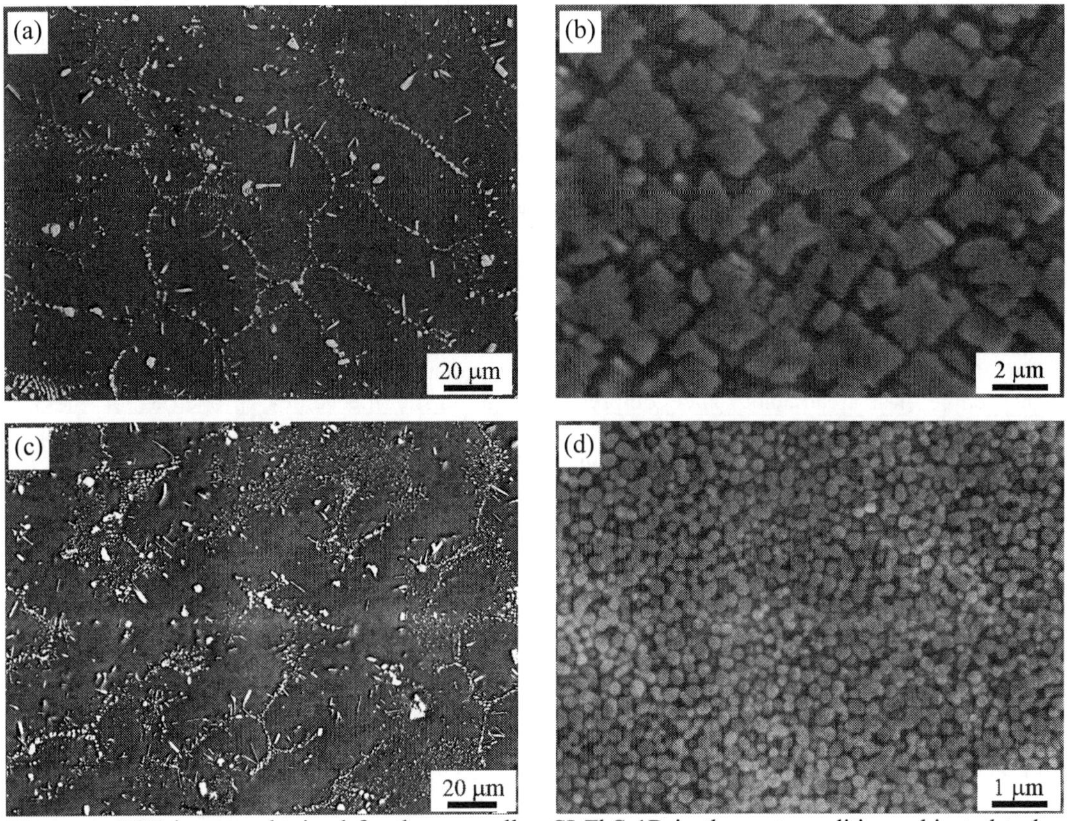

Figure 1. BSE images obtained for the superalloy SLZhS-1R in the cast condition subjected to heat treatments: (a, b) after homogenization and heterogenization annealing, (c, d) after homogenization annealing followed by air cooling and ageing. Arrows show plate-like precipitates of unknown phases. White particles are carbides.

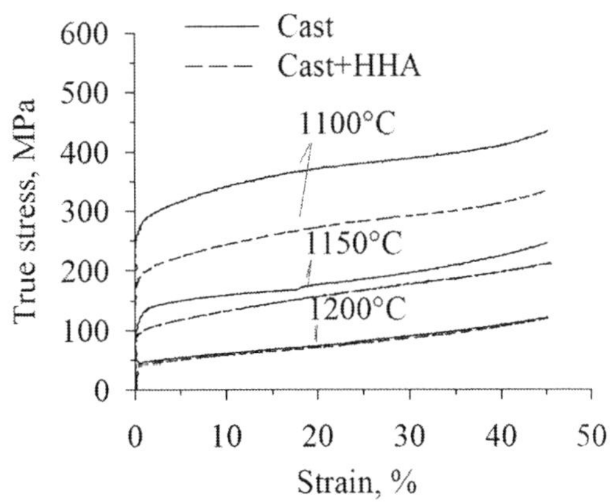

Figure 2. True stress-strain curves obtained for the superalloy SLZhS-1R as a result of compression tests of samples at elevated temperatures (HHA – homogenization and heterogenization annealing).

Table 1. Tensile mechanical properties of the superalloy SLZhS-1R in comparison with those of the Russian nickel base superalloy ZhS6U.

Superalloy	σ_{UTS}, MPa / δ, %			
	20 °C	650 °C	750 °C	850 °C
SLZhS-1R	1015 / 4.5	885 / 1.1	679 / 0.5	604 / 0.3
ZhS6U [5, 6]	920-970 / 3-6	-	-	-
	830 / 3	-	-	-

Note that the obtained properties can be improved via more precise alloying with Co and W. On the other hand, the new superalloy should not be considered as a casting superalloy, and an efficient working process can help in attaining enhanced mechanical properties.

4. Conclusions

A preliminary study of the microstructure and mechanical properties has been performed for the newly developed nickel-based superalloy SLZhS-1R heavily alloyed with Co and W. It was shown that the chemical homogeneity can be improved, and coarse coagulated or fine γ' precipitates can be obtained using simple heat treatments. The compression and tensile mechanical properties of the new superalloy were evaluated. It showed higher strength properties at room temperature that can be ascribed to the efficient solid solution strengthening caused by high alloying with Co and W. Further efforts are to be directed to the development of efficient hot working, which would promote the formation of a finer grained microstructure and properly balanced mechanical properties.

Acknowledgements

The present work was supported by the Russian Science Foundation (Grant No. 18-19-00594). The work was performed using the facilities of the shared services center «Structural and Physical-Mechanical Studies of Materials» at the Institute for Metals Superplasticity Problems of Russian Academy of Sciences.

References

[1] Reed R C 2006 *The Superalloys: Fundamentals and Applications* (Cambridge University Press)
[2] Ford D A, Fullagar K P L, Bhangu H K, Thomas M C, Burkhoider P S, Korinko P S, Harris K and Wahl J B 1999 *Trans. ASME.* **121** 138
[3] Harris K and Wahl J B *Superalloys 2004* (Warrendale, PA: TMS), pp. 45-52
[4] Hobbs R A, Tin S and Rae C M F 2005 *Metall. Mater. Trans. A* **36** 2762
[5] Polyanskii V M, Gavrilyuk V V, Zagorskii V Z, Logunov A V, Polyanskii A M and Silis M I 2004 *Metal Sc. Heat Treat.* **46** 392 (in Russian)
[6] Heat resistant casting alloys obtained by vacuum melting. *Industry standard* (in Russian).

IOP Conf. Series: Materials Science and Engineering **447** (2018) 012046 doi:10.1088/1757-899X/447/1/012046

Features of the surface relief of TiNi alloy in coarse-grained and ultrafine-grained states at room and elevated temperatures

A A Churakova[1,2], E F Gilmanova[2] and D V Gunderov[1,3]

[1]Institute of Physics of Molecules and Crystals UFRC RAS, 151 October Av., 450075, Ufa, Russia
[2]Ufa State Aviation Technical University, 12 K. Marx. St., 450008, Ufa Russia
[3]Saint Petersburg State University, 7/9 University emb. 199334, Saint Petersburg, Russia

E-mail: churakovaa_a@mail.ru

Abstract. The effect of coarse-grained (CG) and ultrafine-grained (UFG) structures processed by equal channel angular pressing on the formation and features of the surface relief during mechanical tensile tests at room and elevated temperatures was studied. The formation of the UFG structure leads to a significant increase in strength, but the ductility decreases. Coarse-grained samples in the quenched state after tensile tests have a developed relief and microcracks. After tensile testing of the annealed CG sample, a similar situation is observed. On samples in the UFG state, no developed relief and cracks were found. The surface of the UFG sample in the annealed state at a temperature of 500 °C shows a very weak relief, but no cracks are observed. Possible reasons of these effects are discussed.

1. Introduction

Alloys of titanium nickelide (TiNi) belong to the class of functional materials with shape memory effects (SME) and superelasticity (SE), caused by thermoelastic martensitic transformations. These alloys are widely used as structural and functional materials in engineering and medicine [1]. A promising trend in improving the service properties of metallic materials, including TiNi alloys, is the formation of an ultrafine-grained (UFG) state in them by methods of severe plastic deformation (SPD) [2-3]. As a result of equal channel angular pressing (ECAP), an ultrafine-grained (UFG) structure with a grain size of about 400 nm is formed in TiNi alloys [2], which leads to a significant increase in the mechanical and functional properties of shape memory alloys. However, the mechanisms of mechanical behavior and deformation of the UFG TiNi are not sufficiently studied, despite years of research by various scientific groups. The aim of this work was to study the influence of the method of ECAP and heat treatment on the mechanical properties, structure and morphology of the alloy surface after tensile testing.

2. Material and experimental methods

The material of the study was a two-component $Ti_{50.2}Ni_{49.8}$ alloy. The $Ti_{50.2}Ni_{49.8}$ alloy is enriched in Ti with respect to stoichiometry at room temperature possesses the B19′ crystal lattice of the monoclinic martensite and the Ti_2Ni with the FCC lattice [1]. The temperatures of the martensitic transformations B2-B19' are M_s = 65 °C, M_f = 38 °C, A_s = 79 °C, and A_f = 107 °C.

The TiNi alloy was homogenized at a temperature of 800 °C for one hour, and then quenched in water to form a supersaturated solid solution of Ti in Ni. The state after quenching was taken as the initial coarse-grained. Thermal treatment of the TiNi alloy was carried out at 500 °C for one hour, followed by cooling in air. The ultrafine-grained structure was formed by the method of equal-channel angular pressing (ECAP) in the regime: n = 6 passes, T = 450 °C on rods with a diameter of 10 mm. The microstructure and surface relief of the alloys were studied using an OLYMPUS GX5 optical microscope and a JEOL JSM 6395 scanning electron microscope. Mechanical tensile tests of small samples were carried out at room temperature at a rate of 0.04 mm/min in a machine for mechanical tensile testing of small specimens developed in IPAM USATU.

3. Results

According to the results of optical microscopy, the initial alloy in a coarse-grained state is characterized by a structure with a grain size of the original austenite of about 170 μm with martensite twins inside the grains (figure 1a). The $Ti_{50.2}Ni_{49.8}$ alloy after ECAP is also in the martensite state. The presence of martensitic boundaries makes it difficult to estimate the austenite grain size after ECAP, but the estimated austenite grain size is about 0.7 μm (figure 1b).

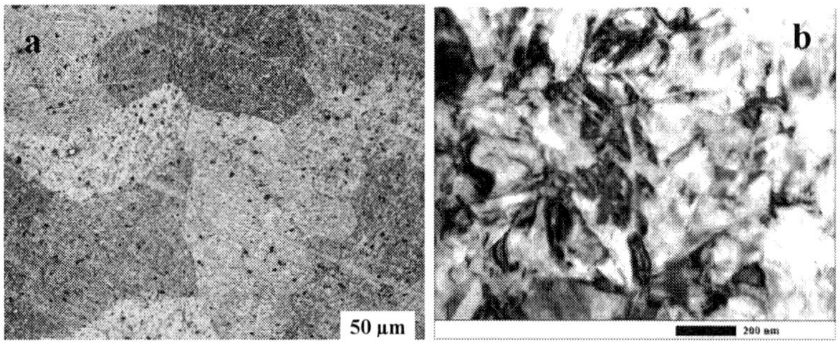

Figure 1. (a) Microstructure of the quenched $Ti_{50.2}Ni_{49.8}$ alloy, optical microscopy; (b) Microstructure of the $Ti_{50.2}Ni_{49.8}$ alloy after ECAP, TEM.

Tensile tests showed that the "σ - ε" curves have the form characteristic of the deformation of Ti-Ni alloys [3]. During the tensile test at room temperature, when the critical stress (σ_m) is reached, an area (5-10% of deformation) appears where the material is deformed due to the deformation-induced martensitic transformation. After this, an elastic deformation of martensite occurs, and then a plastic deformation of martensite takes place. The alloy in a quenched state has the following characteristics: the stress of the beginning of martensitic transformation (σ_m) is 220 MPa, the yield stress (YS) is 580 MPa, the tensile strength (UTS) is 850 MPa, and the elongation is 21%. Mechanical tests at room temperature led to the destruction of samples without localizing the deformation and the formation of a neck. ECAP leads to a significant increase in strength and yield stress (up to 1070 MPa and 930 MPa, respectively), and the elongation decreases to 18%. The ductility of the CG alloy after annealing at 500 °C increases noticeably, possibly due to the annihilation of the quenching stresses. Annealing at 500 °C of the UFG specimens leads to a decrease in the yield stress to 870 MPa, while the ductility varies insignificantly (figure 2).

One of the few effective methods for the direct study of the features of deformation-induced martensitic transformations is the metallographic observation of the surface relief of a sample during tensile testis. The surface was studied in four structural states: a coarse-grained state with a grain size of 170 μm, a coarse-grained state after annealing at 500 °C (d=170±15 μm), an ultrafine-grained state with a grain size of 700 nm, and a UFG state after annealing at a temperature of 500 °C (d=0.8-1.0 μm).

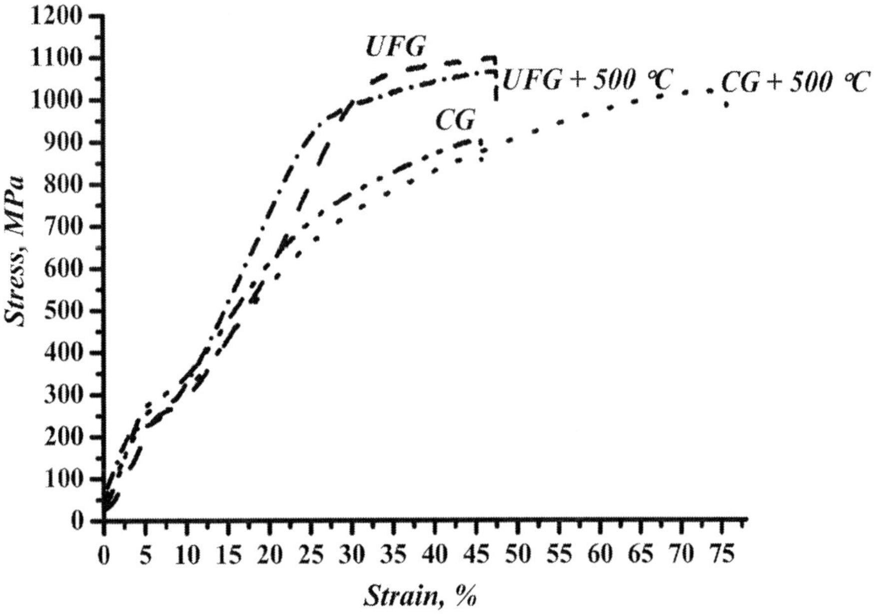

Figure 2. The «stress-strain» curves of the $Ti_{50.2}Ni_{49.8}$ alloy in various states: CG; CG and annealing at 500 °C; UFG; UFG and annealing at 500 °C.

Figure 3a shows a typical image of the surface relief of a quenched alloy. On the surface of the sample there is a developed relief. In addition, there are numerous cracks on the working gage. Cracks are also observed in SEM images. After tensile testing of the annealed sample, a similar pattern is observed (figure 3c). Deformation bands are also detected by scanning electron microscopy (figure 3d), which indicates that the relief has become even more developed.

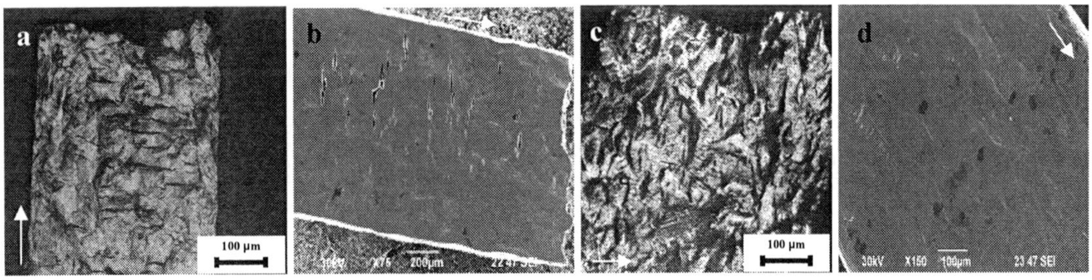

Figure 3. The structure of the surface relief of the $Ti_{50.2}Ni_{49.8}$ alloy in the CG state (a, b) and the annealed state at T = 500 °C (c, g): OM (a, c), SEM (b, d). The arrows indicate the direction of stretching.

The study of the deformation relief on the surface of the samples after the tensile tests of the UFG alloy by the OM method did not reveal a developed relief (figure 4), no cracks were observed. Perhaps this is due to the grain refinement after ECAP. In SEM, the relief is also not observed (figure 4b). The surface of the UFG specimen, annealed at a temperature of 500 °C, has a weak relief. Perhaps this is due to the coarsening of grains and an increase in ductility after annealing.

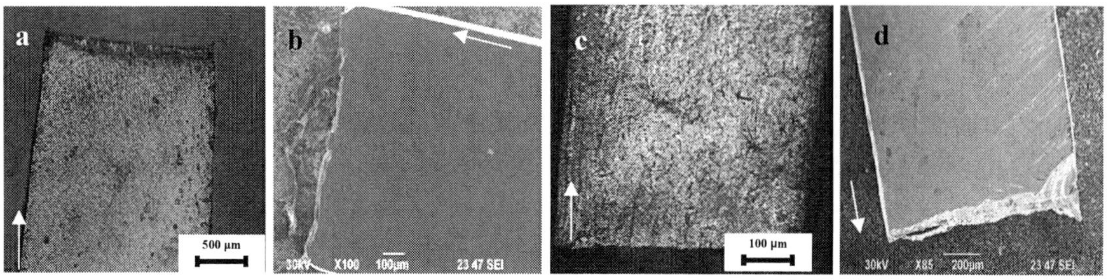

Figure 4. Structure of the surface relief of the $Ti_{50.2}Ni_{49.8}$ alloy in the UFG state (a, b) and annealed state at T = 500 °C (c, g): OM (a, c), SEM (b, d). The arrows indicate the direction of stretching.

In the initial state, the alloy has a high ductility and under tension undergoes a large deformation, but it breaks down without the forming of a neck. Hence, we observe expressed deformation bands and considerable work hardening of the material under tension in the martensitic structure, which leads to the formation of a developed relief and cracks during tensile testing of the sample in the initial state. ECAP leads to a decrease in ductility, and the material is not so significantly deformed and is work hardened when elongated to failure. Martensitic deformation is localized in individual small grains. Therefore, the developed relief and cracks are not formed.

4. Conclusion

The results of mechanical tests showed that annealing of the quenched state leads to an increase in ductility and strength. ECAP leads to an increase in the yield stress by about 40%. Annealing of ECAP states leads to a slight decrease in strength and yield stress. Samples after tension in the quenched state have a developed relief and microcracks. ECAP samples after tensile tests do not demonstrate such relief.

Acknowledgment

This work was supported by grant No. 6.37.204.2016 from Saint Petersburg State University.

References

[1] *Shape Memory Alloys: Fundamentals, Modeling and Applications*. ed. by Brailovski V, Prokoshkin S, Terriault P and Trochu F Montreal: Ecole de technologie superieure (ETS), Universite du Quebec, CANADA, 2003 p 851
[2] Gunderov D, Lukyanov A, Prokofiev E, Churakova A, Pushin V, Prokoshkin S, Stolyarov V and Valiev R 2013 *Mater. Sci. Forum* **738-739** 486
[3] Valiev R, Gunderov D, Lukyanov A and Pushin V 2012 *J. Mater. Sci.* **47** (22) 7848

IOP Conf. Series: Materials Science and Engineering **447** (2018) 012047 doi:10.1088/1757-899X/447/1/012047

Effect of preliminary heterogenization on the structure and hardness of cryorolled aluminum alloy D16

R R Ilyasov, E V Avtokratova, S V Krymskiy, O S Sitdikov and M V Markushev

Institute for Metals Superplasticity Problems, Russian Academy of Sciences, 39 Khalturin st., Ufa 450001, Russia

E-mail: ilyasov@imsp.ru

Abstract. The effect of preliminary heat treatment on transformations of the grain structure of the aluminum alloy D16 under isothermal rolling at the temperature of liquid nitrogen and subsequent annealing was investigated. It was found that heterogenization by intensifying the nucleation of new grains facilitates the nanostructuring of the alloy during cryorolling, as well as the formation of a fine-grained structure upon static recrystallization at post-deformation annealing.

1. Introduction

In recent years, aluminum alloys processed by severe plastic deformation at low (cryogenic) temperatures have become of the increasing scientific and commercial interest. This is reasoned by the fact that such processing results in nanostructuring of the alloy and a noticeable increase in its strength [1-3]. Along with the deformation conditions, their intensity is determined by a number of structural factors, including the degree of heterogeneity of the initial structure of the alloy. However, many aspects relating to this effect on the structure and the related properties, developed upon cryo-straining, still do not have a systematic description and remain unexplored.

The aim of the work is to study the effect of the main strengthening phases on the structural and microhardness changes in the aluminum alloy D16 during cryogenic rolling and post-deformation annealing in a wide temperature range.

2. Material and procedure

A hot-pressed rod of the aluminum alloy D16 of standard chemical composition (Al-4.4Cu-1.4Mg-0.7Mn, wt.%) with a diameter of 60 mm possessing a fibrous non-recrystallized structure (fiber thickness 100-200 μm, subgrain size 2 μm) was cut into plates with thickness of about 5 mm. After water quenching from a temperature of 505 °C, some of the plates were heterogenized for 5 hours at 400 °C. Subsequent isothermal rolling up to a total strain of $e \approx 2$ was performed at the temperature of liquid nitrogen on a laboratory six-roller mill. Therewith, the long axis of the plates coincided with the rolling direction. Subsequent annealing was carried out in air in the temperature range of 190-500 °C with an exposure time of 0.5 hours.

The microstructure of the alloy was studied using a Nikon L150 optical microscope after etching of mechanically grinded and polished templates in a standard Keller's reagent. The volume fraction of secondary particles was estimated on the images obtained using a TESCAN MIRA 3 LMH scanning electron microscope (SEM). Objects for electron backscatter diffraction (EBSD) analysis and transmission electron microscopy (TEM) were prepared using a TenuPol-5 twin-jet polisher at -28 °C in a 20% solution of nitric acid in methanol. The EBSD analysis was carried out using an Oxford Instruments HKL Channel 5 system. On the EBSD maps, low-angle boundaries (with misorientation from 2 to 15°) were highlighted by thin white-, and high-angle boundaries (more than 15°) - by thick black lines. The dislocation structure was analyzed using a JEOL 2000EX transmission electron microscope (TEM) at an accelerating voltage of 160 kV.

Vickers microhardness (HV) was measured by a standard procedure at a load of 1 N at room temperature on an Axiovert – 100 microscope equipped with a MHT-10 microhardometer.

3. Results and discussions

The TEM analysis showed that the matrix of the pre-quenched alloy, supersaturated with the main alloying elements, contained only the compact T-phase ($Al_{20}Cu_2Mn_3$) precipitates with a length of about 340 nm and a thickness of about 70 nm (figure 1 a). In addition, platelet particles of the θ and S phases precipitated and coagulated during subsequent heterogenization annealing (figure 1 e) were formed owing to the decomposition of the aluminum solid solution. The average size of such precipitates located along the grain boundaries reached about 1000 x 250 nm, while in the grain interiors they were much finer, not exceeding 380 x 70 nm. The volume fraction of these phases was about 8%.

Cryorolling of a pre-quenched alloy formed an inhomogeneous deformation structure consisting of arrays of deformation bands with predominantly weakly misoriented cellular structures (figure 1 b, c). Dislocation pile-ups were mainly noticed near the T-phases (figure 1 d), even though individual crystallites of a nanometric size with a TEM contrast typical for highly deformed nanograins were also present. In the pre-heterogenized state, on the contrary, a more homogeneous structure was formed with a more uniform distribution of dislocations, cells and dislocation boundaries, as well as a larger fraction of nano-sized crystallites, preferably located near particles (figure 1 f-h).

The observed effect is, apparently, originated from the interaction of lattice dislocations with secondary phases. It is generally accepted that the precipitates act as obstacles for the dislocation motion and provide their accumulation in adjacent matrix regions, followed by their transformations in the nuclei of new grains. Hence, the acceleration of dynamic recrystallization/deformation nanostructuring in a heterogenized alloy was related to the fact that the main strengthening phases "stimulated" the heterogeneous nucleation and grain refinement [4].

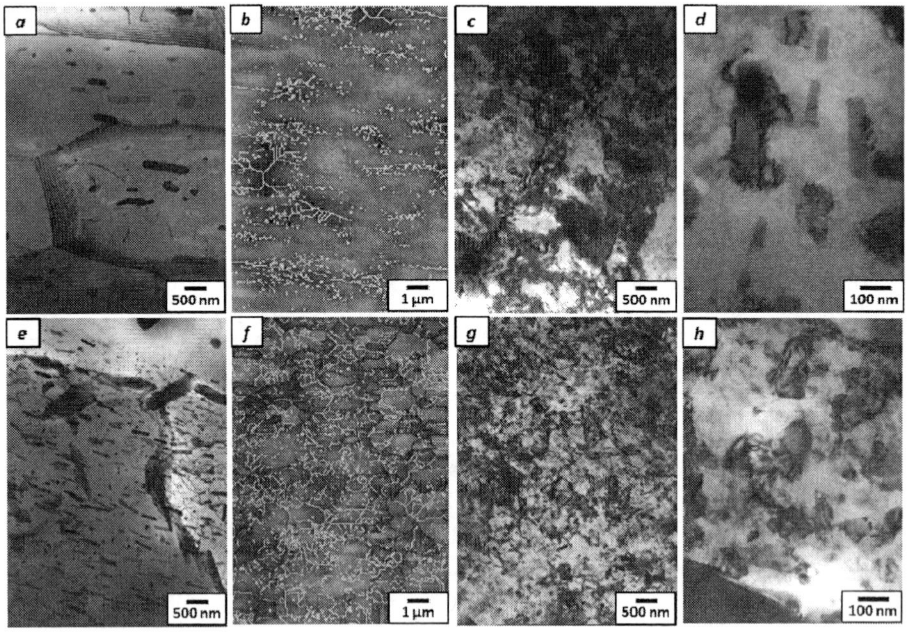

Figure 1. The structure of the alloy D16 after: quenching (a), heterogenization annealing at 400°C, 5 hrs (e), and subsequent cryorolling in the pre-quenched (b-d) and heterogenized (f-h) states. TEM (a,c,d,e,g,h), SEM-EBSD (b,f).

The evaluation of microhardness showed that after quenching the hardness of the alloy was about 115 HV, while heterogenization annealing led to a drop in hardness up to 75 HV; caused by a decrease in the alloying of a aluminum solid solution due to its decomposition with the formation and coarsening of secondary phases (i.e., decreasing the effects of solid-solution hardening and dispersion hardening). After cryogenic rolling, the increase in the microhardness of the alloy was almost the same in both states (about 65 HV) due to the developed deformation structure with signs of nanostructure (figure 1). As a result, the quenched alloy showed higher hardness (about 180 HV), whereas the hardness in heterogenized state was fixed at a level of 150 HV (figure 2).

The subsequent half-hour annealing at 190 °C further increased the hardness of the pre-quenched alloy up to 200 HV owing to the decomposition of the aluminum solid solution with the formation of dispersed products (zones) in a heavily deformed matrix. In contrast, such a response to aging during low-temperature annealing of ф pre-heterogenized alloy was completely lost due to the low degree of supersaturation of its solid solution upon preliminary heat treatment. As a result, the hardness of the deformed alloy decreased due to the occurrence of static recovery and recrystallization, which also took place more readily in the pre-heterogenized alloy (figure 3). With increasing the annealing temperature to 400 °C, the same restoration processes in the matrix, as well as an increase in the precipitate sizes, led to an intensive decrease in the alloy hardness in both states up to the level of the initial hot-pressed rod (about 80 HV) (figure 2). At annealing temperatures of beyond 400 °C, i.e. above the solvus point for the main strengthening phases [3], the alloy matrix was normally saturated by main alloying elements in accordance with the phase diagram and, hence, water-cooling after annealing led again to the formation of a supersaturated solid solution, accompanied by the repeat solution hardening of the alloy at room temperature in both structural states (figure 2).

As mentioned above, increasing the annealing temperature promoted the occurrence of static recrystallization in the present alloy and, hence, an entirely uniform fine-grained structure with an average grain size of 10 μm was evolved in the pre-quenched alloy at 500 °C (figure 4 a, c). Preliminary heterogenization promoted, in turn, the earlier (at lower temperatures) formation of more homogeneous recrystallized structures with a finer grain size (figure 4 b, d). Thus, a more uniform structure with a finer grain size of about 8 μm was processed in a heterogenized alloy (figure 4 d). This suggests that precipitates of the main strengthening phases also served as very effective structure control agents of, which stimulated the grain refinement upon static recrystallization in a heavily deformed alloy.

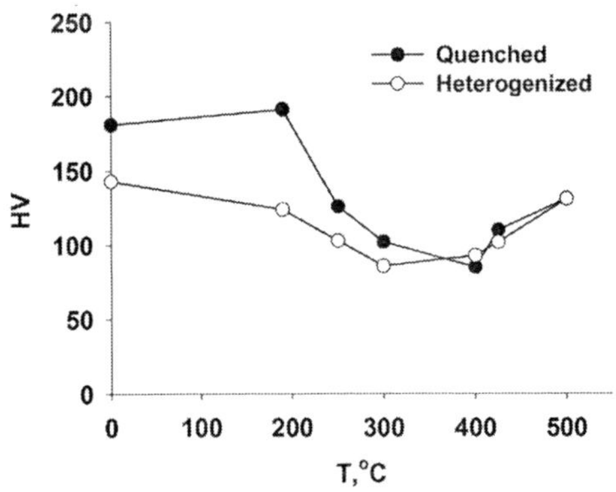

Figure 2. Microhardness of the cryorolled alloy D16 vs temperature of 0.5 hr post-deformation annealing.

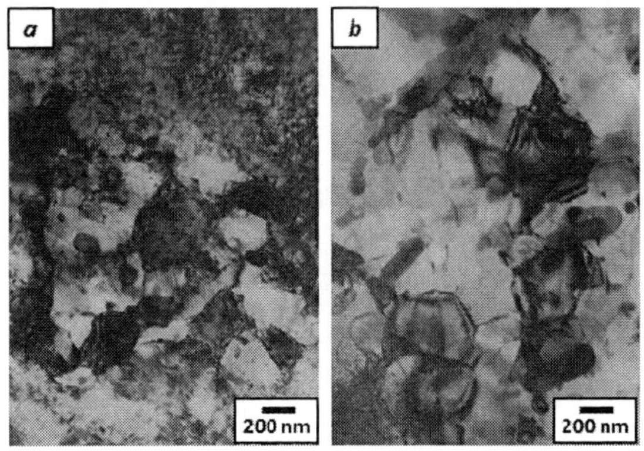

Figure 3. TEM structure of the cryorolled and annealed at 250 °C alloy D16 in the pre-quenched (a) and heterogenized (b) states.

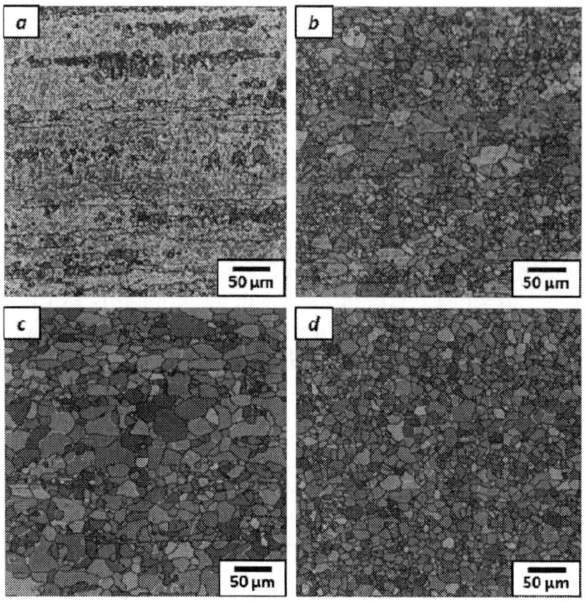

Figure 4. Microstructure of the cryorolled and annealed at 400 (a, b) and 500 °C (c, d) alloy D16 in the pre-quenched (a, c) and heterogenized (b, d) states.

4. Conclusions

1) The analysis showed that preliminary heterogenization of the age-hardenable aluminum alloy D16 facilitates the formation of a nanocrystalline structure during cryogenic rolling by stimulating dynamic recrystallization, as well as the development of a fine-grained structure during subsequent annealing by stimulating static recrystallization.
2) The highest hardness of about 200 HV was achieved in the pre-quenched alloy D16 after cryorolling and the subsequent half-hour annealing at 190 °C. At the same time, cryorolling of the preliminary heterogenization alloy led to a hardness of only 150 HV, which then monotonically decreased upon the subsequent annealing below the solvus point. Post-deformation annealing at higher temperatures led to recovery and recrystallization in both structural states, which reduced the hardness level up to the value of about 80 HV, corresponding to the initial hot-pressed rod.

Acknowledgments
The present research was performed in accordance to the state assignment of IMSP RAS. Experimental work was carried out using the facilities of the Shared Service Center of IMSP RAS «Structural and Physical-Mechanical Studies of Materials».

References
[1] Ma E 2006 *JOM* **4** 49
[2] Markushev M V, Avtokratova E V, Kazakulov I Y, Krymsky S V, Mochalova M Y, Murashkin M Yu and Sitdikov O S 2011 *Russian Metallurgy (Metally)* **4** 364
[3] Krymskiy S V, Avtokratova E V, Sitdikov O S, Markushev M V and Mikhaylovskaya A V 2015 *Phys. Metals Metallogr.* **7** 676
[4] Humphreys F J and Hatherly M 2004 *Recrystallization and Related Annealing Phenomena*: second ed. Elsevier p 658

IOP Conf. Series: Materials Science and Engineering **447** (2018) 012048 doi:10.1088/1757-899X/447/1/012048

Features of the deformation behavior under pulse current and ultrasound in materials with phase transformation

V V Stolyarov[1,2], Yu V Tsarenko[3] and V V Rubanik[3]

[1] Mechanical Engineering Research Institute of Russian Academy of Sciences, 101990, Moscow, Russia
[2] National Research Nuclear University MEPhI, 115409, Moscow, Russia
[3] Institute of Technical Acoustics of National Academy of Belorussia, 210023, Vitebsk, Belorussia

E-mail: vlstol@mail.ru

Abstract. The paper presents the experimental results on tension behavior of the wire under electrical current and ultrasonic wave for coarse-grained pure Ni and TiNi. The materials display electroplastic and acoustoplastic effects that can be responsible for a significant strengthening or softening of the material and may influence the deformability of the material.

1. Introduction

For the production of various semi-finished products, thermomechanical processing, including cold rolling or drawing is used. In this case, there is a problem of simultaneous increase in deformability and functional properties of materials, since the first one is achieved due to high processing temperatures, which can lead to deterioration of the second one. One of the approaches to solving this problem is the use of external energy effects, not accompanied by high temperatures. In alloys with thermoelastic martensitic transformation, the methods of deformation-thermal processing in combination with pulsed current (PC) [1] or ultrasonic (US) action [2] really allow to significantly increase the technological and operational properties, however, the mechanism of such influence is not sufficiently studied. The influence of the combined action of both effects is of particular interest. It was shown that for coarse-grained copper, the combination of electroplastic (EPE) and acoustoplastic (APE) effects can significant influence the dislocation mobility and deformation behavior of the material [3]. The purpose of this work is a comparative study of the deformation behavior in materials with or without phase transformation under conditions of electropulse and / or ultrasonic action.

2. Experimental

Wire samples with a coarse-grained structure of annealed pure nickel (NP2, mm Ø1.6) and the $Ti_{50}Ni_{50}$, $Ti_{49.3}Ni_{50.7}$ shape memory alloy (Ø1.38mm) were used as the starting materials. The tensile test was carried out with a horizontal R-5081-20 machine, at 20 and 200 °C, a ram speed of 1 mm/min) and a vertical Instron machine (at 20 °C, a ram speed of 1.5 mm/min.). The ultrasound frequency was 18 kHz, the amplitude of ultrasonic shifts (A) from 5 to 20 μm, and their duration varied from 1 to 5 s. Regimes of pulse current were: a - for Ni, pulse duration τ=2 x 10^{-4} s, density j = 120 A/mm^2, b - for TiNi, τ = 1 x 10^{-4} s, j = 500 A/mm^2. The difference in the current modes is conditioned by different electrical conductivity of materials.

IOP Conf. Series: Materials Science and Engineering **447** (2018) 012048 doi:10.1088/1757-899X/447/1/012048

3. Results

Figure 1 shows the stress-displacement curves of coarse-grained nickel with a different sequence of external influences: (a) current + ultrasound; (b) ultrasound + current. The impact of the pulsed current, and then ultrasound, leads to stress jumps downward, caused by EPE of 25-50 MPa and APE up to 180-220 MPa (figure 1a). When changing the sequence of effects, an additional decrease in the stress conditioned by the total effect of ultrasonic and current, was observed (figure 1b).

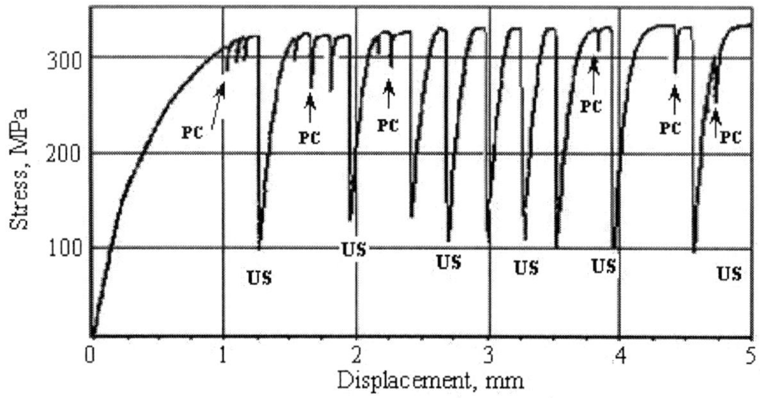

a

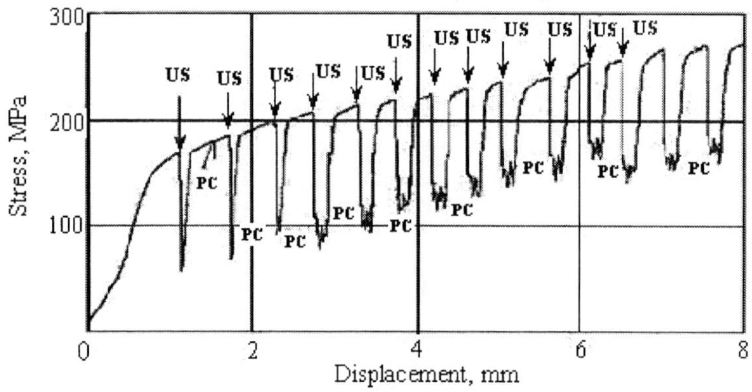

b

Figure 1. Stress-displacement curves of Ni under external influences: (a) current + ultrasound, A=20 μm; (b) - ultrasound + current, A=5 μm.

This result agrees with the data for copper in [3] and complements the paper [4]. Note that with decreasing intensity of ultrasound (A decreases to 5 μm), APE decreases to 100-120 MPa. The noticeable difference in the values of EPE and APE is explained by the difference in the intensity of the impacts, frequency and duration.

The serial introduction of ultrasound and current pulses at ambient temperature in $Ti_{49.3}Ni_{50.7}$ alloy leads to emergence of differently directed stress jumps (figure 2a): up (from current) and down (from ultrasound), but at 200 °C (in steady austenite) all stress jumps are directed (figure 2b) down. The divergence of jumps at ambient temperature testifies to various natures of acoustic and electroplastic effects that require careful study.

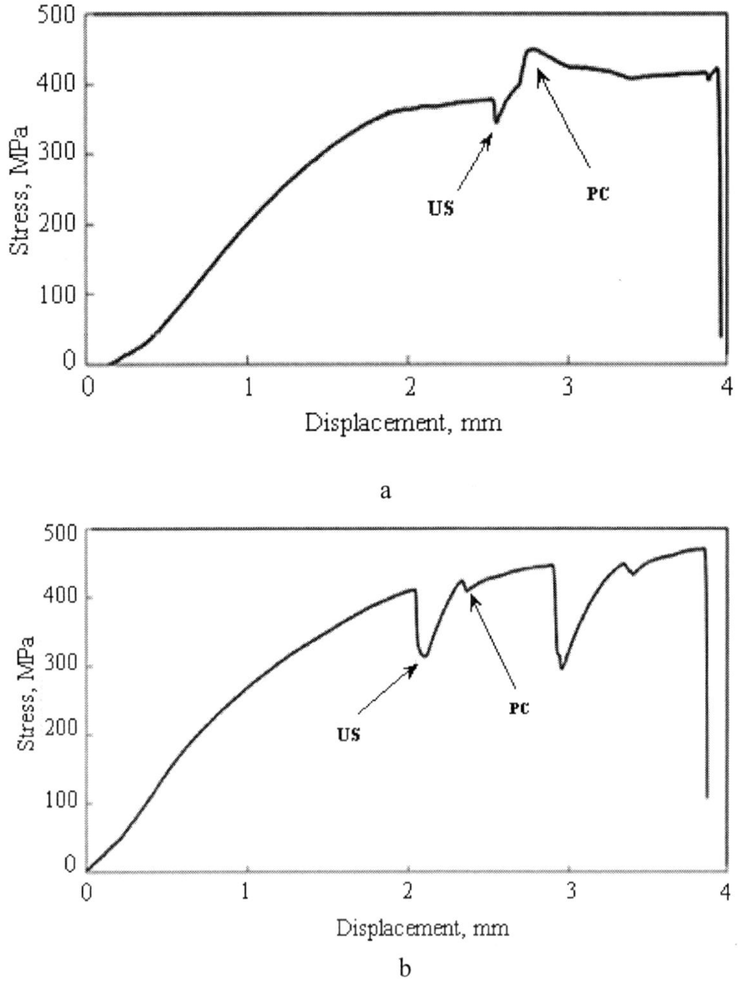

Figure 2. Stress-displacement curves of $Ti_{49.3}Ni_{50,7}$ under sequential introducing of ultrasound and pulse current at (a) 20 °C and (b) 200 °C (b).

The combined effect of ultrasound and pulse current was investigated for the $Ti_{50}Ni_{50}$ alloy (figure 3). If current pulses and ultrasound are injected simultaneously, unidirectional stress jumps markedly different in amplitude are observed. Such a feature may be associated with the inverse phase transformation M→A or with the straining mechanical impacts, however it requires further research.

The features of the jumps observed on the deformation curves for both effects are their complete reversibility and asymmetric type of the drop and rise. Indeed, after an almost instantaneous drop in stress, a noticeably slower rise to the initial value on the stress-strain curve follows. Since both materials were in a recrystallized coarse-grained state and the instantaneous temperatures were much less than the recrystallization threshold (less than 200 °C), the mechanism of the phenomena can be associated with local energy absorption on the crystal defects of the material, called the Blaga - Langenecker effect [5].

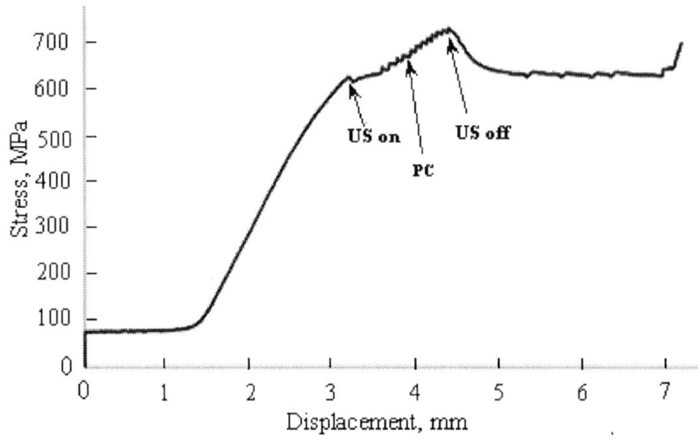

Figure 3. Stress-displacement curve of $Ti_{50}Ni_{50}$ under simultaneous introducing ultrasound and pulse current at 20 °C.

The nature of upward jumps arising only in a shape memory alloy is associated with the implementations of a reverse phase transformation when the temperature rises above As (the starting temperature of the inverse transformation M → A) from the introduction of ultrasound or current pulse [6].

Acknowledgment
The work has been supported by RFBR, grant # 16-58-48001.

References

[1] Potapova A A, Stolyarov V V, Bondarev A B and Andreev V A 2012 *Machine Build. Eng. Education* **2** 33 (In Russian)
[2] Rubanik V V, Rubanik V V, DorodeikoV G and Miliukina S N 2013 *Mater. Sci. Forum* **738-739** 362
[3] Kozlov A V, Mordyuk B N and Chernyashevsky A V 1995 *Mater. Sci. Eng.* A **190** 75
[4] Rubanik V V 2017 *Initiation of Shape Memory Effect in TiNi Alloys Under Ultrasonic Wave* (Doct. Sci. thesis, Minsk) (In Russian)
[5] Langenecker B 1966 *IEEE Trans. Sonics & Ultrasonics* **13** 1
[6] Siua K W, Ngana H W and Jones I P 2011 *Int. J. Plast.* **27/5** 788

IOP Conf. Series: Materials Science and Engineering **447** (2018) 012049 doi:10.1088/1757-899X/447/1/012049

Modeling of the stress-strain state for a disk and a shaft from different Ni-based alloys during welding under shear pressing

A Kh Akhunova[1,2], S V Dmitriev[1,3], V A Valitov[1,4], E V Galieva[1]

[1] Institute for Metals Superplasticity Problems of Russian Academy of Sciences, 39 Khalturin st., Ufa 450001, Russia
[2] Ufa State Aviation Technical University, 12 K. Marx st., Ufa 450008, Russia
[3] National Research Tomsk State University, 36 Lenin Ave., Tomsk 634050, Russia
[4] Bashkir State University, 32 Validy Str., Ufa 450076, Russia

E-mail: akhunova_a@mail.ru

Abstract. The stress-strain state in a disk and a shaft made of different heat-resistant Ni-based alloys during pressure welding is analyzed by the finite element method. The disk and the shaft have a complimentary cone shape, and during welding the shaft is inserted into the disk or inserted and simultaneously rotated. In the first case, a single component shear deformation is realized, while in the second case, one has a two-component shear deformation in the welding zone. The cone angle of 5° can be recommended, because in this case the plastic deformation in the welding zone is about 3%, which is typically sufficient for obtaining a reliable joint.

1. Introduction

Improving the manufacturing technology of rotor structures designed for the 5th generation gas turbine engine is a very important problem [1-16]. The use of bimetallic parts, such as blisk, disk–disk and disk–shaft, made of heat-resistant nickel and titanium alloys, reduces the weight of the engine. Of particular interest for the engine is a disk fixed on the turbine shaft. When creating such units, it is necessary to ensure a reliable connection between the welded surfaces, since the presence of defects in the weld of such structures can lead not only to the destruction of the part, but also to the destruction of the entire engine. Pressure welding (PW) is widely used to obtain permanent joints in such structures. In the process of PW, a solid-phase joining takes place under the relatively small applied pressure between the parts being welded, which provides plastic deformation of the contact surfaces and stimulates diffusion processes. This approach has several advantages, e.g., it is easy to execute, it allows welding of similar materials and materials with very different properties, it allows controlling the structure and properties of the joint. However, this method should be applied with caution, because in some cases the joining can be accompanied by significant pore formation, which leads to heterogeneity of the welded joint and to the deterioration of the quality of welding. To avoid such drawbacks, a combined PW can be used, in which together with pressing, the shear deformation is present in the welding zone. The application of shear during welding helps to increase plastic deformation in the layer close to the surfaces to be welded and does not affect strongly the rest of the material, which is very important for such parts of the aircraft engine as disks and shafts.

In this paper, we consider the method for producing a welded joint by PW, carried out by inserting and/or rotating a shaft in contact with a disk. With the help of finite element simulation, the stress-strain state in the disk and the shaft from the heat-resistant nickel-base alloys EP975 and EK79 is investigated with the aim to estimate the magnitude of shear deformation at the contact surface. It is well known that shear deformation in the welding zone improves the quality of welding.

2. Finite element model

Computer simulation is performed using the DEFORM-2D software package. Pressure welding of the shaft and the disk is modeled in the axisymmetric setting. A schematic view of the disk and the shaft is presented in figure 1 a. Different forms of the shaft and the disk are considered. The shaft is a cylinder with a diameter of 12 mm and a height of 12 mm, coupled with a truncated cone with a height of 20 mm having the angle 0°, 0.5° or 5°. In the case of a cone angle of 0°, the shaft has a simple cylindrical shape. The hole in the disk has a narrowing with the same cone angle as that of the shaft, so that the shaft and the disk have complimentary shapes. The disk height is 14 mm and the diameter is 28 mm.

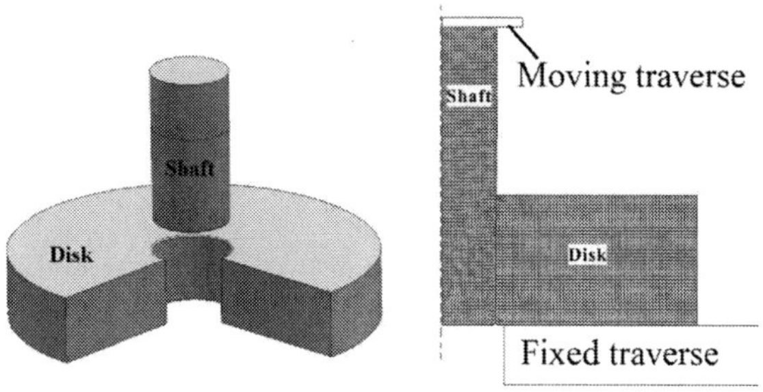

Figure 1. (a) Schematic view of the disk and shaft of a gas turbine engine. Part of the disc is removed for clarity. (b) Finite element model of the shaft and the disk. The disk lies on a fixed traverse, and the shaft is inserted in the disk with or without rotation about its axis.

The finite element model is shown in figure 1 b. When conducting computer simulation, the deforming tool (the upper moving traverse) and the supporting body (the lower fixed traverse) have the properties of an absolutely rigid body. The deformable bodies (the disk and the shaft) are assumed to be elastic-plastic. The disk and the shaft are made of the ultrafine-grained superalloys EP975 and EK79, respectively. The material properties were determined by experimental stress-strain curves obtained under uniaxial compression of the alloys at the welding temperature. Welding was carried out in isothermal conditions at a temperature of 1100 °C.

Deformable bodies were divided into twenty-node isoparametric finite elements with a quadratic approximation of the displacement fields. The number of elements is 2500 for the disk and 1700 for the shaft. For welding, the disk was mounted on a fixed traverse and the conical part of the shaft was placed into the disk hole. Contact conditions at the boundaries of the traverse-shaft and the disk-shaft pairs are described by the Ziebel friction model. The value of the friction coefficient is assumed to be 0.3. The deformation of the shaft and the disk in the course of PW, carried out during the insertion and/or rotation of the shaft in contact with the disk, was simulated. Two different loading schemes are considered:

Scheme 1: the shaft is inserted into the disk by the motion of the upper traverse down by 3 mm with the strain rate of 10^{-3} s^{-1};

Scheme 2: a combination of the insertion of the shaft into the disk and simultaneous rotation of the shaft about the z axis. The shaft is inserted into the disc by 3 mm with a strain rate of 10^{-3} s^{-1} and rotates with a speed of 3 rpm.

3. Simulation results

When analyzing the results of computer simulation, the distribution of the shear deformation is considered, since the normal deformation components have little effect on the quality of the welded joint.

The simulations show that, in welding according to Scheme 1, a single component of shear deformation prevails, which is ε_{rz}, while $\varepsilon_{r\theta}$ is much smaller. In this case one can speak of a one-component shear deformation. On the other hand, welding according to Scheme 2 results in both shear deformation components, ε_{rz} and $\varepsilon_{r\theta}$, to be comparable, i.e., a two-component shear deformation takes place in this case.

The distribution of ε_{rz} in the disk and the shaft after welding according to Scheme 1 is shown in Figure 2 for the cone angle a) 0°, b) 0.5°, and c) 5°. It can be seen that shear deformations in the contact zone increase with an increase in the cone angle. In the cases shown in b) and c) the shear deformation is distributed more evenly, while in a) it is concentrated close to the upper and lower surfaces of the disk. The maximum values of ε_{rz} increase with increasing cone angle. In b) they are 10 times greater than in a) and in c) 100 times greater than in a).

The distribution of ε_{rz} and $\varepsilon_{r\theta}$ in the disk and the shaft at the end of welding according to Scheme 2 are presented in figure 3 and 4, respectively. Both components of shear deformation are of the same order of magnitude in this case. The component ε_{rz} is distributed similarly to the case of Scheme 1. The distribution of $\varepsilon_{r\theta}$ is somewhat different from the distribution of ε_{rz}, but in both cases shear deformation is localized in the welding zone. As the cone angle increases, the maximum values of shear deformation also increase similarly to the case of Scheme 1. The area of maximum values with increasing angle of the shaft is distributed more evenly in the Scheme 2.

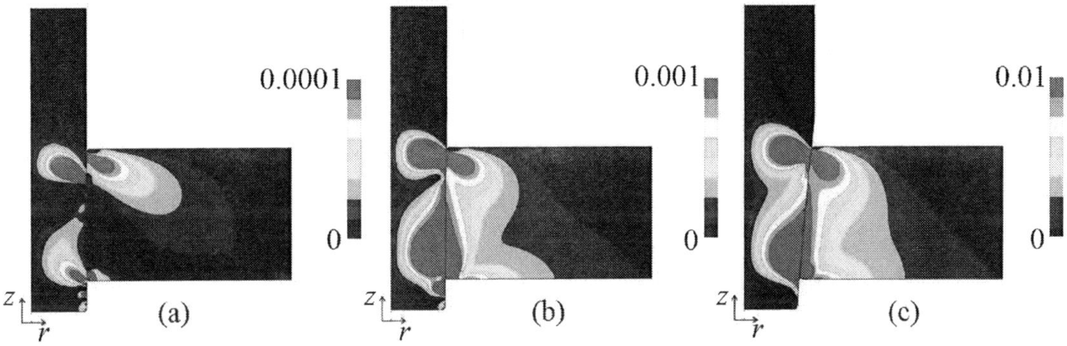

Figure 2. Distribution of shear strain ε_{rz} in the shaft and disk deformed according to Scheme 1 for the taper angle: a) 0°, b) 0.5°, and c) 5°. Shear strain $\varepsilon_{r\theta}$ is nearly zero in this case.

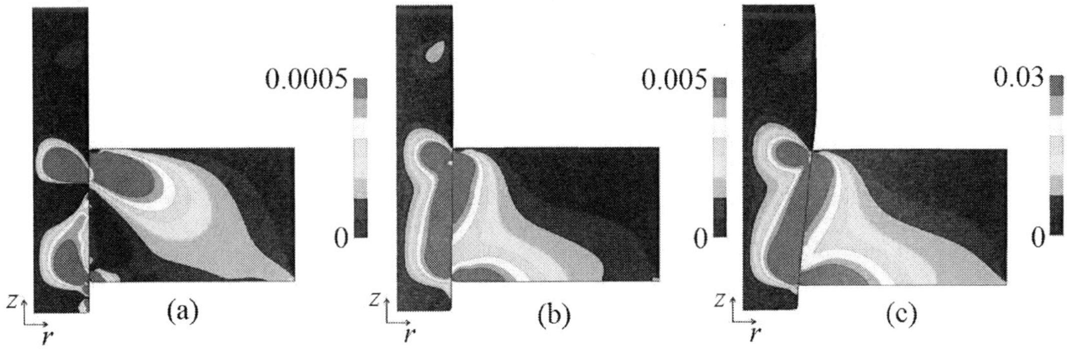

Figure 3. Distribution of shear strain ε_{rz} in the shaft and the disk deformed according to Scheme 2 for the taper angle: a) 0°, b) 0.5°, and c) 5°.

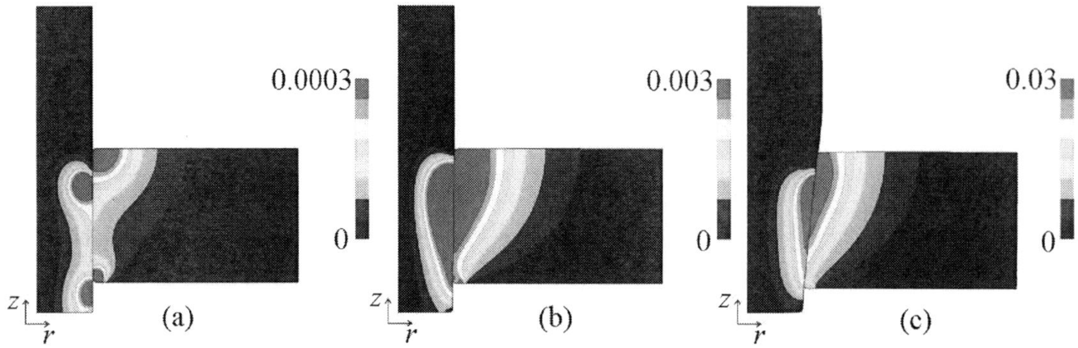

Figure 4. Distribution of shear strain $\varepsilon_{r\theta}$ in the shaft and the disk deformed according to Scheme 2 for the taper angle: a) 0°, b) 0.5°, and c) 5°.

4. Conclusions

In our work we came to the following conclusions:

1) To create a permanent connection between the disk and the shaft, it is preferable to use pressure welding, which is carried out with a combination of insertion and rotation of the shaft in the disk. In this case, a two-component shear deformation is provided, which, as is well known, improves the quality of the welded joint.

2) To improve the quality of the welded joint, the shaft and the disk should have a complimentary cone shape. The cone angle of 0.5° seems to be too small because it produces only 0.3% of shear strain. The angle of about 5° can be recommended because in this case the plastic shear deformation of about 3% is achieved, which is typically sufficient for obtaining a reliable joint.

Acknowledgements

This work was supported by Russian Science Foundation Grant No. 18-19-00685 (in the part devoted to computer simulation). The work was partly supported by the IMSP RAS State assignment No. AAAA-A17-117041310215-4 (in the part devoted to experimental data for the choice of pressure welding conditions).

References

[1] Sims C T, Stoloff N S and Hagel W C 1987 *Superalloys II: High-Temperature Materials for Aerospace and Industrial Power* (New York: Wiley)
[2] Jozwik P, Polkowski W and Bojar Z 2015 *Materials* **8** 2537
[3] Povarova K B, Valitov V A, Ovsepyan S V, Drozdov A A, Bazyleva O A and Valitova E V 2014 *Russian Metallurgy* (*Metally*) **9** 733
[4] Galieva E V, Lutfullin R Ya, Akhunova A Kh, Valitov V A and Dmitriev S V 2018 *Sci. Technol. Weld. Join.* **23** 612
[5] Akhunova A K, Valitova E V, Valitov V A and Dmitriev S V 2017 *Welding Int.* **31** 879
[6] Valitov V A, Akhunova A Kh, Galieva E V, Dmitriev S V, Lutfullin R Y and Zhigalova M Y 2017 *Letters on Materials* **7** 180
[7] Galieva E V, Valitov V A, Zhigalova M Y, Lutfulin R Y and Akhunova A Kh 2017 *Mater. Phys. Mech.* **33** 57
[8] Akhunova A Kh, Galieva E V, Drozdov A A, Arginbava E G, Dmitriev S V and Lutfullin R Y 2016 *Letters on Materials* **6** 211
[9] Akhunova A K, Valitova E V, Dmitriev S V, Valitov V A and Lutfullin R Y 2016 *Welding Int.* **30** 488
[10] Galieva E, Valitov V, Lutfullin R, Dmitriev S, Akhunova A and Mukhametrakhimov M 2016 *Mater. Sci. Forum* **838-839** 350

IOP Conf. Series: Materials Science and Engineering **447** (2018) 012049 doi:10.1088/1757-899X/447/1/012049

[11] Safiullin R V, Dmitriev S V, Akhunova A K and Safiullin A R 2014 *Materialwiss. Werkst.* **45** 835
[12] Safiullin R V, Kruglov A A, Akhunova A K, Safiullin A R and Dmitriev S V 2012 *Materialwiss. Werkst.* **43** 786
[13] Akhunova A, Dmitriev S, Kruglov A and Safiullin R 2010 *Key Eng. Mater.* **433** 319
[14] Galieva E, Valitov V, Lutfullin R and Bikmukhametova A 2018 *Def. Dif.. Forum* **385** 150
[15] Povarova K B, Valitov V A, Drozdov A A, Bazyleva O A, Galieva E V and Arginbaeva E G 2018 *Russian Metallurgy* (*Metally*) **1** 42
[16] Valitov V, Povarova K, Bazyleva O, Drozdov A, Ovsepyan S and Galieva E 2016 *Mater. Sci. Forum* **838-839** 523

IOP Conf. Series: Materials Science and Engineering **447** (2018) 012050 doi:10.1088/1757-899X/447/1/012050

Computer simulation of superplastic forming of a three-sheet structure containing an ultrafine-grained core

V Ganieva[1], A Kruglov[1,2], R Lutfullin[2], O Rudenko[2], F Enikeev[1]

[1]Ufa State Petroleum Technological University, 1 Kosmonavtov St., Ufa 450062, Russia
[2]Institute for Metals Superplasticity Problems RAS, 39 Khalturin St., Ufa 450001, Russia

E-mail: alweld@go.ru

Abstract. Superplastic forming of a three-sheet structure is considered. The focus is on overcoming the main disadvantages of the process of manufacturing of such structures by means of the SPF/DB technique including high temperatures of treatment and the formation of folds on the outer surface of the article manufactured. The application of ultrafine-grained sheets of titanium alloy Ti-6Al-4V has been analyzed in order to avoid these disadvantages. Various combinations of the use of ultrafine-grained and microcrystalline sheets as skin and core sheets are considered by means of the finite element analysis. The standard power law of superplasticity is used for analysis with material constants determined from technological experiments. The conclusion is made that the most attractive way appears to be the usage of ultafine-grained material for the core sheet and microcrystalline material for the skin sheets.

1. Introduction

Modern aerospace industry deals with articles that combine high specific strength and reliability. Such requirements are met by three-layer structures with corrugated filler, obtained by superplastic forming (SPF), concurrent with diffusion bonding (DB) of titanium alloys [1]. The scheme of this process was published in many works [2-5]. Briefly, it should be noted that the structure forms a package of three sheets. Two sheets serve as shells. The inner sheet is selectively connected to the skin by means of DB. In the SPF process, the inner sheet is stretched between the skins. As a result, the face sheets take the form of a die, while the core sheet forms stiffeners.

It should be noted that interest in three-layer structures still does not disappear. The key detail of the 5th generation aircraft engine – a hollow fan blade – has a similar design. For the first time the English company Rolls-Royce began to produce hollow blades for aircraft engines of the new generation. Now hollow fan blades are developed by the French company Snecma and others. The unique face plates of golf club heads also are three-layered [6].

Along with the development of the use of three-layer corrugated structures manufactured by the SPF/DB technique, the problems of their quality and production efficiency continue to be essential. The main of them are as follows:

- high temperatures of treatment 900–925 °C [1];
- oxidation during superplastic forming leads to the formation of the so-called α-layer on the surface of the article, which must then be removed by chemical etching;
- low resistance of the tools made of expensive heat-resistant materials;

- formation of folds on the surface of the product.

Solving the above problems, a filler sheet made of the ultrafine-grained (UFG) titanium alloy can be produced, while the face sheets can be made of rolled sheets having a microcrystalline (MC) structure [7, 8]. The domestic corporation VSMPO-AVISMA has experience in manufacturing commercial sheets with a grain size of less than 1 μm.

The purpose of this study is to show the possibility of solving these problems on the basis of the results of modeling and experimental data.

2. Materials and methods

Figure 1 shows a scheme of a fragment of a three-sheet structure. All three sheets are the commercial titanium alloy VT6 (Ti-6Al-4V). Skin sheets with a thickness of s_s have an MC structure with an average grain size of 3 μm. The core sheet with a thickness of s_c possesses an UFG structure with an average grain size of 1 μm. The pressure p from the inertia gas is applied to the free surfaces of the inner cavities of the sheet package. The height of the structure H is equal to the maximum distance between the plates of the die. The angle between the ribs and the skin is denoting as φ. The temperature of SPF is equal to 750 °C.

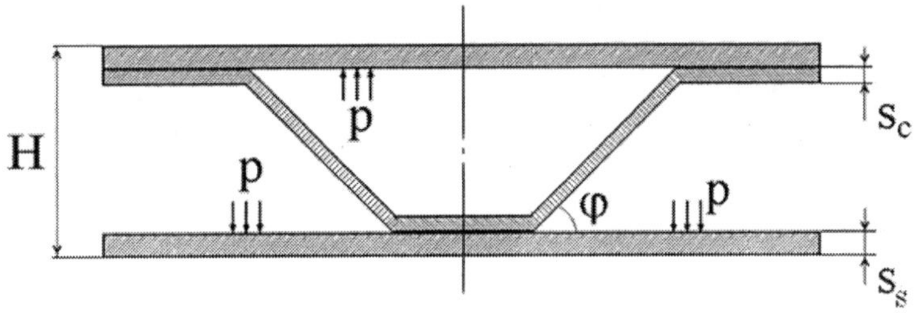

Figure 1. Fragment of a three-sheet structure

3. Results and discussion

The following standard power law of superplasticity has been chosen for analysis

$$\sigma = K\xi^m \quad \text{or} \quad \xi = C\sigma^n, \tag{1}$$

where σ is the flow stress, ξ is the strain rate; K, m, C, and n are the material constants to be determined experimentally. To determine the values of material constants for UFG titanium sheets, technological experiments with constant pressure have been done in accordance with the procedures described in details in [9]. The obtained results are collected in table 1.

Table 1. Results of technological experiments on Ti-6Al-4V

T SPF, [°C]	p, [MPa]	t, [s]	H, [mm]	R, [mm]	r_0, [mm]	s_0, [mm]	Average grain size, mμ
750	2.0	1000	35	35	1	1	~1
	2.5	470					
	3.0	305					

To describe the mechanical behavior of the VT6 sheet alloy with the UFG structure, the following pair of experimental data was treated in accordance with procedures described in [9]: the duration of forming under a constant pressure of 2.5 and 3.0 MPa equals to 470 and 305 s respectively. As a result, it was found that m=0.4216 and K=1181 MPa·s^m. This set of material constants was used in calculations to describe the mechanical response of the UFG material. The constant pressure generated under relatively low pressure resulted in the fact that considerable grain growth influenced the mechanical response of the material under study. Therefore, to describe the rheological behavior of the MC sheet alloy, the following pair was chosen: the duration of forming under a constant pressure of 2.0 and 3.0 MPa equals to 1000 and 470 s respectively. The values of the determined material constants were be as follows: m=0.2955 and K=531 MPa·s^m. This set of material constants was used in calculations to describe the mechanical response of the MC material.

The formulation of a boundary value problem in the mechanics of solids is described in details in [10]. Standard ANSYS software was used to solve it. The calculation area was divided into 397 PLANE82 elements with the Plane strain option. The tool was set as an absolutely rigid body using the TARGE169 element. The boundary conditions at the interface between the tool and the deformable body were set by the contact element CONTA172.

The following numerical data were chosen for the calculations: s_s= 1.6 mm, s_c=0.8 mm, H=10 mm, φ=60°. Computer modeling was carried out in a two-dimensional setting by using the software ANSYS10.0 (ED). In view of a symmetry, only half of the represented fragment was considered. The finite element mesh is shown in figure 2.

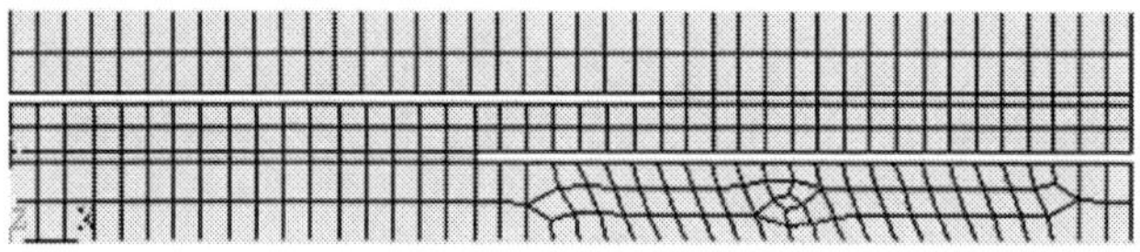

Figure 2. Finite element mesh at the initial position.

The regime of loading (pressure time cycle) has been chosen as follows: at the first stage of forming, the pressure is increased linearly with time up to its maximum value p_{max} to ensure optimum conditions for superplasticity [11]. At this stage, the core sheet is deformed under conditions of superplasticity, stiffening ribs are formed. The first stage continues until the skin sections joined to the core touch the surface of the die. The second stage proceeds under conditions of creep up to smoothing the outer folds.

Four different combinations of microstructures of skin and core sheets and regimes of loading have been considered as listed in table 2.

Table 2. Variants of microstructures and regimes of loading used in calculations

No.	Layer's thicknesses and microstructures	Duration of forming, s / p_{max}, [MPa]	Holding time, s, under constant p_{max}=4 [MPa]	Depth of the groove, [mm]	Equivalent strain of the core (max)
1	1.6+0.8+1.6 UFG+UFG+UFG	900/2	7200	0.11	1.446
2	1.6+0.8+1.6 UFG+UFG+UFG	3600/4	7200	None	1.459
3	1.6+0.8+1.6 MC+UFG+MC	900/2	7200	0.12	1.193
4	1.6+0.8+1.6 MC+UFG+MC	3600/4	7200	None	1.308

The results of calculations for all the considered variants show that if the value of the angle of inclination of the stiffeners to the skin does not exceed 60°, the thickness of the skin sheet should be at least twice as thick as the core sheet. The parameters that control the wrinkle smoothing process are the holding time (which is quite obvious) and the time of forming. When forming according to the law of linear pressure, an increase in time of forming leads to the decrease in the strain rate. As seen in figure 3, the most favorable situation is associated with the regime of loading up to p_{max}=4 MPa for 1 hour and consequent holding under this pressure for 2 hours.

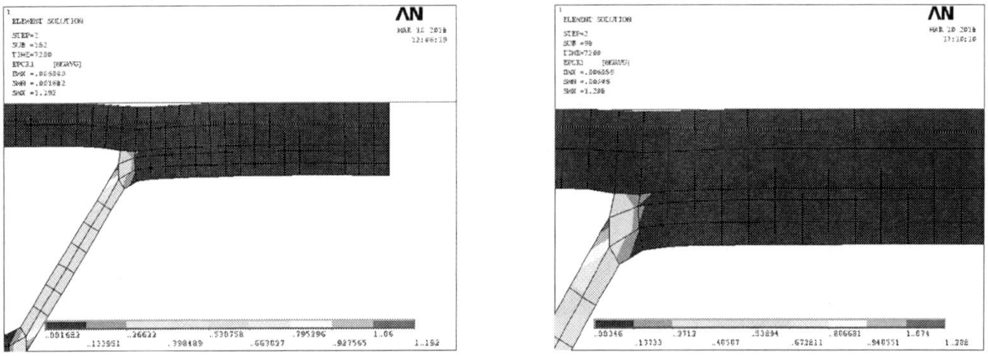

a) b)

Figure 3. Distributions of the first principal creep strain calculated for variants 3 (a) and 4 (b).

4. Conclusions

The durability of the figured dies increases significantly with decreasing temperature, which is important for the stability of the product dimensions. As the simulation results show, the problem of folds is solved both in the UFG+UFG+UFG circuit (Variant 2) and in the MC+UFG+MC circuit (Variant 4). Taking into account the cost of the UFG material, the MC+UFG+MC circuit design is clearly preferable. The relatively long duration of the SPF reduces the efficiency of the process as the productivity of the process decreases. The problem of folds could be solved using skin sheets with a large grain size, thereby reducing the exposure time. However, this technique will cause a problem during welding and will not lower the temperature of the process that reduces the structural strength. Thus, the use of the UFG material as a core sheet in a three-layer structure makes the SPF/DB process economically attractive.

References

[1] *Superplastic Forming of Structural Alloys*: *Proc. Symp*. 1982 ed. Paton N E and Hamilton C H (Warrendale, PA: TMS−AIME) 414 p
[2] Brewer W D, Bird R K and Wallace T A 1998 *Mater. Sci. Eng.* **A243** 299
[3] Cheng J-Ho and Lee S 1994 *J. Mater. Process. Technol.* **45** 249
[4] Shao J, Li Z Q, Xu H, Han X and Zhang R 2016 *Mater. Sci. Forum* **838-839** 585
[5] Kruglov A A, Karimova A F and Enikeev F U 2018 *Letters on Materials* **8** 174
[6] Safiullin A R, Safiullin R V and Kruglov A A 2010 *Rev. Adv. Mater. Sci.* **25** 281
[7] Valiakhmetov O R et al 2010 *Nanotechnol. Russia* **5** 108
[8] Kruglov A, Lutfullin R, Mulyukov R, Mukhametrakhimov M, Rudenko O and Safiullin R 2016 *Mater. Sci. Forum* **838–839** 610
[9] Enikeev F U and Kruglov A A 1995 *Int. J. Mech. Sci.* **37** 473
[10] Enikeev F U 2008 *Russian J. Non-Ferr. Metals* **49** 34
[11] Kruglov A A, Musina R Sh and Enikeev F U 2014 *Russian J. Non-Ferr. Metals* **55** 564

IOP Conf. Series: Materials Science and Engineering **447** (2018) 012051 doi:10.1088/1757-899X/447/1/012051

The effect of holding temperature on the strength of the diffusion bond of Ti-alloy and stainless steel through the ultrafine-grained interlayers of Ni and Ni-2%Cr alloy

R G Khazgaliev[1], I I Ibragimova[2], M F Imayev[1,3], M Kh Mukhametrakhimov[1] and R R Mulyukov[1,3]

[1] Institute for Metals Superplasticity Problems, Russian Academy of Sciences, Ufa, Russia
[2] Ufa State Aviation Technical University, Ufa, Russia
[3] Bashkir State University, Ufa, Russia

E-mail: sloth-usatu@mail.ru

Abstract. The manuscript considers the effect of diffusion bonding pressure, as well as holding at different temperatures, on the strength of the joints between titanium alloy and stainless steel, formed by diffusion bonding through ultrafine-grained interlayers of Ni and the Ni-2%Cr alloy. In samples with an interlayer of nickel after holding at 25 °C, the maximum strength of 380 MPa is achieved after diffusion bonding in the pressure range of 4-8 MPa. In samples with an interlayer of Ni-2%Cr, after holding at 25 °C, the highest strength of 490 MPa is achieved after bonding at a pressure of 12 MPa at 700 °C for 20 minutes. In the case of Ni-2%Cr interlayer, the strength of the joint falls more slowly than in the case of Ni interlayer. The mechanical properties of the joints are discussed in terms of the effect of chromium doping of nickel interlayer on the temperature range of transformation of austenite to martensite of the formed TiNi intermetallic compound.

1. Introduction

The combination of high corrosion and mechanical properties of titanium alloy with the strength of stainless steel in one product will make it possible to expand the capabilities of the engineering, aerospace and chemical industries. Diffusion bonding (DB) is a promising method for obtaining a perfect permanent joint [1,2]. Direct DB of titanium alloy and stainless steel does not allow obtaining the necessary properties due to the formation of brittle intermetallics of the Ti-Fe system. To prevent the formation of Ti-Fe intermetallics, various interlayers are used in the joint zone [3].

The DB technique through nickel interlayer was well studied [4-8]. By means of DB through a nickel layer, a joint of pure Ti and stainless steel was obtained with the strength of 311 MPa [4]. In the study [4-8], coarse-grained nickel was used and the bonding temperature range was 750-950 °C. It was revealed that at the titanium alloy-nickel interface, the Ti_2Ni, TiNi, and $TiNi_3$ intermetallic layers were formed, over which failure occurred [7-10]. The application of nanocrystalline Ni reduces the temperature and duration of bonding while maintaining the strength of the joint and the character and site of fracture [9, 10]. In the study [9], the formation of a regular network of cracks in the Ti_2Ni and $TiNi_3$ layers adjacent to the TiNi layer was observed. It was assumed that the main cause of failure is the anomalous change in the coefficient of thermal expansion (CTE) as a result of austenite to martensite transformation (AMT) in the temperature range of 30-100 °C in the formed TiNi layer [9,

11, 12]. In this connection, in order to increase the strength of the joint, it seems promising to try, by doping the TiNi layer, to reduce the AMT temperature below the operating temperature of the product.

It is known that chromium most strongly decreases the temperature of AMT in TiNi. With the addition of about 1 at.% Cr the temperature of TiNi AMT decreases by approximately 150 °C [12]. Reasoning from the character of the failure, one can assume that chromium alloying of nickel might cause a decrease in the temperature of AMT. In the case of nickel interlayer at the optimal temperature and holding time, the joint failed along the Ti-alloy/interlayer interface, and after DB through the Ni-2%Cr alloy, the joint failed along the stainless steel/interlayer interface [16,17].

An increase in the strength of the joint of the Ti64-type titanium alloys with microduplex or precipitation-hardening stainless steels when using nickel-based alloys with the additions of Cr, Fe, Mo, and Al was found earlier. However, these results were not analyzed in terms of lowering the temperature of AMT [13-15]. The alloying elements used by the authors [13-15] reduce the temperature of AMT, as evidenced by their high strength values. In addition, the effect of holding at low temperatures on mechanical properties at room temperature was not studied in [13-15]. The use of interlayers of nickel alloys with the additions of Cr, Fe, Mo and Al instead of pure nickel during DB of the Ti64-type titanium alloys with microduplex or precipitation-hardening stainless steels resulted in an increase in the strength of the joint [13-15]. The strength value 560 MPa of the joint was obtained using an interlayer of nickel alloy with 15.6% Fe and 4.9% Mo [13]. The strength in the range of 515-525 MPa was achieved after DB through an interlayer of nickel alloy with 8.6% Fe, 16.7% Cr, 0.4% Al [14,15]. The doping of TiNi with these elements reduces the temperature of AMT [12], therefore the increase in strength may be due to the AMT temperature decrease of the. However, the authors did not consider this fact. The assumption of the effect of AMT on strength can be verified by examining the mechanical properties of the joint after holding the samples at temperatures below the temperature range of AMT.

This work is aimed to study the effect of bonding pressure on the strength of diffusion bonding of titanium alloy and stainless steel through the fine-grained interlayer of the Ni-2%Cr alloy, as well as to determine the effect of holding at various temperatures on the strength of the joint.

2. Experimental

DB was performed using a titanium-based PT3V alloy (Ti-2%V-4.5%Al) and 12Kh18N10T stainless steel (0.12%C-18%Cr-10%Ni-1%Ti). The sizes of the joint parts were 4x4x16 mm^3, the interlayers were made of 0.2 mm thick plates of pure Ni and the Ni-2%Cr alloy. To obtain an ultrafine grained structure in Ni and the Ni-2%Cr alloy, they were deformed by torsion under quasi-hydrostatic pressure of 5 GPa in 5 revolutions [18]. The DB regimes were as follows: the pressure range was P = 4-16 MPa, the temperature was T = 700 °C, the duration was t = 20 minutes, the residual gas pressure was did not exceed $2x10^{-3}$ Pa. The microstructure was analyzed using a Tescan Vega scanning electron microscope with an X-act Oxford Instruments detector for energy dispersive analysis with shooting mode in the BSE contrast. Tensile tests of samples with the dimensions of the working part 1x1x5 mm^3 were performed at room temperature with a mobile grip speed of 1 mm/min. Three samples were used to measure the main ultimate tensile strength σ_u.

3. Results and discussion

After DB at a pressure of 4 MPa, the thicknesses of the intermetallic layers are: Ti_2Ni (0.63 ± 0.16 μm), TiNi (1.82 ± 0.14 μm) and $TiNi_3$ (1.22 ± 0.15 μm) (figure 1). The chemical composition at the interface of Ni-2%Cr/stainless steel changes smoothly, and there are no sharp differences in phase contrast. In this layer, the Ti:Cr ratio is close to 1:2, which indicates the formation of $TiCr_2$ intermetallics [16,17].

In the samples with Ni interlayer after holding at 25 °C, the maximum strength of 380 MPa is achieved after DB at the pressure range of 4-8 MPa. A further increase in the pressure of DB results in a decrease in the strength of the joint to 300 MPa (figure 2a). After holding the DB samples at 10 °C, the joint strength drops to 150-250 MPa (figure 2a). In samples with an interlayer of Ni-2%Cr, after

IOP Conf. Series: Materials Science and Engineering **447** (2018) 012051 doi:10.1088/1757-899X/447/1/012051

holding at 25 °C, the highest strength of 490 MPa is achieved after DB at a pressure of 12 MPa (figure 2b). After holding the DB samples at -10 °C, the joint strength drops to 150-300 MPa, while the maximal strength is also observed after DB at a pressure of 12 MPa (figure 2b). As we see, lowering the holding temperature results in a decrease in the strength of the joint when using a layer of both Ni and Ni-2%Cr alloy. However, as the holding temperature decreases, the strength of the joint through Ni-2%Cr interlayer falls more slowly than in the case of Ni. Even after holding at the temperature of -10 °C, i.e. 20 degrees lower than 10 °C, the strength of the joint through Ni-2%Cr interlayer is approximately the same as in the case of Ni interlayer.

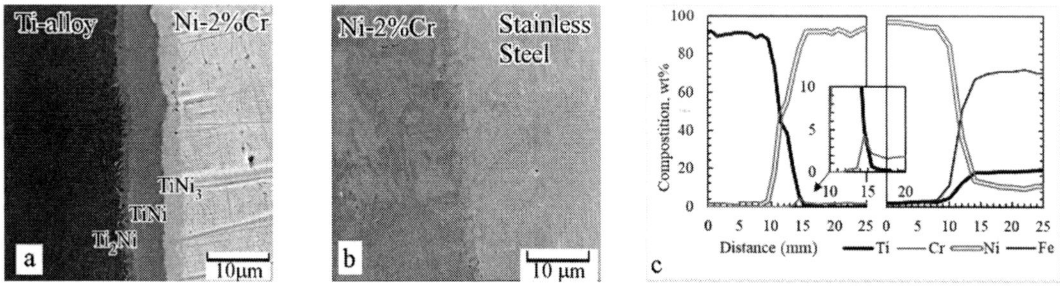

Figure 1. Microstructure (a, b) and chemical composition (c) of joint zones of titanium alloy and stainless steel through Ni-2% Cr interlayer after 4 MPa DB at 700 °C for 20 min.

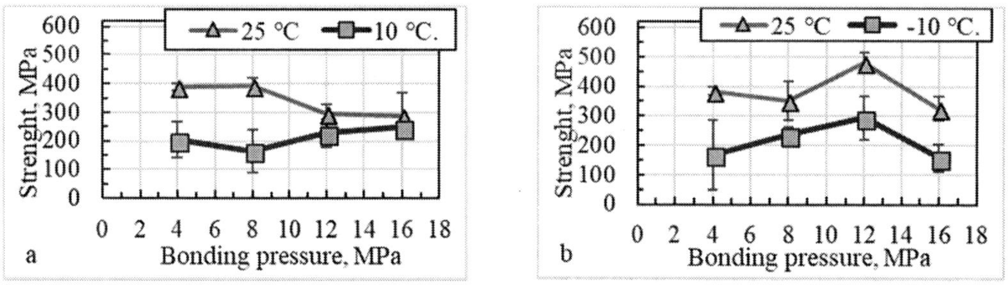

Figure 2. Influence of DB pressure on the joint strength of the titanium alloy and stainless steel through a layer of (a) Ni, (b) Ni-2%Cr alloy. DB was performed at 700 °C for 20 min.

The samples, subjected to DB through Ni interlayer, irrespective of the holding temperature, were always fractured with "participation" of the TiNi layer: either along the Ti_2Ni/TiNi interface or along the TiNi/$TiNi_3$ one [9,16]. The fracture of the samples bonded through Ni-2%Cr interlayer at 4 MPa, 700 °C for 20 minutes occurred along the Ni-2%Cr/stainless steel interface (figure 3a). After DB under a higher pressure and holding at 25 °C, the fracture passed along the Ti_2Ni + TiNi / $TiNi_3$ intermetallic interfaces (figure 3c). After holding at -10 °C, the samples bonded through Ni-2%Cr interlayer were fractured along the Ti_2Ni + TiNi/$TiNi_3$ intermetallic interfaces (figure 3b).

An increase in the strength of the joint through Ni-2%Cr interlayer after holding at 25 °C, as well as a slower decrease in strength with a decrease in the holding temperature, can be explained by shifting the temperature of the onset of the TiNi martensitic transformation to lower temperatures. In the case of Ni-2%Cr, the martensitic transformation of TiNi apparently begins just above room temperature and its negative effect on strength is less than in the case of Ni interlayer, where the austenite to martensite transformation begins much higher than room temperature and at room temperature the degree of transformation is already essential. A further increase in strength at room temperature is possible by increasing the concentration of elements reducing the temperature range of the austenite to martensite transformation of TiNi below the room temperature.

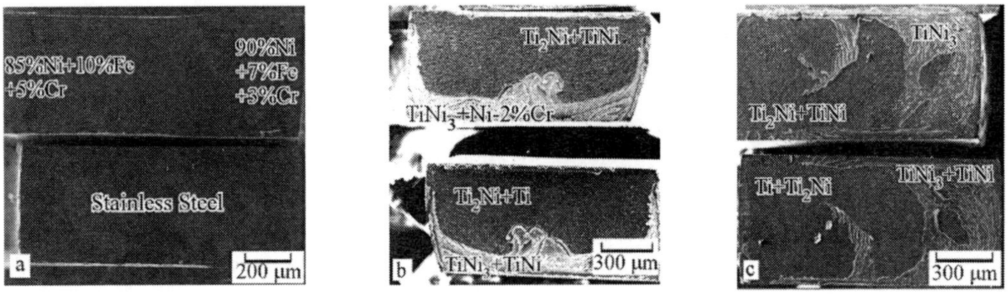

Figure 3. Fracture surfaces of the samples, subjected to DB through Ni-2%Cr interlayer at 700 ° C for 20 min under pressure (a, b) 4 MPa and (c) 16 MPa after holding at (a, c) 25 °C and (b) 10 °C and failed during tensile deformation at 25°C.

Acknowledgements

The present work was accomplished according to the state assignment of IMSP RAS No. AAAA-A17-117041310221-5. Electron microscopic studies and mechanical tests were carried out on the facilities of shared services center of the Institute for Metals Superplasticity Problems of Russian Academy of Sciences «Structural and Physical-Mechanical Studies of Materials».

References

[1] Karakozov E S 1986 *Welding of Metals by Pressure.* (Moscow: Mashinostroenie) p 280 (In Russian)

[2] Petrenko V R, Kireev L S and Peshkov V V 2004 *Welding of Titanium with Steel: Monograph* (Voronezh: Voronezh State Technical University) p 173 (In Russian)

[3] Kornilov I I and Boryskina N G 1956 *Doklady AN SSSR* **108** 1063 (In Russian)

[4] Kundu S, Chatterjee S, Olson D L and Mishra B 2007 *Metall. Mater.Trans. A* **38** 2053

[5] Kundu S and Chatterjee S 2008 *Mater. Charact.* **59** 631

[6] Kundu S and Chatterjee S 2006 *Mater. Sci. Eng. A* **425** 107

[7] Thirunavukarasu G, Kundu S, Mishra B and Chatterjee S 2014 *Metall. Mater.Trans. A* **45** 2067

[8] Thirunavukarasu G, Kundu S, Mishra B and Chatterjee S 2014 *Metall. Mater.Trans. A* **45** 2078

[9] Khazgaliev R G, Mukhametrakhimov M Kh, Imayev M F, Shayakhmetov R U and Mulyukov R R 2015 *Russian Phys. J.* **58** 822

[10] Khazgaliev R G, Mukhametrakhimov M Kh, Mulyukov R R and Lutfullin R Ya 2011 *Perspectivnye Materialy* **12** 529 (In Russian)

[11] Uchil J, Mohanchandra K P, Ganesh Kumara K, Mahesh K K and Murali T P 1999 *Physica B* **270** 289

[12] Otsuka K and Ren X. 2005 *Progr. Mater. Sci.* **50** 511

[13] Sam S, Kundu S and Chatterjee S 2012 *Mater. Design* **40** 237

[14] Kundu S, G. Anand G and Chatterjee S 2013 *Metall. Mater.Trans. A* **44** 2196

[15] Kundu S, Mishra B, Olson D L and Chatterjee S 2013 *Mater. Design* **51** 714

[16] Khazgaliev R G, Imayev M F and Mulyukov R R 2017 *Deformatsiya i Razrushenie Materialov* **5** 18 (In Russian)

[17] Ibragimova I I, Khazgaliev R G, Imayev M F and Mulyukov R R 2017 *Vector nauki TGU* **4(42)** 40 (In Russian)

[18] Khisamov R Kh, Safarov I M, Mulyukov R R and Yumaguzin Yu. M 2013 *Phys. Solid State* **55** 1

IOP Conf. Series: Materials Science and Engineering **447** (2018) 012052 doi:10.1088/1757-899X/447/1/012052

High pressure torsion induced structural transformations in Ti- and Zr-based amorphous alloys

D V Gunderov[1,3], E V Boltynjuk[2], E V Ubyivovk[2], A A Churakova[1,3], G E Abrosimova[4], V D Sitdikov[1], A R Kilmametov[5], R Z Valiev[1,2]

[1]Ufa State Aviation Technical University, 12 K. Marx str., Ufa 450008, Russia
[2] Saint Petersburg State University, 28 Universitetskiy pr., Saint Petersburg 198504, Russia
[3]Institute of Molecule and Crystal Physics - Subdivision of the Ufa Federal Research Centre of the Russian Academy of Sciences, Prospekt Oktyabrya 151, Ufa, 450075, Russia
[4] Institute of Solid State Physics Russian Academy of Sciences, 2 Academician Ossipyan str., Chernogolovka, Moscow District, 142432, Russia
[5]Karlsruhe Institute of Technology (KIT), Institute of Nanotechnology, Hermann-von-Helmholtz-Platz 1, 76344 Eggenstein-Leopoldshafen, Germany

E-mail: dimagun@mail.ru

Abstract. The melt-spun (MS) Ti50Ni25Cu25 alloy and the Zr62Cu22Al10Fe5Dy1 bulk metallic glass (BMG) were subjected to high pressure torsion (HPT). X-ray diffraction (XRD) measurements show a shift of the first diffraction halo to a low angle after HPT processing, which corresponds to an increase in the values of the radius of the first coordination sphere and the free volume. Direct density measurements confirmed an increase in free volume values. A special TEM procedure was used for a detailed study of the microstructure of both amorphous alloys after HPT processing. The study revealed the formation of a large density of shear bands (SBs) in both alloys. Nanocrystals are formed directly in shear bands as a result of strain-induced nanocrystallization. Amorphous nanoclusters with a size of 20 nm are formed in an amorphous matrix surrounding the SBs in the HPT-processed MS alloy Ti50Ni25Cu25. The formation of nanoclusters was not observed in BMG Zr62Cu22Al10Fe5Dy1 after HPT processing.

1. Introduction

High pressure torsion (HPT) seems to be a promising method to form a high density of localized shear bands for structural transformations of amorphous alloys [1-6]. Plastic deformation also leads to changes in the mechanical properties of amorphous alloys [1-10]. However, the features of structural transformations occurring in amorphous alloys during HPT processing are not clear. Preparing lamellae for TEM from the cross section of the HPT-processed specimen using a focused ion beam (FIB) provides a new opportunity to study the structural features of amorphous materials processed by HPT. Important information about the structure of amorphous alloys is given by the free volume, which can be estimated

IOP Conf. Series: Materials Science and Engineering **447** (2018) 012052 doi:10.1088/1757-899X/447/1/012052

by the XRD method [1, 7]. A new unique method [11] provides the ability to measure the density of HPT-processed samples with small sizes and complex shapes.

In this work, the melt-spun (MS) Ti50Ni25Cu25 alloy and the bulk metallic glass (BMG) Zr62Cu22Al10Fe5Dy1 were subjected to HPT processing. Complex structural studies using TEM, XRD and density measurements were utilized to study the effect of HPT on the structure and properties of amorphous alloys of different chemical compositions.

2. Experimental

The $Ti_{50}Ni_{25}Cu_{25}$ ribbons with a thickness of 0.04 mm and a width of 2 mm were produced by the melt spinning technique. Cylindrical rods with a diameter of 5 mm and a length of 50 mm from the $Zr_{62}Cu_{22}Al_{10}Fe_5Dy_1$ BMG (Zr_{62} BMG) were made by copper mould casting. Fragments of the ribbons and BMG were subjected to HPT under an applied pressure of 6 GPa at room temperature for 10 and 5 revolutions, respectively. The structure of the HPT-processed specimens was examined by TEM on a Zeiss Libra 200FE. Data on selected area electron diffraction (SAED) were collected from an area of 80 nm in diameter. To prepare the TEM samples from the cross section of the HPT-processed specimen, the lamella samples were cut out by a focused ion beam (FIB) using a Zeiss Auriga Crossbeam SEM-FIB according to the procedure described in [8]. Lamellae were cut from areas close to the surface of the HPT sample. The structure was also examined by X-ray diffraction (XRD) under Cu radiation employing a Rigaku Ultima IV device. The density measurements were performed in accordance with the precise technique of measuring the bulk density of small solid objects using laser confocal microscopy described recently in [11].

3. Results and discussion

According to the XRD and TEM studies, the initial MS $Ti_{50}Ni_{25}Cu_{25}$ alloys and Zr_{62} BMG have a predominantly amorphous structure. TEM studies of the lamellae revealed a large number of shear bands in HPT-processed states (figure 1). The primary shear bands have an average thickness of 5-10 nm. The shear bands are distorted and assembled in a group or bundles. The average distance between the shear bands in a group ranges from 20 to 50 nm. In the HPT-processed Zr_{62} BMG the groups of shear bands (zones of structural transformation with a banded contrast) have a total width of 200 nm and more (figure 1a). In the HPT-processed MS $Ti_{50}Ni_{25}Cu_{25}$ the groups of shear bands have a slightly smaller total width – up to 100 nm (figure 1b).

Nanocrystals were observed directly in shear bands in MS TiNiCu subjected to HPT (crystalline reflections in a microdiffraction pattern taken from the shear bands in figure 1b). These nanocrystals were formed as the result of strain-induced nanocrystallization, which was previously found in this amorphous alloy [2, 9]. There are areas with a cluster-type amorphous structure around the bundles of shear bands. The average size of clusters is 20-30 nm. According to the SAED pattern (figure 1b), the clusters have an amorphous structure. Probably, the clusters and their boundaries have different contents of free volume. However, the nature of the cluster-type amorphous structure in the HPT-processed MS $Ti_{50}Ni_{25}Cu_{25}$ has not been revealed yet.

The formation of a cluster-type amorphous structure after HPT was not detected in the Zr_{62} BMG. Some fine crystalline reflections are also observed in the microdiffraction pattern taken from individual shear bands (figure 1a), which may also indicate the strain-induced nanocrystallization. However, nanocrystals were not observed on the most shear bands. Apparently, nanocrystallization in a shear band depends on the conditions of shear bands propagation. Thus, it can be concluded that the HPT-induced nanocrystallization in the Zr_{62} BMG is less intensive that in the HPT-processed TiNiCu.

IOP Conf. Series: Materials Science and Engineering **447** (2018) 012052 doi:10.1088/1757-899X/447/1/012052

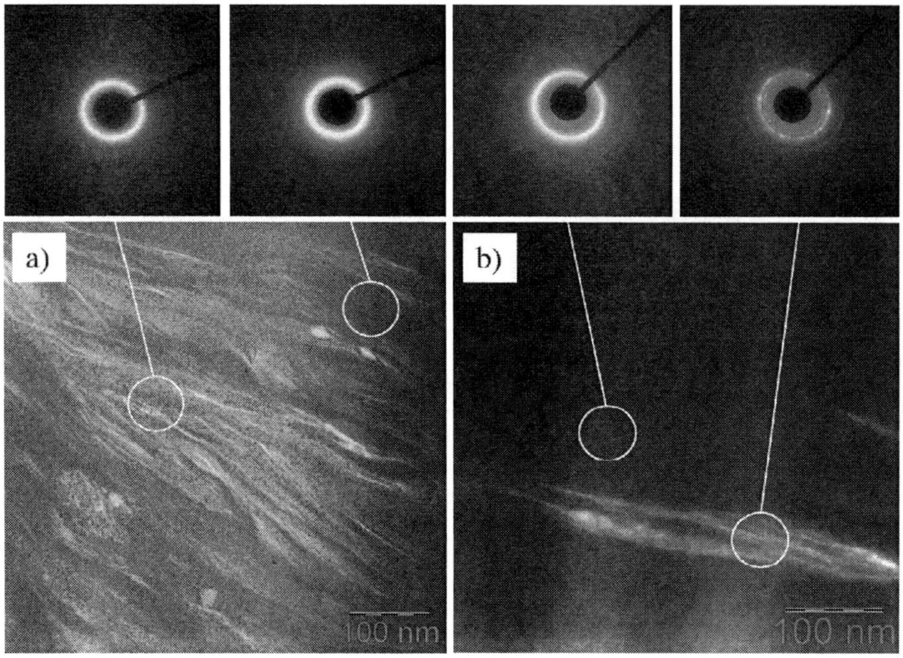

Figure 1. STEM BF image of a) $Zr_{62}Cu_{22}Al_{10}Fe_5Dy_1$ BMG subjected to HPT, on the top – corresponding microdiffraction patterns from different regions of the sample. b) cluster-type amorphous structure and shear bands with nanocrystals of MS $Ti_{50}Ni_{25}Cu_{25}$ subjected to HPT, on the top – corresponding microdiffraction patterns from different regions of the sample.

XRD shows the HPT-induced shift of the first diffraction halo to low angles for both alloys, which means an increase in the radius (R) value of the first coordination sphere. This means an increase in the content of free volume (ΔV) [7]. The increment of free volume was calculated according to the procedure described in [7]. ΔV was increased approximately by 0.5 % in the HPT-processed Zr_{62} BMG compared to the initial state. ΔV was increased approximately by 1.8 % in the HPT-processed MS $Ti_{50}Ni_{25}Cu_{25}$ in comparison with the initial state. However, the analysis of changes in free volume, based on the shift of the first diffraction halo, may contain a considerable error for the MS $Ti_{50}Ni_{25}Cu_{25}$ due to more intensive crystallization in this alloy. It should be noted that an increase in free volume leads to an increase in the ductility of amorphous alloys [12].

Density measurements performed according to the procedure [11] demonstrate that the initial Zr_{62} BMG has a density of ρ equal to 6.98 kg/m^3, which correlates with the data for bulk samples of this BMG obtained by hydrostatic weighing [13]. HPT processing leads to a decrease in the density values of the Zr_{62} BMG and the MS $Ti_{50}Ni_{25}Cu_{25}$ ($\Delta\rho$) approximately by 2%. The value of ΔV=2 % for the HPT-processed Zr_{62} BMG determined by the direct method, is much larger than ΔV determined by XRD (0.5%). This is probably due to the fact that the HPT processing leads to the formation of nano-sized pores or cracks in the HPT-processed sample. Previously, nano-sized pores and nano-voids in the SPD-processed alloy were discovered in [14]. It should be noted that the increase in ΔV = 0.5% obtained by XRD in this work is close to ΔV observed in other works on the HPT processing of BMGs [15].

4. Conclusions

The direct density measurements and analysis of first diffraction halo shift demonstrate an increase in the free volume content in both amorphous MS $Ti_{50}Ni_{25}Cu_{25}$ and Zr_{62} BMG. The formation of high density of shear bands in amorphous MS $Ti_{50}Ni_{25}Cu_{25}$ and Zr_{62} BMG was detected on lamella samples using TEM. However, the structural transformations of these two amorphous alloys occur in different ways. HPT leads to the formation of groups of shear bands (zones of structural transformation with a banded contrast) with total width of 200 nm and more in Zr_{62} BMG, whereas groups of shear bands have a slightly smaller width – up to 100 nm in MS $Ti_{50}Ni_{25}Cu_{25}$. HPT-induced nanocrystallization in shear bands and the formation of a cluster-type amorphous structure were observed in MS $Ti_{50}Ni_{25}Cu_{25}$. The formation of nanoclusters was not observed in Zr_{62} BMG after HPT processing and strain-induced nanocrystallization is less prominent in this alloy. Additional studies are required for a deep understanding of the strain-induced structural transformations in amorphous alloys.

Acknowledgments

The authors acknowledge Saint-Petersburg State University for a research grant 6.65.43.2017. The work was partially supported by Deutsche Forschungsgemeinschaft (project number HA 1344/32-1. The scientific investigations were performed at the Interdisciplinary Resource Centre for Nanotechnology of the Research Park of St. Petersburg State University and center for collective use "Nanotech", Ufa State Aviation Technical University.

References

[1] Meng F, Tsuchiya K, Seiichiro I and Yokoyama Y 2012 *Appl. Phys. Lett.* **101** 121914
[2] Valiev R Z, Gunderov D V, Zhilyaev A P, Popov A G and Pushin V G 2004 *J. Metast. Nanocryst. Mater.* **22** 21
[3] Wang X D, Cao Q P, Jiang J Z, Franz H, Schroers J, Valiev R Z, Ivanisenko Y, Gleiter H and Fecht H-J 2011 *Scripta Mater.* **64** 81
[4] Gunderov D, Slesarenko V, Lukyanov A, Churakova A, Boltynjuk E, Pushin V, Ubyivovk E, Shelyakov A and Valiev R 2015 *Adv. Eng. Mater.* **17** 1728
[5] Gunderov D V, Boltynjuk E V, Ubyivovk E V, Lukyanov A V, Churakova A A, Kilmametov A R, Zamula Y S and Valiev R Z 2018 *J. Alloys Comp.* **749** 612
[6] Boltynjuk E V et al 2018 *J. Alloys Comp.* **747** 595
[7] Shao H, Xu Y, Shi B, Yu C, Hahn H, Gleiter H and Li J 2013 *J. Alloys Comp.* **548** 77
[8] Chen Y M, Ohkubo T, Mukai T and Hono K 2009 *J. Mater. Res.* **24** 1
[9] Glezer A M, Sundeev R V and Shalimova A V 2011 *Dokl. Phys.* **56** 476
[10] Abrosimova G and Aronin A 2018 *J. Alloys Comp.* **747** 26
[11] Kilmametov A, Gröger R, Hahn H, Schimmel T and Walheim S 2017 *Adv. Mater. Technol.* **2** 1600115
[12] Murali P and Ramamurty U 2005 *Acta Mater.* **53** 1467
[13] Churyumov A Y, Bazlov A I, Zadorozhnyy V Y, Solonin A N, Caron A and Louzguine-Luzgin D V 2012 *Mater. Sci. Eng. A* **550** 358
[14] Ribbe J, Schmitz G, Gunderov D, Estrin Y, Amouyal Y, Wilde G and Divinski S V 2013 *Acta Mater.* **61** 5477
[15] Edalati K, Yokoyama Y and Horita Z 2010 *Mater. Trans.* **51** 23

New carbon membrane for water desalination via reverse osmosis

R I Babicheva[1], S V Dmitriev[2,3], A A Kistanov[4], M Dahanayaka[4], AW-K Law[1,5] and K Zhou[4]

[1] Environmental Process Modelling Centre, Nanyang Environment and Water Research Institute, Nanyang Technological University, Singapore 639798, Singapore
[2] Institute for Metals Superplasticity Problems, Ufa 450001, Russia
[3] National Research Tomsk State University, Tomsk 634050, Russia
[4] School of Mechanical and Aerospace Engineering, Nanyang Technological University, Singapore 639798, Singapore
[5] School of Civil and Environmental Engineering, Nanyang Technological University, Singapore 639798, Singapore

E-mail: ri.babicheva@gmail.com

Abstract. The recent development of membrane-based water desalination and purification technologies shows new approaches to meeting global demand for freshwater. Carbon-based nanomaterials have shown great potential in various desalination technologies. Such materials are suitable for membrane separation, offering advantageous structure, mechanical properties and porosity. In the present molecular dynamics study, water desalination via reverse osmosis is simulated using a new sp^2-carbon membrane. It is found that the membrane under consideration demonstrates a very good rate of ion rejection and a reasonably good water flux, which grows with increasing temperature.

1. Introduction

The lack of freshwater resources, unevenly distributed in the world, makes researchers move towards the development of new good membranes for the desalination of sea water [1]. Obviously, for a good salt rejection rate, the pore size of a membrane must be smaller than the diameters of Na^+ (~7.5 Å) and Cl^- (~6.5 Å) ions and at the same time must be large enough for a good permeate water flux. The latter can be observed only when the pore diameter is larger than the size of water molecule (~4.0 Å). It is important to find single-layer structures with high strength and proper hole sizes. It is believed that due to strong covalent bonding, graphene membranes can be widely used for water desalination. Unfortunately, there is a drawback of such structure, namely, its hexagonal pore size is too small for the penetration of a water molecule. Therefore, researchers are looking for membranes capable to overcome this problem. For example, in a recent molecular dynamics study by Liu et al. [2] the desalination performance of a single-layer C_2N membrane during forward osmosis was addressed. An unstrained membrane showed a relatively poor water flux due to the insufficiently large size of pores. The authors improved the water flux through the membrane, applying a bi-axial tensile strain up to 12%. Another method is the introduction of holes using techniques such as graphene perforation by ion beams [3]. It is important to look for membranes that can be used for water desalination, avoiding difficulties such as application of tensile stress or perforation.

In a series of recent theoretical works, new single-layer carbon structures with different hole sizes were proposed [4,5]. We believe that some of these structural carbon modifications may be promising candidates for water desalination. In the present work, we study the water desalination performance of a new sp^2 two-dimensional carbon structure using the density functional theory (DFT) and molecular dynamics (MD).

2. Simulation setup

2.1. Material of study

For the seawater desalination through reverse osmosis, the carbon membrane having a particular new structure is chosen. Figure 1 a shows the membrane structure obtained after optimization of the DFT geometry using the VASP software. This single-layer material is characterized by an intrinsic pore size of 5.2÷5.9 Å. The first-principle calculation using the Bader analysis showed that the charge density is evenly distributed within the structure indicating that the charge of C atoms is equal to zero.

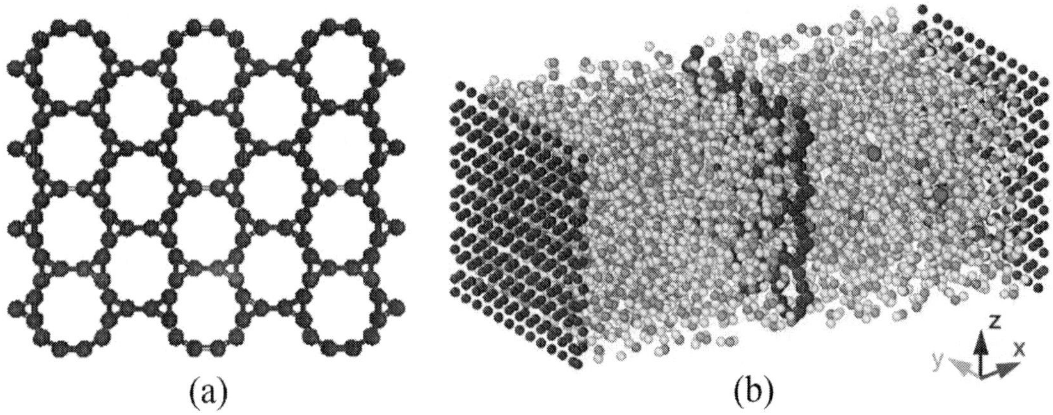

Figure 1 (a) Structure of carbon membrane after DFT geometry optimization. (b) MD model for water desalination via reverse osmosis. The single-layer carbon membrane is located between the feed solution having Na^+ and Cl^- ions (on the right) and the pure water box (on the left). Two graphene pistons are placed on the two ends of the model. Here, water molecules are indicated in blue shadow, Na^+ in pink, Cl^- in yellow, while C atoms are given in grey. In order to distinguish the ions and C atoms of the membrane, their sizes are enlarged.

The simulation model for the water desalination process consists of pure water and the feed solution at two sides of the single-layer carbon membrane, which is located in the middle of the model at x=0. In addition, there are two graphene pistons at the two ends of the model, which are treated as rigid bodies (figure 1 b). To model the seawater, the concentration of Na^+ and Cl^- ions in the feed solution equal to 0.8 mol/L is chosen. The dimensions of the membrane, the solution boxes and the pistons along the y and z directions are 29 and 23 Å, respectively. The lengths of the boxes for the feed solution and pure water along the x-direction are equal and are set to 50 Å. In order to separate the left and right parts of the model and to avoid atomic interactions between them, a vacuum space having a length of 30 Å along the x-direction was introduced between the pure water box and the feed solution.

During simulation, atomic interactions between particles of a system are considered as electrostatic and van der Waals ones. The latter interactions are described by the Lennard-Jones (LJ) potential with parameters taken from the AMBER03 force field [6], while the particle-particle particle-mesh (PPPM) method is used to characterize long-range electrostatic interactions. The LJ parameters for the interaction of atoms between different types of atoms are calculated using the Lorentz-Berthelot mixing rule. In order to create water, the so-called Transferable Intermolecular Potential 3P (TIP3P) water model [7] and the SHAKE constraint algorithm [8] were adopted for the simulation.

2.2. Modeling

The MD simulations are performed using the atomic/molecular massively parallel simulator (LAMMPS) program package [9]. Periodic boundary conditions are applied along the three orthogonal directions x, y and z of the simulation model. Initially, the system is equilibrated at a corresponding reverse osmosis temperature, which can be 280, 300, 320, 340 or 360 K in the NVT ensemble for 20 ps. Further, in order to create an osmotic pressure of 100, 200 or 300 MPa, the corresponding forces are imposed on the right graphene piston for 3 ns to compress the seawater box. For visualization of the atomic structure and phase composition at different temperatures, the OVITO and VMD packages are used [10,11].

3. Results and discussion

Figure 2 a shows the permeation water flux through the membrane as a function of temperature for the three considered applied pressures. It was revealed that at a pressure of 300 MPa, the water flux monotonously increases from 30 to 140 L/(h cm^2). For pressures of 100 and 200 MPa, similar to 300 MPa, the flux increases from 2 to 70 L/(h cm^2) and from 20 to 95 L/(h cm^2), respectively. It should be noted that in these two cases, the general trend of increasing water flux with increasing temperature slows down in the temperature range close to 340 K. The permeation water flux for 100 MPa at 340 K is even lower than at 320 K. The reason for such non-monotonic behavior of water flux with increasing temperature at relatively low pressures remains to be explained in the next more detailed study. At this stage of the study, we can speculate that a decrease in the permeate water flux at 340 K can be explained by the temperature effect on the pressure of the feed solution. It is known that the latter decreases with increasing temperature. This behavior obviously leads to the reduction of the overall reverse osmosis pressure and the reduction of the flux that is especially noticeable at a lower applied pressure (100 MPa), while in the case of high external pressures the temperature effect is not so pronounced (figure 2 a). Normally, reverse osmosis systems are designed in such a way that the feed pressure is adjusted to compensate the water flux changes caused by an increase in temperature.

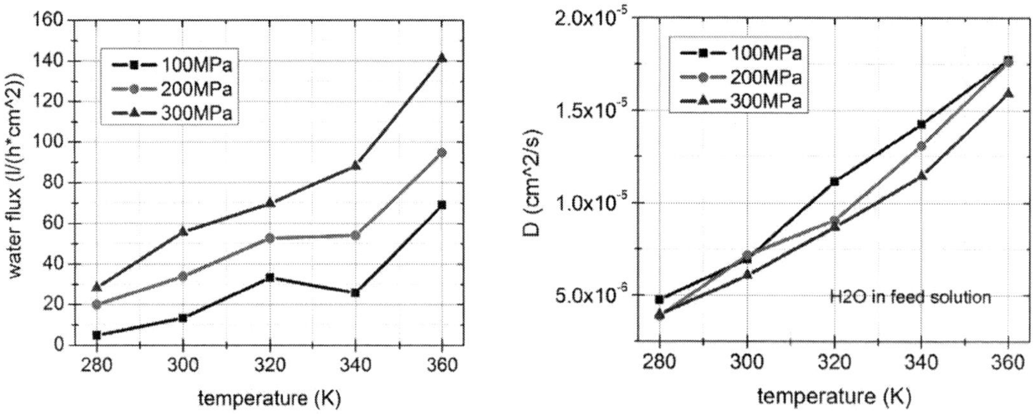

(a) (b)

Figure 2 (a)Temperature dependence of permeate water flux through sp^2 carbon membrane shown in figure 1 a. (b) Temperature dependence of diffusion coefficient D for water molecules in feed solution.

It was revealed that in all the considered cases the studied carbon membrane demonstrates 100% salt rejection. The Na^+ and Cl^- ions cannot go through the membrane pores.

In order to investigate the diffusion behavior of water molecules in the feed solution, their mean square displacement (MSD) $\langle \Delta s^2(t) \rangle$ was calculated for the considered temperatures and reverse osmosis pressures. This parameter reflects the deviation of the position of the particles with respect to the reference position over time. From the obtained MSD curves, the diffusion coefficient D can be found as $\langle \Delta s^2(t) \rangle \approx 6Dt$, where t is the simulation time. Figure 2 b shows the dependences of D for water molecules in the salt water on temperature of reverse osmosis process for differed pressures. Overall, the diffusion coefficient increases with increasing temperature from $\sim 5.0\times10^{-6}$ cm^2/s at 280 K up to $\sim 1.75\times10^{-5}$ cm^2/s at 360 K. The applied reverse osmosis pressure affects insignificantly the diffusion coefficient. Thus, the curves for 100 MPa and 200 MPa are very close to each other, while D values for the case of 300 MPa are slightly lower especially at higher temperatures.

Temperature determines the rate of diffusion of water molecules. Its increase leads to activation of diffusion (see figure 2 b) and, obviously, to an increase in water flux through the membrane, as can be seen in figure 2 a.

4. Conclusions

The obtained results demonstrate that the carbon membrane having the structure shown in figure 1 a can be successfully used for water desalination via reverse osmosis. The material demonstrates a high water flux and perfect salt rejection rate, which can be explained by the fact that the size of its intrinsic pores are permeable to water molecules, unlike those of the conventional graphene structure, and at the same time, the pores are not big enough for the Na^+ and Cl^- ion penetration. Such a membrane does not require the application of tensile deformation or the introduction of holes (defects) to increase the water flux.

In the following studies it is important to analyze the theoretical strength of the considered carbon membrane, e.g., in line with the MD study performed earlier for graphene [12] or using the DFT approach.

Acknowledgements
The work of R.I.B. was supported by the Russian Science Foundation, grant No. 17-79-10410. S.V.D. was partially supported by the state assignment of IMSP RAS. The work of A.A.K. was partly supported by the Russian Science Foundation, grant No. 14-13-00982.

References

[1] Hebbar R S, Isloor A M, Inamuddin and Asiri A M 2017 *Environ. Chem. Lett.* **15** 643
[2] LiuB, LawA W-K and Zhou K 2018 *J. Membrane Sci.***550** 554
[3] Buchheim J, Wyss R M, Shorubalko I and Park H G 2016 *Nanoscale* **8** 8345
[4] BelenkovE A and Greshnyakov V A 2013 *New Carbon Mater.* **28** 273
[5] BelenkovE A and Kochengin A E 2015 *Phys. Solid State* **57** 2126
[6] Duan Y, Wu C, Chowdhury S, Lee M C, Xiong G M, Zhang W, Yang R, Cieplak P, Luo R, Lee T, Caldwell J, Wang J M and Kollman P 2003 *J. Comput. Chem.* **24** 1999
[7] JorgensenWL, ChandrasekharJ, MaduraJD, ImpeyRW and KleinML 1983*J. Chem. Phys.* **79** 926
[8] Ryckaert J P, Ciccotti G and Berendsen H J C 1977 *J. Comput. Phys.* **23** 327
[9] Plimpton S 1995 *J. Comput. Phys.* **117** 1
[10] Stukowski A 2010 *Modelling Simul. Mater. Sci. Eng.* **18** 015012
[11] Stukowski A, BulatovVV and Arsenlis A 2012 *Modelling Simul. Mater. Sci. Eng.* **20** 085007
[12] Baimova JA, Dmitriev SV, Zhou K and Savin AV 2012 Phys. Rev. B **86** 035427

IOP Conf. Series: Materials Science and Engineering **447** (2018) 012054 doi:10.1088/1757-899X/447/1/012054

Structural transformations in the NiAl alloys with deviations from the stoichiometric composition during stepwise cooling

A A Chaplygina, P A Chaplygin, M D Starostenkov

Department of Physics, Altai State Technical University, 46 Lenin Ave., 656038 Barnaul, Russia

E-mail: alesya_ch@mail.ru

Abstract. The Monte Carlo simulation method shows the formation of B2 superstructure domains in the course of structural phase transitions during stepwise cooling of the Ni-Al intermetallic compound with a deviation from stoichiometry. The effect of the deviation from stoichiometry on the structure and energy characteristics of the alloy during stepwise cooling is analyzed in this article. It is shown that the highest percentage of the ordered phase after stepwise cooling has the alloy with a stoichiometric composition.

1. Introduction

The NiAl intermetallic alloy has the B2 (CsCl) ordering with the Pm3m space group based on a bcc lattice. The Ni–Al system is characterized by large differences in the size of atoms and the electronic structure. Beta-alloys of the Ni–Al system were used in the investigations of low-stability pre-transitional states [1]. It is well known that mono-nickel aluminide is characterized by a high melting temperature (1638 °C) and heat of formation. The crystals of NiAl exhibit stronger elastic anisotropy, and the related anisotropy of properties is compared to the structure with a disordered bcc-lattice [1].

There are two primary methods of computer simulation in condensed matter physics. Currently, the most common method is molecular dynamics, rather than the Monte Carlo method. The method of molecular dynamics is most adequate in the investigation of dynamic processes in materials. However, it has a very important disadvantage: the order-disorder phase transformations are relatively slow to get a certain practical result and require a long experiment time. The use of the Monte-Carlo method can significantly accelerate research. For this reason, the phase transformation in the NiAl alloy is investigated using the Monte Carlo method.

The purpose of this work is to study the structural and phase characteristics of the intermetallic compound NiAl with deviations from the stoichiometric composition during stepwise cooling from the solid solution by the Monte Carlo method.

2. Computer simulation model

The atoms of nickel aluminide are located at the points of the bcc lattice, ordered in accordance with the B2 superstructure (figure 1). The computational cell includes 32×32×32 atomic layers (65536 atoms). Periodic boundary conditions are used for an effective modelling of an infinite system. The starting configuration is a disordered bcc solid solution. To simulate diffusion in a crystal, a single vacancy is introduced in the computational cell; the corresponding concentration of vacancies is 1.53×10^{-5}. The diffusion of atoms occurs through the mechanism of vacancies, which was previously applied in a similar way in the works [2-11]. The structure of the alloy changes at discrete points in time with an act of diffusion corresponding to the atomic jump in the vacant site. Only atoms from the

1st, 2nd, or 3rd coordination sphere can jump into the vacant site. The probability of the hopping to the vacant site is determined by the Metropolis algorithm.

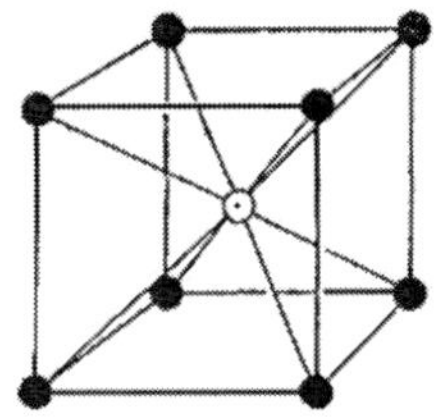

Figure 1. B2 superstructure

The interatomic interactions are described by the Morse potential $\varphi(r_{ij}) = D_{KL}\beta_{KL}e^{-\alpha_{KL}r_{ij}}\left(\beta_{KL}e^{-\alpha_{KL}r_{ij}} - 2\right)$, where $\alpha_{KL}, \beta_{KL}, D_{KL}$ are the potential parameters of bonds between atoms of the K and L sorts and r_{ij} is the distance between the atoms. The configuration energy of the system is calculated as $E = 1/2\sum_{i=1}^{N}\sum_{j=1}^{M}\varphi\left(r_i - r_j\right)$, where $r_i - r_j$ are the radius-vectors of atoms i and j, N is the number of atoms in the system, M is the number of nearest neighbours, including atoms of three coordination spheres of interaction.

3. Results and discussion

In this study, the structure evolution of Ni-Al alloys having a deviation of ± 5% from stoichiometry is analyzed during stepwise cooling. In the process of stepwise cooling of the alloy, the average configuration energy per atom (figure 2), the long-range order parameter (figure 3 a) and the short-range order parameter (figure 3 b) are analyzed.

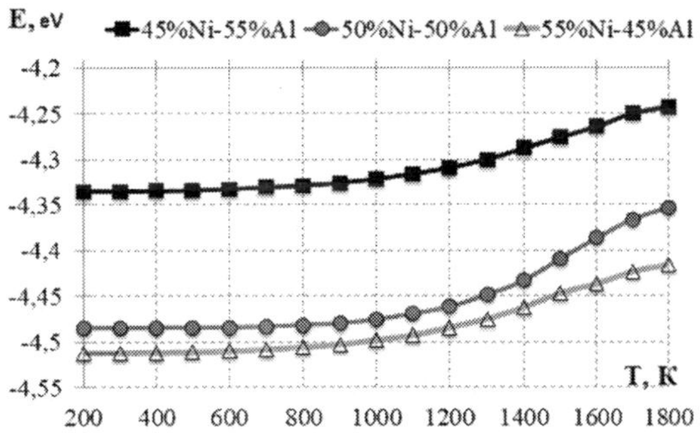

Figure 2. Temperature dependence of the average configuration energy of NiAl alloys without and with deviations from stoichiometry

The lowest configuration energy value has the alloy with 45% Al and the highest value - the alloy with 55% Al, as it is clearly seen from figure 2. The temperature ranges of the disorder-order phase transition are close for the three alloys, but the alloy with a stoichiometric composition moves to the ordered state faster and at a higher temperature.

The behavior of the short-range order parameter in the first coordination sphere (figure 3 a) suggests a tendency of the alloy towards ordering upon cooling, as evidenced by an increase in negativity of the short-range order parameter. At higher temperatures, the degree of ordering is smaller. A system without deviation from stoichiometry shows a stronger tendency towards ordering than systems with deviations from stoichiometry. Changes in the short-range order parameter for all the three studied alloys (figure 3 a) correlate with changes in the configuration energy (figure 2).

The temperature dependence of the long-range order parameter is of particular interest. In the alloy with an equiatomic composition, the long-range order parameter reaches 0.9, and in the alloys having deviations from stoichiometry it reaches 0.89 for 45% Al and 0.72 for 55% Al. In the NiAl alloy without deviation from stoichiometry, the value of the long-range order parameter grows faster than in non-stoichiometric alloys. The maximum value of the long-range order parameter corresponds to the largest volume of the ordered structure.

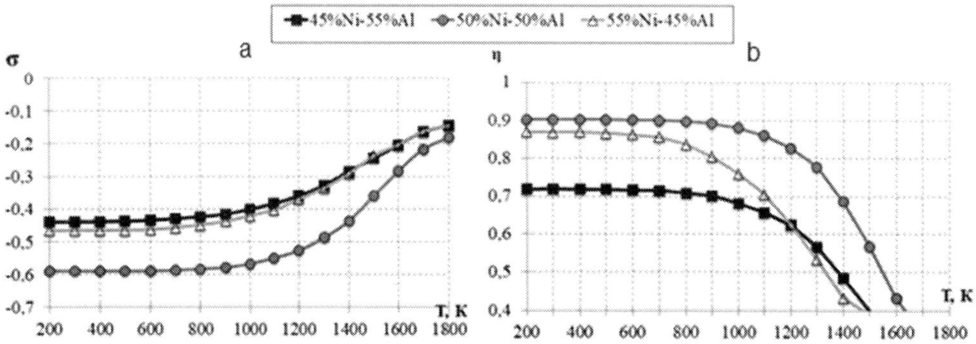

Figure 3. Temperature dependence of the short-range (a) and long-range (b) order parameters in NiAl alloys without and with deviations from stoichiometry

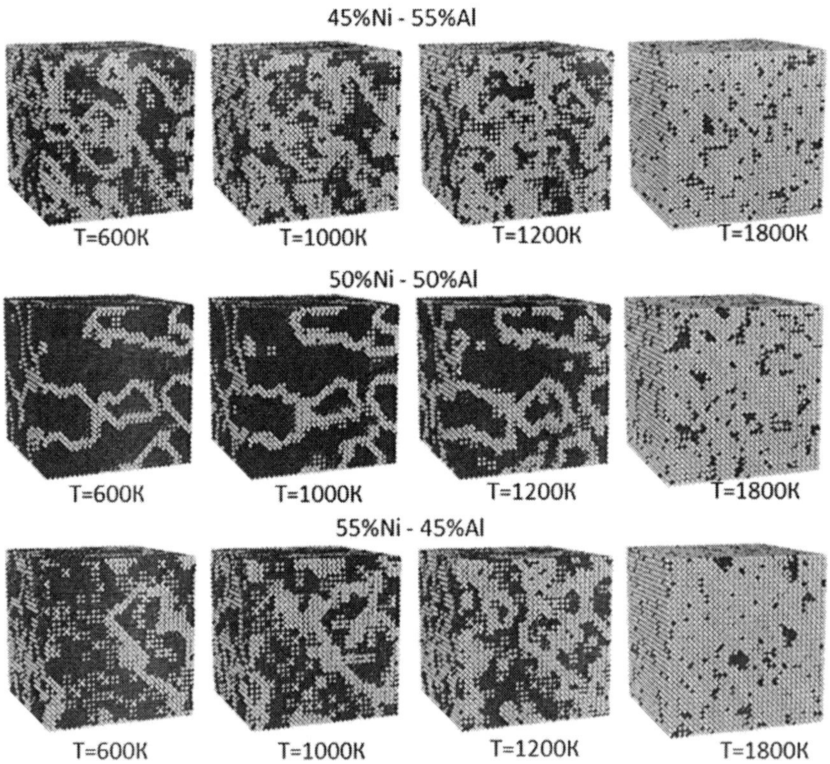

Figure 4. Domain structure evolution in NiAl intermetallic alloy with deviation from stoichiometry during stepwise cooling

Let us take a look at the domain structure changes of NiAl with various compositions during stepwise cooling, as presented in figure 4 for different temperatures. At high temperature (1800 K) only small-size domains of the B2 superstructure are observed for all configurations. The size of the antiphase domains with the B2 superstructure increases when cooled down to 1600 K. In an equiatomic alloy, at about 1500 K, the domain structure begins to form with two domains within the computational cell. The process of ordering continues with a decrease in temperature down to 1000 K, keeping disordered areas close to antiphase boundaries. In the alloy with 45% Al, the number of disordered regions decreases with decreasing temperature. The formation of domains is suppressed in the alloy with 55% Al.

4. Summary

Using the Monte Carlo simulation, we have shown the formation of the B2 superstructure antiphase domains after the completion of the disorder - order phase transition in NiAl alloys. Domains are separated by antiphase boundaries in the directions <110> and <100>. The analysis has demonstrated that deviation from stoichiometry exerts a noticeable effect on the structure and energy characteristics of the system. After stepwise cooling, the most ordered phase corresponds to the equiatomic composition and the most disordered one to the 45% Ni-55% Al alloy.

Acknowledgments
This research was supported by the grant of the President of the Russian Federation MK-5094.2018.2.

References

[1] Kositsyn S V and Kositsyna I I 2008 *Uspekhi Fiziki Metallov* **9** 195 (In Russian)
[2] Potekaev A I, Kulagina V V, Chaplygina A A, Chaplygin P A and Starostenkov M D 2017 *Russ. Phys. J.* **60** 1
[3] Kulagina V V and Potekaev A I 2011 *Izvestiya VUZov. Fizika* **54** 369 (In Russian)
[4] Potekaev A I, Kulagina V V, Chaplygina A A, Chaplygin P A and Starostenkov M D 2017 *Russ. Phys. J.* **59** 1532
[5] Chaplygina A A, Chaplygin P A, Starostenkov M D, Potekaev A I and Kulagina V V 2016 *Russ. Phys. J.* **59** 605
[6] Potekaev A I, Grinkevich L S, Chaplygin P A, Starostenkov M D and Chaplygina A A 2015 *Russ. Phys. J.* **58** 485
[7] Starostenkov M D, Chaplygina A A and Romanenko V V 2014 *Key Eng. Mater.* **592-593** 321
[8] Potekaev A I and Kulagina V V 2011 *Russ. Phys. J.* **54** 5
[9] Potekaev A I, Kulagina V V, Chaplygina A A, Starostenkov M D and Klopotov A A 2013 *Russ. Phys. J.***55**1248
[10] Chaplygina A A, Starostenkov M D, Kulagina V V and Potekaev A I 2012 *Russ. Phys. J.* **55** 814
[11] Dmitriev S V, Ovcharov A A, Starostenkov M D and Kozlov E V 1996 *Phys. Solid State* **38** 996

IOP Conf. Series: Materials Science and Engineering **447** (2018) 012055 doi:10.1088/1757-899X/447/1/012055

Microstructure and mechanical properties of a novel polycrystalline rhenium-containing nickel-based disc superalloy subjected to hot forging

Sh Kh Mukhtarov [1], R V Shakhov [1], A A Ganeev [1], R A Gaisin [1], A V Logunov [2], V M Imayev [1], R M Imayev [1]

[1] Institute for Metals Superplasticity Problems, Russian Academy of Sciences
39 Khalturin st, Ufa 450001, Russia
[2] PJSC "UEC-Saturn", 163 Lenin Ave, Rybinsk 152903, Russia

E-mail: vimayev@mail.ru

Abstract. A newly developed heavy alloyed polycrystalline nickel-based superalloy SDZhS-15 containing rhenium and intended as a structural material for turbine discs in gas turbine engines has been studied. Multiple homogenization and heterogenization heat treatment was developed for the superalloy to improve its hot workability. The heat treated workpieces were subjected to single-step hot forging under quasi-isothermal conditions at temperatures slightly below the γ' solvus temperature followed by air cooling and ageing. As a result, sound forgings with an inhomogeneous, partially recrystallized microstructure were produced. The obtained forgings were used for preparation of specimens for mechanical testing. The new superalloy showed superior high strength and high temperature capability, while retaining reasonable ductility in comparison with currently known disc superalloys.

1. Introduction

Gas turbine engines (GTE) provide a challenging environment for the development of novel materials and associated processing methods. Particularly, the increase of GTE efficiency requires the use of novel materials and processes that would meet the growing operational requirements due to the increase in temperature and loading of components [1]. To meet these requirements, novel nickel-based superalloys are designed towards heavier alloying to achieve properly balanced mechanical properties, i.e. higher creep resistance along with higher strength, fatigue strength and other service properties.

In respect of polycrystalline nickel-based superalloys, the following alloy designing approaches can be distinguished:

i) an increase in the volume fraction of the γ' phase precipitates by increased concentrations of precipitate forming elements such as Al, Ti, Nb and Ta. In this case, the γ' solvus temperature increases approaching the melting temperature of the superalloy;

ii) alloying with elements such as rare-earth yttrium, cerium and lanthanum forming dispersed precipitates different from the γ' phase and providing additional dispersion hardening;

iii) alloying with elements providing solid-solution strengthening, such as molybdenum, tungsten, rhenium, ruthenium, etc. In this regard, there is a risk of forming topologically close packed (TCP) phases during long-term high temperature exposure that can reduce the high temperature capability of a superalloy. Thus, the alloying additions should be precisely chosen to avoid the formation of undesirable phases.

In the present study, a hot forging procedure was developed for the newly designed polycrystalline superalloy SDZhS-15 containing rhenium, followed by microstructure characterization and evaluation of mechanical properties. The obtained mechanical properties are to be compared with those of currently known high γ' containing nickel-based superalloys.

2. Experimental

The experimental nickel-based superalloy SDZhS-15 with a nominal chemical composition of Ni-28(Cr, Co)-12.5(Al, Ti, Nb, Ta)-9(Mo, W, Re)-0.17(C, La, Y, Ce, B) (in wt. %) was taken as a starting material. The ingots of the superalloy with a size of Ø 100×180 mm were manufactured by vacuum induction melting. Small pieces were cut out of the initial ingot and remelted in a laboratory arc-melting furnace under argon atmosphere to ensure good chemical homogeneity. The ingot dimensions used in the present study were approximately Ø 45×15 mm.

To enhance the hot workability, long-term homogenization and heterogenization anneals were carried out in the temperature range of (T_s-120 °C)÷(T_s+20 °C), where T_s is the solvus temperature; each annealing stage was performed in a time interval of τ=2-20 h and followed by slow furnace cooling. The heat treated workpieces with dimensions of 35×22×15 mm^3 were subjected to forging under quasi-isothermal conditions. To do this, the workpieces were put into a thick-walled can made of a more ductile nickel base superalloy in contrast to SDZhS-15 and having nearly the same solvus temperature as SDZhS-15. The canned workpieces were heated up to a temperature 20÷40 °C lower than T_s and forged at a strain rate of $\dot{\varepsilon}\approx10^{-2}$ s^{-1} to an engineering strain of $\varepsilon\approx$80% using the die tool preheated to T=930 °C. The forged workpieces were left to cool in air then subjected to aging at T=T_s-360 °C (6 h) and T=T_s-470 °C (32 h) followed by air cooling.

The microstructural examination was carried out for the cross sections of the forged workpieces. Scanning electron microscopy (SEM) in backscattering electron (BSE) mode was used for this. The electron backscattered diffraction (EBSD) analysis was performed with a scan-step size of 0.5 μm. In doing so, the γ' precipitates were assumed as a γ phase and the microstructure was assumed as a quasi-single-phase. The EBSD analysis was conducted using the CHANNEL 5 processing software. Grain boundaries having a misorientation angle less than 2° were excluded from consideration taking into account the measurement accuracy. Grain boundaries having a misorientation angle more than 15° were assumed as high-angle boundaries. X-ray diffraction (XRD) measurement was made using Cu-K_α radiation.

The obtained forgings were used for preparation of samples for mechanical testing. Tensile and long-term strength tests were carried out using flat samples with a gauge section of 15×3×1.5 mm^3. All specimens for mechanical testing were prepared by electrospark cutting followed by fine grinding of working surfaces.

3. Results and discussion

3.1. Initial as-cast material

The X-ray diffraction analysis of the as-cast alloy revealed only peaks corresponding to the γ and γ' phases (figure 1 a). Thus, any undesirable phases, including TCP ones, were not detected in the superalloy (within the sensitivity of the XRD analysis).

Figure 1 b represents the BSE images of SDZhS-15 in the as-cast condition. A typical dendritic microstructure was obtained. Nonequilibrium eutectic colonies resulted from dendritic segregation were clearly distinguished (figure 1 b). The size of the elongated dendrites was in the range of (10-20)×(50-100) μm, the arm spacings were about 20 μm. The γ' phase precipitates had a size mostly in the range of 0.1-0.2 μm and predominantly cuboidal morphology; there were also coarse coagulated γ'

particles with a size of 1-2 μm. White particles were carbides, which had a size of 0.5-5 μm (figure 1 b). The volume fraction of the γ′ phase was visually 65-70%.

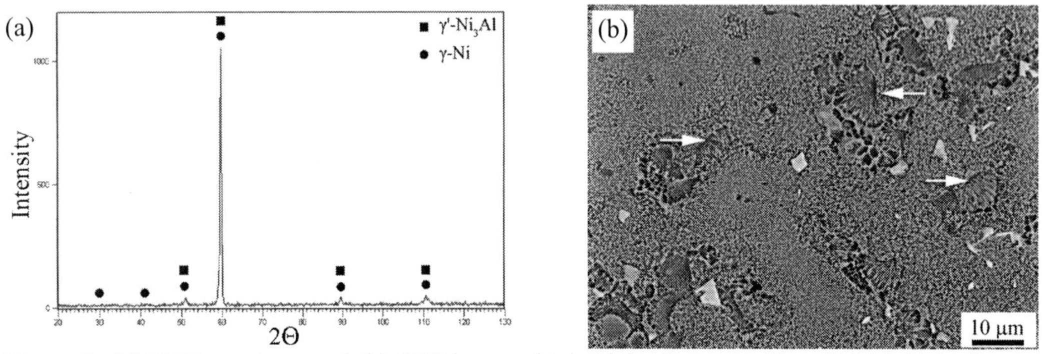

Figure 1. (a) XRD spectrum and (b) BSE image obtained for the superalloy SDZhS-15 in the as-cast condition: (a) TCP phases were not detected in the superalloy, (b) arrows show nonequilibrium eutectics; white particles are carbides, dark particles are γ′ precipitates formed during cooling of the ingot.

3.2. Effect of hot working on microstructure and mechanical properties

In order to improve the hot workability, the as-cast superalloy prior to hot forging was subjected to heat treatment, which included multiple homogenization and heterogenization anneals. This reduced the level of dendritic segregation and led to coagulation and spheroidization of the γ′ precipitates.

To ensure nearly isothermal conditions and all-round compression, hot forging was fulfilled in a thick-walled can made of a nickel-based superalloy, which was more ductile and had a solvus temperature similar to that of SDZhS-15. The heat treated canned workpieces were preheated at a temperature 20÷40 °C lower than T_s and forged under quasi-isothermal conditions to a strain of ε≈80%, then removed from the cans and aged. Sound forgings almost free of any lateral cracks were produced using this technique.

Figure 2 shows the EBSD orientation map and BSE images obtained from the central part of the forged workpiece. One can see that coarse non-recrystallized γ grains occupied about 70 vol.%. A large number of low-angle boundaries were observed in these grains suggesting that the dynamic recovery occurred very extensively (figure 2 a). This is confirmed by the fact that the fraction of high-angle boundaries was only 45.2%. Nevertheless, fine recrystallized grains with a size of $d_γ$=3-20 μm, occupying about 30 vol.% were formed that was the result of dynamic recrystallization (figure 2a). Thus, a partially recrystallized microstructure was obtained after canned forging, in spite of the higher strain rate and quasi-isothermal forging conditions. Apparently, this was promoted by all-round compression provided during canned forging. The microstructure of the forged material contained coarse and dispersed precipitates of the γ′ phase with a size of 1-3 μm and about 0.15-0.2 μm, respectively. Large precipitates were not dissolved at the forging temperature, whereas the dispersed γ′ precipitates appeared during cooling of the forged workpiece. Note that the fine γ′ precipitates had a rounded morphology. This can be ascribed to the effect of a higher strain rate and a slower cooling rate in the case of canned forging. White carbide particles appeared after ageing. They were mostly located near the coarse γ′ precipitates along the γ grain boundaries and had a size of 0.2-2 μm.

The forged and aged workpieces were used to prepare specimens for mechanical testing. Table 1 represents the mechanical properties of the superalloy SDZhS-15 after multiple heat treatment, canned forging and ageing in comparison with those of other industrial superalloys applied as disc materials for GTE. It is seen that the strength properties and creep resistance obtained for SDZhS-15 are higher than those of the known disc superalloys. Note that the specimens tested at T=650 °C and a loading of P=1200 MPa for 100 hours were not fractured. This indicates a significant potential of the new superalloy in terms of creep resistance. It is worth noting that the superalloy showed fairly high strength, while retaining lower but quite reasonable ductility (table 1).

Thus, preliminary heat treatment improving hot workability was developed for the new nickel base superalloy SDZhS-15. Note that the composition of the superalloy is close to the composition of some single-crystal superalloys and non-deformable cast superalloys [1-4]. This suggests that heavy alloying of nickel-based superalloys with refractory elements, including rhenium is not an obstacle to the following wrought processing. Hot workability was improved due to the reduction of dendritic segregation, coagulation and spheroidization of the γ' precipitates. This allowed forging the superalloy under quasi-isothermal conditions with relatively high strain rates, which provided a partially recrystallized microstructure as a result of dynamic recrystallization. The forged material was cooled in air from the forging temperature and then aged. As a result, both fine and coarse γ' precipitates were obtained. The material with the microstructural state obtained demonstrated superior strength and long-term strength at T=650 °C, while retaining a fairly reasonable ductility compared to the currently known disc superalloys. As expected, further optimization of hot working and strengthening heat treatment might additionally enhance the mechanical properties of the new developed superalloy.

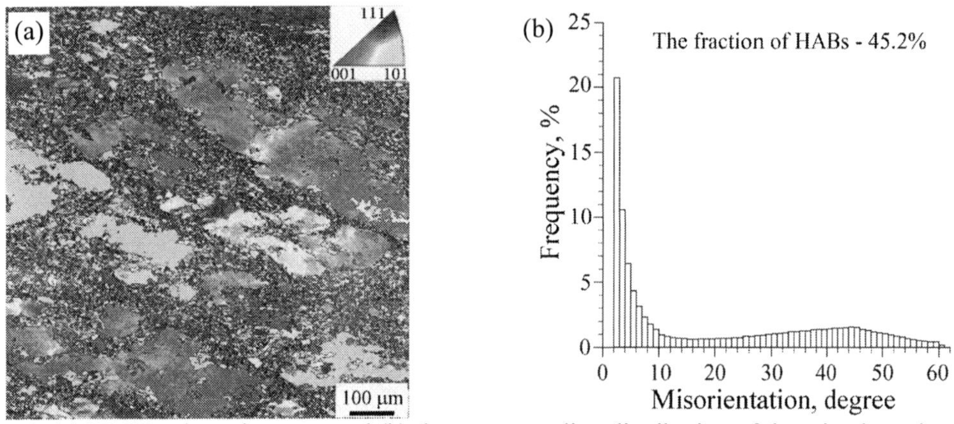

Figure 2. (a) EBSD orientation map and (b) the corresponding distribution of the misorientation-angle for grain boundaries obtained from the central part of the workpiece, subjected to preliminary multiple heat treatment and canned forging; (a) high- and low-angle grain boundaries are indicated by black and white lines, respectively.

Table 1. Mechanical properties of the superalloy SDZhS-15 in comparison with those of other nickel base disc superalloys (RT – room temperature).

Superalloy	UTS, MPa	YS, MPa	δ, %	Long-term strength at T=650 °C, 100 hours, MPa
SDZhS-15 (present work) RT T=650 °C	1708 1486	1250 1080	11 8	>1200
N18 RT [5]	1620	1080	>20	>1000
Alloy 10 RT [6]	1650	1210	12-17	-
AF115 RT [7]	1646	1190	20	1030
TMW-4 RT [8]	1700	1210	9	-
ME3 (Rene104) RT [9]	1580 1650	1150 1150	9-18 21	-
FGH100 RT [10] T=705 °C	1600 1345	1172 1090	19 20	T=705 °C, 897 MPa, the creep life τ=42.8 h

4. Conclusions

The preliminary multiple heat treatment has been developed for the new ingot-metallurgy high γ' containing nickel-based superalloy SDZhS-15, which improved its chemical homogeneity, provided coagulation and spheroidization of γ' precipitates and improved hot workability of the as-cast superalloy. Successful hot forging of the new superalloy using a can made of a nickel-based superalloy under quasi-isothermal conditions was fulfilled at temperatures slightly below the solvus temperature. As a result, an inhomogeneous and partially recrystallized microstructure was produced in the forged workpiece. In spite of the fact that the obtained microstructure was far from being optimal, the new superalloy subjected to canned hot forging followed by ageing demonstrated superior mechanical properties. Further efforts should be directed at optimizing the processes of hot working and strengthening heat treatment in order to produce a more homogeneous and properly designed microstructure.

Acknowledgements

The present work was supported by the Russian Science Foundation (Grant No. № 18-19-00594). The work was performed using the facilities of the shared services center «Structural and Physical-Mechanical Studies of Materials» at the Institute for Metals Superplasticity Problems of Russian Academy of Sciences.

References

[1] Reed R C 2006 *The Superalloys: Fundamentals and Applications* (Cambridge University Press)
[2] Ford D A, Fullagar K P L, Bhangu H K, Thomas M C, Burkhoider P S, Korinko P S, Harris K and Wahl J B 1999 *Trans. ASME,* **121** 138
[3] Harris K and Wahl J B 2004 *Proceedings of the 10th Inter. Symp. Superalloys 2004* (Warrendale, PA: TMS), pp. 45-52
[4] Hobbs R A, Tin S and Rae C M F 2005 *Metall. Mater. Trans. A* **36** 2762
[5] Guedou J Y, Lautridou J C and Honnorat Y 1992 *Proceedings of the 7th Inter. Symp. Superalloys 1992* (Warrendale, PA:TMS), pp. 267-276
[6] Rice D, Kantzos P, Hann B, Neumann J and Helmink R 2008 *Proceedings of the 11th Inter. Symp. Superalloys 2008* (Warrendale, PA: TMS), pp. 139-147
[7] Merrick H, Benn R C and Bhowal P R High strength powder metallurgy nickel base alloy: *US Patent No. 6468368 B1* 2002-10-22
[8] Gu Y F, Cui C, Harada H, Fukuda T, Ping D, Mitsuhashi A, Kato K, Kobayashi T and Fujioka J 2008 *Proceedings of the 11th Inter. Symp. Superalloys 2008* (Warrendale, PA: TMS), pp. 53-61
[9] Gao Y, Stoelken J S, Kumar M and Ritchie R O 2007 *Acta Mater.* **55** 3155
[10] Jia C L, Ge C C and Yan Q Z 2016 *Mater. Sci. Eng. A* **659** 287

IOP Conf. Series: Materials Science and Engineering **447** (2018) 012056 doi:10.1088/1757-899X/447/1/012056

Temperature- induced structure evolution of Ag nanoparticles

S L Gafner, D A Bashkova, Yu Ya Gafner

Katanov State University of Khakassia, 90 Lenin Av., Abakan 655000, Khakassia, Russia

E-mail: ygafner@khsu.ru

Abstract. The molecular dynamics method with a modified tight-binding (TB-SMA) potential has been used to study the thermal stability of the initial fcc phase in perfect silver clusters to 2.0 nm in diameter, including the "magical" icosahedral size of N = 55 and 147 atoms. It has been shown that the temperature factor can cause a transition from the initial fcc phase to other structural modifications in small Ag clusters (up to 200 atoms). Thus, the fcc structure of clusters with N > 200 atoms is thermally stable to the melting point. For silver nanoparticles of smaller size, the situation is much more complicated, and numerous cases of thermally induced changes in the cluster structure have been observed, often occurring in different scenarios. Thus, to use such small silver clusters, it is necessary to study in detail the problems of the thermal stability of the cluster structure, apparently taking into account the influence of various "magic" numbers.

1. Introduction

Metal clusters with sizes of several nanometres can be used in future technologies, such as biosensors, in catalysis or in medicine [1]. An ultra-small cluster is a nanoparticle that shows a significant in size, it is a nanoparticle of less than 2.0 nm. The structural, electronic, and optical behavior of the nanoparticles with such dimensions can demonstrate new properties. Novel, ultrafast and instant methods of synthesis facilitate the development of a representative number of applications based on silver nanoparticles and their functionalization and make these applications available to the scientific and technical community. The main quality of some of these applications is based on the morphology of the nanoparticles.

In particular, the properties of silver clusters open up a wide field of applications due to their optical activity and antimicrobial effect [2]. However, the rapid oxidation/sulfidation from the ambient atmosphere dramatically reduces all bonuses of silver and causes complications in terms of practical applications. For example, sensors containing Ag usually require storage in inert gases, and the operating time of such devices in the ambient atmosphere is rather limited. One of the existing solutions is to make a protection layer that can prevent Ag structures from degradation due to reactions with environmental species. However, these treatments are not always applicable or rather difficult to use on an industrial scale [3].

The first type of stabilizers is an inorganic scaffold/substrate. Due to the strong cohesive energy between the metal atoms and between the atoms of a metal and another inorganic material, for example, silicon, the metal clusters on the surfaces must be passivated by organic molecules or separated by an inactive matrix. For silver clusters, molecules with thiol groups are often used to passivate cluster surfaces. A natural alternative is to passivate the bulk surface instead of the cluster [4]. Several types of Ag nanoparticles were experimentally synthesized using various protective ligands, such as proteins, peptides and dendrimers. However, experiments with thiol groups were the most successful, since it was here that high stability, a small size of Ag clusters, and a narrow distribution of them in size were observed [5, 6].

While the majority of the hosts/scaffolds employed for Ag nanoclusters are liquid, polymer or organics based materials, it has only very recently been discovered that some bulk, oxyfluoride glasses can also host Ag nanoclusters. The obvious advantages of the glassy state, such as the possibility of fiber, film, and arbitrary bulk shape manufacturing, open up a novel application [7].

A possible approach to solve this problem could be the formation of very pure particles with a perfect crystalline structure. They should be more stable against the above-mentioned phenomena of degradation. A very promising approach for preparation of such nanoparticles (NPs) is the aggregation of metal atoms into clusters from very pure sources in a vacuum. These clusters can be collimated into beams, size selected and deposited or implanted on/in different substrates, which provides great opportunities for controlling the structure and properties of materials on the nanoscale [3].

Indeed, in mono-crystalline NPs with fcc structure, only atoms present on the surface would be reactive due to reduced coordination numbers and the presence of dangling bonds. In the case of poly-crystalline particles, there is higher probability that the reactive species can attack silver atoms at the grain boundaries, thus, converting them into compounds in expense of the pure metal. We believe that the use of silver nanoparticles with a size equal to the "magic" numbers of different structures can become another unaccounted opportunity to increase the stability of the structure. This position is confirmed by experimental analysis of the distribution of experimentally obtained particles in terms of size (mass spectrum). It showed that clusters with a certain number of atoms that correspond to "magic" numbers are much more stable than clusters of a different size. At the same time, the degree of defectiveness of the surface of clusters with "magic" numbers of atoms is minimal, while in others it can be very significant.

Due to the new applications of sub-nanometric material, it is important to analyze the structural, electronic, and optical behavior on several units of atoms, as well as to show the stabilization mechanisms of these sub-nanometric materials. Since the different internal structure of metallic nanoparticles means different physico-chemical properties, from the point of view of using silver nanoclusters, it is very important to determine the limits of the thermal and dimensional stability of the initial crystal structure of nanoclusters.

2. The computer model

Many various methods have been developed to imitate the behavior of nanoclusters. First of all, we should mention the molecular dynamics method (MD). It is most appropriate for determining the structures of metal clusters of a certain diameter, whichare stable in a wide temperature range. The initial objects by computer simulations were spherical fcc silver clusters obtained by cutting a sphere from a perfect fcc lattice with a size of up to 2.0 nm. The choice of clusters of such small size is due to the fact that it is in these sizes that the so-called "magic" numbers play the most important role in the stability of the structure [8 -10].

Clusters were simulated within a canonical ensemble using a Nóse thermostat. In this case, the number of particles N, volume V, and temperature T remain unchanged (NVT ensemble), and the total momentum p of the system is zero. The temperature during the simulation was determined from the average kinetic energy of the atoms, which was calculated based on the Verlet velocity algorithm with a time step h = 1 fs. The simulation was performed with the MDNTP program developed by Dr. Ralf Meyer (University of Duisburg, Germany). After analyzing various representations of potential energy, Ag nanoclusters were simulated using well-established tight-binding (TB-SMA) potentials [11] with a fixed cutoff radius corresponding to the fifth coordination shell inclusive. Structural transitions were determined using visualizers and from the graphs of the radial distribution function and the temperature dependence of the potential energy.

All clusters at T = 50 K were subjected to the obligatory procedure of thermal relaxation to optimize the internal structure and shape. Then the simulated systems were gradually heated from 50 to 1000 K using the Nóse thermostat. The lower temperature limit of 50 K is set so as not to miss the point of the change in the cluster structure, which can be observed below 100 K. The upper limit of 1000 K is sufficient to melt the simulated particles, since the melting temperature for Ag clusters decreases significantly with decreasing particle size compared with the case of a bulk material (T_m = 1235.1 K). To determine the most stable cluster structure, an array of equally sized nanoparticles was taken. During heating, the temperature was changed with a step of 20 K; in the region of structural transitions, the step was 1 K; at each fixed temperature, the clusters were held for 1 ns.

3. Results and discussion

As the most interesting examples, let us consider the Ag_{55} and Ag_{147} silver nanoclusters, i.e. clusters with the number of atoms exactly corresponding to geometric "magic" icosahedral numbers. The simulation results showed that the smallest of the simulated clusters behaved quite predictably under heating. Already at the preliminary stage of thermal relaxation, a polytype transition of the fcc → Ih occurred and an ideal icosahedral structure was formed. That is, as expected, with such a small cluster size and the number of atoms equal to two complete Ih shells, the gain in surface energy and the absence of structural defects immediately lead to the formation of an extremely stable icosahedral modification, which is fixed by many authors for a variety of metals. No pre-melting effects were observed here, the dependence of the potential energy of the Ag_{55} cluster on the temperature had a strictly linear character up to the melting point, after which a very rapid destruction of the long-range order occurred in the cluster (figure 1).

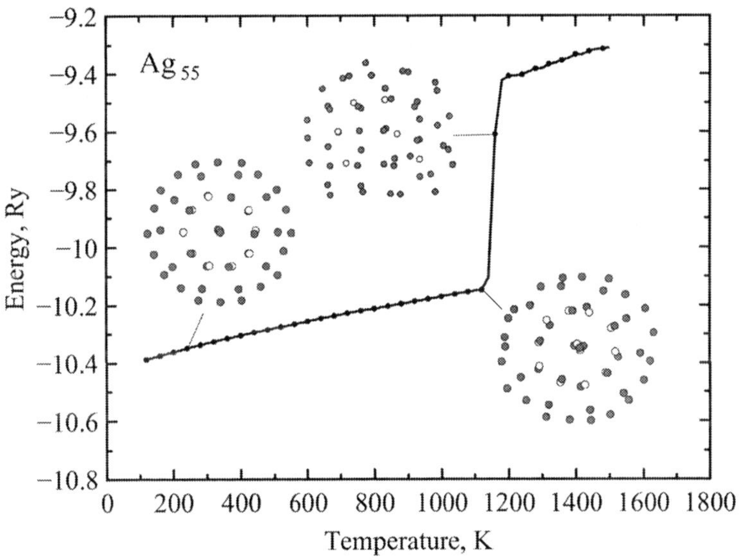

Figure 1. Temperature dependence of the potential energy of the fcc Ag_{55} cluster and structural changes in the nanoparticle.

The second considered cluster of the group of ideal icosahedral ones under consideration was already completely filled with the third Ih shell. In this case, the number of atoms becomes much larger (N = 147) than in the Ag_{55} cluster, which can be decisive for such small subnanometer clusters. A typical dependence of the potential energy of the cluster Ag_{147} on the temperature is shown in figure 2. In this figure we will dwell in more detail, but first we give some preliminary arguments.

As already mentioned, in the bulk state, silver has a face-centered cubic lattice, but the competition between the bulk and surface energies in the nanometer range can lead to the formation of several different isomers, the stability boundaries of which have not yet been determined. Many theoretical studies were mainly focused on searching for the total energy of a cluster at very low temperatures in various structural modifications as a measure of their relative stability, and the data obtained were directly affected by the chosen potential. Apparently, we must recognize that even insignificant defects in silver, especially at small sizes, can play a very important role in determining the equilibrium structure of a particle.

Another cause of such serious discrepancies consists in very close energies calculated for different structural modifications. For example, the MD simulation of gold clusters close to silver in the processes of structuring performed in [12] with the Gupta potential at N = 38 atoms gives the following free cluster energies in its ground state: –3.4405 eV/atom in the amorphous state, –3.44 eV/atom in the fcc state, and –3.431 eV/atom in the icosahedral (Ih) state. Such small differences between these three isomers were retained up to room temperature. At T > 250 K, thermal energy is sufficient to overcome potential barriers between these structures; as a result, all three isomers begin to appear randomly. Such significant changes are believed to be mainly caused by the fact that, with such cluster sizes, almost all atoms lie on the cluster surface, and an insignificant change in their arrangement induced by thermal diffusion is sufficient for a spontaneous formation of another structural modification. This conclusion was supported experimentally: the cluster structure and shape fluctuations were clearly visible in gold clusters with a diameter of 2.0 nm in using an electron microscope. The larger size of the fluctuating particles is due to the fact that the cluster structure and shape changes were induced by both thermal processes and electron irradiation.

The authors of [12] applied MD simulation with the Gupta potential to perform similar estimates for larger gold clusters. The binding-energy difference between the amorphous and Ih structures was found to be 9.4 meV/atom at N = 55, and the five-particle Dh modification at N = 57 turned out to be more stable than the amorphous structure by only 5.7 meV/atom. To verify this result, the authors of [12] performed an additional ab-initio study of the relative stability of gold clusters with crystalline and amorphous structures. The binding energy difference per atom between the most stable amorphous and crystalline modifications for clusters of the sizes given above was found to be smaller than 0.01 eV/atom. The only specific feature found was the distribution of the electron charge density: it was slightly broader in amorphous clusters.

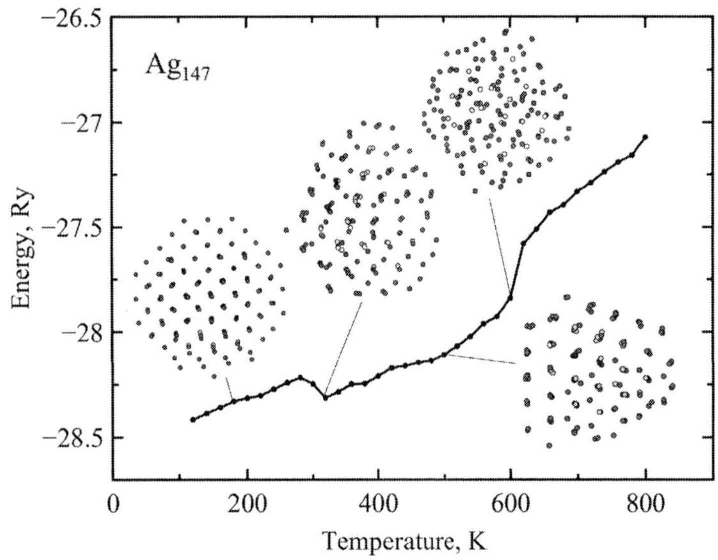

Figure 2. Temperature dependence of the potential energy of the fcc Ag_{147} cluster and structural changes in the nanoparticle.

All this can be seen well from the analysis of figure 2. Here, unlike the Ag_{55} cluster, the structural transition fcc → Ih was not observed in the preliminary stage of thermal relaxation of Ag_{147}. This fact indicates that the gain in surface energy at such a size is no longer so significant, and to overcome the energy barrier between polytypic modifications, a larger value of thermal energy kT is required. This result is confirmed by the work [10], where it was shown that the energy stability of the Ag_{147} cluster decreases by approximately 30-35% compared to Ag_{55}, becoming practically identical with the decahedral (Dh) cluster Ag_{146}. The reason for this phenomenon is that in the case of Ag_{55}, about 76% of the atoms lie on the cluster surface, whereas in the Ag_{147} cluster it is only 62%. It is clear that due to the smaller coordination number and, consequently, a lower value of the binding energy, surface atoms more easily move around the cluster, forming an energetically more advantageous structure.

In full accordance with our theoretical considerations, the cluster Ag_{147} retained its original fcc structure to temperatures of the order of 280 K. Further, the structure was rearranged and the cluster acquired a decahedral or icosahedral structure, which gradually disintegrated during further heating. The complete destruction of the crystal structure of Ag_{147} nanoclusters is observed at temperatures above 620 K. Thus, in spite of the fact that the Ag_{147} cluster strictly corresponded to the geometric "magical" icosahedral number, during heating approximately the same proportion of both icosahedral and decahedral modifications could be formed, which indicates close values of the energies of these five-particle structures [10].

4. Conclusions

When studying nanoclusters at the end of the 20th century, the fact seemed unusual that their properties were significantly different from the properties of bulk compounds of the same composition. This was is not surprising for the smallest clusters, whose electronic and atomic structure was fundamentally different from the structure of the bulk body. However, even fairly large clusters with internal structures similar to the bulk body may have properties different from those characteristic of it.

When a metal particle decreases in size to several hundred atoms, the density of states in its conduction band drastically changes. The continuous density of states of the bulk body is replaced by a set of discrete levels, the intervals between which can be greater than the thermal energy k_BT, which results in the formation of gaps. Clusters of different sizes and internal structures have different electronic structures, hence, different distances between the levels, which can be used in the design of nanomaterials and the development of various technical devices. For example, light-induced transitions between energy levels control the color of a material. The ability to interact with other materials also depends on the size and structure of the cluster.

It was previously believed that small-sized metal clusters should mostly have a structure with five-particle symmetry, since it is these structures that provide surface energy growth. Hence, the main goal was to determine the size boundary at which the metal cluster changes the structure from the icosahedral one to the structure corresponding to the bulk material. However, as shown in this study, this concept is too simplified; at least for some metals, the problems of thermal stability of the cluster structure should be studied in more detail, taking into account the effect of different “magic” numbers.

Acknowledgments

This work was supported by the Russian Foundation for Basic Research (project no. 17-48-190320 and 18-42-190001).

References

[1] Hua D Hongtao Y U 2015 *Adv. Natur. Sci.* **8** 1
[2] Roese S Engemann D Hoffmann S Latussek K Sternemann C and Hövel H 2016 *J. Phys.: Conf. Ser.* **712** 012068
[3] Novikov S M, Popok V N, Evlyukhin A B, Hanif M, Morgen P, Fiutowski J, Beermann J, Rubahn H-G and Bozhevolnyi S I, 2017 *Langmuir* **33** 6062
[4] Alkis S, Krause J L, Fry J N and Cheng H-P 2009 *Phys. Rev. B.* **79** 121402(R)
[5] Padmos J D, Boudreau R T M, Weaver D F and Zhang P 2015 *J. Phys. Chem. C.* **119** 24627
[6] Copp S M, Schultz, D Swasey S M, Faris A and Gwinn E G 2016 *Nano Lett.* **16** 3594
[7] Velázquez J J, Tikhomirov V K, Chibotaru L F, Cuong N T, Kuznetsov A S, Rodríguez V D, Nguyen M T and Moshchalkov V V 2012 *Optics Express* **20** 13582
[8] Akbarzadeh H and Yaghoubi H 2014 *J. Coll. Interface Sci.* **418** 178
[9] Chiu Y-P, Wei Ch-M and Chang Ch-S 2008 *Phys. Rev. B.* **78** 115402
[10] Liu D, Wen Z and Jiang Q 2011 *Curr. Nanosci.* **7** 463
[11] Cleri F and Rosato V 1993 *Phys. Rev. B.* **48** 22
[12] Garzon I L, Michaelian K, Beltan M R, Posada-Amarillas A, Ordejon P, Artacho E, Sanchez-Portal D and Soler J M 1999 *Eur. Phys. J. D* **9** 211

Non-symmetric kinks in Klein-Gordon chains free of the Peierls-Nabarro potential

Zh G Rakhmatullina[1], **P G Kevrekidis**[2] **and S V Dmitriev**[1,3]

[1] Institute for Metals Superplasticity Problems, Ufa 450001, Russia
[2] Department of Mathematics and Statistics, University of Massachusetts, Amherst, Massachusetts 01003, USA
[3] National Research Tomsk State University, Tomsk 634050, Russia

E-mail: dmitriev.sergey.v@gmail.com

Abstract. There exist several approaches to discretize Klein-Gordon field equations in a way that static kink solutions can be obtained from a two-point map, iteratively, starting from any admissible initial value. Such kinks can be placed anywhere with respect to the lattice, in other words, they do not experience the Peierls-Nabarro potential (PNP). So far, to the best of our knowledge, only symmetric PNP-free kinks have been analyzed for the ϕ^4 and sine-Gordon equations. It is interesting to construct PNP-free discretisations for the ϕ^6 equation supporting asymmetric kinks. In the present study, two versions of such discretisations are derived.

1. Introduction

Discrete Klein-Gordon equations supporting kinks free of the Peierls-Nabarro potential (PNP) have been considered in a number of studies [1-9]. Such discretizations are interesting for a number of reasons. One of them is the mobility of such kinks, possessing the zero-frequency Goldstone mode, even in the regime of strong discreteness. So far, only symmetric ϕ^4 and sine-Gordon kinks have been analyzed. Recently, kinks in the continuum ϕ^6, ϕ^8, ϕ^{10} and ϕ^{12} models have been studied [10-13]. It is tempting to derive PNP-free discrete analogues of such fields, and in the present study we do this for the ϕ^6 case.

With the use of the discretized first integral approach [9], we derive the Hamiltonian PNP-free discrete version of the ϕ^6 equation, which is essentially the Klein-Gordon model proposed by Speight and Ward [1]. Then we obtain the momentum-conserving PNP-free discrete ϕ^6 model suggested in [2].

Kink solutions of the discrete Klein-Gordon models find their applications, e.g., in the solid state physics, describing the dynamics and interaction of dislocations, dislocation kinks and jogs, domain walls, crowdions, crowdion clusters [14-17], etc. Kink mobility is a crucial issue in all such applications.

2. ϕ^6 field

We consider the Hamiltonian of the ϕ^6 field,

$$H = E_K + E_P, \tag{1}$$

with kinetic and potential energies given by

$$E_K = \frac{1}{2}\int_{-\infty}^{\infty} \phi_t^2 \mathrm{d}x, \tag{2}$$

and

$$E_P = \frac{1}{2}\int_{-\infty}^{\infty} \left[\phi_x^2 + \frac{k}{4}\phi^2(1-\phi^2)^2\right] \mathrm{d}x, \tag{3}$$

respectively. Here $\phi(x,t)$ is the unknown function of spatial, x, and temporal, t, variables. The last term in (3) gives the energy of the six-order on-site potential, which we denote as

$$V(\phi) = k\phi^2(1-\phi^2)^2/8. \tag{4}$$

From the Hamiltonian (1)-(3) the following equation of motion can be derived

$$\phi_{tt} = \phi_{xx} - V'(\phi) = \phi_{xx} - k\phi(1-4\phi^2+3\phi^4)/4 \equiv D(\phi(x,t)) . \tag{5}$$

In the following we will discretize the equation of motion (5) on the lattice $x=nh$, where $n=0, \pm1, \pm2...$, and h is the lattice spacing. The general form of the constructed discrete models will be as follows

$$\ddot{\phi}_n = \Delta_2\phi_n + F(\phi_{n-1}, \phi_n, \phi_{n+1}) \equiv D(h, \phi_{n-1}, \phi_n, \phi_{n+1}), \tag{6}$$

where

$$\Delta_2\phi_n = \left(\frac{1}{h^2}\right)(\phi_{n-1} - 2\phi_n + \phi_{n+1}), \tag{7}$$

and, in the continuum limit ($h\rightarrow0$),

$$F(\phi_{n-1}, \phi_n, \phi_{n+1}) \rightarrow k\phi(1-4\phi^2+3\phi^4)/4. \tag{8}$$

Note that the standard discretization reads

$$\ddot{\phi}_n = \Delta_2\phi_n + k\phi_n(1-4\phi_n^2+3\phi_n^4)/4, \tag{9}$$

but in this model kinks typically have only two possible equilibrium center positions with respect to the lattice, one of them being stable and another one unstable.

PNP-free discretisations can be derived with the help of the first integral of the static version of (6), taken with zero integration constant (C=0), which can be written either as

$$U(x) \equiv \phi_x^2 - k\phi^2(1-\phi^2)^2/4 = 0, \tag{10}$$

or as

$$u(x) \equiv \pm\phi_x - \sqrt{k}\phi(1-\phi^2)/2 = 0. \tag{11}$$

2.1. Discretization conserving energy

We consider a discrete version of the first integral (11) in the form

$$u(h, \phi_{n-1}, \phi_n) = \frac{\phi_n - \phi_{n-1}}{h} - \frac{\sqrt{k}}{2}\frac{\phi_{n-1}+\phi_n}{2}\left(1 - \frac{\phi_{n-1}^2+\phi_{n-1}\phi_n+\phi_n^2}{3}\right) = 0. \tag{12}$$

The Hamiltonian (1)-(3) can be discretized as follows

$$\mathcal{H} = \frac{h}{2}\sum_n(\dot{\phi}_n^2 + u^2), \tag{13}$$

from which the following discrete equations of motion can be derived,

$$\ddot{\phi}_n = -u(\phi_{n-1}, \phi_n)\frac{\partial}{\partial\phi_n}u(\phi_{n-1}, \phi_n) - u(\phi_n, \phi_{n+1})\frac{\partial}{\partial\phi_n}u(\phi_n, \phi_{n+1}). \tag{14}$$

Substituting (12) into (14) one obtains the equations of motion of the discrete ϕ^6 model conserving energy (13),

$$\begin{aligned}\ddot{\phi}_n = \Delta_2\phi_n &+ \frac{\sqrt{k}}{12h}[(\phi_{n-1}^3 - \phi_{n+1}^3) - 3\phi_n^2(\phi_{n-1} - \phi_{n+1})] + \\ &+ \frac{k}{144}\left[-2(\phi_{n-1}^5 + \phi_{n+1}^5) - 8\phi_n(\phi_{n-1}^4 + \phi_{n+1}^4) + \right. \\ &+ (9 - 15\phi_n^2)(\phi_{n-1}^3 + \phi_{n+1}^3) + (24\phi_n - 16\phi_n^3)(\phi_{n-1}^2 + \phi_{n+1}^2) + \\ &\left. + (27\phi_n^2 - 10\phi_n^4 - 9)(\phi_{n-1} + \phi_{n+1}) - 6\phi_n^5 + 24\phi_n^3 - 18\phi_n\right].\end{aligned} \tag{15}$$

From the construction, it is clear that static solutions of (15) can be found from (12), which is the cubic two-point map. Since in the discretized first integral (11) the integration constant was set to be zero, only kink solution can be found from (12). For this, one can take any initial value of the two-point map in the range $0<\phi_n<1$ and find iteratively the equilibrium kink solution. The arbitrariness of the choice of the initial value means that the static equilibrium kink can be placed anywhere with respect to the lattice, meaning that the PNP is zero for the kink.

2.2. Discretization conserving momentum

Now we consider the first integral in the form of (10) and discretise it as follows,

$$U(h,\phi_{n-1},\phi_n) = \frac{(\phi_n-\phi_{n-1})^2}{h^2} - \frac{k}{4}\phi_{n-1}\phi_n(1-\phi_{n-1}\phi_n)^2 = 0. \quad (16)$$

The equation of motion (5) then can be discretized as follows, in the spirit of [2],

$$\ddot{\phi}_n = \frac{U(h,\phi_n,\phi_{n+1}) - U(h,\phi_{n-1},\phi_n)}{\phi_{n+1}-\phi_{n-1}}. \quad (17)$$

Substituting (16) into (17) one finds the discrete ϕ^6 model of the form

$$\ddot{\phi}_n = \Delta_2\phi_n - \frac{k}{4}[\phi_n - 2\phi_n^2(\phi_{n+1}+\phi_{n-1}) + \phi_n^3(\phi_{n+1}^2 + \phi_{n+1}\phi_{n-1} + \phi_{n-1}^2)]. \quad (18)$$

This model can be written as a conservation law:

$$\mathrm{d}/\mathrm{d}t\,[\dot{\phi}_n(\phi_{n+1}-\phi_{n-1})] = J(h,\phi_n,\phi_{n+1},\dot{\phi}_n,\dot{\phi}_{n+1}) - J(h,\phi_{n-1},\phi_n,\dot{\phi}_{n-1},\dot{\phi}_n), \quad (19)$$

where

$$J(h,\phi_n,\phi_{n+1},\dot{\phi}_n,\dot{\phi}_{n+1}) = U(h,\phi_n,\phi_{n+1}) + \dot{\phi}_n\dot{\phi}_{n+1}, \quad (20)$$

and consequently, as a result of direct summation, conserves the linear momentum [2],

$$P = \sum_n \dot{\phi}_n(\phi_{n+1}-\phi_{n-1}). \quad (21)$$

Again, it is clear from the construction that static solutions of (18) can be found from the cubic two-point map (16). The integration constant in the discretized first integral (10) was set to be zero, so that only kink solution can be found from (16). Equilibrium kink solutions are found from the two-point map, iteratively, starting from any initial value in the range $0<\phi_n<1$. Such kinks do not experience PNP, since they can be placed anywhere with respect to the lattice.

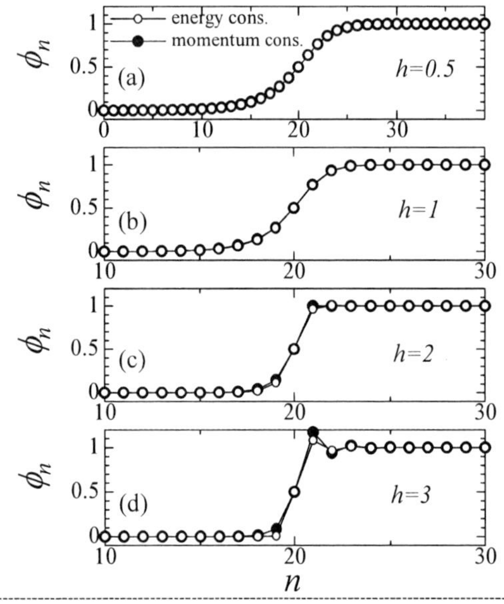

Figure 1. Static kink profiles shown by open and filled dots for the energy- and momentum-conserving models, respectively, for k=2 in (4). Panels (a)-(d) present the results for different degrees of discreteness, h=0.5, 1.0, 2.0, and 3.0, respectively. For large h, the kink profiles in the two models are noticeably different, and they become closer to each other for smaller h, because both models have the same continuum limit. Note the asymmetry with respect to the central point in the discrete ϕ^6 kink profile.

3. Numerical results

In figure 1, profiles of static kinks are plotted for the energy-conserving model (15) (open dots) and the momentum-conserving model (18) (filled dots). The case of k=2 is considered. Kink solutions were obtained for these models from the cubic two-point maps (12) and (17), respectively, using ϕ_n=0.5 as the initial value. The results for the discreteness parameter h=0.5, 1.0, 2.0, and 3.0 are shown in (a)-(d), respectively. The asymmetry of the kinks with respect to the central point can be clearly seen. For large h, the kink profiles in the two models are noticeably different, and they become closer to each other for smaller h, as it should be, as both models have the same continuum limit.

4. Conclusions

Two discrete ϕ^6 models, supporting asymmetric kinks free of PNP were derived, one of them, (15), conserves energy (13) and the other one, (18), conserves momentum (21).

The obtained results demonstrate that, in principle, discrete kinks can be highly mobile in that sense that they can be accelerated by any weak external force. Discrete kinks (or domain walls [18]) play an important role in nanomaterials transporting mass, energy, momentum, or/and electric charge, depending on the application.

As a continuation of this work, one can study collisions between slowly moving kinks boosted with the use of the zero-frequency Goldstone mode. It is also important to study the effect of the discreteness parameter h on the linear phonon spectra of the two vacuums, $\phi_n=0$ and $\phi_n=\pm1$, and on the existence and frequencies of the kink's internal modes. It would be also interesting to analyze the interaction of moving kinks with PT-symmetric perturbations similar to what was done for the continuum case [19,20].

Acknowledgements

The work of ZhGR was supported by the Russian Foundation for Basic Research, grant no. 17-02-00984-a. SVD thanks the Russian Science Foundation, grant no. 16-12-10175, for their financial support.

References

[1] Speight J M and Ward R S 1994 *Nonlinearity* **7** 475; ibid. 1997 **10** 1615; 1999 **12** 1373
[2] Kevrekidis P G 2003 *Physica* D **183** 68
[3] Dmitriev S V, Kevrekidis P G and Yoshikawa N 2006 *J. Phys. A* **39** 7217
[4] Barashenkov I V, Oxtoby O F and Pelinovsky D E 2005 *Phys. Rev. E* **72** 035602(R)
[5] Dmitriev S V, Kevrekidis P G and Yoshikawa N 2005 *J. Phys. A* **38** 7617
[6] Cooper F, Khare A, Mihaila B and Saxena A 2005 *Phys. Rev. E* **72** 036605
[7] Speight J M and Zolotaryuk Y 2006 *Nonlinearity* **19** 1365
[8] Oxtoby O F, Pelinovsky D E and Barashenkov I V 2006 *Nonlinearity* **19** 217
[9] Dmitriev S V, Kevrekidis P G, Yoshikawa N and Frantzeskakis D J 2006 *Phys. Rev. E* **74** 046609
[10] Belendryasova E and Gani V A 2017 *J. Phys. Conf. Ser.* **934** 012059
[11] Marjaneh A M, Gani V A, Saadatmand D, Dmitriev S V and Javidan K 2017 *J. High Energy Phys.* **2017** 28
[12] Gani V A, Lensky V and Lizunova M A 2015 *J. High Energy Phys.* **2015** 147
[13] Khare A, Christov I C and Saxena A 2014 *Phys. Rev. E* **90** 023208
[14] Braun O and Kivshar Y 2004 *The Frenkel–Kontorova Model: Concepts, Methods, and Applications* (New York: Springer)
[15] Derlet P M, Nguyen-Manh D and Dudarev S L 2007 *Phys. Rev. B* **76** 054107
[16] Dmitriev S V, Medvedev N N, Chetverikov A P, Zhou K and Velarde M G 2017 *Phys. Status Solidi RRL* **11** 1700298
[17] Dmitriev S V, Korznikova E A and Chetverikov A P 2018 *J. Exp. Theor. Phys.* **126** 347
[18] Evazzade I, Lobzenko I P, Saadatmand D, Korznikova E A, Zhou K, Liu B and Dmitriev S V 2018 *Nanotechnology* **29** 215704
[19] Saadatmand D, Borisov D I, Kevrekidis P G, Zhou K and Dmitriev S V 2018 *Commun. Nonlinear Sci.* **56** 62
[20] Saadatmand D, Dmitriev S V, Borisov D I and Kevrekidis P G 2014 *Phys. Rev. E* **90** 052902

IOP Conf. Series: Materials Science and Engineering **447** (2018) 012058 doi:10.1088/1757-899X/447/1/012058

Dependence of the microstructure and microhardness of the AA2024-O alloy on the thermal and mechanical action on the weld during friction stir welding

V V Karmanov, A L Kameneva

Department of Innovative Technologies of Mechanical Engineering, Perm National Research Polytechnic University, 29 Komsomolsky pr., Perm 614990, Russia

E-mail: annkam789@mail.ru

Abstract. The microstructure of the weld from the AA2024 alloy is highly dependent on the heat input in the weld, and therefore it is necessary to understand the relationship between these characteristics to improve the mechanical properties of the weld. The effect of the microstructure in various zones of the weld on its microhardness has been ascertained. The structure of various weld zones was studied depending on the heat input to the weld. Therefore, the experiment was performed at different values of the external axial force and the values of the constant speed of the rotation of the tool and welding. The dependence of the microhardness and microstructure of the weld zones on the temperature gradient in both normal and traverse directions has been established.

1. Introduction

Alloys with low melting temperature (e.g. Al-Cu-Mg) are not weldable by melting processes, but they are easily welded by friction stir welding (FSW). The main sources of heat during FSW are the work of the sliding friction force and the rolling friction torque generated between the tool and the welded surfaces, as well as the strain energy of the work pieces. During FSW, the unit thermal power is determined by the dependence:

$$q = q_1 + q_2 = \frac{4M_{r.f.} \cdot \omega}{\pi D^2} + \frac{4F_{s.f.} \cdot V_W}{\pi D^2} \qquad (1)$$

where q is the total specific heat capacity, q_1 is the specific thermal power under rotational motion of the tool about its axis, q_2 – is the specific thermal capacity when the tool is moving along the interface of the work piece, $M_{r.f.}$ is therolling friction torque, N·m; ω is the tool rotation speed, s^{-1}; D is thetool diameter, m, $F_{s.f.}$ is the sliding friction force, N; V_W is the linear speed of movement of the tool or welding speed, $m{\cdot}s^{-1}$ [1, 2].

Intensive transfer of metal in combination with high thermal conductivity of aluminum leads to a rapid temperature equalization on the surface of welded parts. Therefore, the tool moving along the connection line at the initial stage of weld formation constantly draws a colder metal into the plastic strain. When welding proceeds at high traverse speeds, as well as a slow tool rotation speed, the material receives less work per unit length of the weld, i.e. fewer tool rotations per millimetre. Under such conditions, the plasticised material may not reach a sufficiently high temperature. Aluminium alloys can withstand only a certain shear strain rate, which is dependent on temperature. Insufficient axial force on the tool and, as a consequence, insufficient mechanical action and frictional warming of the weld do not contribute to achieving an evenly welded joint [3].

At the present time, there are a large number of papers devoted to the analysis of the microstructure evolution of the weld and the width of the heat affected zone (HAZ) depending on the tool rotation speed (w), welding speed (V_w) and forces (P) occurring during FSW [4]. The purpose of this work is to study the influence of the external axial force acting on the tool on the change in microstructure and microhardness in various zones of the aluminum alloy AA2024-O, welded at values of constant speed of rotation of the tool and welding.

2. Material and Experimental Test Procedures

Sheets of the AA2024-O alloy (12 mm thick) with technological plating were investigated. The chemical composition is given in table 1. The welding process was carried out by friction stir welding equipment I-ISTIR PDS-5. The tool made from MP159 material (Cr-Ni-Co alloy) consisted of a shoulder with a diameter of 25.4 mm, along with a threaded pin with a diameter of 12.7 mm and a lengh of 12 mm. Sheets were welded with constant values of w=300 rpm, V_w=300 mm/min, and different heat and mechanical input due to the change in force on the tool: P=23 kN (low), 26 kN (medium), and 29 kN (high). The maximum depth of the tool plunge into the workpiece and the angular deviation from the vertical axis were maintained at 11.3 mm and 1.50, respectively. The welding parameters were kept the same for all samples in order to obtain comparable conditions.

Table 1. The chemical composition of the AA2024-O alloy, wt%

Element:	Si	Fe	Cu	Mn	Mg	Cr	Zn	Ti	others	Al
AA2024-O	0.5	0.5	3.8-4.9	0.3-0.9	1.2-1.8	0.10	0.2	0.15	0.20	Bal.

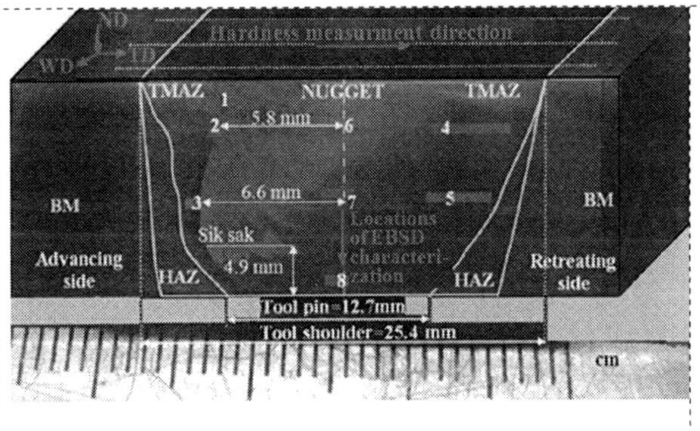

Figure 1. Macrograph of the weld cross-section. The weld was obtained at the high heat input.

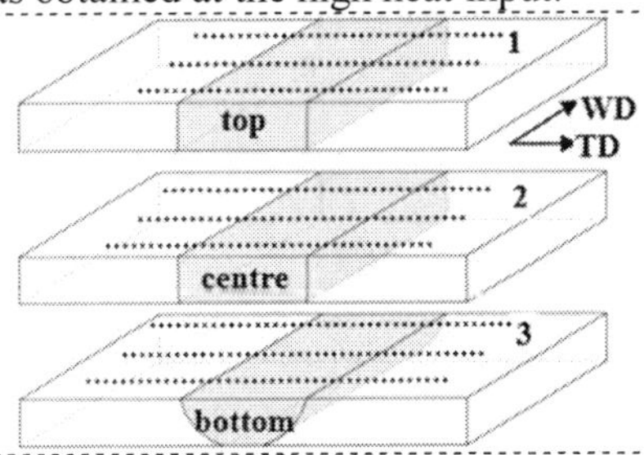

Figure 2. Upper, central and bottom samples cut in transverse direction

Rectangular samples of 50x25x12 mm in size were cut from the welding zone using a Discotom-50 (Struers) machine. One part of the samples was cross-sectioned in the normal direction of the welded sheets (figure 1), the second part was slitted in the transverse direction of the welded sheets (figure 2). Then each sample was cut into three portions to study the microstructure and defects in the normal and transverse directions. EBSD analysis was accomplished by an EDX 55 Ultra energy-analyzer of an ULTRA 55 electron microscope using OIM software. A macrograph of the weld cross-section is shown in figure 1. Points 1-9 are the locations of the EBSD characterizations. For the SEM and EBSD analysis, polished surfaces were etched with Keller's reagent (100 ml HCl, 100 ml HNO_3, 25 ml HF, and 100 ml distilled water). The etched specimens were rinsed in ethanol, and then blow dried and observed by the ULTRA 55. Vickers hardness ($HV_{0.2}$) was measured in the transverse direction on surfaces 1, 2, 3 of the top, centre and bottom portions of each sample.

DuraScan 70 microhardness tester (EMCO-TEST) with a microanalysis system was used to measure microhardness in 0.3 mm increments. The microhardness of the three parallel tracks was averaged to refine the measurements and obtain a single microhardness map for the top, centre, and bottom portions of each sample (figure 2). The distance between the tracks was 9 mm.

The weld zones temperature was determined by the procedure of Ref. [5].

$$T = T_m \cdot k\left(\frac{\omega^2}{V \cdot 10^4}\right)^{\alpha} = 502 \cdot 0.7\left(\frac{300^2}{300 \cdot 10^4}\right)^{0.05} = 295^0 C, \tag{2}$$

where: T is the FSW temperature (°C), T_m is the melting temperature of the sheet material (°C); α is reported to range from 0.04 to 0.06; k is constant, it is between 0.65 to 0.75; ω is the tool rotation speed; V is the welding speed; 502 ^{0}C is the temperature of AA2024 solidus.

3. Results and discussion

Figure 3 presents IPX maps of tested areas, which are marked in figure 1 in the transverse direction and in the normal direction along the weld central axis (WCA). Microhardness profiles are presented in figure 4. Depending on the welding thermal cycle and vicinity to the nugget, several microstructural changes occured within all zones . The nugget and the TMAZ (thermo-mechanically affected zone) on the advancing and retreating sides differ in size and shape of the grains. The upper part of the nugget, located under the tool shoulde,r experiences ahigher plastic strain and temperature than other weld points. The maximum microhardness values of the given part decrease along the WCA in the normal direction depending on the heat input: high – from 140-145 HV to 105 HV, medium - from 135 HV to 100 HV, and low - from 130 HV to 95 HV (figure 4). The microhardness decrease towards the bottom occurs due to the deterioration of the mixing degree of the material and a decrease in temperature and intensity of the plastic strain development during the FSW.

The side of the nugget on the advancing side with respect to the transverse direction, regardless of the heat input, is characterized by a fine-grained structure free of defects such as discontinuities, cracking, porosity etc (figure 3 a, b). This means that on the advancing side, the degree of plastic strain is higher. The grain diameter is about 1-3 μm. In addition to fine and equiaxial grains with predominantly (111) and (101) crystal orientations, the nugget contains fine rounded black particles of the strengthening phase Al_2CuMg (~ 300 nm) typical for the AA2024 aluminum alloy (figure 3 e, f). A significant displacement of the microhardness map at the weld bottom relative to its axis on the advancing side occurs due to an insufficient intensity of plastic strain, if the heat input is minimal. The wide nugget transition zone (NTZ) with alternating layers of elongated grains and fine crystallites between the TMAZ and the nugget is formed on the advancing side (figure 3 a, b). High microhardness is retained in this zone (figure 4). The fine-grained structure in the nugget on the retreating side is observed only with high heat input (figure 3 c). From figure 3 c, d it is clear that there is no smooth narrow interface between the nugget, NTZ, and TMAZ on the retreating side.

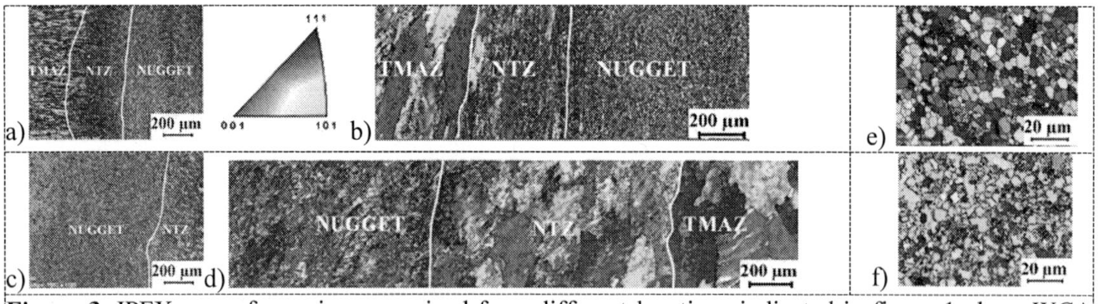

Figure 3. IPFX maps for regions examined from different locations indicated in figure 1 along WCA and in transverse direction: 3 (a,b); 6 (c,d); 7 (e); 8 (f). Heat input: high (a,c,e,f), low (b,d). Advancing side (a,b), retreating side (c,d). The color triangle shows poles aligned with the sheet normal direction.

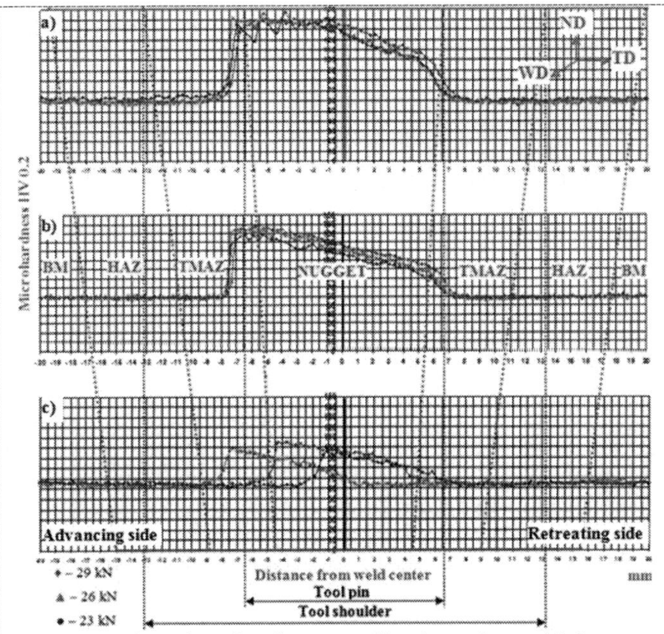

Figure 4. The micro-hardness profile along the weld for top, centre, and bottom samples marked in figure 2a: top (a), centre (b), bottom (c).

The decrease in microhardness to 90 V on the retreating side is due to coagulation and misorientation of the grains as well as to the appearance of discontinuities and an increase in the dissolution of particles (figure 3 b, d; figure 4). Only a small amount of fine crystallites among the coarse grains appears in the NTZ on the retreating side (figure 3 d). This means that on the retreating side, the flow is greater, and the peak temperature is higher, because the metal in front of the rotating tool shifts to this side.

In both TMAZ, the grains are bent upwards indicating an additional displacement relative to the parent sheets, caused by the advancing pin somewhat similar to the flow around it (figure 3 a,b,d). It has been found that the grains within the TMAZ contain a high density of subgrains and dislocations (figure 3 b,d).

Significant coarsening of grains was observed in the TMAZ due to high-temperature exposure during FSW. The microhardness of the TMAZ decreases sharply to the microhardness of the base material (60 HV), corresponding to the AA2024-O alloy (figure 4).

4. Conclusions

The advancing side, irrespective of the thermal and mechanical effects, is characterized by finely dispersed and equiaxed grains with a diameter of about 1-3 μm, in which the crystallographic orientations (111) and (101) predominate. The maximum microhardness (140-145 HV) of this weld part is explained by the presence in the grains of fine rounded black particles of the Al_2CuMg strengthening phase (~ 300 nm), typical for the aluminum alloy AA2024. The displacement of microhardness maps at the weld bottom relative to its axis on the advancing side is due to an insufficient intensity of plastic strain if the mechanical and thermal impact is minimal. Increaseing the temperature on the retreating side leads to coarsening and misorientation of the grains, dissolution of the strengthening phase of Al_2CuMg and, as a consequence, decreasing the weld microhardness on this side to 105 HV. This means that the degree of plastic strain is higher on the advancing side, and the peak temperature is higher on the retreating side. The reason for the weld microhardness decrease in the TMAZ to the base material of 60 HV is a significant coarsening of the grain due to the high temperature impact during the FSW process. The optimal welding parameters are: w = 300 rpm, V = 300 mm / min and P = 29 kN.

References

[1] Kotlyshev R R, Shuchev K G, Kramskoy A V 2010 *Vestnik DGTU* **10** 648 (In Russian)
[2] Reznikov A N and Reznikov L A 1990 *Thermal Processes in Technological Systems* (Moscow: Mechanical Engineering) 288 p.
[3] Balasubramanian V 2008 *Mater. Sci. Eng. A* **A480** 397.
[4] Sato Y S, Yamashita F, Sugiura Y, Park S H C and Kokawa H 2004 *Scripta Mater.* **50** 365.
[5] Mishra R S, Ma Z Y 2005 *Mater. Sci. Eng. R* **50** 1.

IOP Conf. Series: Materials Science and Engineering **447** (2018) 012059 doi:10.1088/1757-899X/447/1/012059

Au nanoclusters in the plasmonic applications

Yu Ya Gafner, S L Gafner

Katanov State University of Khakassia, pr. Lenina 90, Abakan, 655000 Republic of Khakassia, Russia

E-mail: ygafner@khsu.ru

Abstract. One of the new applications of gold nanoparticles is the use of the plasmon effect, which occurs due to the interaction of an electromagnetic wave with free electrons in nanostructures. Particles of gold with a size smaller than the wavelength of a visible electromagnetic wave can strongly absorb light due to the collective oscillation of conduction electrons. Bulk gold has a high-symmetry fcc structure, but controlling the collection of metal atoms in solution can also lead to the growth of many anisotropic forms. The use of these different structures has a significant effect on the plasmonic properties, since defects at twin boundaries can serve as scattering centers of conduction electrons in a metal. Thus, on the basis of controlling the size, shape, and structure of Au nanoparticles, it is possible to control light with unprecedented accuracy, and therefore the determination of the morphology of individual gold nanoparticles is of great interest for studying their possible optical interaction.

1. Introduction

Silver and gold nanoparticles are currently two of the most studied nanostructured nanomaterials. Such studies involve the prediction of various physical properties of nanoparticles, novel methods of synthesis, sophisticated functionalizations and applications that motivate the scientific community to constantly improve these nanomaterials [1].

At present, gold nanoparticles are extensively studied due to their possible widespread use in microelectronics, sensors, catalysis, nonlinear optics, and biomedicine [1–3]. Moreover, gold nanoclusters serve as the basis for many synthesized molecular nanocrystalline materials. For example, when placed on a fullerene base, gold nanoclusters can form cluster composite superstructures. This information is sufficient for a comprehensive theoretical and experimental study of the structural, electronic, dynamic, and other physical and chemical properties of both isolated and passivated gold clusters.

It is well known that all metallic nanoparticles exhibit unusual optical, thermal, chemical, and physical properties mainly due to the high percentage of surface atoms and the free path of electrons on a nanometer scale (10–100 nm for many metals at room temperature). Critical parameters affecting these properties are the size and shape of nanoparticles; therefore, these parameters should be controlled by nanotechnologies.

The fact that the properties of many clusters depend, firstly, on the number of atoms in them and, secondly, on their structure, means that to determine the structure of nanoparticles, it is very important to identify the structure and properties relationship. Although bulk gold has a fcc structure, small atomic groups can form configurations that can violate standard crystallographic laws: when the size of the Au cluster reaches the nanometer scale, a number of highly symmetric structures become energetically favorable for the system, which leads to the formation of icosahedral (Ih) and decahedral (Dh) modifications with a pentagonal symmetry [4].

The light interaction with nanostructures, especially those fabricated of noble metals, causes various fascinating optical phenomena. One of the current directions of research in the field of nano-optics is the development of metal nanostructures that efficiently interconvert the propagating and localized optical fields and thereby facilitate the generation of strongly enhanced local electrical fields. Field enhancements (FEs) occur due to resonantly excited surface plasmons, which represent collective electron oscillations in metals coupled to electromagnetic fields in neighboring dielectrics. Strong FEs are extremely important for practical applications, for example, in sensing and catalysis, as well as in surface-enhanced Raman scattering (SERS). Noble metal nanostructures are also very promising for the formation of surface plasmon resonators, which have demonstrated enormous increase in the fluorescence of quantum emitters, thus facilitating the development of ultra-bright and stable single-photon sources. In addition, they can be used as real-time sensors, plasmonic solar cells, building blocks for light-energy guiding devices and metamaterials with unique optical properties, *etc* [2].

Gold exhibits excellent plasmonic properties and can be advantageously used in most of the above-mentioned applications. The frequency and intensity of plasmon resonance depend on the distribution of polarization charge across the nanostructure, which, in turn, is determined by its shape. Thus, to control the shape of the Au nanostructure means to control the wavelengths of light that it scatters and absorbs, as well as the extent to which it amplifies local electric fields [3].

2. Using of Au nanoparticles in plasmonics

Coinage metals, such as Au, Ag, and Cu, have been important materials throughout history. Although they were admired in ancient cultures primarily for their ability to reflect light, their applications became much more sophisticated with our deeper understanding and control of the atomic world. Today, these metals are widely used in electronics and catalysis, as well as structural materials, but when they are fashioned into structures with nanometer-sized dimensions, they also become tools for a completely different set of applications that involve light. These new applications go far beyond merely reflecting light and have renewed our interest in maneuvering the interactions between metals and light in a field known as plasmonics.

In plasmonics, metal nanostructures can serve as antennas for converting light into localized electric fields (E-fields) or as waveguides for directing light to the desired locations with nanometer precision. These applications become possible due to a strong interaction between incident light and free electrons in the nanostructures. With a tight control over nanostructures in terms of size and shape, light can be effectively manipulated and controlled with unprecedented accuracy [5].

Although there are many varieties of plasmonic Au nanostructures, they can be differentiated based on the plasmonic modes they support: localized surface plasmons (LSPs) or propagating surface plasmons (PSPs). In LSPs, the time-varying electric field associated with light (E_o) exerts a force on the gas of negatively charged electrons in the conduction band of the metal and drives them to oscillate collectively. At a certain excitation frequency, this oscillation will be in resonance with the incident light, resulting in a strong oscillation of surface electrons, commonly known as the localized surface plasmon resonance (LSPR) mode. Structures that support LSPRs experience a uniform E_o when excited by light as their dimensions are much smaller than the wavelength of light.

In contrast, PSPs are supported by Au structures that have at least one dimension that approaches the excitation wavelength. In this case, the E_o is not uniform across the structure and other effects must be considered. For example, in a structure such as nanowires, SPs propagate back and forth between the ends of the structure. This can be described as a Fabry-Perot resonator with the resonance condition $l = n\lambda_{sp}$, where l is the nanowire length, n is an integer, and λ_{sp} is the wavelength of the PSP mode. One should also consider the reflection from the ends of the structure, which can change the phase and resonant length. The propagation lengths can be in the tens of micrometers (for nanowires), and the PSP waves can be manipulated by controlling the geometrical parameters of the structure.

It has now become clear that the size, shape, and structure of nanoparticles determine their optical properties, including resonance frequencies. Consequently, by adjusting the size, appearance of the Au nanostructure and environmental parameters, it is possible to control the light in various ways. Therefore, it is not surprising that the successes in the synthesis and assembly of Au nanostructures provided new opportunities for precise control of the interaction of Au nanoparticles with a light wave and, therefore, with discovered applications that were previously considered impossible. Thus, based on controlling the size, shape and structure of nanoparticles, light can be effectively controlled with unprecedented accuracy, and therefore the determination of the morphology of individual gold nanoparticles is of great interest for studying their possible optical interactions.

Gold have a highly symmetrical, face centered cubic crystal structure, but many anisotropic shapes can be grown by controlling the assembly of metal atoms in solution. Au nanostructures synthesized by chemical methods usually have one of four possible forms: an ideal nanoparticle; a nanoparticle with single twinning; a nanoparticle with multiple twinning (often with a symmetry axis of the fifth order); interconnected nanoparticles stacked with packaging defects. The use of these different structures has a significant effect on the properties of a plasmon, since defects at the boundaries of twins can serve as scattering centers for conduction electrons in a metal. But even in the case of the same type of synthesized nanostructure, the peak and shape of the plasmon resonance can vary greatly [5].

3. Results and discussion

Since the different internal structure of a metal nanoparticle means different physico-chemical properties, from the point of view of using gold nanoclusters in plasmon applications, it is very important to determine the boundaries of the thermal and dimensional stability of the initial crystal structure of nanoparticles. For this purpose, possible configuration changes in gold clusters when heated to the melting temperature were considered and an attempt was made to determine the dimensional boundaries of clusters in which a structural transition can occur. To analyze the clusters, the MDNTP computer program developed by Dr. Ralf Meyer from the University of Duisburg (Germany) will be used. The interatomic interaction forces were calculated using the modified tight-binding potential TB-SMA [6].

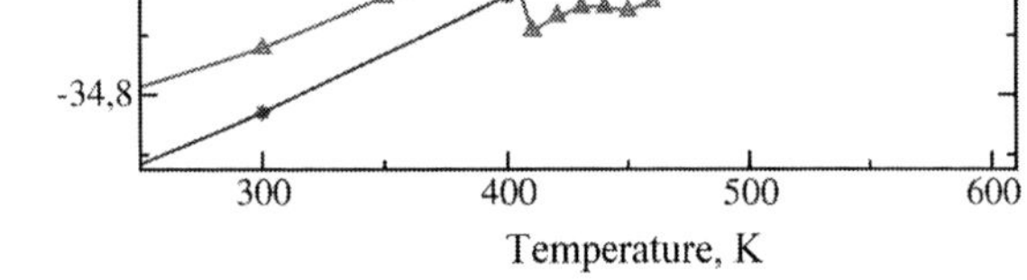

Figure. 1. Potential energy E_p vs. absolute temperature T during melting and crystallization of an Au_{135} cluster.

The simulation was performed in terms of the NVT ensemble; clusters with diameters D up to 3.5 nm were investigated. All clusters had the same initial internal structure, which corresponds to a perfect fcc lattice; to optimize their internal structure and shape, they were subjected to the procedure of thermal relaxation at 100 K.

To determine the cluster structure, we considered an ensemble of clusters of the same size and gradually heated them to 1300 K, which significantly exceeded their melting point. During heating, the temperature was varied stepwise in steps of 10–20 K on average; at each fixed temperature, the clusters were kept for 0.4 ns. The results of this study showed that, with a particle diameter of more than 3.0 nm, the initial fcc structure was retained in all cases up to the melting temperature. This fact suggests the thermal stability of the fcc gold clusters with a diameter of more than 3.0 nm; however, with a smaller cluster size, the situation is not that simple.

As a typical example, let us consider a gold cluster consisting of 135 atoms (D = 1.6 nm), which is cut from a perfect fcc lattice. With an increase in temperature of about 400 K, a structural transition from the initial fcc phase occurs, which is accompanied by a sharp change in the potential energy (figure 1). As the number of atoms in a cluster increases, the point of the polytypic transition shifts toward the melting temperature. It was shown that with an increase in cluster size, the structural transition occurs at a higher temperature. For example, in the case of a gold cluster with $N = 201$ atoms (D = 2.0 nm), substantial jumps in the potential energy are observed only near the melting point. This change in the potential energy corresponds to the transition of a cluster from a close-packed state (fcc lattice) to a state with an icosahedral structure. A similar picture of a stepwise change in the potential energy of the system before melting was also observed for a cluster consisting of 1157 atoms (D = 3.33 nm). However, the analysis showed that in this case there are no structural transformations, and energy jumps are associated with fluctuations near the phase transition.

Thus, with an increase in the number of atoms in a cluster, the following tendency is observed. At low temperatures, for a gold cluster with N = 55, a fcc–icosahedral structure transition occurs. However, with a further increase in temperature to 350–400 K, the icosahedral structure is destroyed with the formation of an amorphous-like state; for a cluster with the number of atoms $N = 135$, the structures alternate, and the amorphous phase is fixed in the medium temperature range for a short time. For a cluster with the number of atoms N = 201, the amorphous phase was not observed up to the melting point.

4. Conclusions

The facts that we have established have demonstrated that for small gold nanoclusters the process of structure formation is more complex than it follows from previous experiments. Until now, it was believed that with a decrease in the cluster diameter, the ratio of the cluster surface to its volume sharply increased, which was responsible for the most significant contribution of only the surface component to the total energy. As a result, the cluster tends to acquire the maximum spherical shape; for nanoparticles of fcc metals, this leads to the appearance of a cubooctahedral shape. In small clusters of metals, such as Au, Ag, Cu, Ni, and some others, with sizes below the critical size, the appearance of other crystal modifications with icosahedral or decahedral symmetry was observed.

However, in the case of gold nanoparticles, it cannot be definitely stated that the transition from the fcc configuration occurs only in the pentagonal phase and is always accompanied by a decrease in energy. Near this phase transition, the cluster is in a very unstable state, and the presence of structural defects in nanostructures leads to fluctuations of the potential energy near the point of the polytypic-transition. In all likelihood, it should be accepted that, for gold, especially for small particle sizes, even rather small defects can play a very important role in the formation of the equilibrium structure of particles. Thus, the mechanisms responsible for the formation of the structure in small gold particles require a more comprehensive study, since in many cases the observed picture remains quite unclear.

Acknowledgments
This work was supported by the Russian Foundation for Basic Research (project no. 17-48-190320 and 18-42-190001).

References

[1] Horta-Fraijo P, Cortez-Valadez M, Hurtado R B, Vargas-Ortiz R A, Perez-Rodriguez A and Flores-Acosta M 2018 *Physica E* **97** 111
[2] Novikov S M, Popok V N, Evlyukhin A B, Hanif M, Morgen P, Fiutowski J, Beermann J, Rubahn H-G and Bozhevolnyi S I 2017 *Langmuir* **33** 6062
[3] Wiley B J, Chen Y, McLellan J M, Xiong Y, Li Z-Y, Ginger D and Xia Y 2007 *Nanolett.* **7** 1032
[4] Gafner Yu Ya, Goloven'ko Zh V and Gafner S L 2013 *J. Exp. Theor. Phys.* **116** 252
[5] Rycenga M, Cobley C M, Zeng J, Li W, Moran Ch H, Zhang Q, Qin D and Xia Y 2011 *Chem. Rev.* **111** 3669
[6] Cleri F and Rosato V 1993 *Phys. Rev. B* **48** 22

IOP Conf. Series: Materials Science and Engineering **447** (2018) 012060 doi:10.1088/1757-899X/447/1/012060

Stability of in-plane delocalized vibrational modes in triangular Morse lattice

D U Abdullina[1,2], M N Semenova[3], A S Semenov[3], D S Ryabov[4], G M Chechin[4], E A Korznikova[1], J A Baimova[1,2] and S V Dmitriev[1,5]

[1] Institute for Metals Superplasticity Problems, 39 Khalturin st., 450001 Ufa, Russia
[2] Bashkir State University, 32 Validy St. 450076 Ufa, Russia
[3] North-Eastern Federal University n.a. M.K. Ammosov, Polytechnic Institute (branch) in Mirny, 5/1 Tikhonova st., 678170 Mirny, Sakha (Yakutia), Russia
[4] Southern Federal University, 105 Bolshaya Sadovaya st., 344006 Rostov-on-Don, Russia
[5] National Research Tomsk State University, 36 Lenin Ave., 634050 Tomsk,, Russia

dina.abdullina25@gmail.com

Abstract. Delocalized nonlinear vibrational modes (DNVMs) play a very important role in the study of the dynamics of a nonlinear lattice in solid state physics. Such modes are exact solutions to the equations of motion of atoms dictated by the lattice space symmetry. If the amplitude of DNVM is above the threshold value, it is modulationally unstable. In the present study, we consider the stability of DNVMs in a two-dimensional (2D) triangular lattice with atoms interacting via the Morse potential. Critical exponents are calculated numerically as functions of the DNVM amplitude. Extrapolation to the zero value of the critical exponent gives an estimation of the DNVM amplitude, below which it is stable.

1. Introduction

Spatially localized nonlinear excitations, called discrete breathers (DBs), are actively studied in various crystals and model nonlinear lattices by means of molecular dynamics [1-21] and *ab initio* simulations [22,23]. Existing attempts to experimentally identify DBs in crystals were summarized in the review [2]. It has recently been demonstrated that delocalized nonlinear vibrational modes (DNVMs), synonymously called bushes of nonlinear normal modes [24], have a close relationship to DB. Particularly, it has been shown that DBs can be obtained by superimposing a bell-shaped function upon a DNVM having a frequency outside the phonon spectrum [4-6]. DBs can also emerge spontaneously from a DNVM having a frequency outside the phonon spectrum as a result of modulational instability [18-21]. From this it is clear that the stability of DNVM is a very important issue.

DNVMs are the exact solutions to the nonlinear equations governing atomic motion, and this is true for any interatomic potential used in simulations. This is so because DNVMs are found considering only the lattice space symmetry using the group theory [24]. For a two-dimensional (2D) triangular lattice, Chechin and Ryabov have recently derived eight DNVMs with in-plane atomic motion [25]. Here we study the stability of all eight DNVMs by estimating critical exponents as functions of the initial atomic displacements that define the amplitude of DNVM. Extrapolating the results to the zero value of the critical exponent, we find the maximal initial displacement for a stable DNVM.

2. Simulation details

A two-dimensional (2D) close-packed triangular lattice is considered. The interatomic interactions are described by the classical Morse potential

$$V(r) = A\{\exp[-2\beta(r - r_m)] - 2\exp[-\beta(r - r_m)]\}, \quad (1)$$

where r is the distance between the considered pair of atoms, A, β, r_m are the potential parameters. The function $V(r)$ has a minimum at $r=r_m$. A is the bond breaking energy (depth of the potential) and β defines the bond stiffness. With the proper choice of units of energy, distance, and time we set $A=1$, $r_m=1$, and the atom mass $m=1$. We take $\beta=5$ and the cut-off radius equal to $5r_m$. Then the equilibrium interatomic distance is $\rho=0.9881329$. The computational cell includes 24×24 atoms. Periodic boundary conditions are applied.

In figure 1 eight DNVMs with in-plane displacements of atoms are shown [22]. All arrows have equal length D_0. The initial conditions are set by setting the initial displacements according to these patterns with zero initial velocities of atoms. The choice of D_0 specifies the DNVM amplitude. To facilitate the development of the instability, random in-plane initial displacements are added to all atoms with a maximal deviation from the equilibrium position equal to 10^{-9}.

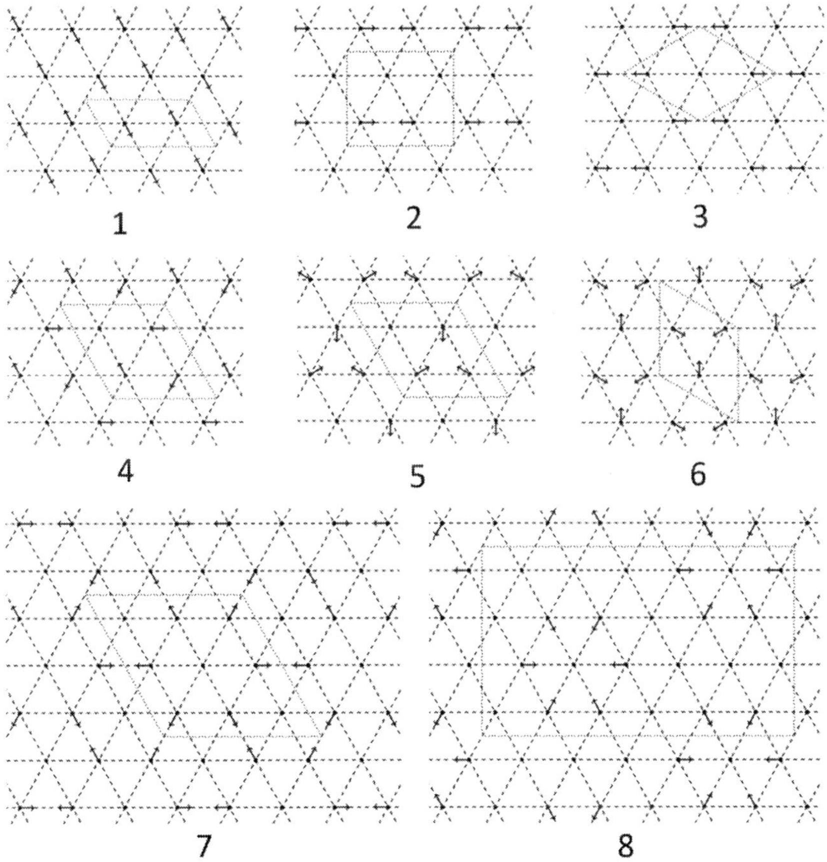

Figure 1. Eight in-plane DNVMs supported by a triangular lattice, as found by Chechin and Ryabov [22]. The arrows have length D_0, which defines the initial amplitude of DNVM. Dotted lines show the translational cells of the displacement patterns.

3. Numerical results and discussion

In figure 2 we plot, as an example, the numerical results obtained for DNVM 2. Note that in this DNVM all moving atoms vibrate in the horizontal (x) direction. As a measure of the instability of this mode, one can analyse the appearance of displacements in the vertical (y) and out-of-plane (z) directions. It was found that the z-component of the displacements remained nearly zero but the y-component firstly remained at the level of the initial perturbation, and then started to grow exponentially with time, $\max|\Delta y|\sim e^{\alpha t}$, with the critical exponent α. This is illustrated in figure 2 a for three values of the initial atomic displacements, D_0=0.03, 0.04, and 0.05. Note that the logarithmic scale is used for the ordinate. The straight line segments show the slopes from which critical exponents were estimated. In figure 2 b the critical exponent as a function of D_0 is shown for DNVM 2. The straight line represents a linear least square fit $\alpha = a + bD_0$ with a=-0.641 and b=45.3. This shows that the critical exponent vanishes at D_0=0.0142. For smaller values of D_0, DNVM 2 is modulationally stable. The coefficient b indicates the sensitivity of the critical exponent to the amplitude of DNVM 2.

In table 1 the coefficients a and b are given for all DNVMs studied.

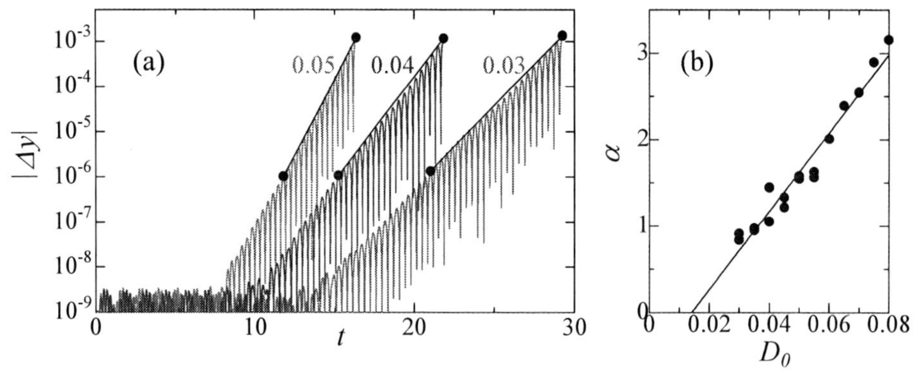

Figure 2. Results for DNVM 2. (a) Absolute value of displacement $|\Delta y|$ as a function of time for a moving atom. Three curves give the results for three different values of the initial displacements of atoms, D_0=0.03, 0.04, and 0.05, as indicated in the figure. Straight lines show the slopes from which critical exponents were estimated. (b) The critical exponent as a function of the initial displacements of atoms. The straight line shows the linear least square fit.

Table 1. Coefficients of the least square fit $\alpha = a + bD_0$ for eight DNVMs studied DNVMs.

DNVM	1	2	3	4	5	6	7	8
a	-0.383	-0.641	-1.75	-0.811	-3.92	-1.49	-0.170	-0.055
b	13.3	45.3	70.1	23.8	125	73.3	23.4	11.8

4. Conclusions

Using molecular dynamics, the modulational instability of eight DNVMs with in-plane atomic vibrations for a triangular Morse lattice was studied. It is found that the critical value of the DNVM amplitude, above which it becomes unstable, is of the order of 1% of the interatomic distance ρ. In the future studies, it would be important to analyze the effect of the computational cell size on the development of modulational instability. The stability of transverse DNVMs, as well as two-dimensional DNVMs should also be analyzed.

Acknowledgements

The work of A.S.S. (discussion of the numerical results) was supported by the Russian Foundation for Basic Research, grant no. 18-32-00171_mol_a. D.S.R and G.M.Ch. have worked under state assignment no. 3.5710.2017/8.9 (derivation of DNVM and discussion of the results). E.A.K. thanks the Russian Foundation for Basic Research, grant no. 17-02-00984_a (writing the paper). Work of J.A.B. was supported by the grant of the President of the Russian Federation for state support of young Russian scientists - doctors of sciences MD-1651.2018.2 (discussion of the results). S.V.D. thanks the Russian Science Foundation, grant no. 16-12-10175, for their financial support (design of the research). This work was partly supported by the state assignment of IMSP RAS.

References

[1] Flach S and Gorbach A V 2008 *Phys. Rep.* **467** 1
[2] Dmitriev S V, Korznikova E A, Baimova Yu A and Velarde M G 2016 *Phys. Uspekhi* **59** 446
[3] Baimova J A, Korznikova E A, Lobzenko I P and Dmitriev S V 2015 *Rev. Adv. Mater. Sci.* **42** 68
[4] Korznikova E A, Fomin S, Soboleva E and Dmitriev S V 2016 *JETP Lett.* **103** 277
[5] Barani E, Lobzenko I P, Korznikova E A, Soboleva E, Dmitriev S V, Zhou K and Marjaneh A 2017 *Eur. Phys. J.* B **90** 38
[6] Korznikova E A, Kistanov A A, Sergeev K S, Shepelev I A, Davletshin A R, Bokij D I and Dmitriev S V 2016 *Letters on Materials* **6** 221 (in Russian)
[7] Murzaev R T, Bachurin D V, Korznikova E A and Dmitriev S V 2017 *Phys. Lett.* A **381** 1003
[8] Evazzade I, Lobzenko I P, Korznikova E A, Ovid'ko I, Roknabadi M and Dmitriev S V 2017 *Phys. Rev.* B **95** 035423
[9] Barani E, Korznikova E A, Chetverikov A P, Zhou K and Dmitriev S V 2017 *Phys. Lett.* A **381** 3553
[10] Zakharov P V, Korznikova E A, Dmitirev S V, Ekomasov E G, Zhou K 2019 *Surf. Sci.* **679** 1
[11] Haas M, Hizhnyakov V, Shelkan A, Klopov M and Sievers A J 2011 *Phys. Rev.* B **84** 144303
[12] Murzaev R, Babicheva R I, Zhou K, Korznikova E A, Fomin S, Dubinko V I and Dmitriev S V 2016 *Eur. Phys. J.* B **892** 168
[13] Murzaev R T, Kistanov A A, Dubinko V I, Terentyev D A and Dmitriev S V 2015 *Comp. Mater. Sci.* **98** 88
[14] Terentyev D A, Dubinko A V, Dubinko V I, Dmitriev S V, Zhurkin E E and Sorokin M V 2015 *Model. Simul. Mater. Sc.* **23** 085007
[15] Bachurina O V, Murzaev R T, Korznikova E A and Dmitriev S V 2017 *Mater. Phys. Mech.* **33** 49
[16] Kiselev S A, Bickham S R and Sievers A J 1993 *Phys. Rev.* B **48** 13508
[17] Dmitriev S V 2016 *Letters on Materials* **6** 86
[18] Burlakov V M, Kiselev S A and Rupasov V I 1990 *JETP Lett.* **51** 544
[19] Cretegny T, Dauxois T, Ruffo S and Torcini A 1998 *Physica* D **121** 109
[20] Khadeeva L Z and Dmitriev S V 2010 *Phys. Rev.* B **81** 214306
[21] Dmitriev S V, Korznikova E A, Bokij D I and Zhou K 2016 *Phys. Status Solidi* B **253** 1310
[22] Lobzenko I P, Chechin G M, Bezuglova G S, Baimova Y A, Korznikova E A and Dmitriev S V 2016 *Phys. Solid State* **58** 633
[23] Chechin G M, Dmitriev S V, Lobzenko I P and Ryabov D S 2014 *Phys. Rev.* B **90** 045432
[24] Chechin G M and Sakhnenko V P 1998 *Physica* D **117** 43
[25] Semenova M N, Semenov A S, Bebikhov Yu V, Ryabov D S, Chechin G M, Rakhmatullina Zh G, Korznikova E A and Dmitriev S V 2018 *Fundamental'nye Problemy Sovremennogo Materialovedenia* **15** 257 (in Russian)

IOP Conf. Series: Materials Science and Engineering **447** (2018) 012061 doi:10.1088/1757-899X/447/1/012061

Mechanisms of nanoscale structure formation during electron beam treatment of silumin

V D Sarychev[1], Yu F Ivanov[2], S A Nevskii[1], Yu S Serenkov[1], E A Vysotskaya[1], V E Gromov[1]

[1] Siberian State Industrial University, Novokuznetsk, Russia 654007

[2] Institute of High Current Electronics SB RAS, Tomsk, Russia 634055

E-mail: nevskiy.sergei@yandex.ru

Abstract. The effect of low-energy high-current electron beams on the structure of silumin has been studied. It has been found that the electron beam treatment leads to the formation of a heterogeneous multi-layer structure in the material. A columnar crystallization structure has been observed in the surface layer. The transverse dimensions of the columns are between 400 and 600 nm, whereas the second phase interlayer dimensions are between 80 and 100 nm. The transverse dimensions of the columns at a depth of 80 to 100 μm are 1.0-1.2 μm. The cause of the columnar structure is the thermoconcentration-capillary instability, which leads to the formation of vortices and the displacement of the second phase particles to the columnar boundaries. The considerable thickness of the columnar crystallization layer is caused by the fact that silicon and other alloying elements influence the dependence of liquid aluminum surface tension on temperature by introducing nonlinearity. As a result, there is an effect of the "thermal drill", i.e. the formation of vortices. The formed vortices stir the melt throughout the depth, and the downward flow of the liquid moves to the center of the bath at the cooling stage, which is manifested in an increase in the thickness of the columnar crystallization layer. It manifests itself in an increase in the thickness of the columnar crystallization layer. At depths over 100 μm, the fracture of silicon particles is observed as a result of differences in the elastic moduli and linear expansion coefficients of the aluminum matrix and the silicon plate.

1. Introduction

Aluminum and silicon based alloys are increasingly used in a number of industries, for example, in the engine building industry for the manufacture of pistons. Studies of the structure of piston materials after their operation, shows that damaged layers with a thickness from 10 to 20 μm [1] are formed there. In this regard, it is necessary to develop technologies for surface modification of aluminum alloys, which would provide thickness of the hardened layer larger than that of the damaged layers formed during operation. Recently, different ways and methods of surface modification have been suggested, and one of the prospective methods is the low energy high current electron beam treatment [2]. The advantage of this treatment method is that its periodic pulsed character allows, on the one hand, increasing the time interval of the molten state of surface layers and, on the other hand, providing hardening effects leading to the formation of sub-micro and nanoscale structures [3]. The disadvantage of the electron beam treatment is the formation of craters on the surface of the material, which become dangerous stress concentrators under cyclic loading [4, 5]. Therefore, it is necessary to choose treatment modes that minimize the influence of craters. To find such modes, one needs information on the mechanisms of the influence of electron beam treatment on the structure of materials. As was shown in earlier studies [4-9], the effect of the electron beam on materials has a diverse character. It involves heating, melting and vaporization of the surface of materials. Electron beams also generate thermoelastic waves that affect the mechanical properties of the radiation-exposed material. During the melting of metals under the action of electron beams, convective flows are formed in the melt. The intensity and character of these flows determine the features of heat and matter transfer, which affect the physical and mechanical properties of materials. The present article aims at the study of the mechanisms of the effect of electron beam treatment on the structure of silumin.

2. Material and methods of investigation

The Al–12% Si silumin, which is widely used in the engine building industry for the manufacture of pistons, was studied. The sample for studies had dimensions of 20×20×10 mm^3. Surface modification was accomplished by means of the electron-beam treatment technique. The energy density of the beams was 35 J/cm^2, the pulse duration was 150 μs, the number of pulses was 3, and the pulse repetition frequency was 0.3 s^{-1}.. Transmission electron microscopy (TEM) was used to study the structural and phase states of the modified silumin at different distances from the irradiated surface.

3. Results and discussion

Analysis of the TEM images of the cross-section of the samples has shown that the electron beam treatment leads to the formation of a gradient structure. At a depth of 10 to 100 μm from the radiation-exposed surface, the structure of columnar crystallization is observed. The layers of the second phases are located on the boundaries of the columns. The transverse dimensions of the columns are from 400 to 600 nm; the transverse dimensions of the second-phase interlayers are from 80 to 200 nm. Apparently, the columnar crystallization structure appears due to thermoconcentration-capillary instability, which leads to the formation of a vortex and the displacement of the second-phase particles to the boundaries [10-12]. At depths from 80 to 100 μm, one can observe a transition layer, extending from the crystallization zone to the thermal effected one. The transverse dimensions of the columns at this depth are 1.0-1.2 μm. Such an increase in the size of the columns can be explained by the transformation of the interlayers into sphere-shaped particles. The large depth of the columnar crystallization layer is due to the fact that silicon and other alloying elements affect the dependence of the surface tension of liquid aluminum on the temperature [13]. As a result, the "thermal drill" effect appears, which consists in the formation of a vortex structure that mixes the melt throughout the depth, as well as of a downward flow of liquid moving towards the center of the bath during the cooling stage. This explains the increase in the thickness of the columnar crystallization layer. To confirm the proposed mechanism, the mathematical problem of the temperature distribution in a silumin sample in the form of a plate has been solved. Following [12, 14], the enthalpy approach can be used for this purpose. In this approach, the heat conduction equation can be written as follows:

$$\rho\frac{\partial H}{\partial t}=\frac{\partial}{\partial z}\left(k\frac{\partial T}{\partial z}\right) \tag{1}$$

where z is the coordinate, ρ is the density, k is the coefficient of heat conductivity, and H is the enthalpy. Phase transitions under the effect of electron beams are taken into account as follows:

$$\rho H(T)=\begin{cases}\rho_s C_s T,\ T<T_L;\ \rho_L\dfrac{L_L}{\Delta T_L}T,\ T_L\le T\le T_L+\Delta T_L;\rho_L C_L T,\ T_L+\Delta T_L\le T<T_V, \\ \rho_V\dfrac{L_V}{\Delta T_V}T,\ T_V\le T\le T_V+\Delta T_V;\ \rho_V C_V T,\ T_V+\Delta T_V\le T\end{cases} \tag{2}$$

where C_p is the coefficient of heat capacity, indices S, L, V stand for solid, liquid and vaporous states. On the surface of the sample z =0 the heat flow is given by

$$-k\frac{\partial T}{\partial z}=q_0(t),\ q_0(t)=\begin{cases}\frac{E_S}{t_0},0\le t\le t_0\\ 0,t>t_0\end{cases} \tag{3}$$

On the bound of plate $z = h$ there is no heat flow

$$\partial T/\partial z=0 \tag{4}$$

The initial temperature is equal to $T(0,z)=T_0$ at any depth of the plate 0<z<h, h is the thickness of plate. The numerical solution of equations (1)-(4) was obtained by means of an implicit difference scheme using the double-sweep method. The thermal and physical constants of silumin were computed using the mixture principle. The computations were carried out for beams with a power density ranging from 15 to 35 J/cm^2 and a pulse duration t_0 =150 μs. Figure 1 shows the temperature distribution in the sample at different time moments for materials exposed to an electron beam with an energy density of 35 J/cm^2. One can see that at t = 150 μs, i.e. at the end of the pulse, the temperature of the material surface is equal to T = 900 K (curve 2), which is above the melting temperatures of aluminum and silicon, but lower than the temperature of vaporization. This is due to the high thermal conductivity and vaporization heat of silumin. At the end of the pulse, the processes of cooling and crystallization are started. Calculations of the fusion penetration depth (table 1) show that the thickness of the melted layer increases with the energy density of the electron beam. Its values agree with the experimental data [15]. At t =200 and 300 μs (curves 3 and 4, respectively), gradual temperature equalization across the depth is observed. At distances from 80 to 100 μm from the surface, the temperature increases. As mentioned above, at these distances, the process of transformation of the second-phase interlayers to sphere-shaped particles, i.e. spheroidization process, dominates. According to the literature data [16], this process has a diffusion origin, though recently mechanisms alternative to diffusion have been proposed as well.

For instance, in [17] a mechanism was proposed according to which the spheroidization occurs as a result of failure of silicon plates due to the difference in the coefficients of linear thermal expansion of the matrix and inclusion. Since the volume fraction of silicon plates is small compared to that of the aluminum matrix, the latter makes the main contribution to thermal expansion. The linear expansion coefficient of aluminum is four times higher than that of silicon. Therefore, thermal expansion (contraction) of the two phases in most cases is incompatible, which leads to the inevitable appearance of mechanical stresses between the phases. Silicon inclusions can accept only a quarter of the thermal expansion (contraction) from the aluminum matrix by means of their own thermal expansion (contraction). Cracks are formed at the non-homogeneities of the surfaces of the inclusions. They play the role of capillaries for aluminum atoms. Mechanical stresses generated by the cracks cause capillary forces that move the atoms of the matrix into the gaps formed between inclusions. The flows of vacancies and silicon atoms occur in the opposite direction.

Table1. Dependence of the thickness of the melted layer in aluminum and silicon samples

E_s, J/cm^2	Silumin *d*, μm	Aluminum *d*, μm	Experiment *d*, μm
15	23	18	23
20	39	39	30
25	54	57	55
35	80	-	80

4. Conclusions

The mechanisms of the effect of electron beams on the structure of silumin have been established. It has been demonstrated that electron beams with an energy density of 35 J/cm^2, and a pulse duration of

150 μs lead to the formation of a heterogeneous multi-layer structure. The surface layer has a columnar crystallized structure. Its depth is 80 μm. Mathematical modeling of the heating operation shows that the depth of fusion is also 80 μm. One of the reasons for the formation of a columnar crystallized structure is thermoconcentration-capillary instability appearing as a result of the temperature gradient occurrence, which, in turn, leads to the formation of vortices and the second-phase particles expulsion to the boundaries of columns. At depths over 100 μm, the process of spheroidization of silicon plates is observed.

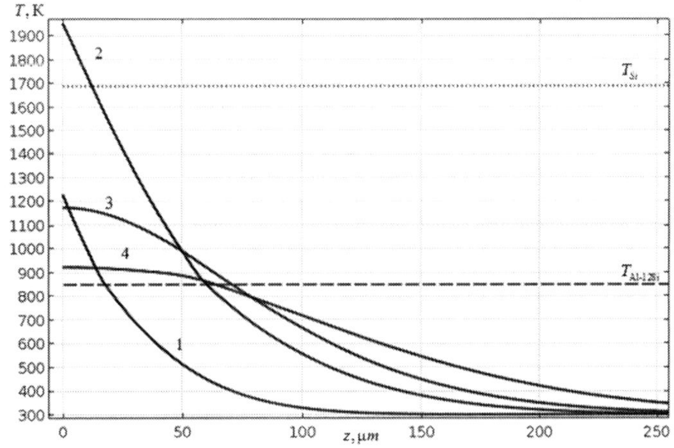

Figure1. Distribution along the depth at different instants of time t 1 - 50 μs, 2 - 150 μs, 3 - 200 μs, 4 - 300 μs. $T_{Al-12Si}$- eutectic temperature for silumin, T_{Si} – melting temperature for unblended silicon.

Acknowledgement
The study was carried out with the financial support of Ministry of Education and Science of Russian Federation within the framework of the state task # 3.1283.2017/4.6 and Russian Foundation of Basic Research (project # 18-48-420001).

References
[1] Fadeev A V, Bazhenov V E, Koltygin A V 2015 *Russian J. Non-Ferr. Metals* **56** 26
[2] Shulov V A, Gromov A N, Teryaev D A and Engel'ko V I 2016 *Russian J. Non-Ferr. Metals* **57** 256
[3] Novikov A S, Shulov V A and Engel'ko V I 2012 *Sil'notochnye Impul'snye Elektronnye Pushki dlya Aviatsionnogo Dvigatelestroeniya* (Pulsed High-Current Electron Beams for Aviation Propulsion Engineering) (Moscow: DIPAK) (In Russian)
[4] Markov A B, Meisner L L, Yakovlev E V, Meisner S N, Gudimova E Y, Petrov V I 2015 *Izvestiya Vuzov. Fizika* **58** 9/3 173 (In Russian)
[5] Ozur G E, Proskurovsky D I and Rotshtein V P 2016 *Techn. Phys. Lett.* **42** 328
[6] Bleicher G A, Krivobokov V P, Pashchenko O V 2014 *Erosiya Poverkhnosti Tverdogo Tela pod Deistviem Moshnykh Puchkov Zrayagennykh Chastits* (Erosion of the surface of solids at power charged particles beams action) (Novosibirsk: Nauka) (In Russian)
[7] Khaimzon B B, Sarychev V D, Soskova N A and Gromov V E 2013 *Steel in Trans.* **43** 55
[8] Sarychev V D, Khaimzon B B, Gromov V E and Soskova N A 2012 *Titan* **1 (35)** 4 (In Russian)
[9] Volkov N B, Leivi A Ya, Talala K A and Yalovets A P 2010 *Tech. Phys.* **55** 484
[10] Nevskii S, Sarychev V, Konovalov S, Kosinov D, Komissarova I 2016 *Mater. Sci. Forum* **870** 34

[11] Berg J, Acrivos A 1965 *Chem. Eng. Sci.* **20** 737
[12] Konovalov S, Chen X, Sarychev V, Nevskii S, Gromov V, Trtica M 2017 *Metals* **7** 1
[13] Traidia A, Roger F *2011 Int. J. Heat Mass Transfer* **54** 2163
[14] Samarskii A A, Vabishchevich P N 2003 *Vychislitel'naya Teploperedacha (Computational Heat Transmission)* (Moscow: URSS) (In Russian)
[15] Ivanov Y F, Petrikova E A, Ivanova O V, Ikonnikova I A, Tkachenko A V 2015 *Russian Phys. J.* **58** 478
[16] Courtney T H, Kampe J C M 1989 *Acta Metall.* **37** 1747
[17] Liu X, Beausir B, Zhang Y et al. 2018 *J. All. Comp.* **730** 208

IOP Conf. Series: Materials Science and Engineering **447** (2018) 012062 doi:10.1088/1757-899X/447/1/012062

The influence of temperature-strain rate conditions on hot workability and microstructure of powder metallurgy nickel-based superalloy EP741NP

A A Ganeev, V A Valitov, F Z Utyashev, V M Imayev

[1] Institute for Metals Superplasticity Problems, Russian Academy of Sciences, 39 Khalturin st, Ufa 450001, Russia

E-mail: vimayev@mail.ru

Abstract. The influence of temperature-strain rate conditions on the compression properties, hot workability and changes in the microstructure after the uniaxial single-step isothermal compression and the three-step isothermal compression with intermediate anneals has been studied for the powder metallurgy (PM) nickel-based superalloy EP741NP. Compression tests were carried out in the temperature range of T=900-1190 °C to a total strain of $e\approx1.05$. The study of the microstructure was carried out for the cross sections of the deformed samples. The optimal strain temperature in terms of hot workability and recrystallization kinetics was defined as T=1140-1160 °C, i.e. 45÷25 °C below the γ' solvus temperature T_s. Three-step compression experiments with intermediate anneals were performed at the optimal temperature of deformation with a strain rate in the range of $\dot{\varepsilon}=10^{-4}$-$10^{-1}$ s^{-1} to obtain improved hot workability and fast recrystallization kinetics. BSE observations and EBSD analysis showed that the three-step uniaxial compression at T=1140-1160 °C with $\dot{\varepsilon}=5\times10^{-3}$ s^{-1} and intermediate anneals provided reasonable hot workability and the formation of a recrystallized fine-grained microstructure with predominantly high-angle grain boundaries.

1. Introduction

Powder metallurgy (PM) nickel-based superalloys are widely used for manufacturing of various parts in gas turbine engines (GTE). They are typically produced from powder, which is consolidated by hot isostatic pressing (HIP) into the desired parts. It is known that the mechanical properties of PM consolidated superalloys can be significantly improved by hot working [1-4]. However, low hot workability can impede the fabrication of high-quality wrought products from PM superalloys. In addition, hot working should provide fast recrystallization kinetics and the formation of a fine-grained microstructure that is the key to attaining enhanced mechanical properties. Therefore, the temperature-strain rate conditions of hot working should be precisely selected to ensure the most reasonable hot workability and extensive occurrence of recrystallization processes throughout the material volume.

The present study was aimed to develop the temperature-strain rate conditions of the uniaxial hot forging procedure in order to obtain improved hot workability and the formation of a fine-grained recrystallized structure in the PM nickel-based superalloy EP741NP. This Russian alloy is known as the most widely used PM superalloy for the manufacture of GTE parts.

2. Experimental

The nominal chemical composition of the alloy EP741NP is as follows: Cr - 9.3; Ti - 1.8; Al - 5.2;

Mo - 4; Nb - 2.7; Co – 15.7; W - 5.4; Hf – 0.2, C – 0.04, B – 0.014 (wt. %). The initial material was supplied as workpieces with a size of Ø90×60 mm. The material was produced from a powder with a size of particles up to 140 μm, which was consolidated by HIP. The γ′ solvus temperature (T_S) was determined by differential scanning calorimetry and quenching experiments. It was defined as T_S=1180±5 °C. The obtained PM superalloy was subjected to heterogenization annealing to coagulate and spheroidize the γ′ precipitates. Small samples with a size of Ø10×15 mm were cut from the heat treated alloy to carry out compression tests.

At first, single-step compression tests were performed in the temperature range of T=900-1190 °C with an initial strain rate of $\dot{\varepsilon}=10^{-3}$ s^{-1} to a total strain of $e\approx1.05$. The true stress-true strain curves were built taking into account the gradual increase in the cross section of the sample during compression. Hot workability (deformability) was evaluated by measuring the maximal crack depth (L) on the lateral surfaces of the deformed samples. The strain temperature range, providing the most reasonable combination of hot workability, volume fraction and recrystallized grain size, was chosen for subsequent three-step compression tests with varying strain rate ($\dot{\varepsilon}=10^{-4}$-$10^{-1}$ s^{-1}). To increase the volume fraction of recrystallized grains, intermediate recrystallization anneals after each annealing step were additionally performed at the deformation temperature. All the deformed samples were finally cooled in air.

The study of the microstructure was carried out for the central part of the cross sections of the deformed samples. Optical microscopy and scanning electron microscopy (SEM) in backscattering electron (BSE) mode were used for this. Electron backscattered diffraction (EBSD) analysis was performed with a scan-step size of 0.5 μm. In this case, the γ′ precipitates were assumed as a γ phase, and the microstructure was assumed as a quasi-single-phase. The EBSD analysis was conducted using CHANNEL 5 processing software. Grain boundaries having a misorientation angle of less than 2° were excluded from consideration taking into account the measurement accuracy. Grain boundaries having a misorientation angle of more than 15° were assumed as high-angle boundaries.

3. Results and discussion

3.1. Single-step compression tests

Figure 1 represents the true stress-true strain curves, obtained as a result of compression testing of the samples. With increasing the test temperature the flow stresses decreased, and the hot workability improved reaching its maximum at around the γ′ solvus temperature. Table 1 shows the dependences of workability and microstructural parameters of the deformed samples on the test temperature. All compressed samples had lateral cracks, which became smaller with increasing strain temperature: one can see that the value of L decreased with increasing strain temperature (table 1). A microstructural examination (not presented here) showed that the volume fraction ($V_{r.g.}$) and the size of the recrystallized γ grains (d) increased with increasing test temperature. After compression at T=1175 and 1190 °C, the mean size of recrystallized grains reached d=14 and 65 μm, respectively. One can see that the size of recrystallized grains sharply increased near the γ′ solvus temperature that is associated with the dissolution of the γ′ phase. Taking into account the hot workability, the volume fraction and the size of recrystallized grains depending on the test temperature, the optimal strain temperatures were defined in the range of T=1140-1160 °C.

3.2. Three-step compression tests with intermediate anneals

Three-step compression tests with intermediate anneals were carried out at T=1140 °C, varying the strain rate. Figure 2 represents the appearance of samples and the corresponding EBSD orientation maps obtained from the central parts of the samples, deformed at a strain rate ranging from $\dot{\varepsilon}=10^{-4}$ to 5×10^{-2} s^{-1}. Table 2 shows the dependences of workability (maximal crack depth on the lateral surfaces) and microstructural parameters of the deformed samples on the strain rate after a three-step compression with intermediate anneals. All compressed samples had lateral cracks. However, it should be noted that the L value, characterizing the workability, has changed non-monotonically depending

IOP Conf. Series: Materials Science and Engineering **447** (2018) 012062 doi:10.1088/1757-899X/447/1/012062

on the strain rate, which reaches the smallest value after deformation at $\dot{\varepsilon}=5\times10^{-3}$ s^{-1}. At the same time, the volume fraction of recrystallized grains monotonically increased with increasing strain rate (table 2). Most likely, compression with a higher strain rate imparted a higher strain energy, which promoted the recrystallization processes during intermediate anneals. The fact that the size of recrystallized γ grains decreased with increasing strain rate indicates that compression with a higher strain rate created a larger number of recrystallization centers and the formation of new recrystallized grains predominated over the growth of recrystallized grains during anneals after each compression step. The efficient occurrence of recrystallization processes under the three-step compression with intermediate anneals at T=1140 °C is also confirmed by the fact that predominantly high-angle grain boundaries were obtained (table 2).

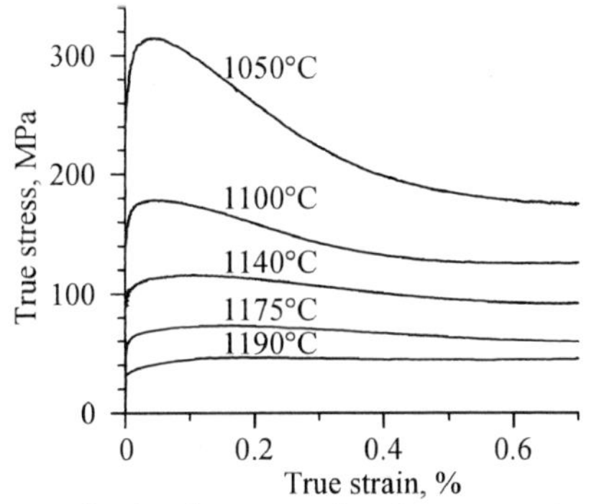

Figure 1. True stress-strain curves obtained as a result of compression tests of the samples of the alloy EP741NP at different temperatures ($\dot{\varepsilon}=10^{-3}$ s^{-1}).

Table 1. The maximal crack depth on the lateral surfaces and the microstructural parameters obtained for samples of the alloy EP741NP subjected to single-step compression depending on the strain temperature ($\dot{\varepsilon}=10^{-3}$ s^{-1}, e=1.05).

The maximal crack depth and the microstructural parameters	Deformation temperature, °C			
	1100	1140	1175	1190
The maximal crack depth, L, mm	4	3.2	2.6	2.2
The fraction of recrystallized grains, $V_{r.g.}$, %	48	55	60	≈100
The mean size of recrystallized grains, d, μm	1.2	3.3	14	65

Figure 2. (a-c) The appearance of lateral surfaces of the samples and (d-f) EBSD orientation maps obtained from the central part of the samples of the EP741NP alloy subjected to three-step compression tests at T=1140 °C with intermediate anneals: (a, d) $=10^{-4}$ s^{-1}, (b, e) $=5\times10^{-3}$ s^{-1}, (c, f) $\dot{\varepsilon}=5\times10^{-2}$ s^{-1}.

Comparing tables 1 and 2, one can see that the three-step compression at T=1140 °C with intermediate anneals was more effective from the viewpoint of hot workability and recrystallization behavior. Indeed, the predominantly fine-grained recrystallized microstructure ($V_{r.g.}$=70-100%) was obtained after the three-step compression with intermediate anneals at T=1140 °C, in contrast to the single-step compression at the same temperature ($\dot{\varepsilon}=10^{-3}$ s^{-1}), which provided only $V_{r.g.}$=55%. It is worth noting that the three-step compression with intermediate anneals and $\dot{\varepsilon}=5\times10^{-3}$-$5\times10^{-2}$ s^{-1} provided appreciably better hot workability than the single-step compression (tables 1 and 2). At the same time, the size of recrystallized γ grains after the single-step compression and the three-step compression with intermediate anneals at T=1140 °C was found to be similar (d=3.3 and 3.7-6 μm, respectively). This may be due to the fact that a significant volume fraction of the γ′ phase di not dissolve at T=1140 °C and stabilized the microstructure against grain growth during intermediate recrystallization anneals.

Note that the L value slightly increased with an increase in the strain rate from $\dot{\varepsilon}=5\times10^{-3}$ to 5×10^{-2} s^{-1} indicating a decrease in hot workability (figure 2 b, c, table 2). Most likely, an increase in the strain rate led to impeding dynamic recrystallization and dynamic recovery. Despite the wider development of recrystallization processes during intermediate anneals after straining with $\dot{\varepsilon}$= 5×10^{-2} s^{-1}, this promoted the formation of cracks on the lateral surfaces of the samples.

Table 2. The maximal crack depth on the lateral surfaces and the microstructural parameters obtained for the samples of the alloy EP741NP subjected to a three-step compression with intermediate anneals depending on the strain rate (T=1140 °C, e=1.05).

The maximal crack depth and the microstructural parameters	The strain rate $\dot{\varepsilon}$, s^{-1}		
	10^{-4}	5×10^{-3}	5×10^{-2}
The maximal crack depth, L, mm	3.3	1	1.6
The fraction of recrystallized grains, $V_{r.g.}$, %	70	84	≈100
The mean size of recrystallized grains, d, μm	6	5.1	3.7
The fraction of high-angle grain boundaries, %	70	79	90

Thus, multiple hot working with intermediate recrystallization anneals at temperatures 45÷25 °C below T_s was found to be much more effective for the PM superalloy EP741NP than single-step hot working at the same temperatures. In the former case, recrystallization processes occurred more extensively that provided the formation of a fine-grained microstructure with predominantly high-angle grain boundaries. As has been recently shown [5], the positive influence of multiple hot working and intermediate recrystallization anneals on the recrystallization behavior and hot workability can be used for manufacturing of parts made of the alloy EP741NP with a gradient microstructure and mechanical properties.

4. Conclusions

The influence of temperature-strain rate conditions on compression properties, hot workability and microstructure changes after the uniaxial single-step isothermal compression and the three-step isothermal compression with intermediate anneals was studied for the powder metallurgy (PM) nickel-based superalloy EP741NP. Compression tests and microstructural examination showed that the three-step uniaxial compression at T=1140-1160 °C with $\dot{\varepsilon}=5\times10^{-3}$ s^{-1} to a strain of e≈1.05 and intermediate anneals at the same temperature provided improved hot workability and the formation of a more uniform recrystallized fine-grained microstructure with predominantly high-angle grain boundaries in contrast to the single-step uniaxial compression under the same conditions.

Acknowledgements

The present work was supported by the Russian Science Foundation (Grant No. № 18-19-00685). The work was performed using the facilities of the shared services center «Structural and Physical-Mechanical Studies of Materials» at the Institute for Metals Superplasticity Problems of Russian Academy of Sciences.

IOP Conf. Series: Materials Science and Engineering **447** (2018) 012062 doi:10.1088/1757-899X/447/1/012062

References

[1] Reed R C 2006 *The superalloys: Fundamentals and Applications* (Cambridge University Press)
[2] Fatkhullin O X, Eremenko V I, Vlasova O N, Griz N M, 2001 *Technology of Light Alloys* **No. 5-6** 149 (in Russian)
[3] Kononov S A, Perevozov A S, Kolachev B A, 2007 *Metals* **No. 5** 86 (in Russian)
[4] Beresneev A G, Logunov A V, Logacheva A I, 2008 *Vestnik MAI* **15(13)** 83 (in Russian)
[5] Ganeev A A, Valitov V A, Utyashev F Z, Imayev V M, 2018 Effect of thermomechanical treatment on formation of gradient structure and mechanical properties in a disk made of PM nickel based superalloy *Phys. of Metals & Metallogr.* In press

Study of the microstructure of the pipe steel after rolling in the plate mill 5000

V A Nekit[1], S I Platov[1], M L Krasnov[2], V N Urtsev[3], M V Ivanushkin[1]

[1] Nosov Magnitogorsk State Technical University, 38, Lenin Str., Magnitogorsk 455000, Russia
[2] PJSC "MMK", 93, Kirov St., Magnitogorsk 455000, Russia
[3] Research and Technology Centre «Ausferr», 18, Gorky St., Magnitogorsk 455000, Russia

E-mail: psipsi@mail.ru

Abstract. The article is devoted to an experimental study of the influence of thermo-mechanical processing parameters on the microstructure of pipe steel. Rational temperature-time modes of hot rolling and parameters of accelerated cooling, which allow obtaining a structural state that ensures the achievement of the required complex of consumer properties were determined.

1. Introduction

Currently, the main trend in the ferrous metallurgy is to reduce alloying and achieve the required properties of products due to precise control of the material structure [1-3]. The distribution of velocities and modes of reduction, rolled stock cooling and other factors are little used for normalization during the production of ordinary grades of steel in the hot rolling mills. This normalization is important because the processes of structure formation occur in almost all parts of the rolling mill, and for the implementation of the defined structural states, strictly regulated thermo-mechanical processing of the material is required [4-11].

The controlled low-temperature rolling is used to increase the blow toughness and to improve the surface quality of pipe steels. At a regulated rolling process, pauses are widely used in rolling modes. The use of regulated cooling between deformations allows approaching to the modes of thermomechanical processing and two-phase rolling in the finishing block of deformations, which significantly expands the capabilities of the rolling mill in obtaining high consumer properties of products. According to the set of the required consumer properties of pipe steels, it is necessary to have microscopic structures in the finished product obtained at high cooling rates: upper and lower bainites and marten site. Since the temperature field of the rolled product is very uneven during cooling, the task of obtaining the required structural state in certain parts of the finished product, for example, in the surface layers, can be set.

The aim of the work is to study the possibilities of the equipment of a rolling mill 5000 to achieve temperature and deformation modes, allowing obtaining the required properties of the final product by ensuring a rational cooling path.

2. Methodology

A series of experimental rolling of 5 slabs to a final thickness of 27 mm were performed. For further research samples were taken from each of the rolled sheets. Mechanical properties were determined and the structural condition of the specified volumes of finished rolled product was certified. Mill process analysis system operates with two main electronic documents containing technological information about rolling [15]. To check the reliability of the data included in the passport, the information stored in the Level 1 database of the APCS system was used. This information is the initial, formed by processing the data of the first level on the basis of specialized algorithms. The Level 1 database records the system information about the metal tracking. The dependence of the coordinates of the head part of the metal on time for each unit located on the mill line is being recorded.

The temperature and time modes of experimental rolling on a thick plate rolling mill 5000, starting with the issuance of the metal from the furnace and ending with its exit from the accelerated cooling line, are presented in the table. The process of the pipe steel rolling includes the following operations: heating of the plates to a temperature of 1172-1178 °C.; their excerpt at the roller table during 80-160 s.; 6-pass roughing rolling on a reversing rolling mill; pause between deformation (BDP) during 500-1750 s.; finis rolling with 18-20 passes during 170-300 s; exposure of the workpiece on a roller conveyor 2 and accelerated cooling in a specialized line within 30-60 s.

Table 1. Temperature-time modes of experimental rolling

Operation	The surface temperature of the rolled metal, °C		Duration, s	Cooling rate, °C/c
	Beginning	End		
Mode 1				
Warming	15	1177	4597	0,25
Rough rolling	1037	990	63,30	0,74
Pause			1662,60	0,15
Finish rolling	748	764	337,35	0,05
Accelerated cooling	740	548	33,62	5,71
Mode 2				
Warming	15	1175	5331	0,22
Rough rolling	1044	992	61	0,85
Pause			1716	0,15
Finish rolling	742	780	200	0,19
Accelerated cooling	736	521	60,50	3,55
Mode 3				
Warming	15	1178	4452	0,26
Rough rolling	1036	977	63,40	0,93
Pause			534,57	0,17
Finish rolling	885	918	177,55	0,19
Accelerated cooling	867	679	40,08	4,69
Mode 4				
Warming	15	1172	4285	0,27
Rough rolling	1045	984	66,23	0,92
Pause			1198,77	0,15
Finish rolling	799	853	178,87	0,30
Accelerated cooling	805	637	33,59	5,00
Mode 5				
Warming	15	1172	4328	0,27
Rough rolling	1025	969	60,55	0,92
Pause			1112,83	0,15
Finish rolling	800	831	185,83	0,17
Accelerated cooling	777	602	61,60	2,84

3. Results and discussion

The structure of the rolling sheets for 90-95% consists of ferrite (F), upper and lower bainites (UB and LB) and martensite (M). The remaining volume is perlite (P), carbides, nitrides, carbonitrides and marten site-austenite. The volume fractions of M and LB decrease from the surface to the centre of the sheets, while the volume fractions of F and P increase in this direction.

The metal of all five experimental modes is characterized by a different-grain ferrite component, which may be due to the partial implementation of the conditions of the process of recrystallization of austenite in the rolling mill line. A common feature of the structural state is that the average size of F grains in the metal layers located at a distance of up to 4 mm from the upper and lower surface of the sheet is 1.2–1.4 times smaller than the average size of ferrite grains located in deeper layers.

The metal, rolled in all modes, except the third, is characterized by a striped structure, expressed in significant elongation areas, containing M, UB, LB and carbides. These regions have a visually distinct boundary with a ferrite matrix. It should be noted that the striped structure is much less pronounced in the central layers of the rolled products. The sheet rolled according to regime No.3 is characterized by a higher temperature of finish rolling; the striped structure is weakly expressed already in the near-surface layers, and in the central layer it is practically absent.

An important experimental result is that the increase (about 2 times) of the time of passage of the line of accelerated cooling, with a constant mode of heat removal from the metal surface (the scheme of inclusion of nozzles) leads to an increase in the volume fractions of UB and, especially, LB throughout the cross section of the sheet, thereby increasing the average value of microhardness (see figure 1).

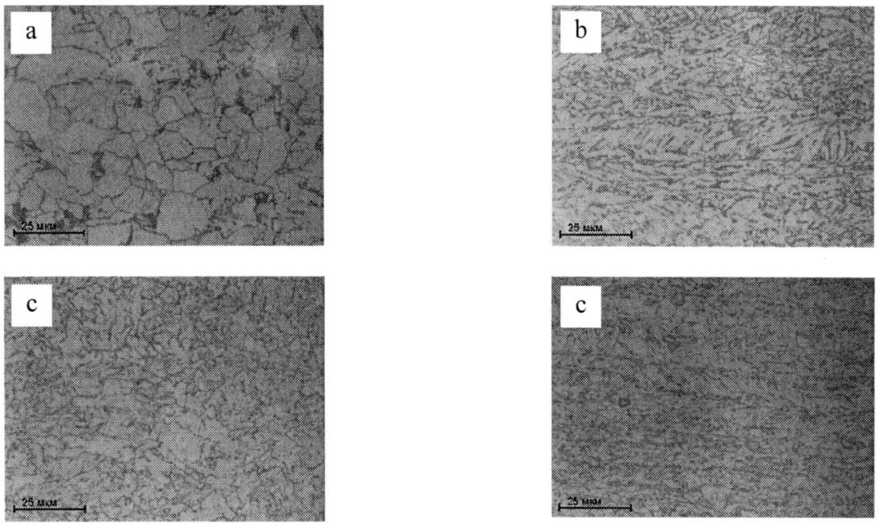

Figure 1. Structural state of rolled sheets mode No. 3, 1 mm from the bottom surface of the sheet (a), Hv_{100}=206, F+P; mode No. 5, 1 mm from the bottom surface of the sheet (b), Hv_{100}=244, F+P+UB+LB; mode No. 3, the central layer (c), Hv_{100}=214,F+P+UB; mode No.5, the central layer (d), Hv_{100}=214, F+P+UB.

The increase in the volume fraction of LB is due to the more complete flowing of the process of austenite disintegration during accelerated cooling. This is confirmed by the dimension of the values t

$$t = \frac{\Delta T_o}{\tau_o},$$

where ΔT is the decrease in the temperature of the layer surface as a result of accelerated cooling, °C/s; τ_o is the time of passage of the accelerated cooling line, s.

For mode No. 5 t=2.82, while for mode No. 2 with almost the same τ_0=60c (see table), t=3.55. The decrease in the cooling rate in the case of mode 5 is explained by the heating of the metal layers as a result of the thermal effect of the process of austenite decay [12].

4. Summary

According to the results of a series of experimental rolling of plates to the final thickness of 27 mm, the values of a number of significant technological parameters are set, compliance with which allows obtaining the required mechanical properties of the strips:

- the surface temperature of the metal at the beginning of the finish rolling is 800-820 °C;
- the time of passage of the line of accelerated cooling is 60–63 s;
- the surface temperature of the sheets at the exit of the line of accelerated cooling is 600-620 °C.

Compliance with these parameters leads to an increase in the volume shares of the upper and, in particular, the lower bainite throughout the section of rolled products.

References

[1] Shmakov A V, Salganik V M, Denisov S V, Gareev A R and Pustovoitov D O 2012 *Steel* **42** **2** 148 (In Russian)
[2] Gareev A R, Shmakov A V, Murikov S A, Urtsev V N and Devyatov D K 2012 *Metallurgist* **55** **11-12** 935 (In Russian)
[3] Gareev A R, Murikov S A, Platov S I, Urtsev V N and Shmakov A V 2015 *Rolling Stock Production* **5** 14 (In Russian)
[4] Gareev A R, Murikov S A, Platov S I, Urtsev V N and Shmakov A V 2015 *Rolling Stock Production* **2** 30 (In Russian)
[5] Platov S I, Nekit V A, Ogarkov N N and Zhelezkov O S 2016 *Modern Metallurgy of New Millenium. Proc. II Int. Scientific-Practical Conf.* (Lipetsk, Lipetsk State University) p. 206 (In Russian)
[6] Platov S I, Ogarkov N N and Nekit V A 2016 *Mater. Eng. and Technol. for Prod. Process. II, Ed A A Radionov, G G Mikhailov and D A Vinnik* (TransTech Publ.) p. 620
[7] Platov S I, Nekit V A, Ogarkov N N and Zhelezkov O S 2016 *Processing Solid and Laminate Materials* **2** **45** 45 (In Russian)
[8] Gareev A R, Murikov A S, Platov S I, Urtsev V N and Shmakov A V 2014 *Rolling Stock Production* **7** 20 (In Russian)
[9] Urtsev V N, Khabibulin D M, Kaptsan A V, Platov S I and Voronkov S N 2002 *Method of Production of Cold Rolled Strips of Ultra-Low Carbon Steel* (Patent RUS 2212457) (In Russian)
[10] Morozov A, Nikiforov B, Salganik V, Pesin A and Platov S 2001 *Simulation of Materials Processing: Theory, Methods and Applications* (Toyohashi, Toyohashi Unive of Technol.) p. 555 (In Russian)
[11] Kaptsan F V, Fomichev A V, Pronin A A and Voznesensky A A 2006 *Technology Improvements at OJSC "MMK". Proceedings Central Lab. OJSC "MMK"* **10** 378 (In Russian)
[12] Gareev A R 2015 Improvement of thermo mechanical modes of deformation of continuously cast slabs and design of the cooling line SHSGP in order to improve the quality *Cand Sci. Thesis* (Magnitogorsk, Magnitigorsk State Univ.) 136 p. (In Russian)

IOP Conf. Series: Materials Science and Engineering **447** (2018) 012064 doi:10.1088/1757-899X/447/1/012064

Structure of carbon steel under plastic deformation

G V Shlyakhova [1,2], L B Zuev [1,3], A V Bochkareva [1]

[1] Institute of Strength Physics and Materials Science, Siberian Branch, Russian Academy of Sciences, 3/4 Akademichesky ave., Tomsk 634055, Russia
[2]Seversk Technological Institute, MEPhI National Research Nuclear University, 65 Kommunistichesky ave., Seversk 636036, Russia
[3]Tomsk State University, 36 Lenin ave., Tomsk 634050, Russia

E-mail: avb@ ispms.tsc.ru

Abstract. Results of studies on the structure of carbon steel 1045 under severe plastic deformation are presented. The atomic force microscopy method was used to estimate the metallographic structure of steel damage after cyclic impact. It was found that the plastic flow localization process occurs in narrow micro-volumes. During the process of cyclic loading, fatigue strength increases, while plastic deformation decreases, and vice versa.

1. Introduction

During the operating of equipment and engineering production, in a large number of structures the fracture occurs at loads below the yield point. As a result of repeated effects of cyclic loads on the material, there is a change in its structure, due to which different microscopic discontinuities form. These discontinuities can subsequently grow in size and eventually result in the development of a main crack. This process is particularly facilitated by a multiple alternation of stresses in local areas of the materials. The majority of products and structures, such as bridge structures, road and railway transport, metalworking machines and tools, presses, aircrafts, lifting mechanisms, pipelines and others, operate under the conditions of cyclic loads [1].

The complexity of studying and predicting the fatigue failure of materials is caused by the fact that the formation and propagation of a crack occur in small local regions of a material. Such regions are determined by the structural components of the material and their orientation relative to operating loads. This significantly increases the probabilistic nature of the occurrence of a micro-crack and its ability for further development [2].

With increasing the frequency of cyclic loading, the deformation rate increases. As the time period for softening of materials decreases and the distortion of crystal lattice increases, the intensity of grain fragmentation into blocks and their mutual disorientation increases [3].

It is well known that the resistance to fractures in metals and alloys under cyclic loading depends on many factors, such as the chemical composition of the metal, operating temperature, stress concentration, cyclic loading asymmetry, cyclic loading frequency, etc. [4]. To carry out such investigations, atomic-force microscopy and optical microscopy are widely used. These methods of comprehensive analysis of metal surfaces allow obtaining the metallographic images with high spatial resolution [5-9].

The aim of the work is to study the degree of damage of surface and structural changes in carbon steel 1045, which is widely used in engineering and metallurgy under different cyclic loads.

2. Materials and methods

The studies were carried out with flat samples of hot-rolled carbon steel 1045. The chemical composition of steel 1045 is presented in table 1.

The specimens for the experimental studies having a dog-bone shape were cut from a hot-rolled 4-mm steel sheet. The obtained metallographic test samples were etched with 4% alcohol solution of nitric acid, after that they were subjected to fatigue fracture tests using a Biss UTM 150 fatigue testing machine with the number of cycles *(N)* 1×103, 5×103, 20×103 and 90×103. The maximum load reached 150 kN and the peak displacement was 150 mm. Microstructure studies of the surface of the deformed and unloaded samples were carried out using an Neophot-21 (OHM) optical microscope and an atomic force microscope (AFM) (Solver PH47-PRO, manufactured by JSC "Nanotechnology-MDT", Zelenograd, Russia). The Solver PH 47-PRO device software, as well as a set of cantilevers, allows us to use different modes of atomic force analysis. Therefore, all sorts of structural features of a solid surface can be observed. The mechanical properties of carbon steel were determined using a portable hardness tester TPP-10.

Table 1. Chemical composition of carbon steel 1045.

C	Si	Mn	Ni	S	P	Cr	Cu	As
0.42÷0.5	0.17÷0.37	0.5÷0.8	< 0.3	< 0.04	< 0.035	< 0.25	< 0.3	< 0.08

3. Experimental results

Metallographic investigations of the surface of carbon steel samples in the initial state are shown in figure 1. As it is seen, a significant part of the image of the sample presents a Widmanstatten structure, consisting of large grains without a regular geometric shape, wheras the other part show a normal structure. The ferrite and pearlite ratios are 45% and 55%, correspondingly, as can be seen in figure 1. The average grain size, measured by interception method, is about 21.9±1.2 μm. For ferrite and pearlite, this is 18.6±1.1 μm and 25.1±1.3 μm, respectively. The results obtained are comparable with the AFM data.

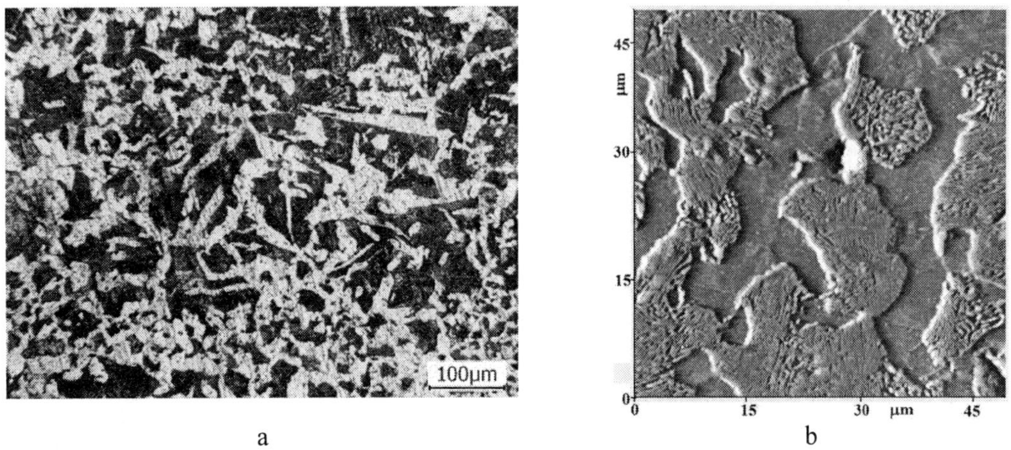

Figure 1. Microstructure of low-carbon steel 1045 according to optical microscopy (a) and atomic-force microscopy (b) of a 50×50 μm area

One can see in figure 1 b that the cementite lamina has almost the same direction within a certain pearlite grain and this direction is markedly different from this orientation within consecutive grains. Several nodular pearlites are segregated at the boundaries of pearlite grains.

The results of metallographic studies of low-carbon steel 1045 under cyclic loading are presented here. The changes in the microstructure of pearlite and the extent of the surface damage were obtained for a certain number of cycles (N), such as 1×10^3, 5×10^3, 20×10^3, 90×10^3, see figure 2. Analysis of surface scan images of steel 1045 at N=1×10^3 did not revealed any structural changes in pearlite grains, see figure 2 a. However, at N=5×10^3, it can be seen that some cementite plates are bent or even partially crushed without formation of micro-cracks, see figure 2 b. Like cementite, a brittle carbide, is prone to micro-cracks formation, especially if it has an elongated shape when N=20×10^3. The cracks are located mainly across the plates and less along them, see figure 2 c. Within a number of cycles N=95×10^3, the crushed cement plates are separated, the distance between them becomes greater, and the cavity can be filled with a softer phase – ferrite. In some cases, the destruction of cementite plates is carried out on a number of closely spaced planes, which leads to the formation of defective zones. In such zones there are numerous fragments of cement plates that are displaced relative to each other, see figure 2 d.

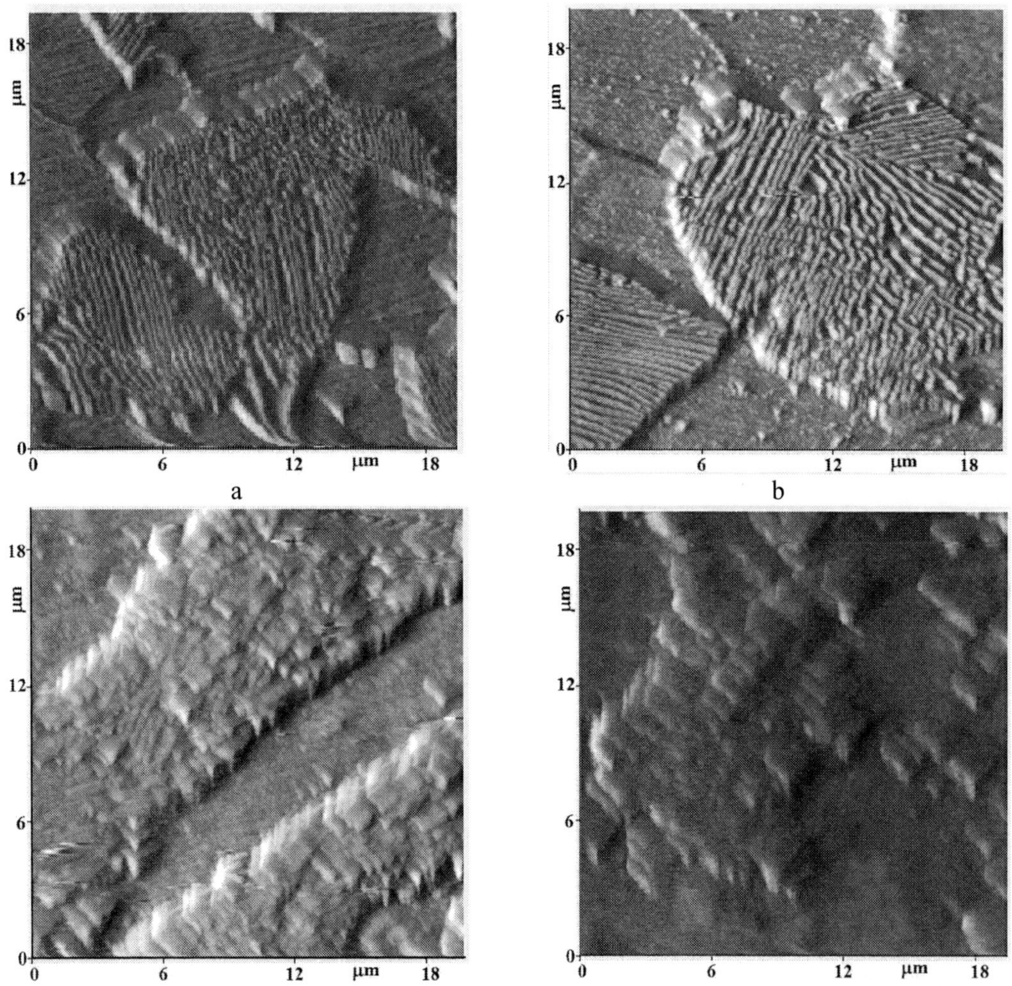

Figure 2. Scanning-probe images of the lamellar pearlite structure at the interface under different cyclic loading: (a)N=1×10^3, (b) N=5×10^3, (c) N=20×10^3, (d) N=95×10^3.

Experimental results obtained for low carbon steel under cyclic loading showed that for the number of cycles *(N)* in the interval from 5×10^3 to 20×10^3 the hardness of steel increases. But the future increment of N leads to a reverse decrease to the initial values (see table 2).

Table 2. The dependence of hardness *(HV)* low-carbon steel 1045 on the number of cycles *(N)*

Number of cycles (N)	Hardness, HV (MPa)
Initial state	1735
1×10^3	1735
5×10^3	2178
20×10^3	2178
95×10^3	1735

4. Summary
The experimental study of carbon steel 1045 revealed special features of lamellar pearlite appeared during plastic deformation. Localization of the plastic flow occurs in narrow microvolumes. The lamellar perlite decay starts from the impaired subcolonies and the critical concentration of damage gives rise to the destruction of cementite laminae.

Acknowledgements
This work was performed within the frame of the Fundamental Research Program of the State Academies of Sciences for 2013-2020, line of research III.23. and the Russian Foundation for Basic Research (grant no. 17-08-00751-a).

References
[1] Proceedings of the International Conference VHCF-5, Edited by C. Berger, H.-J. Christ, DVM, June 28-30, 2011, Berlin, Germany, 980
[2] Honeycombe R W K 1984 *The Plastic Deformation of Metals* (London: Edward Arnold Ltd.)
[3] Mylnikov V V, Shetulov D I and Chernyshov E A 2013 *Russ. J. Non-Ferr. Metals* **54 (3)** 229
[4] Shetulov D and Myl'nikov V V 2014 *Rus. Metally* **3** 241
[5] Giessible F J 1994 *Jpn. J. Appl. Phys*. 1994 **33(1) 6B** 3726
[6] Sizova O V, Shlyakhova G V, Kolubaev A V, Kolubaev E A, Psakhie S G, Rudenskii G, Chernyavsky A and Lopota V 2014 *Adv. Mat. Res*. **872** 174
[7] Danilov V I, Shlyakhova G Vand Semukhin B S 2014 *Appl. Mech. Mater.* **682** 351
[8] Barannikova S A, Shlyakhova G V and Zuev L B 2014 *Appl. Mech. Mater.* **682** 3
[9] Lunev A G, Nadezhkin M V, Shlyakhova G V, Barannikova S A and Zuev L B 2014 *Appl. Mech. Mat*er **1623** 371

IOP Conf. Series: Materials Science and Engineering **447** (2018) 012065 doi:10.1088/1757-899X/447/1/012065

The effect of Cr and Zr content on the microstructure and properties of the Cu-Cr-Zr system alloy

E A Sarkeeva[1, 2], V D Sitdikov[1], G I Raab[1], W Wei[2,3,4], I V Alexandrov[1,2]

[1] Ufa State Aviation Technical University, 12 K. Marx, 450000 Ufa, Russia
[2] School of Materials Science and Engineering, Changzhou University, 1 Gehu Road, Changzhou 213164, P.R .China
[3] Jiangsu Key Laboratory of Materials Surface Science and Technology, Changzhou University, Changzhou 213164, P.R. China
[4] Joint Laboratory of Functional Nanostructured Metals, Changzhou University, Changzhou 213164, P.R. China

E-mail: sarkeeva.e@inbox.ru, benjamin.wwei@163.com

Abstract. This article investigates the effect of various contents of Cr and Zr on the microstructure and properties of the alloy of the Cu-Cr-Zr system. It has been defined that an increase in the content of Zr leads to a decrease in the grain size both in the initial state and after severe plastic deformation. The increase in the content of Zr significantly enhances the strength of the alloy.

1. Introduction

Cu-based alloys are considered to be good conductors and are widely used in different branches of industry: in heat exchange units, in resistance welding machines (electrodes and other fittings), in the electrical engineering, electronics engineering, instrument engineering and etc. [1, 2]. However, coarse-grained Cu-Cr-Zr alloys exhibit low strength that limits their application.

The traditional way to improve the strength properties of metallic materials is the application of plastic deformation and/or the addition of alloying elements. For example, the addition of Cr significantly reinforces the Cu alloy and increases the level of high-temperature strength. The inclusion of a small percentage of Zr leads not only to an increase in strength properties, but also to a considerable improvement of ductility at elevated temperatures. Alloys of the Cu-Cr-Zr system are dispersion-strengthened materials [2]. The strength of Cu alloys can be further increased due to the formation of an ultrafine-grained structure by the methods of severe plastic deformation. [3-6].

The aim of this work is to investigate the influence of Cr and Zr additions on the structure and properties of the alloys of the Cu-Cr-Zr system during severe plastic deformation.

2. Experimental methods

In this work, the experiments were carried out on Cu-0.5%Cr-0.2%Zr, Cu-1.8%Cr-0.8%Zr, Cu-0.8%Cr-0.08%Zr (wt. %) alloys. The alloys were subjected to high-temperature treatment at T=1000 °C during τ=30 minutes with a subsequent quenching in 5% NaCl solution to form a supersaturated solid solution, thereby forming the initial state. Hereafter, the samples with a size of 12×12×80 mm in the form of rectangular parallelepipeds were deformed by equal channel angular pressing at room temperature with a rate of 0.2 mm/sec with an interior angle of the fitting ψ=120° and an exterior angle φ=0° for equal channel angular pressing via route A. The number of passes was 4. The degree of deformation for each pass was 0.8.

To study the microstructure, we used an OLIMPUS GX51 optical microscope, a JEOL JSM-6490LV scanning electron microscope, and a JEOL 2100 transmission electron microscope X-ray structural analysis of the samples was performed by a Rigaku Ultima IV (Bragg-Brentano geometry) diffractometer using the $Cu_{K\alpha}$ monochromatic radiation at the reflected beam (parabolic graphite monochromator). The measurements were conducted along the area of 2×2 mm^2 located in the center of the samples.

The microhardness of the investigated samples was measured by a "Micromet" with a load of 100 g and a holding time under load of 10 seconds. The investigations of electrical conductivity were performed at the temperature T=23 °C by the eddy current method. The measurement error was ±2%.

3. Results and discussion

The structure of the alloys was investigated in the initial state and after severe plastic deformation. Figure 1 a, b, c shows rounded inclusions with a size of several microns (table 1) in the initial state against the matrix of the α-solid solution. These inclusions are excessive Cr and Zr, which are formed during solidification of the alloy and are not dissolved during quenching of the solution, since the solubility of Cr at a temperature of 1000 °C in the copper matrix is only 0.4-0.45 wt. %, and Zr – 0.15 wt. % [2, 7]. The size of large inclusions is 4.3±0.1 μm in the Cu-0.5Cr-0.2Zr alloy, 3.2±0.2 μm in the Cu-0.8Cr-0.08Zr alloy, 2.6±0.2 μm in the Cu-1.8Cr-0.8Zr alloy. It can be said that an increase in the content of Zr leads to a decrease in the size of the inclusions. An increase in the degree of alloying by Zr leads to a decrease in the grain size. In particular, the grain size is 260±11 μm in the Cu-0.8Cr-0.08Zr alloy, 140±8 μm in the Cu-0.5Cr-0.2Zr alloy, 110±4 μm Cu-1.8Cr-0.8Zr alloy. This is due to the particles of the non-dissolved Zr. Thus, the alloys in the initial state are characterized by a coarse-grained microstructure, in which with increasing Zr content the average grain size decreases.

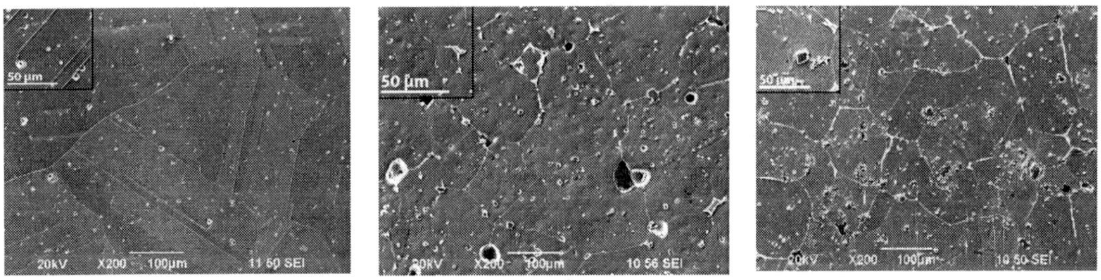

a b c

Figure 1. Scanning electron microscopy. The microstructure of the alloys in the initial state: the Cu-0.8Cr-0.08Zr alloy (a), the Cu-0.5Cr-0.2Zr alloy (b), the Cu-1.8Cr-0.8Zr alloy (c).

The implementation of severe plastic deformation leads to a decrease in the size of the inclusions (d_{av} in table 1). Besides, the average distance (L_{av}) between the inclusions decreases resulting in an increase in the density of their distribution. The latter could be explained by mechanical fragmentation of particles.

Figure 2 demonstrates the microstructure of the alloy after severe plastic deformation in the longitudinal section of the samples. Severe plastic deformation leads to the formation of an elongated fragmented structure with an ultrafine size in the transverse direction. The longitudinal axis of the elongated fragments is oriented at an angle of 35 degrees in respect to the axis of the sample. Due to the high elongation, only the transverse dimension of this structure can be defined. Its average transverse size is 170±10 nm in the Cu-0.8Cr-0.08Zr alloy, 160±7 nm in the Cu-0.5Cr-0.2Zr alloy, 155±10 nm in Cu-0.2Cr-0.8Zr. As reported above, in the initial state of these alloys, the same, but more pronounced tendency was observed.

Along with large inclusions observed in images of scanning electron microscopy, observations by the method of transmission electron microscopy revealed small particles about 7-11 nm in size (figure 2). The paper [8] reports the presence of Cr and Cu_5Zr phases in these alloys. Small particles are generated as a result of dynamic ageing implemented during severe plastic deformation. This is demonstrated by the change in the lattice parameter in the alloys (table 1) towards the parameter of the pure Cu lattice (0.3615 nm) associated with the impurity depletion of the main matrix. On the other hand, severe plastic deformation considerably increases the concentration of defects (table 1) that leads to acceleration of diffusion processes, i.e. to the kinetics of ageing.

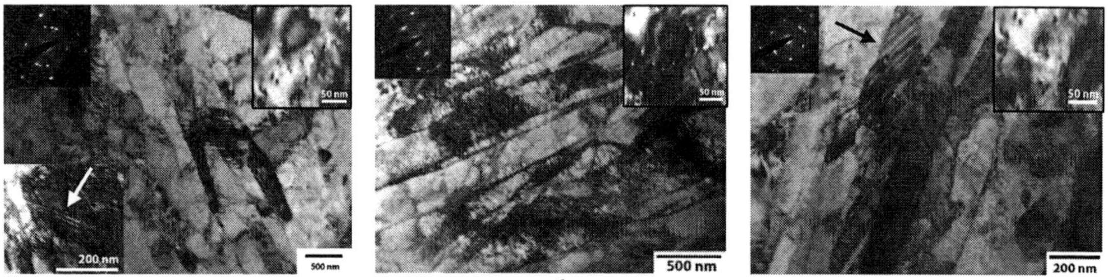

a b c

Figure 2. Transmission electron microscopy. The microstructure of the alloys after severe plastic deformation: the Cu-0.8Cr-0.08Zr alloy (a), the Cu-0.5Cr-0.2Zr alloy (b), the Cu-1.8Cr-0.8Zr alloy (c)

Figures 2 a and 2 c demonstrate the presence of twins (indicated by arrows). With an increase in the degree of alloying, the number of twins grows due to the decrease in the stacking fault energy, which is consistent with [9].

Table 1. Microstructure parameters of the alloys

State	Cu-0.8Cr-0.08Zr		Cu-0.5Cr-0.2Zr		Cu-1.8Cr-0.8Zr	
	Solid solution	Severe plastic deformation	Solid solution	Severe plastic deformation	Solid solution	Severe plastic deformation
H_v, MPa	750±20	1460±30	950±20	1500±40	1030±60	1650±40
IACS, %	32.0±0.7	33.0±0.7	29.0±0.6	30.0±0.6	30.0±0.6	29.0±0.6
D_{av}, mkm	260±11	0.17±0.01	140±8	0.160±0.007	110±4	0.155±0.010
d_{av}, mkm	3.2±0.2	1.0±0.1	4.3±0.1	2.5±0.2	2.6±0.2	1.9±0.1
L_{av}, mkm	38	11	34	21	20	21
Lattice parameter, nm	-	3.615208 ±0.000001	3.62167 ±0.00005	3.62025 ±0.00002	3.618412 ±0.000029	3.61669 ±0.00015
Lattice strain, %	-	0.23 ±0.02	0.074 ±0.005	0.2400 ±0.0005	0.037 ±0.001	0.20 ±0.02
Crystallite size, nm	-	39±3	145±22	32±3	229±59	34±4
Disl.dens, 10^{15} m^{-2}	-	6.7±0.6	-	4.6±0.3	-	3.84±0.6

Table 1 shows the increase in microhardness after severe plastic deformation. In particular, the microhardness value of all three ultrafine-grained alloys dramatically increased in comparison with the initial state (table 1). Besides, the highest microhardness is achieved in the Cu-1.8Cr-0.8Zr alloy with a high content of alloying elements.

It is known that Cr and Zr atoms dissolved in the Cu matrix scatter the electrons responsible for the conductivity, and, consequently, significantly degrade the electrical conductivity. Table 1 shows that the electrical conductivity of the Cu-0.5Cr-0.2Zr and Cu-1.8Cr-0.8Zr alloys is about 30% IACS (International Annealed Copper Standards). This indicates that almost the same number of atoms is dissolved in the matrix. However, the electrical conductivity in the Cu-0.8Cr-0.08Zr alloy is slightly higher after severe plastic deformation. Firstly, this is due to the fact that the Cu-0.8Cr-0.08Zr alloy contains a smaller amount of Zr, and, secondly, the dynamic aging and the formation of new precipitates take place during severe plastic deformation. Nevertheless, a small amount of Cr and Zr elements still remains in the matrix leading to low conductivity.

4. Conclusions

1. The increase in the content of Zr in the alloy and the implementation of equal channel angular pressing allow achieving the structure with the smallest transverse cross section of fragments equal to 155 nm.

2. The increase in the content of Zr in the alloy and the implementation of equal channel angular pressing activate twinning processes.

3. The process of dynamic ageing is implemented in the alloy during severe plastic deformation. As a result, the particles of Cr and Cu_5Zr precipitate in the matrix mainly in the form of a sphere and with a mean radius of 7-11 nm.

4. The increase in the content of Zr in the alloy and the implementation of equal channel angular pressing significantly increase the microhardness. The highest microhardness of 1650 MPa is achieved in the Cu1.8Cr-0.8Zr alloy that is 1.6 times higher compared to the initial state

Acknowledgments
Authors are grateful to the financial support of the Ministry of Education and Science within the framework of the project No. 16.1969.2017/4.6. W. Wei is grateful to be supported by the National Natural Science Foundation of China under Grant No. 51561001, the Priority Academic Program Development of Jiangsu Higher Education Institutions(PAPD), the Top-notch Academic Programs Project of Jiangsu Higher Education Institutions (TAPP),and the Science and Technology Bureau of Jiangsu Province, P.R. China under grant No. BY2016029-19.

References

[1] Mishnev R, Shakhova I, Belyakov A and Kaibyshev R 2015 *Mater. Sci. Eng. A* **629** 29.
[2] Osintsev O E and Phyodorov V N 2004 *Copper and Copper Alloys, Russian and Foreign Brands* (Moscow: Mashinostroenie) p. 215 (In Russian)
[3] Purcek G, Yanar H, Saray O, Karaman I and Maier H 2014 *Wear* **311** 149
[4] Valiev R Z, Alexandrov I V 2007 *Bulk Nanostructured Metal Materials: Preparation, Structure and Properties* (Moscow: Akademkniga) p. 271 (In Russian)
[5] Alexandrov I V, Sitdikov V D, Abramova M M, Sarkeeva E A, Wei K X and Wei W 2016 *J. Mater. Eng. Perf.* **25** 4303
[6] Sarkeeva E A, Abramova M M, Oreshkina R A, Sitdikov V D, Wei K X, W Wei, Musin F F and Alexandrov I V 2015 *Mater. Phys. Mech.* **24** 211
[7] Batra I S, Dey G K, Kulkarni U D and Banerjee S 2002 *Mater. Sci. Eng. A* **356** 3236.
[8] Zhang S, Li R, Kang H, Chen Z, Wang W, Zou C, Li T and Wang T 2017 *Mater. Sci. Eng. A* **680** 108
[9] Pang Y, Xia C, Wang M, Li Z, Xiao Z, Wei H, Sheng X, Jia Y and Chen C 2014 *J. Alloys. Comp.* **582** 786

IOP Conf. Series: Materials Science and Engineering **447** (2018) 012066 doi:10.1088/1757-899X/447/1/012066

The effect of speed asymmetry on the strain state in aluminium bimetals during accumulative rolling

A M Pesin, D O Pustovoitov and O D Biryukova

Nosov Magnitogorsk State Technical University, 38 Lenina ave., Magnitogorsk 455000, Russian Federation

E-mail: fimapatisonchik@inbox.ru

Abstract. Methods of severe plastic deformation are one of the most modern and cost-effective methods for obtaining high values of mechanical properties in metals and alloys. It is possible to obtain an ultrafine-grained structure due to such processes. The shear component has a significant effect on the value of the true deformation in case of using the methods of severe plastic deformation. In this paper, the strain state of aluminium alloys Al5083, Al1070, and Al2024 in the process of bimetal asymmetric rolling and accumulative rolling was investigated. Symmetrical and asymmetrical rolling of a sample was simulated in the Deform 2D software package. The thickness of the workpiece was 2 mm, the reduction by one pass was 50%, the friction coefficient on the contact of the metal and the rolls ranged from 0.1 to 0.4, the velocity of the rolls was 1.047 rad / s in the symmetric case. The discrepancy between the speeds of the rolls was 10-50%. The presented data agree with the known experimental data on the processing of aluminium alloys, thus the result obtained provide for obtaining the UFG structure in these alloys.

1. Introduction

Methods of severe plastic deformation are one of the most modern and cost-effective methods for obtaining high values of mechanical properties in metals and alloys. It became possible to obtain an ultrafine-grained structure with their help. Such structural changes lead to a significant increase in strength - this parameter increases in 2 times in pure metals and by 40-80% in alloys [1]. The development of severe plastic deformation technologies for industrial and commercial applications is a complex scientific task.

Processing of an aluminium alloy by the severe plastic deformation method makes it possible to refine the metal structure. This leads to an increase in the dislocation density. It is important that the yield stress and tensile strength also show an increase. The crushed grain in such conditions is an indispensable factor for maintaining the ductility of the metal at a proper level. Severe plastic deformation extends the temperature range at which superplasticity is manifested. It should be noted that if the temperature at which the material is processed is low, the smaller will be the grain growth in the structure, which naturally improves the mechanical properties of the material. Also, a high reduction of the strip, which contributes to an increase in the shear strain, leads to a refinement of the grain in the structure.

2. Method of research and materials

In the present research, Al 1070, Al 2024 and Al 5083 alloys were chosen as the material for simulation in the Deform 2D software package. Deform 2D is a forceful system for modeling processes of metal forming with the possibility of analyzing a two-dimensional metal flow with the use of the assumption of plane strain. The main goal was to obtain an ultrafine-grained structure in aluminium materials by different means: symmetrical and asymmetrical rolling of bimetals and accumulative rolling. Endeavours of obtaining the ultrafine-grained structure in such bulk materials are currently unsuccessful [2-4].

The mechanical properties of Al 5083 make it possible to use it under extreme conditions in areas such as shipbuilding, aircraft and automotive industries; it has high hardness and strength. Alloy Al 1070 is characterized by high electrical conductivity and corrosion resistance. Alloy Al 2024 is the most famous and most widely used "aircraft" alloy. The chemical composition of the investigated Al 1070, Al 2024 and Al 5083 alloys is given in tables 1, 2, 3 [5,6].

Table 1. Chemical composition of the Al 1070 alloy (wt%).

Si	Fe	Cu	Mn	Mg	Ti	Al
0.15	0.16	0.01	0.03	0.02	0.01	balance

Table 2. Chemical composition of the Al 2024 alloy (wt%).

Si	Fe	Cu	Mn	Mg	Cr	Al
0.422	0.178	0.02	0.019	0.473	0.001	balance

Table 3. Chemical composition of the Al 5083 alloy (wt%).

Si	Fe	Cu	Mn	Mg	Cr	Al
0.40	0.40	0.10	1.00	4.90	0.25	balance

All the parameters that were taken into account as the initial data are given in table 4.

Table 4. Initial data for simulation processes in the soft package Deform 2D.

	Accumulative rolling	Rolling of bimetals
Material	Al5083	Al1070, Al2024, Al5083
The thickness of the workpiece	2 mm	2 mm
The reduction per one pass	50 %	50 %, 60 %
The friction coefficient	0.1-0.4	0.1-0.4
The velocity of the rolls	1.047 rad/s	1.047 rad/s
The distinction of the velocities	10-40 %	10-50 %
Rolls radius	250 mm	250 mm

3. Results

Figure 1 a-b show the simulation results: in the case of aluminium bimetal rolling (top layer is Al5083 and bottom layer is Al2024) with high-speed asymmetry, which had different distinction of velocities, the value of equivalent deformation increases with a friction coefficient of 0.3 and 0.4, it favourably influences the formation of an ultrafine-grained structure. In this case, these values equally increase in the centre (Al5083-0.5 and Al2024-0.5) of the sample and on its surfaces (Al5083-0.0 and Al2024-0.0).

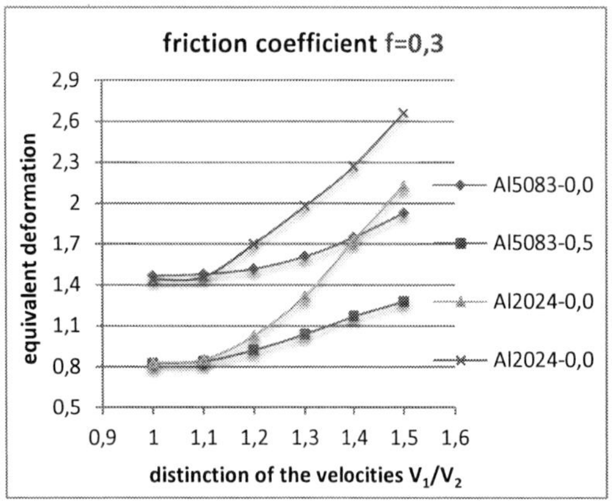

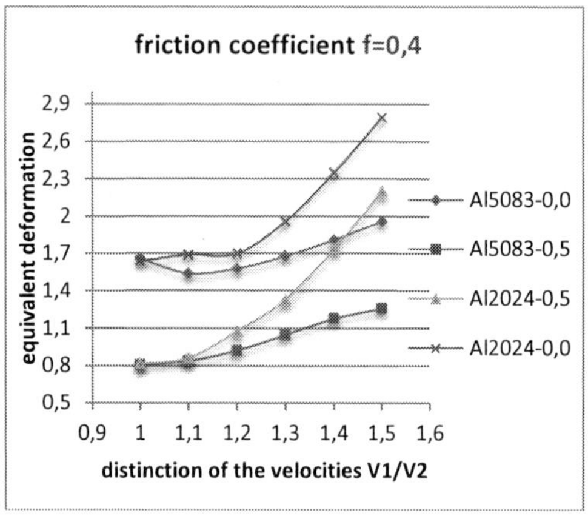

a b

Figure 1. The dependence of the equivalent deformation and the distinction of the velocities of the rolls at a friction coefficient 0.3(a) and 0.4 (b).

In the case of bimetal processing (the top layer is Al5083 and the bottom layer is Al1070) with identical conditions, the situations changes: with increasing speed asymmetry, the value of equivalent deformation rises with increasing friction coefficient. Al5083 is deformed evenly, while Al1070 shows an abrupt increase in the equivalent strain on the metal contact surface, as shown in figure 2 a-b. It is impossible to unequivocally describe the reasons of the upcoming mechanism. The angle of shear in different metal layers differs, the largest angle takes a value of more than 80 degrees, and the structure is not homogeneous.

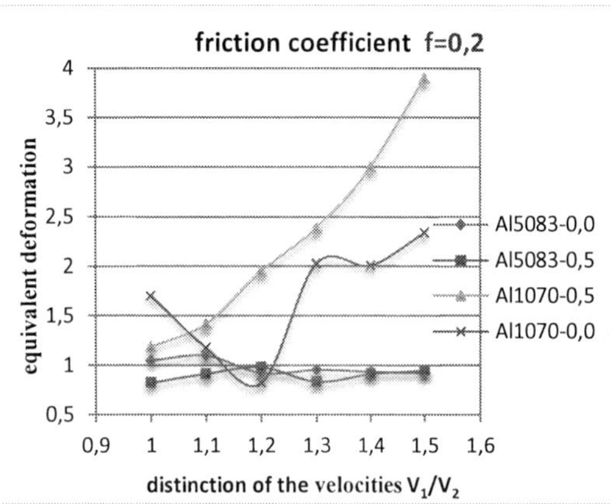

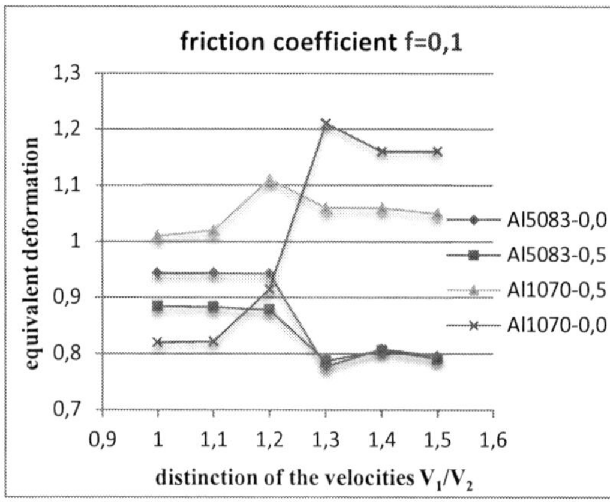

a b

Figure 2. The dependence of the equivalent deformation and the distinction of the velocities of the rolls at a friction coefficient 0.2(a) and 0.1 (b).

The bending of the strip (both the front and rear curtains) varied depending on the velocity of the top work roll, the friction coefficient and the type of the aluminium alloy. This indicates that the reductions along the strip edge are uneven, and the strip alignment occurs if the longitudinal velocity of the top and bottom surfaces of the strip is not the same, which is confirmed by figure 3 a-b.

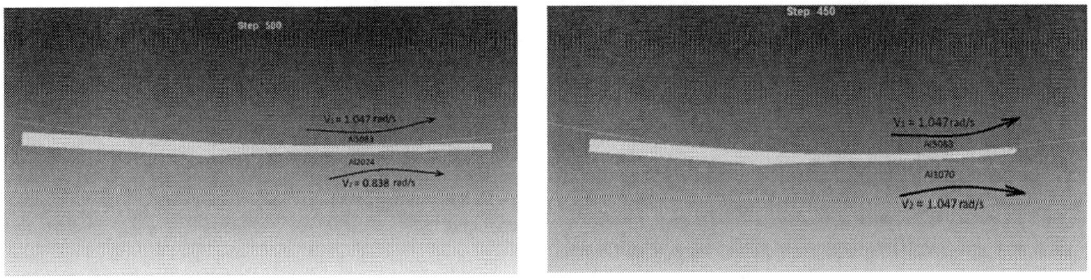

a b

Figure 3. The nature of the metal flow at the asymmetrical rolling of the bimetal Al5083/Al2024 (a) and at the symmetrical rolling of the bimetal Al5083/Al1070 (b).

4. Conclusions

In this paper, the role of equivalent deformation during the asymmetrical rolling of bimetals Al5083/Al2024 and Al5083/Al1070 and accumulative rolling of Al5083 with a distinction of velocities of the rolls was considered. It is shown that the value of the accumulated deformation increases due to the nonmonotonic metal flow, especially during rolling of the bimetal (up to 7.5-9). The value of the shear angle φ for asymmetrical rolling of the bimetals is more than 80°, which ensures that the structure enters the zone of severe plastic deformation.

Acknowledgments

The research was supported by a grant from the Russian Science Foundation (project № 15-19-10030).

References

[1] Hallberg H 2013 *Int. J. Mech. Sci.* **66** 210
[2] Markushev M V 2011 *Letters on Materials* 1 36 (In Russian)
[3] Sedighi M 2016 *Int. J. Adv. Design Manuf. Technol.* **9** 13
[4] Sedighi M 2017. *Canadian Metall. Quart.* **57** 160
[5] Jin-Sung L 2013 *Int. J. Nav. Archit. Ocean Eng.* **5** 580
[6] Shahsavari F 2015 *Procedia Mater. Sci.* **11** 84

IOP Conf. Series: Materials Science and Engineering **447** (2018) 012067 doi:10.1088/1757-899X/447/1/012067

Simulation of the structure formation in alloyed silicon-carbon materials

M S Svirepova[1], E V Vorob'ev[1], N K Plugotarenko[1], T A Bednaya[2]
[1]South Federal University, 105/42 Bolshaya Sadovaya st., Rostov-on-Don 344006, Russia
[2]Don State Technical University, 1 Gagarin square, Rostov-on-Don 344000, Russia

E-mail: berlukova@mail.ru

Abstract. The formation of alloyed thin-film coatings on the basis of a dielectric matrix of silicon-carbon compounds in terms of the assumption on the fractal nature of the formed structures was simulated. Having studied the evolution of two- and three-layer aggregates with increasing temperature by quantum chemistry methods, it is determined that at the initial stage of the structure growth, the hydrogenated aggregates are looser, which contributes to the penetration of metal ions into the interlayer space and access to oxygen and nitrogen atoms. The formation of oxides and nitrides of metals is possible with the addition of a system for obtaining more energy while maintaining the sp^3-hybridization of most of the carbon atoms. The possibility of reorganization of the primary structure and analysis of the formed structures for the presence of percolation clusters was implemented when modelling the growth process of a doped silicon-carbon film. It was established that the concentration of the alloying component of 0.2 mol % is sufficient to form a percolation cluster.

1. Introduction

Films based on a silicon-carbon matrix are amorphous materials. Nanocomposites with the specified electrophysical, optical, and mechanical properties can be created by alloying silicon-carbon films with metal atoms. The concentration of metal atoms in films can reach tens of percent. By varying the composition and concentration of metal atoms, the values of electrical conductivity in films in the range 10^{-4}-10^{14} Ω·cm can be specified [1]. Materials based on a silicon-carbon matrix are used to create gas sensors [2] and electrodes of supercapacitors [3].

The production of new silicon-carbon materials with the desired properties is possible using experimental methods and computer simulation of various crystalline and composite structures at the molecular level [4].

Various quantum mechanical methods of calculation are used to determine the possibility of embedding metal atoms into a silicon-carbon matrix, to obtain the spatial characteristics of particles (molecules) and Hartree-Fock energy [5-6]:

– non-empirical methods (Hartree-Fock method with different basis sets, multiparticle theory, discrete Fourier transform, inverse discrete Fourier transform);

– semi-empirical methods (MNDO, PM3, AM1, TM1, ZINDO, etc.).

Computer simulation based on certain algorithms of particle motion and interaction allows creating a virtual fractal surface and analysing the properties and nature of the growth of real systems.

The formation of the structure is determined by the nature of the process, of the motion and of the particle or cluster pooling. The diffusion-limiting model (DLA) allows one to explore the aggregation of clusters and particles limited by diffusion by implementing the Witten–Sander model, which assumes the presence of an aggregation centre and a free particle that carries out a Brownian motion from infinity and adheres to the aggregation centre when collides with it. The cluster-cluster aggregation model (SSA) accounts for the diffusion controlled cluster-cluster aggregation and implements the possibility of movement of clusters themselves, not only free particles. The reaction limited cluster aggregation model (RLCA) is based on a chemically limited aggregation of clusters with clustes. This model describes the interaction of clusters with a small coefficient of adhesion. There are also "cluster-cluster" aggregation models taking into account the long-range effect of attraction and the "cluster" model with extended detailing of the growth mechanism [7].

2. Methodology

The process of creating silicon-carbon coatings by electrochemical deposition was chosen as the object of modeling. Tetraethoxysilane (TEOS) and hexamethyldisiloxane were used as the source of silicon. Ethanol and methanol were used as carbon sources. Manganese, nickel, zirconium and copper were used as alloying additives to set the functional properties of materials for gas sensors and electrodes of supercapacitors [8]. These metals, according to preliminary experiments, enable the materials to acquire the necessary functional properties. The concentration of dopant varied in the range from 0.1 to 0.5 mole %.

Quantum chemistry methods were used to determine the energy and geometric parameters of the precursor particles. HyperChem software products (preliminary *ab-initio* calculations by the PM3 method) and Gaussian 03 (semi-empirical calculations by the HF method in the 3-21G basis) were used in the calculation of complex nanostructures and the dynamics of their behavior depending on time and temperature.

To model the growth of the dielectric matrix structure, the Witten-Sander diffusion–limited aggregation model and the RLCA ("cluster" aggregation limited chemically) model were used, taking into account the possibility of rotation of the clusters relative to the contact point and aggregate movement [9]. The Monte Carlo method [10] for the implementation of the Wang-Landau algorithm was adopted as the basis for modelling [7].

The modeling of film structures was carried out in the MatLab software environment. In this work, a four-connected square lattice with the number of initial particles up to 10^6 was used for modeling. A certain number of free nodes were determined, which are randomly distributed over a given bounded surface. Among the free reachable nodes, a certain initial set was formed. For each of the initial nodes, a set of neighbors was allocated (the perimeter of the set) and those nodes that were reachable and did not belong to any other set were attached to the initial one. Then the perimeter was built for the new initial set, and the process was repeated until the free nodes were exhausted or until the node from the given set was reached. Each initial particle was labeled as metal-containing or containing only a polymer base. The ratio of such particles was chosen based on experimental capabilities. IN addition, in order to take into account the influence of the chemically aggressive medium, the pH of the medium was set in the form of so-called "interfering" points.

3. Results

Geometrical and energy parameters of complexes with the selected metals are calculated by quantum chemistry methods (figure 1). The use of manganese was the most profitable, therefore at the second stage of modelling manganese was taken as a metal.

Figure 2 shows the simulation results for two-layer (figure 2a) and three-layer (figure 2b) structures. Information on the distances between atoms is summarized in table 1.

The Si-C bond length in two-layer aggregates increases in the presence of hydrogen atoms to 1.87 Å , while in three-layer aggregates the Si-C bond length slightly increases from 1.94 to 1.98 Å with an increase in the number of hydrogen atoms. The length of the C-C bond is approximately the same for two-layer and three-layer aggregates and is 1.36 Å for less hydrogenated structures, which corresponds to the value of the double bond and 1.53 Å for hydrogenated structures (single Sigma bond). Between the layers, the C-C bond ranges from 1.53 to 1.68 Å. The Si-Si bond length in three-layer aggregates is approximately the same and is 2.66 ± 0.02 Å. For bilayer aggregates this relationship is shorter than for the hydrogenated structures (of 2.53 Å). The distance between the atoms in the layers for two-layer structures reaches 3.79 Å, for three-layer 3.85 Å, in three – layer hydrogenated - 3.45 Å.

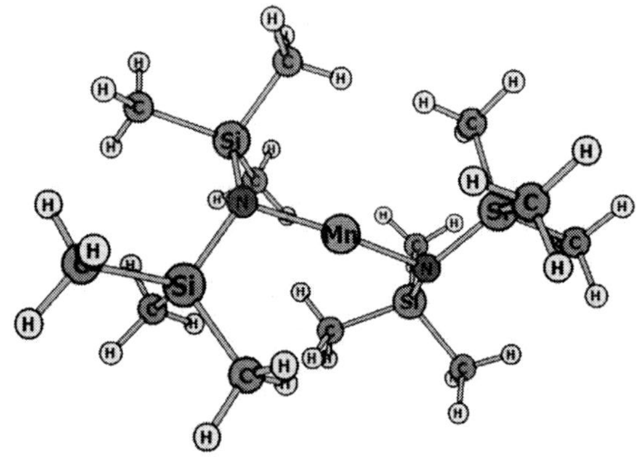

Figure 1. Hexamethyldisilane complex with manganese

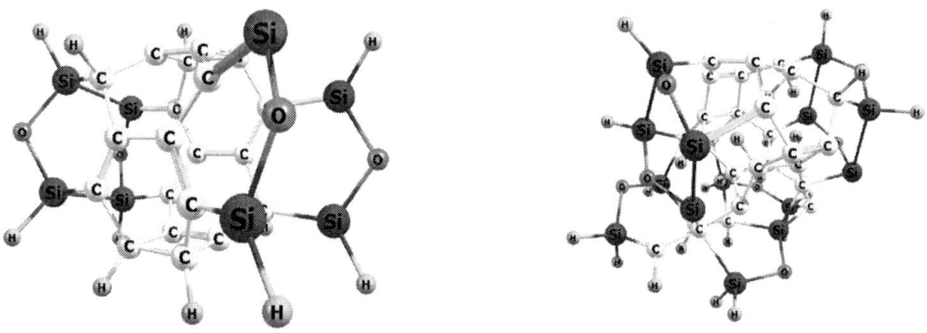

a b

Figure 2. The computer model of the structure of the unit: a) $Si_8C_{20}H_{12}O_4$; b) $Si_{14}C_{27}H_{26}O_7$

Table 1. The distance between the atoms in three-layer units.

The distance between the atoms (Å)	$Si_{14}C_{27}N_7H_{22}$	$Si_{14}C_{27}N_7H_{43}$	$Si_{14}C_{27}H_{26}O_7$	$Si_{14}C_{27}H_{36}O_7$
Si-N(O)	1.66219	1.64510	1.78600	1.69457
Si-C	1.93675	1.95313	1.94512	1.97750
C-C (between layers)	1.52171	1.63092	1.68332	1.62945
C-C (in one layer)	1.35792	1.50653	1.35631	1.54063
Si-Si (between layers)	2.66568	2.64633	2.64826	2.68012
N…N (between layers)	3.40688	3.16203	-	-
O…O (between layers)	-	-	3.28573	3.15433

IOP Conf. Series: Materials Science and Engineering **447** (2018) 012067 doi:10.1088/1757-899X/447/1/012067

At the second stage of modeling, the possibility of reorganization of the primary structure and analysis of the formed structures for the presence of percolation clusters was realized. The search for the lowest concentration at which the formation of percolation cluster is possible is carried out. The presence of a percolation cluster ensures the conductivity of the coating, but high concentrations of the additive lead to a deterioration of adhesion to the substrate.It is determined that the concentration of the alloying component 0.2 mol % is sufficient for the formation of the percolation cluster.

Figure 3 shows the simulation results for different substrate dimensions (blue color indicates particles containing metal, redcolor indicates a silicon-carbon matrix). These results show that as the dimension of the substrate increases, the visual perception of the structure of the material is blurred, it is perceived as a continuous film, details are not visible.

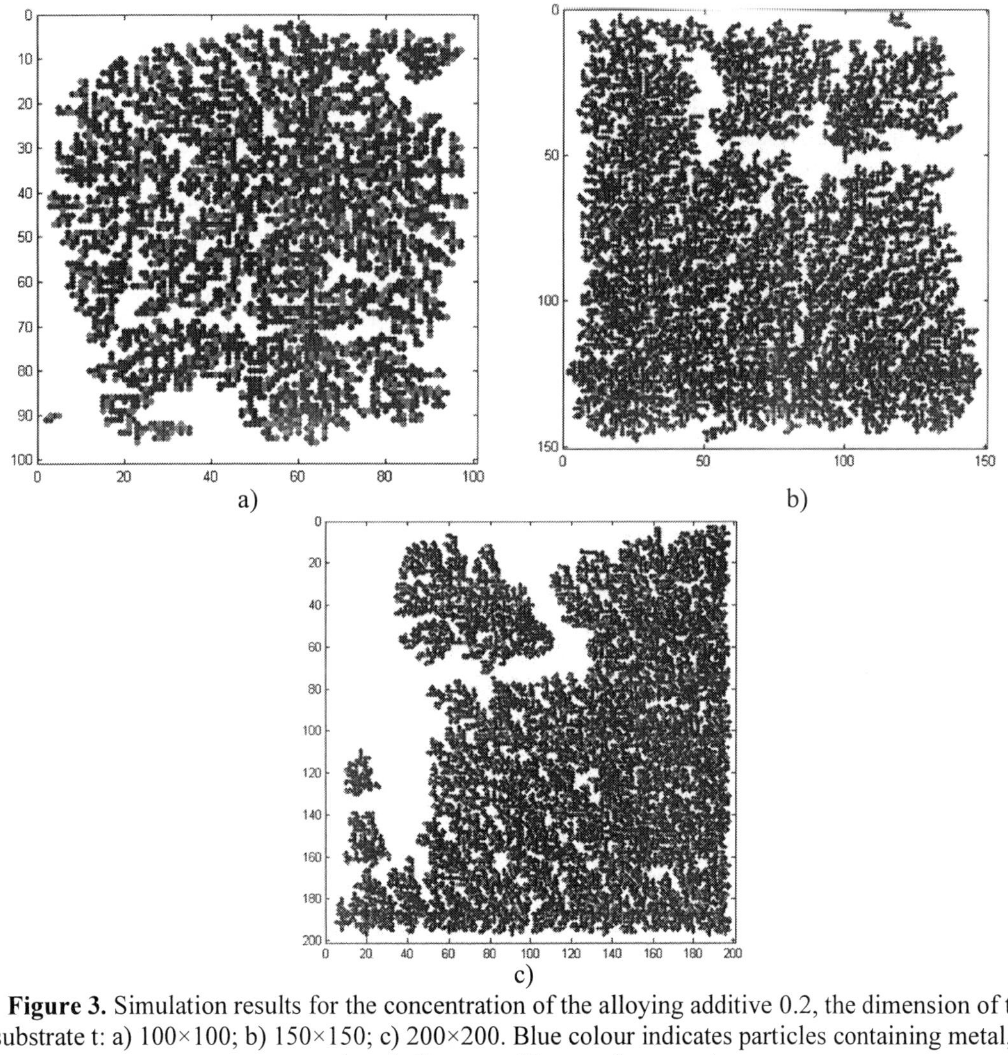

Figure 3. Simulation results for the concentration of the alloying additive 0.2, the dimension of the substrate t: a) 100×100; b) 150×150; c) 200×200. Blue colour indicates particles containing metal, red colour indicates a silicon-carbon matrix

The experiment on the calculation of dimension was carried out on a four-connected lattice of size $n*n$, at n=100 and more. According to the simulation results, with an increase in the number of particles per unit surface, the value of the fractal dimension increases uniformly, which may be due to an increase in the number of particles distributed over the surface area.

The modeling was implemented taking into account three different concentrations of ions hindering the growth of clusters. The concentration of ions interfering with the growth of clusters was chosen in relation to the total number of particles (for particles in the amount of 100n the concentration of interfering ions was 15n, 25n, 35n). The results obtained allow us to speak about the dependence of the onset of percolation, i.e. the probability of cluster formation, on the number of particles at different pH. According to experimental data, it can be concluded that the probability of percolation does not depend on the pH of the medium at the number of initial particles 20n. At a lower concentration of the initial particles with increasing pH, the formation of a percolation cluster occurred earlier than in the case when the medium is not taken into account.

4. Conclusion

Quantum chemical studies have shown the possibility of embedding metal atoms in a silicon-carbon matrix, which will subsequently determine the modes of the process in the formation of doped silicon-carbon coatings.

After studying the evolution of two- and three-layer aggregates with increasing temperature using quantum chemistry methods, it was determined that at the initial stage of the structure growth, hydrogenated aggregates are looser, this will contribute to the penetration of metal ions into the interlayer space and access to oxygen and nitrogen atoms. Further, the communication of additional energy to the system allows the formation of metal oxides and nitrides while maintaining sp^3-hybridization of most carbon atoms than in the absence of an alloying component.

On the basis of standard models and algorithms for determining fractal and percolation clusters, the influence of input data on the final structure of the films is determined. It was also found that the pH of the medium significantly affects the density of the film structure, the growth of the percolation cluster and the fractal dimension.

Acknowledgements

This work was granted in part by the Ministry of Education and Science of the Russian Federation; under Contract №14.575.21.0126 (unique identifier of the contract is RFMEFI57517X0126).

References

[1] Malinkovich M D, Parkhomenko Yu N, Polyakov D S and Shupegin M L 2011 *Russian Microel.* **40** 591
[2] Kim S, Choi J, Jung M, Joo S and Kim S 2013 *Sensors* **13** 13575
[3] Kohsuke Yoshinaga et al 2013 *J. Phys.: Conf. Ser.* **441** 012040
[4] Zyubin A S, Zyubina T S, Dobrovol'skii Y A and Volokhov V M 2016 *Russ. J. Inorg. Chem.* **61** 48
[5] Dykstra C, Frenking G, Kim K and Scuseria G 2005 *Theory and Applications of Computational Chemistry: The First Forty Years* (Elsevier: Amsterdam)
[6] Gerber R B, Shemesh D, Varner M E, Kalinowski J and Hirshberg B 2014 *Phys. Chem. Chem. Phys.* 16 9760
[7] Lyubartsev A P, Martsinovskii A A, Shevkunov S V and Vorontsov-Velyaminov P N 1992 *J. Chem. Phys.* 96 1776
[8] Shmatko V A, Yalovega G E and Myasoedova T N 2015 *Adv. Mater. Res.* **1091** 15
[9] Moruzzi R B, Oliveira A L, Conceicao F T, Gregory J and Campos L C 2017 *Sci. Total Env.* **609** 807
[10] Song Y and Zheng Q 2016 *Progr. Mater. Sci.* **84** 1

IOP Conf. Series: Materials Science and Engineering **447** (2018) 012068 doi:10.1088/1757-899X/447/1/012068

Preparation of graphene nanostructured films for photovoltaics

N Ibrayev, E Seliverstova, A Zhumabekov

Institute of Molecular Nanophotonics, Buketov Karaganda State University, Universitetskaya str. 28, Karaganda 100028, Kazakhstan

E-mail: niazibrayev@mail.ru

Abstract. The results of the development of technology for the preparation of graphene films for solar photocells by the Langmuir-Blodgett method are presented. A comprehenvive study of the dependence of structural, optical, electrophysical and photovoltaic properties on the chemical composition and conditions for the preparation of films based on graphene oxide, reduced graphene oxide and N-doped graphene oxide is carried out. It is shown that using the technology of applying graphene nanostructures to solid substrates by the Langmuir-Blodgett method leads to a decrease in the amount of platinum catalysts used for electrodes of solar devices.

1. Introduction

Modern nanotechnologies have become an integral science that includes all aspects of the preparation, fabrication and application of nanomaterials. Recently, in this field, success has been achieved in the synthesis of many materials, including two-dimensional (2D) nanostructures, which represent a unique target for research.

High conductivity and a high ratio of surface area to volume are the most significant characteristics of layered graphene nanostructures [1]. In addition, it was shown [2, 3] that graphene has a higher oxidation potential than Pt, which is a traditional material used in dye-sensitized solar cells (DSSC). Due to the variety of existing forms of graphene, it is possible to improve the necessary parameters of the counter electrodes in solar cells.

Typically, chemical vapor deposition (CVD) technology [1], aerosol spraying [2] and polyione assembling [3] are used for the deposition of graphene layers. Polyione assembling and CVD are a complicated procedure that requires additional preparation of substrates and a template for growing of graphene films. Furthermore, there is the problem of controlling the thickness of the synthesized coatings, which is a negative factor.

Langmuir-Blodgett technique allows you to control the thickness of the films, as well to predict the orientation of the particles within it. For this project, this means that the combination of possible variations in the number of layers and the distance between the individual graphene sheets enable to control not only the conductivity and resistance of the films, but also to adjust the surface area of the film. This opens up the possibility of creating a highly efficient catalyst graphene layer with high conductivity for using in photovoltaic cells.

This paper presents the results of the development of technology for the production of graphene films for solar cells by the Langmuir-Blodgett method (LB). A comprehensive study was carried out on the dependence of the structural, optical, electrophysical, and photovoltaic properties of films on the chemical composition and preparation conditions. It is shown that the use of the technology of deposition of graphene nanostructures on solid substrates by the Langmuir-Blodgett method leads to a decrease in the amount of platinum catalysts used for electrodes of solar devices.

2. Experiment

Single-layer graphene oxide (SLGO, Cheaptubes), reduced graphene oxide (rGO, Cheaptubes), and nitrogen-doped graphene oxide (NGO) (Sigma-Aldrich) were used as the objects of the study. Monolayers of graphene derivatives were formed by spreading from a solution in the Langmuir trough (KSV Nima). The isotherms of monolayer compression at the water-air interface were measured and investigated. To stabilize the monolayer, it was held for 40 minutes. As a subphase, deionized water was used, purified using the AquaMax water purification system. The specific resistance of water was equal to 18.2 MΩ/cm. The surface pressure was equal to 72.8 mN/m at pH = 5.6 and a temperature of 22 °C.

From measurements of π-A isotherms, surface pressures were chosen as the pressure of monolayer transfer onto a solid substrate, when the sheets of the corresponding graphene derivative contact each other, but do not form significant folds and "wrinkles" on the surface of the film. In this case the monolayer is in a condensed liquid state [4-6].

For the SLGO, the method of preparation of films was used, which was described in detail in [4, 6]. The solid films were obtained by the Y-type transfer at 19 mN/m, since such films are more homogeneous in structure than the Z-type films [4]. The velocity of the substrate motion through the monolayer was equal to 8 and 19 mm/min, and the compression rate of the monolayers of 5 mm/min.

For the preparation of NGO films, data on the phase state of monolayers and their stability, published in [5], were used. In particular, it was established from the isotherms of compression of monolayers that the monolayer is predominantly in the gaseous state in the pressure range from 0 to 2 mN/mm. With further compression of the monolayer, the particles come closer to each other and the film becomes liquid. When comparing the curves recorded for NGO and SLGO, it can be concluded that the behavior of the monolayers for both derivatives of graphene is the same. Solid films of NGO were obtained by the LB method by the Y-type transfer at 10 mN/m. The monolayer was compressed at a barrier velocity of 5 mm/min, the substrate rate was equal to 2 mm/min, and the monolayer stabilization time was equal to 30 minutes.

The rGO films were transferred at a surface pressure of 10 mN/m. The thickness of the films was equal to 1 monolayer, transferred along the Z-type. The monolayer was compressed at a barrier velocity of 5 mm/min, the substrate rate was equal to 2 mm/min, and the monolayer stabilization time was equal to 30 minutes.

For the deposition of graphene films, glass plates coated with an FTO layer were used, on which 1 monolayer of SLGO, rGO or NGO was deposited, as well as layered films based on Pt catalysts+graphene oxides. A platinum catalyst was electrochemically applied from an ethanolic solution of H_2PtCl_6 (Sigma Aldrcih).

The morphology of the films was studied with a Mira 3 (Tescan) scanning electron microscope (SEM). The absorption and transmission spectra of the films were measured with a Cary-300 (Agilent) scpetrophotometer.

The preparation, assembly and measurement of DSSC cell parameters with graphene electrodes was carried out according to the procedure detailed in [7]. Measurements of the electrophysical properties were carried out on a Z-500PRO impedance meter (Elins, Russia) under irradiation of cells with standard simulated solar radiation (Air Mass (AM) 1.5).

3. Results and discussion

For the SLGO it was shown [4, 6] that the packing density of particles within monolayers of a single-layer graphene oxide decreases in the series of DMF-acetone-THF dispersions, although the average particle size in solutions of acetone and THF differs insignificantly. The structural heterogeneity of the surface of graphene oxide films (figure 1 a) can be explained by the result of the simultaneous influence of electrostatic interactions between graphene oxide particles and Van der Waals forces.

SLGO 8 mN/m a NGO 5 mN/m b rGO 10 mN/m c

Figure 1. SEM images of the graphene oxide LB films

When studying the effect of transfer pressure of a monolayer of graphene oxide on the surface of solid substrates and the structural features of LB films, it was found [4] that the film obtained at low surface pressure is more homogeneous in structure, but its optical density was on 30% smaller than that of the film, obtained at high pressure.

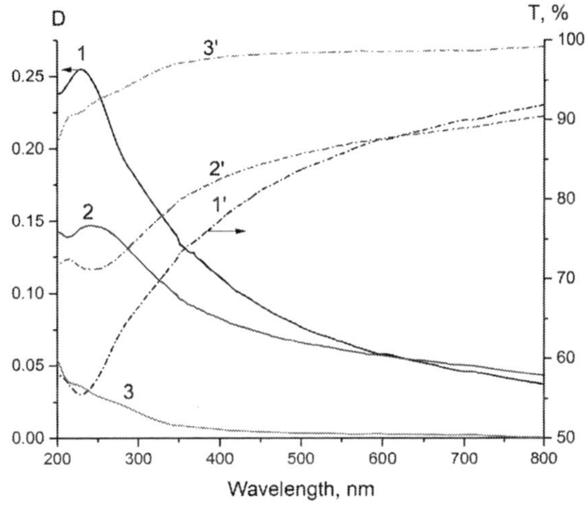

Figure 2. Absorption (1 – 3) and transmission (1' – 3') spectra of LB films based on SLGO (1, 1'), NGO (2, 2') and rGO (3, 3')

Also, the technology of films preparation by the Langmuir-Blodgett method on the basis of graphene oxide doped with nitrogen atoms was developed. SEM measurements showed (figure 1 b) that NGO forms island films. It is possible to distinguish multilayer structures formed by the imposition of individual sheets of NGO in the structure of films transferred at low surface pressures. The synthesized NGO films, like SLGO, have high transparency in the visible spectral range – from 80 to 90% (figure 2), which makes them attractive for various technological applications.

Studying the influence of the preparation conditions of LB films of reduced graphene oxide have shown that rGO forms uniform island films with clearly discernible clusters, whose density increases with increasing surface pressure of monolayer transfer onto solid substrate (figure 1 c). An increase in the surface pressure leads to an increase in the optical density of the absorption band of the reduced graphene oxide. The synthesized rGO films have high transparency in the visible spectral range - from 87 to 96%.

The study of the electrophysical properties of LB films based on SLGO, rGO, and NGO, have shown that LB films prepared from rGO and SLGO with transfer pressures of 10 and 19 mN/m, respectively, have the best electrophysical and conductive parameters.

The obtained LB films were used as counter electrodes in DSSCs. As shown by the measurements (table 1), the use of composite electrodes Pt+graphene oxide is more promising compared to the counter electrodes based on pure derivatives of graphene oxide.

Table 1. Photovoltaic parameters of DSSC with various counter electrodes

Counter electrode	U_{oc}, V	J_{sc}, mA/cm^2	FF	Efficiency, %
Pt	0.67	9.7	0.67	4.4
SLGO	0.41	5.3	0.13	0.31
rGO	0.52	6.0	0.38	1.28
NGO	0.39	3.2	0.10	0.2
Pt+SLGO	13.61	0.262	0.400	3.72
Pt+rGO	13.72	0.404	0.753	3.94
Pt+NGO	13.70	0.505	0.484	2.44

Among the graphene oxide derivatives studied, the photocatalytic activity of composite electrodes decreases in the series of Pt –Pt+rGO – Pt+SLGO – Pt+NGO, which is associated with a low recombination coefficient and a long lifetime of electrons in composite electrodes.

4. Conclusion

Thus, the studies have shown that a graphene oxide film in combination with a platinum catalyst is more promising for photovoltaic cells, than neat graphene oxide electrodes. In our work, as well as in the works [8, 9] it was shown that a graphene-coated Pt composite film provides a large and superior conductive interface for significant improvement of Pt utilization efficiency and charge transfer, which, in turn, leads to a higher power conversion efficiency of the DSSC. As it was compared to a conventional Pt-coated counter electrode, the Pt loading of a composite electrode can be reduced by more than 60% while achieving even better performance.

Graphene oxide LB films, despite the advantages in electrotransport properties, have a sufficiently high resistance, which is reflected in the number of charge carriers injected into the semiconductor layer from dye molecules and in the complex decrease in the efficiency of solar cells.

When studying the photovoltaic properties of solar cells with counter electrodes based on SLGO LB films, it was clarified that the use of Pt + rGO composite electrodes is more promising than the counter electrodes based on pure SLGO or rGO. Among the graphene oxide derivatives represented, the photocatalytic activity of composite electrodes decreases in the series of Pt –Pt+rGO – Pt+SLGO – Pt+NGO, which may be due to the low recombination coefficient and the long lifetime of electrons in the composite electrodes.

Acknowledgments

This work was performed at financial support of Ministry of Education and Science of the Republic of Kazakhstan.

References

[1] He M, Jaehan J and Feng Q 2012 *J. Mater. Chem.* **22** 24254
[2] Cruz R, Pacheco Tanaka D and Mendes A 2012 *Solar En.* **86** 716
[3] Ting-Ting M, Ze-Bao Zh and Ke-Zhi W 2013 *Langmuir* **29** 14314
[4] Seliverstova E, Ibrayev N Kh and Dzhanabekova R Kh 2017 *Rus. J. Phys. Chem. A* **91** 1761
[5] Dzhanabekova R Kh, Seliverstova E V and Ibrayev N Kh 2017 *Bull. Karaganda Univer. Phys.*

series **1(85)** 8
[6] Seliverstova E V, Ibrayev N Kh and Dzhanabekova R 2016 *Nanosystems: Phys., Chem., Math.* **1** 65 (In Russian)
[7] Ibrayev N Kh, Seliverstova E, Nuraje N and Ishchenko A A 2015 *Mat. Sci. Semicond. Proces.* **31** 358
[8] Guai G H, Song Q L, Guo Ch X, Lu Zh S, Chen T, Mang Ch and Li Ch M 2012 *Solar Energy* **86** 2041
[9] Gentian Y, Jihuai W, Yaoming X, Miaoliang H, Jianming L, Leqing F and Zhang L 2013 *Electrochimica Acta* **92** 64

IOP Conf. Series: Materials Science and Engineering **447** (2018) 012069 doi:10.1088/1757-899X/447/1/012069

Synthesis of iron oxide based nanostructures with antimicrobial activity

A N Fomenko, O V Bakina, S O Kazantsev and A M Kondranova

Institute of Strength Physics and Materials Science SB RAS, Tomsk, Russia

E-mail: alserova@ispms.tsc.ru

Abstract. This paper explores the antimicrobial activity of AlOOH and FeOOH nanostructures, and FeOOH/AlOOH composite nanostructures. The synthesized nanostructures have been examined by transmission electron microscopy, X-ray diffraction, low-temperature nitrogen adsorption, and microelectrophoresis. Antimicrobial activity is assessed by a 96-well plate method. It is shown that the synthesized nanoparticles have varying degrees of antibacterial effect on bacterial strains of E.coli and methicillin-resistant Staphylococcus aureus (MRSA). The FeOOH/AlOOH nanostructures possess the highest antibacterial activity, which indicates a synergistic antimicrobial effect of composite nanostructures.

1. Introduction

Today, the problems associated with the emergence of resistant strains of microorganisms and uncontrolled use of antibiotics stimulate the search for new approaches to prevent microbial contamination and its consequences [1]. In view of the pronounced geno- and cytotoxicity of metal nanoparticles, more attention is given to metal oxide nanostructures with low toxicity, such as iron, aluminum, magnesium, and others [2]. Owing to the developed surface and unusual crystalline morphologies [3], such structures can efficiently eliminate pathogens.

The use of iron oxide nanostructures as an antibacterial material is attractive due to their non-toxicity, biocompatibility, and unique magnetic properties. Irshad et al. [4] showed that the inhibitory activity of iron oxides is based on their electrostatic interaction with the bacterial cell membrane, which causes the cell damage in the contact region and induces toxic oxidative stress due to the production of intracellular reactive oxygen species. However, a persistent antimicrobial effect can be achieved at high concentrations of iron oxide nanostructures (> 10 mg/mL), which limits their use. Low antimicrobial activity is associated, first of all, with the agglomeration of nanostructures owing to their high surface energy and magnetization. There are two approaches to enhance the antibacterial activity of iron oxide nanostructures:

1. Surface modification by highly effective antibacterial drugs, e.g., antibiotics [5];
2. Development of antibacterial nanocomposites that would stabilize and preserve the original structure of iron oxide nanostructures [6].

The use of alumina as a component of antibacterial nanocomposites is attractive due to its high adhesion to cell membranes [7]. It adsorbs well both microorganisms and endotoxins [8]. Therefore, we can expect a synergistic antimicrobial effect of the composite, which contains alumina and iron oxide nanostructures. Earlier, we obtained effective antibacterial composite nanostructures AlOOH/Zn [9] and AlOOH/Cuby electric explosion of conductors [10]. The aim of this work is to synthesize FeOOH/AlOOH composite nanostructures and to study their physicochemical characteristics and antimicrobial activity against gram-positive and gram-negative microorganisms, including the antibiotic resistant strain MRSA.

2. Materials and methods

2.1 Synthesis of AlOOH nanostructures

AlOOH nanostructures were synthesized by a one-step route via the facile reaction of AlN/Al composite nanoparticles with water using the procedure detailed in Ref. [11]. AlN/Al nanoparticles were produced by electric explosion of aluminum wire in a nitrogen atmosphere [12].

2.2 Synthesis of FeOOH nanostructures
FeOOH nanostructures were synthesized by the chemical deposition method, which is easily scaled up and does not require the presence of additional toxic solvents, which is crucial for biomedical applications [13]. CP grade reagents were used. The FeOOH synthesis procedure was the following: 90 mL of 1 M sodium sulfate (Na_2SO_4) solution was mixed with 20 mL of 0.5 M $FeCl_3$solution with constant stirring and stirred for 20 minutes; then 80 mL of distilled water was added. The resulting suspension was placed into an autoclave with a Teflon insert and heated to 140 °C for 6 hours. The resulting precipitate was washed with deionized water and dried at 120 °C.

2.3 Synthesis of FeOOH/AlOOH composite nanostructures
FeOOH/AlOOH composite nanostructures were synthesized by adding 6 g of AlOOH at the stage of mixing $Na_2SO_4$and $FeCl_3$, produced according to Paragraph 2.1. After that, the synthesis was carried out according to Paragraph 2.2.

2.4 Nanoparticle characteristics
The synthesized nanostructures were examined using X-ray diffraction methods in Cu Kα radiation (XRD-6000, Shimadzu, Japan), microelectrophoretic analysis (Zetasizer Nano ZSP, UK), low-temperature nitrogen adsorption (Sorbtometer-M, Katakon, Russia), and transmission electron microscopy (JEM-100, JEOL, Japan).

2.5 Antibacterial activity assay
The antibacterial activity of the synthesized nanostructures was studied using the 96-well plate method [14]. The bacterial strains used were *E.coli ATCC 25922* and *clinical strain MRSA*, purchased at the Research Institute of Genetics and Selection of Industrial Microorganisms of the National Research Center Kurchatov Institute (Russia). Microorganisms were grown on meat-peptone agar and Mueller–Hinton broth at a temperature of 37±1 °C. The concentration of bacterial suspensions was approximately 10^7 CFU/mL. Each well of a 96-well plate was inoculated with 100 μL of Mueller–Hinton broth and 20 μL of thebacterial suspension. Suspensions of the investigated nanostructures in deionized sterile water in concentrations of 0.1, 0.075, 0.05, 0.025, and 0.0125 mg/mL were prepared in advance. The resulting suspensions were treated by a HielscherUP-100M ultrasonic homogenizer at a frequency of 30 kHz for 1 minute. The powder suspensions were added to wells (50 μL/well). Wells without nanostructures were used as controls. The plates were incubated at 37±1 °C for 12 hours. The growth kinetics of bacteria was estimated by the change in the optical density of the medium in wells relative to the control wells. The optical density was measured at 600 nm. Statistical processing of the results was carried out using parametric methods at a significance level of $p \leq 0.05$.

3. Results and discussion
Figure 1 shows the electron microscopic images and the main physicochemical characteristics of the synthesized nanostructures. TEM images of nanostructures (figure 1) demonstrate that the nanostructures synthesized by precipitation from $FeCl_3$ are porous agglomerates with a developed surface consisting of 2–5 nm iron oxide flakes (Sample 2). The precipitation of FeOOH nanostructures in the presence of AlOOH preserves the structure of AlOOH nanosheets, and FeOOH is evenly distributed over the nanosheet surface (Sample 3). The study of the samples by the thermal nitrogen desorption method showed that Samples 1 and 3 mostly have mesopores, whereas Sample 2 is microporous. The precipitation of FeOOH on AlOOH sheets increases the specific surface area (SSA) of the sample to 343 m^2/g.

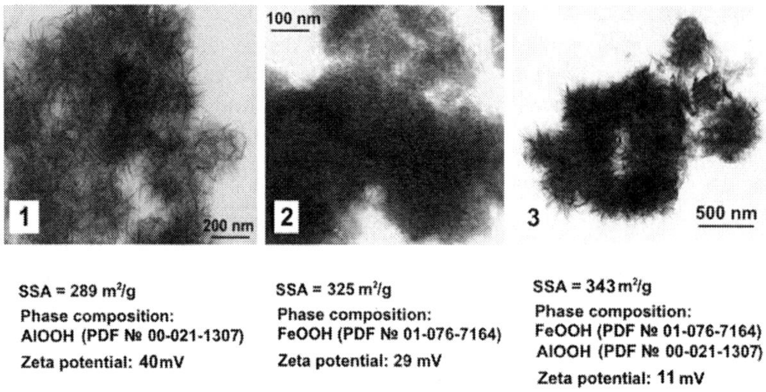

Figure 1. TEM images and characterization of nanostructures: 1 –AlOOH nanostructures; 2 – FeOOH nanostructures; 3 –FeOOH/AlOOH nanostructures.

The antibacterial activity of the AlOOH and FeOOH nanostructures has been separately investigated in a large number of works. It can be expected that composite nanostructures will exhibit higher activity. Our studies revealed that all samples in concentrations up to 0.075 mg/mL have weak antimicrobial effect on *MRSA* (figure 2). The activity of the samples increaseswith increasing concentration, and a pronounced bactericidal effect is achieved at a concentration of 0.1 mg/mL of FeOOH/AlOOH composite nanostructures (figure 2 c).

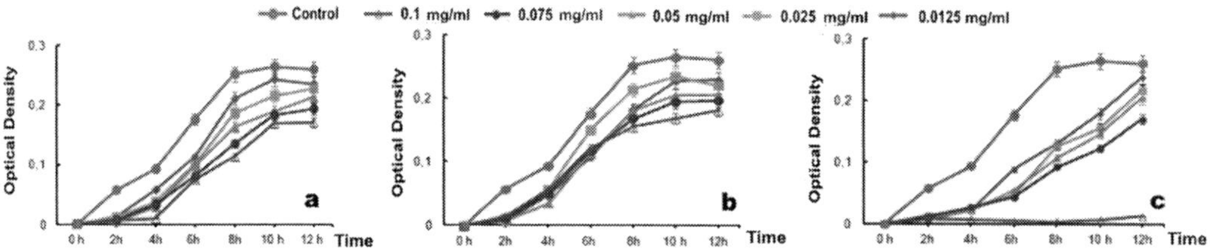

Figure 2. Effect of nanostructures on *MRSA*: a - AlOOH nanostructures; b - FeOOH nanostructures; c - FeOOH/AlOOH nanostructures.

The *E. coli* culture was more sensitive and less selective with respect to all three types of nanostructures (figure 3).

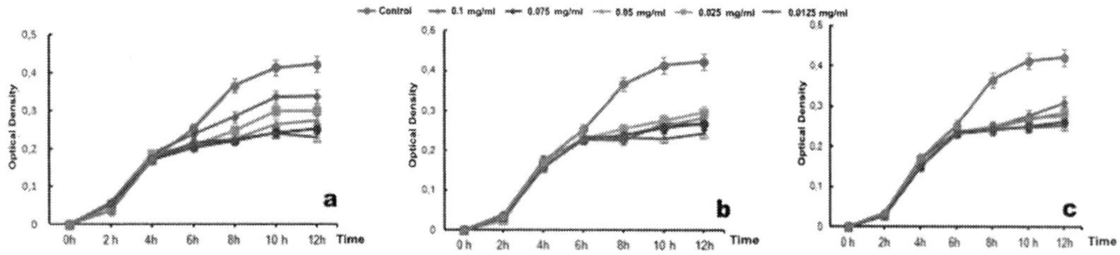

Figure3. Effect of nanostructures on *E.coli*: a - AlOOH nanostructures; b - FeOOH nanostructures; c - AlOOH/FeOOH nanostructures.

AlOOH nanostructures at concentrations of 0.0125 and 0.025 mg/mLslightly inhibit the growth of living *E. coli* cells, but this effect is enhanced with increasing concentrations to 0.1 mg/mL (figure 3 a). The antibacterial effect of FeOOH and FeOOH/AlOOH nanostructures is almost independent of the concentration. After 6 hours of strain incubation with nanostructures, the growth of *E. coli* stops.

4. Conclusions

FeOOH/AlOOH composite nanostructures with high antibacterial activity against gram-positive (*MRSA*) and gram-negative (*E. coli*) bacterial strains were synthesized. Low toxicity and antibacterial efficacy makes them promising candidates for biomedical applications.

Acknowledgements

The work was carried out with the financial support of the Ministry of Education and Science of Russia (Agreement No. 14.604.21.0156, Project Identifier RFMEFI60417X0156).

References

[1] Frieri M, Kumar K and Boutin A 2017 *J. Infect. Public Health* **10** 369
[2] Hoseinzadeh E, Makhdoumi P, Taha P, Hossini H, Stelling J and Amjad Kamal M 2017 *Curr. Drug Metab.* **18** 120
[3] Thukkaram M, Sitaram S and Subbiahdoss G 2014. *Int. J. Biomater.* **13** 213
[4] Irshad R, Tahir K, Li B, Ahmad A, Siddiqui A R and Nazir S 2017 *J. Photochem. Photobiol. B: Biology* **170** 241
[5] Bhattacharya P, Neogi S 2017. *Mat. Res. Expr.* **4(9)** 095005
[6] Nehra P, Chauhan R P, Garg N and Verma K 2018 *Brit. J. Biomed. Sci.* **75(1)** 13
[7] Svarovskaya N V, Bakina O V, Glazkova E A, Fomenko A N and Lerner M I 2017 *Progr. Natur. Sci.: Mat. Int.* **27(2)** 268
[8] Manyasree D,Kiranmayi P andRavi Kumar R V S S N 2017. *Int. J. Pharm. Pharmaceut. Sci.* **10(1)** 32
[9] Lozhkomoev AS, Glazkova EA and Svarovskaya NV 2014 *AIP Conf. Proc.* **1623** 367
[10] Svarovskaya N V, Berenda A V, Bakina O V, Glazkova E A, Lozhkomoev A S, Khorobraya E G and Fomenko A N 2015 *Progr. Natur. Sci.: Mat. Int.* **25(1)** 1
[11] Bakina O V, Svarovskaya N V, Glazkova E A, Lozhkomoev A S, Khorobraya E G and Lerner M I 2015 *Adv. Powder Tech.* **26(6)** 1512
[12] Lerner M I, Glazkova E A, Lozhkomoev A S, Svarovskaya N V, Bakina O V, Pervikov A V and Psakhie S G 2016 *Powder Tech.* **295** 301
[13] Unsoy G, Yalcin S, Khodadust R, Gunduz G and Gunduz U 2012. *J. Nanopart. Res.* **14(11)** 964
[14] Balouiri M, Sadiki M and Ibnsouda S K 2016 *J. Pharmaceut. Anal.* **6(2)** 71

Synthesis and properties of porous nanostructured iron oxide

S O Kazantsev and A M Kondranova

Institute of Strength Physics and Materials Science SB RAS, Tomsk 634055, Russia

E-mail: kzso@ispms.tsc.ru

Abstract. Iron oxide nanoparticles have been synthesized using various chemical methods. The morphology, specific surface area, zeta potential, and acid-base properties of the surface of synthesized nanoparticles are studied. It is shown that the produced samples have a different specific surface area and a positive zeta potential. Brönsted basic sites with pKa = 10.4 and pKa = 12.8 are predominant on the surface of the synthesized nanoparticles.

1. Introduction

Nanodispersed metal oxides are promising materials for the development of new functional coatings, composites, catalysts, and biomedical materials [1–3] because of their specific morphology [4] and specific sorption properties of their surface [5]. For example, magnetic iron oxide nanoparticles are very important for scientific and clinical applications owing to their magnetic properties, low toxicity, and biocompatibility. Iron oxide nanoparticles are used in various industries [6, 7]. The literature contains data on the removal of various pollutants from liquids using magnetic nanoparticles, as well as nanoparticles with a modified surface for better adsorption capacity [8, 9]. The material most often used for these purposes is iron oxide, not pure iron, because iron oxides are more stable to oxidation as compared to metallic iron. Therefore, their magnetic properties and surface properties remain unchanged for a long time. There are various methods for synthesizing nanoparticles , such as microemulsion [10], thermal decomposition [11], co-precipitation method [12], hydrothermal method [13], ionothermal synthesis [14], etc. The variety of methods allows producing nanoparticles with different morphologies: nanosheets [15], nanoflowers [16], octahedral particles [17], microrings [18], etc.

In this paper, various chemical methods were used to synthesize nanostructured iron oxide particles. The morphology, composition, textural and surface properties of the obtained oxides has been studied.

2. Materials and methods

Iron oxide nanoparticles were derived from the inorganic salt using various methods.

Precipitation from salt solution. 50 ml of 0.5 M $FeCl_3{\times}6H_2O$ salt solution was mixed with 1 M NH_4OH solution, added in portions with constant stirring until a pH of about 9-10 was reached. The solution was stirred after adding each portion to bring about equilibrium. Then the solution was filtered, washed, and dried at 120 °C for 2 hours.

Hydrothermal oxidation. 90 mL of 0.1 M Na_2SO_4 solution was placed in a 250 mL beaker, and 20 mL of 0.5 M $FeCl_3$ solution was added to it with constant stirring and stirred for 15-20 minutes. Then 70-80 mL of distilled water was added to make a homogeneous medium. The obtained solution was placed in a sealed autoclave and heated to 140 °C for 4 hours. After that, the autoclave was cooled to room temperature; the resulting precipitate was filtered off, washed, and dried at 120 °C for 2 hours.

The reaction products were examined using X-ray diffraction analysis in Cu Kα radiation (XRD-6000, Shimadzu, Japan), microelectrophoretic analysis (Zetasizer Nano ZSP, Malvern Instruments Ltd, England), low-temperature nitrogen adsorption (Sorptometer-M, Katakon, Russia), transmission electron microscopy (JEM-100, JEOL, Japan), and IR spectroscopy (Nicolet 5700, Thermo Electron, USA).

The acid-base properties of the surface were determined by an indicator method based on the adsorption of Hammett indicators on the surface of solid particles with subsequent spectrophotometric determination of the indicator concentration.

3. Result and discussion

Precipitation from the ferric chloride solution by means of alkali results in the formation of 5–10 nm nanoparticles arranged in agglomerates of various irregular shapes (figure 1 a) from 100 to 500 nm in size.

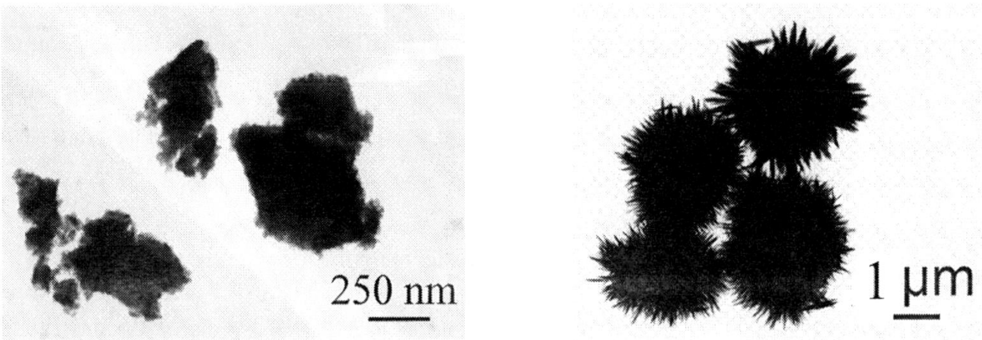

(a) (b)

Figure 1. TEM images of iron oxides synthesized by: a – precipitation from salt solution; b – hydrothermal oxidation.

Hydrothermal oxidation leads to the formation of rods with a size of 1–5 μm and a thickness of about 50 nm agglomerated into spherical particles from 1 to 5 μm in size.

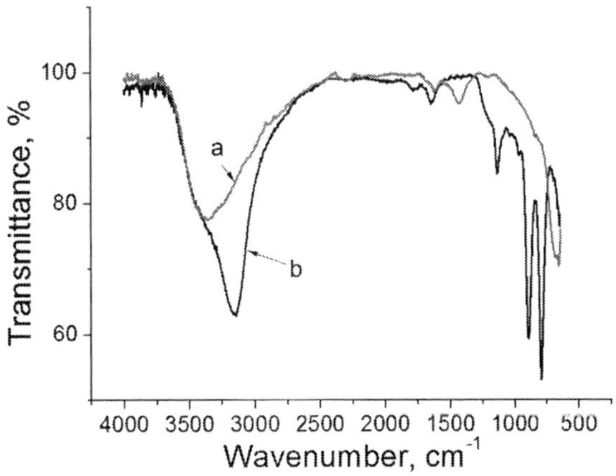

Figure 2. IR spectra of samples synthesized by: a – precipitation from salt solution; b – hydrothermal oxidation.

According to X-ray diffraction analysis, precipitation from salt solution by alkali produces hematite, and hydrothermal oxidation produces goethite (table 1).

IR spectroscopy for hematite Fe_2O_3 (figure 2a) demonstrates the presence of absorption bands of the Fe-O bond at 631 cm^{-1} characteristic for Fe_2O_3, H_2O deformation vibrations at 1622 cm^{-1}, and a broad band with an absorption maximum at 3422 cm^{-1} characteristic for the O-H valence vibrations. The IR spectrum of goethite FeO(OH) (figure 2 b) has a narrow band at 631 cm^{-1} characteristic for the Fe-O bond, OH vibrations at 797, 893 and 1787 cm^{-1}, H_2O deformation vibrations at 1645 cm^{-1}, and a wide absorption band at 3158 cm^{-1} characteristic for the O-H valence vibrations. In addition, the sample has a weak absorption band at 1160 cm^{-1}, which can be attributed to the SO_4^{2-} valence vibrations.

The study of the textural characteristics of the synthesized samples showed that nanoparticles precipitated from the ferric (III) chloride solution by ammonia have the largest specific surface (table 1). All synthesized iron oxides have positive but differing in magnitude zeta potentials in water (table 1).

Table 1. Properties of the synthesized iron oxide nanoparticles.

Synthesis method	S_{sp}, m^2/g	Zeta potential, mV	Phase composition
Precipitation from salt solution	325	29.2	Hematite Fe_2O_3 (01-076-8402)
Hydrothermal oxidation	45	6.2	Goethite FeO(OH) (01-076-7164)

The acid-base properties of the synthesized samples were studied (figure 3) to reveal that the surface of hematite nanoparticles exhibits Brönsted basic sites of different strength (figure 3 a) with pK = 7.4, pK = 10.5, pK = 12.8, and a small number of Brönsted acid sites predominantly at pK = 3.46. The FeO (OH) goethite surface mainly exhibits Brönsted basic sites with pKa = 10.5. The number of basic site of different strength depends on the type and environment of the OH^- groups on the sample surface. Weak Brönsted acid centers can be formed by water molecules that are adsorbed on metal cations or oxygen vacancies on the oxide surface.

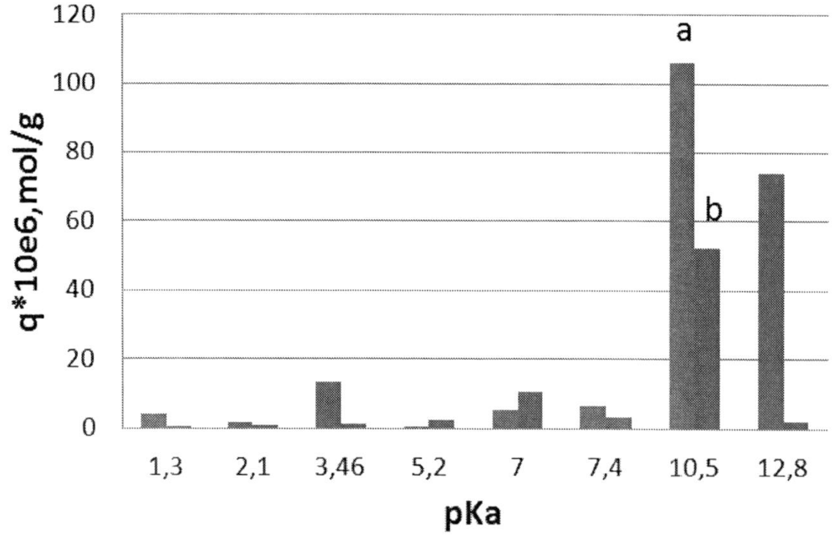

Figure 3. Change in the concentration of active sites on the surface of iron oxides: a – hematite; b – goethite.

Similar acid-base centers are present on the surface of goethite (figure 3 b), as on the hematite surface. However, their number is much smaller. This is most likely due to the phase composition of the reaction products and the morphology of the obtained samples.

4. Conclusions

Nanostructured iron oxide particles with different morphology, phase composition, and textural characteristics were synthesized. It was shown that iron oxide precipitation from the ferric chloride solution results in the formation of a powder with a high specific surface area. The study of the acid-base properties of the surface showed that water molecules and OH^- groups form active Brönsted sites of different strength on the surface depending on the environment. Brönsted basic sites with pKa = 10.4 and pKa = 12.8 are predominant on the surface of the synthesized samples. The obtained data suggest that the synthesized nanostructures can be used as carriers of drugs, including antibiotics, for the development of biomedical materials.

Acknowledgments
The work was carried out with the financial support of the Ministry of Education and Science of the Russian Federation (Agreement No. 14.604.21.0156, Project Identifier RFMEFI60417X0156).

References

[1] Badoga S, Sharma R V, Dalai A K and Adjaye J 2015 *Appl. Catal. A:Gen.* **489** 86
[2] Ciccotti L, do Vale L A S, Hewer T L R and Freire R S 2015 *Catal. Sci. Technol.* **5** 1143
[3] Lozhkomoev A S, Lerner M I, Tsukanov A A, Kazantsev S O, Bakina O V and Psakhie S G 2017 *Phys. Mesomech.* **20** 134
[4] Lozhkomoev A S, Glazkova E A, Bakina O V, Lerner M I, Gotman I, Gutmanas E Y and Psakhie S G 2016 *Nanotechnology* **27** 205603
[5] Tsukanov A A and Psakhie S G 2017 *Phys. Mesomech.* **20** 43
[6] Hu Z H , Kurien U , Murwira K, Ghoshdastidar A, Nepotchatykh O and Ariya P A 2016 *ACS Sustain. Chem. Eng.* **4** 2150
[7] Wang L, Yu J, Dong X T, Li X, Xie Y Z, Chen S H, Li P, Hou H Q and Song YH 2016 *ACS Sustain. Chem. Eng.* **4** 1531
[8] Moeser G D, Roach K A, Green W H, Laibinis P E and Hatton T A 2002 *Ind. Eng. Chem. Res.* **41** 4739
[9] Ponder S M, Darab J G, Bucher J, Caulder D, Craig I, Davis L et al 2001 *Chem. Mater.* **13** 479
[10] Lu T, Wang J H, Yin J, Wang A Q, Wang X D and Zhang T 2013 *Colloid Surf. A-Physicochem. Eng. Asp.* **436** 675
[11] Wang D, Ma Q and Yang P 2012 *J. Nanosci.Nanotechnol.* **12** 6432
[12] Wu S, Sun A Z, Zhai F Q, Wang J, Xu W H, Zhang Q and Volinsky A A 2011 *Mater. Lett.* **65** 1882
[13] Han C L, Zhao D F, Deng C H and Hu K H 2012 *Mater. Lett.* **70** 70
[14] Chen F, Xie S, Huang X and Qiu X 2017 *J. Hazard. Mater.* **322** 152
[15] Zhuang L, Zhang W, Zhao Y X, Shen H, Lin H and Liang J H 2015 *Sci. Rep.* **5** 9320
[16] Zhong L S, Hu J S, Liang H P, Cao A M, Song W G and Wan L J 2006 *Adv. Mater.* **18** 2426
[17] Ooi F, DuChene J S, Qiu G, Graham J O, Engelhard M H, Cao G X, Gai Z and Wei W D 2015 *Small* **11** 2649
[18] Zhong S L, Song J M, Zhang S, Yao H, Xu A W, Yao W T and Yu S H 2008 *J. Phys. Chem. C* **112** 19916

IOP Conf. Series: Materials Science and Engineering **447** (2018) 012071 doi:10.1088/1757-899X/447/1/012071

Nanocomposite material based on nanoporous oxide of aluminum with additives of silver and gold nanoparticles

N Kh Ibrayev, A K Aimukhanov and J B Usupova

E.A. Buketov Karaganda State University, Karaganda100028, Republic of Kazakhstan

E-mail: a_k_aitbek@mail.ru

Abstract. Porous films of anodized aluminum oxide with pore sizes of 50-60 nm were obtained. Silver nanoparticles in porous alumina have been synthesized by the method of controlled single-jet crystallization. The synthesis of gold nanoparticles in the pores was carried out by the reduction of $HAuCl_4$ with NaOH alkali in the pores of aluminum oxide.

1. Introduction

Due to theunique porous structure, the parameters of which (diameter, length and distance between neighboring pores) can be varied during the synthesis, porous alumina films can be used as inorganic membranes [1–3], calibration grids, a templating material for the synthesis of nanowires or nanotubes with a controlled diameter and high geometric anisotropy [4–6], as well as 2D photonic crystals [7, 8].

At present, the greatest use of porous aluminum oxide films has been found as matrices for the synthesis of ordered arrays of anisotropic nanostructures of various compositions. Magnetic materials, such as Fe, Co, Ni, obtained in the channels of porous Al_2O_3, are promising for the creation of a magnetic information storage medium. In this case, the possibility of changing the geometric anisotropy of nanostructures by varying the charge missed during the electrodeposition of the metal makes it possible to control the magnetic properties of nanowires [9].

Membranes of anodic aluminum oxide are a promising material for the creation of high-tech devices operating at elevated temperatures [10]. The coefficient of thermal conductivity of aluminum oxide films is 1.6 W/m×K, which is much higher than the thermal conductivity of polymeric materials and glasses used as solid-state matrices for dye lasers [11–13]. If the pores are filled with laser dyes, then such a system can be used as an active medium for a tunable laser.

2. Method of experiment

The following reagents were used to prepare porous alumina films: Al (0.5 mm thick sheet, 99.99%), oxalic acid $(COOH)_2$, H_3PO_4 conc., CrO_3, $CuCl_2$, HCl, ethanol C_2H_5OH (96%). In order to increase the size of Al crystallites, remove microstresses, and then achieve better ordering of the pores, aluminum substrates were annealed in air for 10 hours at 550 °C in a muffle furnace. For the preparation of solutions, double filtrated and deionized water is used on the AquaMax 360 Basic water purifier. The specific resistance of water was 18.2 MΩ / cm.

An important characteristic of metallic aluminum, which affects the process of ordering of a porous structure, is the roughness of its surface. The alignment of the aluminum surface was carried out by electrochemical polishing in a pulsed mode. Before polishing, the surface of the aluminum plate is cleaned with acetone. For the electrochemical polishing of the aluminum surface, an electrolyte is used, which has the following composition: CrO_3 (185 g/l) and H_3PO_4 (1480 g/l.). The aluminum plates were immersed in a 500 ml heat-resistant dish with the prepared solution at an electrolyte temperature of 80 °C and mixed with a magnetic stirrer (IKAMAG RET control-vise). For electrochemical polishing the following parameters were chosen: voltage - 11-13 V, current - 1.2 A. The polishing time of the aluminum surface lasted 15-20 minutes. Quality control of electrochemical polishing was evaluated by the appearance of the aluminum surface. The temperature of the mixer was 80 ° C, and the temperature of the solution was controlled by a thermocouple. A plate of aluminum (of the same composition and larger in area) was attached to the negative electrode, and a plate of aluminum for polishing was placed on the positive electrode. The current source was Agilent technologies N8740A 150V, 22A, 3300W. The following parameters were set: voltage limits of 11-12 V, current 1.2 A. After that, the samples were washed in distilled water and dried in air. It is noted that an increase in the duration of electropolishing leads to smoothing of the unevenness, but is accompanied by a significant dissolution of aluminum and a decrease in the thickness of the metal plate. Anodic oxidation of aluminum was carried out in a two-electrode electrochemical cell using direct current sources. The resulting porous aluminum oxide matrices were separated from the aluminum base by selective dissolution of the latter in a solution of 0.25 M $CuCl_2$ in water.

Synthesis of silver nanoparticles in porous alumina was carried out by the method of controlled single-jet crystallization. The method is based on the reduction of the silver ammoniac complex by a reducing agent, which was formic acid. To 25 ml of an aqueous solution of silver nitrate with a concentration of 0.4%, 1 pea of NaOH is added. The solution becomes muddy. After deposition of silver on the bottom of the glass, gently drain the top of the solution and wash 2-3 times with distilled water. To the resulting precipitate add 5 ml of deionized water. Then 400 μl of concentrated ammonia are added to the resulting solution and mixed, and the solution becomes clear as a result. In the resulting solution, preliminarily heated to 60 °C aluminum oxide film is dropped and 400 μl of formic acid is added. The reaction is carried out for 4-5 minutes, then the liquid is drained, the alumina film is washed with distilled water and dried at a temperature of 60 °C. The synthesis of different concentrations of nanoparticles in the film was carried out by varying the volume of formic acid added during the reaction.

For the synthesis of gold nanoparticles (AuNPs), we used solutions of $HAuCl_4$ (1 ml of Au solution per 9 ml of deionized water) and NaOH (1g NaOH per 25 ml of water). For the synthesis of AuNPs in aluminum oxide, the pH of the gold solution (initial pH = 1.5-1.95) it is necessary to bring to 10, adding a solution of NaOH (1N) dropwise. Then, the porous alumina film is immersed in a solution of $HAuCl_4$ (pH = 10) and heated at 80 °C for 2 hours. After this time, the color of the film changes from white to raspberry-pink (if the $HAuCl_4$ solution is concentrated, respectively, violet). This is a sign that the gold nanoparticles have formed in the pores. The synthesis of different concentrations of nanoparticles in the film was carried out by varying the volume of NaOH solution added during the reaction.

The synthesized nanostructures were studied using a Tescan Mira 3 LMU microscope at a 15 kV accelerating voltage and a working distance of 7 mm. Information on the local chemical (element) composition was obtained using an energy-dispersive detector, excited by fast electrons of characteristic X-ray radiation

3. Result and discussion

Figure 1 shows the current characteristic for the production of an oxide layer in oxalic acid. The formation of the first pores in the oxide film is most likely to occur in some defective places, for example, at the boundary of the crystallite, in the film cracks. Simultaneously with the growth of pores, the alumina layer begins to grow from the outer part of the film at the film-metal interface, the individual cells of which have the form of a hemisphere.

From these studies, it was found that changes in the current density are observed throughout the anodization process. There are three stages of changes in the current dependence:

Stage 1. When the current is turned on, there is a sharp increase in the current density due to the presence of a thin oxide film on the aluminum surface. At this stage, the thickness of the oxide layer increases.

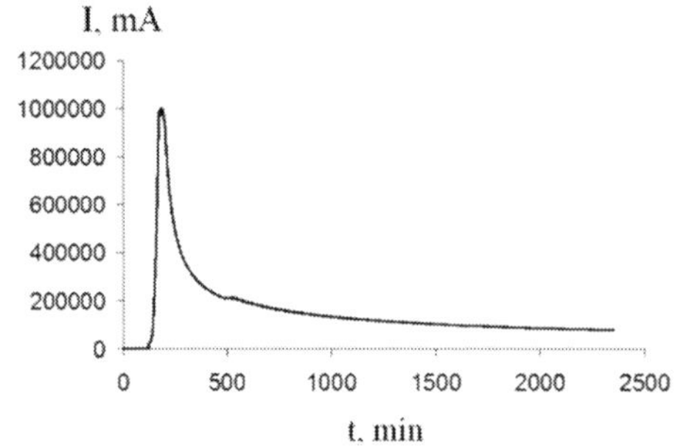

Figure 1. Current characteristic of the process of anodizing aluminum.

Stage 2. The value of the current density starts off fall. At this stage, the porous structure of anodized aluminum is formed, followed by an increase in the pore depth.

Stage 3. The value of the current density passes through a maximum and the current decreases. This occurs in connection with the achievement of the ultimate thickness of the porous alumina layer and the rupture of the upper pore layers.

SEM images of transverse cleavages of the samples and the results of determining the content of chemical elements on the surface of aluminum oxide are shown in figures 2 and 3, respectively. It can be seen that nanoparticles of silver (figures 2 a, 3 a) and gold (figures 2 b, 3 b) are formed on the walls of pores of nanoporous alumina.

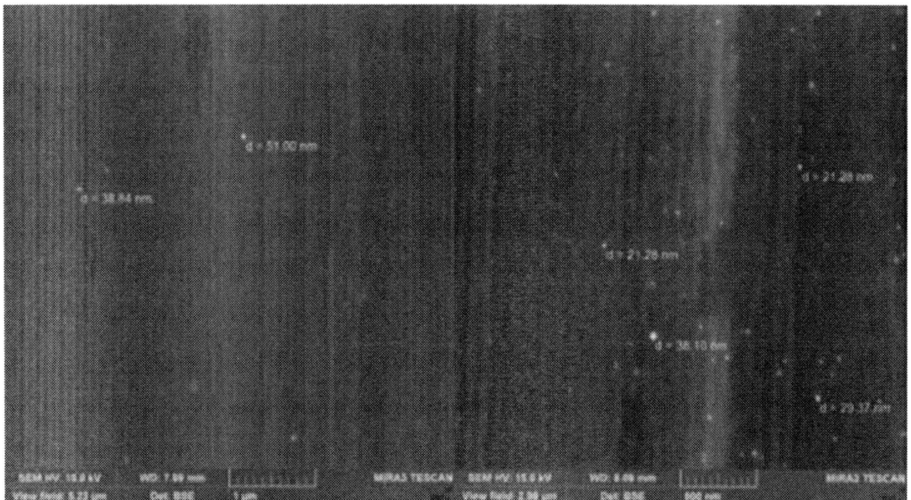

(a)(b)

Figure 2. SEM image of the transverse cleavage of an aluminum oxide film.

Figure 4 a shows absorption spectra of silver nanoparticles of different concentrations in a film of nanoporous alumina. It is seen from the figure that the absorption band of silver nanoparticles in the matrix has a maximum at a wavelength of 387 nm. The appearance of an absorption band with a maximum at wavelength 387 is indicative of the formation of nanoparticles in pores. The half-width of the spectrum is 43 nm, with a change in the concentration of nanoparticles in the film it was 36 nm.

The same figure (figure 4 b) shows the absorption spectra of gold nanoparticles of different concentrations in a film of nanoporous alumina. The appearance of an absorption band in porous aluminum with a maximum at a wavelength of 534 nm is indicative of the formation of gold nanoparticles in pores. It can be seen from the figure that the absorption band of gold nanoparticles in the matrix has a maximum at a wavelength of 534 nm. The half-width of the spectrum is 52 nm, with a change in the concentration of nanoparticles in the film, the half-width of the absorption spectrum was 42 nm.

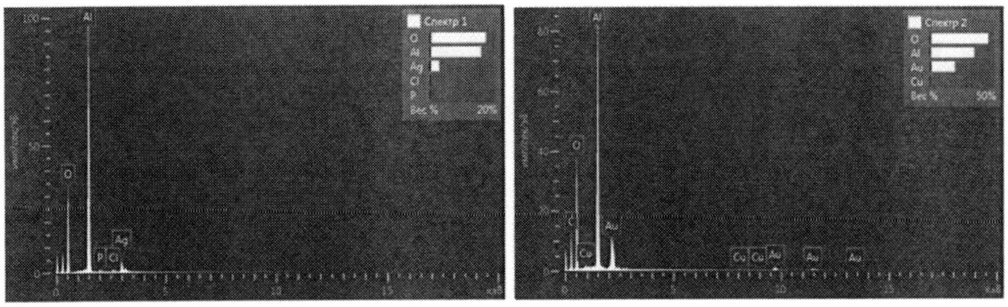

(a) (b)

Figure 3. Energy dispersive spectrum of elements on the surface of an aluminum oxide film.

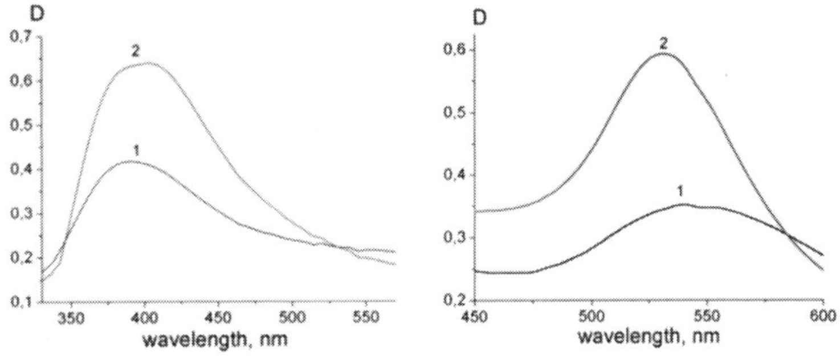

1-C_{Ag}=2•10^{12}mol/m^2;
2-C_{Ag}=7•10^{13}mol/m^2

1-C_{Au}=6•10^{11}mol/m^2;
2-C_{Au}=2•10^{13}mol/m^2

(a) (b)

Figure 4. Absorption spectrum of silver (a) and gold (b) nanoparticles in the matrix of nanoporous alumina.

4. Conclusion

Thus, as a result of the experiments, porous films of anodized aluminum oxide with pore sizes of 50-60 nm were obtained. The synthesis of silver nanoparticles in porous alumina was carried out by the method of controlled single-jet crystallization. The synthesis of gold nanoparticles in pores was carried out by restoring $HAuCl_4$ with alkali NaOH in the presence of a porous alumina film.

References

[1] Choi J, Wehrspohn R B and Gosele U 2005 *Electrochim. Acta* **50** 2591
[2] Inada T, Uno N, Kato T and Iwamoto Y 2005 *J. Mater. Res.* **20** 114
[3] Stair P C, Marshall C, Xiong G, Feng H, Pellin M J, Elam J W, Curtiss L, Iton L, Kung H, Kung M and Wang H H 2006 *Top. Catal.* **39** 181
[4] Strijkers G J, Dalderop J H J, Broeksteeg M A A, Swagten H J M and de Jonge W J M 1999 *J. Appl. Phys.* **86** 5141

IOP Conf. Series: Materials Science and Engineering **447** (2018) 012071 doi:10.1088/1757-899X/447/1/012071

[5] Napolskii K S, Barczuk P J, Vassiliev S Yu, Veresov A G, Tsirlina G A and Kulesza P J 2007 *Electrochim. Acta* **52** 7910
[6] Inguanta R, Piazza S and Sunseri C 2008 *Electrochim. Acta* **53** 5766
[7] Masuda M Y 2006 *Adv. Mater.* **18** 213
[8] Choi J, Luo Y, Wehrspohn RB, Hillebrand R, Schilling J and Gosele U 2003 *J. Appl. Phys.* **94** 4757
[9] Shingubara S 2003 *J. Nanopart. Res.* **5** 17
[10] Kukhta A V, Gorokh G G, Kolesnik E E, Mitkovets A I, Taoubi M I, Koshin Y A and Mozalev A M 2002 *Surface Sci.* **50** 7593
[11] Inada T, Uno N, Kato T and Iwamoto Y 2005 *J. Mater. Res.* **20** 114
[12] Mayer G V, Kopylova T N and Svetlichny V A 2007 *Quantum Electronics* **37** 53
[13] Knobbe E T, Dunn B, Fuqua P D and Nishida F 1990 *Appl. Optics* **29** 2729

IOP Conf. Series: Materials Science and Engineering **447** (2018) 012072 doi:10.1088/1757-899X/447/1/012072

Estimation of the influence of porous nanostructured materials on blood chemistry values

A S Lozhkomoev and S O Kazantsev

Institute of Strength Physics and Materials Science SB RAS, Tomsk, Russia

E-mail: asl@ispms.tsc.ru

Abstract. Nanostructured oxides and hydroxides of iron and aluminum have been synthesized. Their influence on the hematologic and blood chemistry values is examined. The synthesized samples are found to change the hematologic values in different ways. The boehmite nanostructures produced by water oxidation of AlN/Al nanoparticles exert the greatest influence on the blood chemistry values. The obtained results are important for a general understanding of how nanostructured metal oxides interact with blood components, and can be used in the development of new hemostatic agents on the basis of micro-mesoporous metal oxides.

1. Introduction

Blood clotting is a combination of biophysical and biochemical processes that allow the transformation of blood from a liquid to a clot. As a result of the trauma, a three-dimensional network of fibrin filaments is formed under the influence of activators (blood clotting factors). It holds blood under high pressure, being the main mechanism to arrest bleeding in humans.

The formation of afibrin clot depends on a number of variables, such as pH, the ionic strength, and the concentration of calcium ions, fibrinogen, and thrombin [1]. The effect of ions of the biological medium (K^+, Na^+, Cl^-, HCO_3^-, HPO_4^-, SO_4^{2-}, Ca^{2+}, Mg^{2+}) and pH values on the kinetics of fibrin clot formation was studied elsewhere [2]. It was shown that monovalent ions and SO_4^{2-} exert no influence on the formation of the clot, while calcium and magnesium ions substantially increase the viscosity of the clot. The study of the effect of ionic strength and pH on the rate of fibrin clot formation showed that their growth decreases the viscosity of the formed fibrin network.

Platelets and platelet clots aggregate much faster in the presence of calcium, while in the absence of calcium clots appear almost transparent. This is explained by the fact that calcium is crucial for the assembly of procoagulant complexes and the generation of thrombin in vivo. As a result, coagulation begins earlier, and therefore the formed fibrin fibers are thicker [3].

When nanoparticles come into contact with blood plasma, they are immediately covered with a protein corona, which can affect the morphology and abundance of human blood corpuscles [4]. The functioning of coagulation cascades and the complement system can be disturbed by interactions of nanoparticles with protein factors and a reduction of the intact protein in plasma. It is known that the concentrations of coagulation factors in plasma are rather low, e.g., 20, 10, and 200 nM for factors V, VII, and X, respectively, while the fibrinogen concentration is much higher, 9 μM. Therefore, even a slight adsorption of one of the coagulation factors on the surface of nanoparticles can noticeably shift the equilibrium of cascade reactions.

The use of materials whose surface would be able to change the ionic balance of the biological medium, as well as capable of acting as aggregators of coagulation proteins, is promising for the development of micro-mesoporous nanomaterials with a hemostatic effect. Such nanomaterials include oxides and/or hydroxides based on aluminum and iron. They have a developed surface [5, 6], a positive charge [7, 8], and possess selective sorption properties [9–11].

2. Materials and methods

2.1. Synthesis of porous nanostructured metal oxide

Porous aluminum hydroxides and iron oxides were synthesized for the investigation. Aluminum hydroxide was produced by water oxidation of AlN/Al nanoparticles (Sample 1) using the procedure described in reference [5] and hydrolysis of aluminum isopropoxide (Sample 2) according to the following procedure: 10 g of aluminum isopropoxide was heated to 135 °C and kept in a closed flask until complete dissolution (1 hour), then it was added dropwise with constant stirring to deionized water heated to a temperature of 40 or 90 °C. The resulting precipitate was filtered off, washed with 1000 mL of deionized water, and dried at 100 °C for 2 hours. Iron oxides with different morphology and structural characteristics were synthesized under hydrothermal conditions (Sample 3) using Fe nanoparticles (Sample 4). According to the first method, 90 mL of Na2SO4 solution of 1 M concentration was placed in a 250 mL beaker, and 20 mL of FeCl3 solution of 0.5 M concentration was added to it with constant stirring and stirred for 15-20 minutes. Then 70-80 mL of distilled water was added to make a homogeneous medium. Further, the obtained solution was placed in a sealed autoclave, heated to 140 °C, and held for 6 hours. After that, the autoclave was cooled to room temperature; the resulting precipitate was filtered off, washed with 1000 mL of distilled water and dried at 120 °C for 2 hours. According to the second method, 100 ml of a 2.2 M solution of CH3COONa were placed in a 500 mL glass flask and 100 mL of Fe nanopowder suspension of 1 wt. % concentration was added to it with constant stirring and stirred for 15-20 minutes. Then the mixture was heated to boiling and boiled for 6 hours with constant stirring under reflux. 100 mL of distilled water was added to the resulting sol and boiled for 6 hours. The reaction mixture was cooled to room temperature. The resulting precipitate was filtered off, washed with distilled water (at least 1000 mL) and dried at 120 °C for 4 hours.

2.2. Characterization of synthesized porous nanostructured metal oxide

The synthesized samples were examined by transmission electron microscopy using a JEM 2100 microscope (Jeol, Japan) and X-ray diffraction analysis using a Shimadzu XRD-6000 diffractometer (Shimadzu, Japan). The nitrogen adsorption/desorption isotherms were measured at 77 K on a Sorbtometr-M analyzer (Katakon, Russia). The specific surface areas were estimated by the Brunauer–Emmett–Teller (BET) method. Zeta potential measurements were performed using a Zetasizer Nano ZSP (Malvern Instruments, UK). Biochemical analysis was performed using a Cobas c 311 biochemical analyzer (Roche, Switzerland). The studies were performed on whole blood with the synthesized powders added at a concentration of 1.0 mg/mL.

3. Results and discussion

TEM images of the synthesized oxides and hydroxides are shown in figure 1. As can be seen from figure 1 a, water oxidation of AlN/Al nanoparticles results in the formation of agglomerates of crumpled nanosheets up to 200 nm in size and 2–5 nm thick. Aluminum isopropoxide hydrolysis produces similar nanostructures, but the size of nanosheets does not exceed 100 nm (figure 1 b). Iron oxide, synthesized under hydrothermal conditions, had the form of 5 μm agglomerates consisting of rods with a thickness of 20–50 nm and a length of up to 1 μm (figure 1 c). The reaction of iron nanoparticles with sodium acetate and subsequent precipitation during boiling results in the formation of layered iron oxide structures. The layer thickness is 10–20 nm, and the length is up to 1 μm (figure 1 d).

The synthesized nanostructures have a positive surface charge and a large specific surface area. The zeta potential and the specific surface area were respectively 30.5 mV and 286 m2/g for Sample 1, 35.4 mV and 307 m^2/g for Sample 2, 25.7 mV and 101 m^2/g for Sample 3, and 22.3 mV and 143 m^2/g for Sample 4.

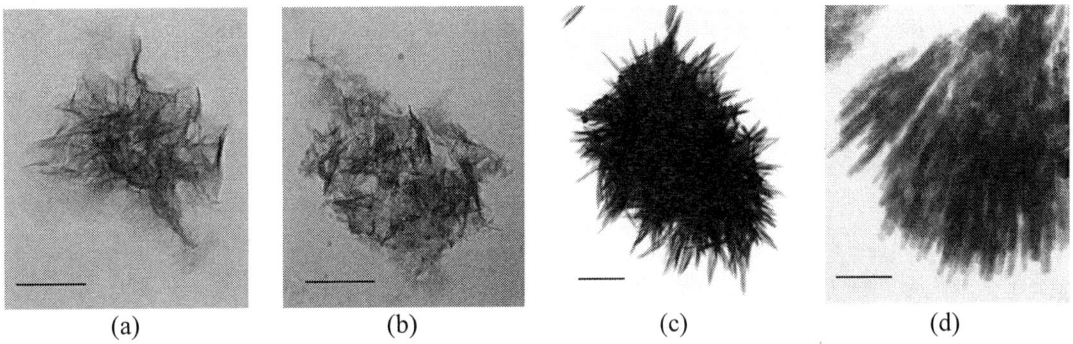

Figure 1. TEM images of nanostructures synthesized by: water oxidation of AlN/Al nanoparticles (a); aluminum isopropoxide hydrolysis (b), under hydrothermal conditions (c), and using Fe nanoparticles (d). All bars are 100 nm.

According to X-ray diffraction analysis (figure 2 a), water oxidation of AlN/Al nanoparticles and aluminum isopropoxide hydrolysis leads to the formation of boehmite. The maxima of the (020) absorption band are shifted to low angles with respect to crystalline boehmite (PDF No. 00-021-1307). This pattern is typical for the AlOOH compound with the pseudoboehmite structure. The sample obtained using iron nanoparticles (figure 2 b, curve 3) is represented by the Fe_2O_3 hematite phase (PDF No. 01-073-3825). The sample produced by the hydrothermal method (figure 2 b, curve 4) is represented by the FeO (OH) goethite phase (PDF No. 01-076-7164).

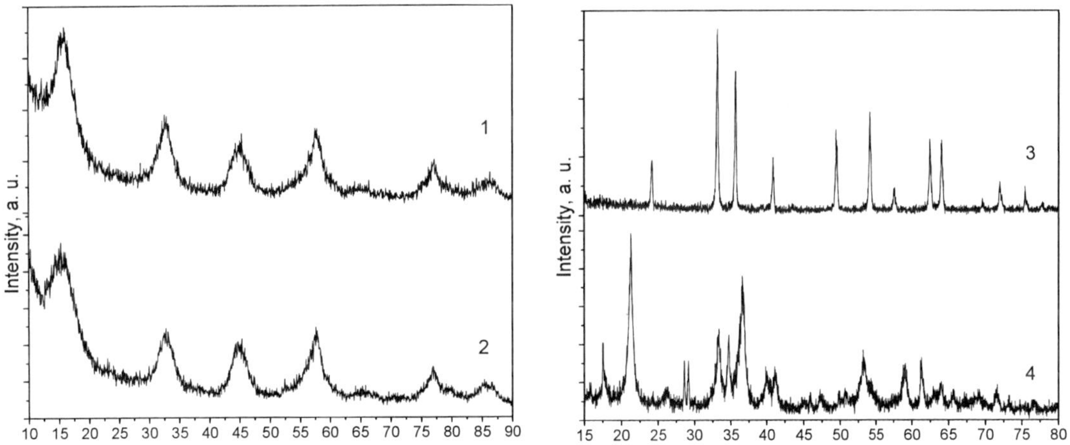

Figure 2. XRD pattern of nanostructures synthesized by: water oxidation of AlN/Al nanoparticles (a, curve 1), aluminum isopropoxide hydrolysis (a, curve 2), hydrothermal method (b, curve 3), and using Fe nanoparticles (b, curve 4).

Table 1 gives the results on the hematologic and blood chemistry values after interaction with the synthesized nanostructures. In blood contact with Sample 1, all blood chemistry values demonstrate the maximum decrease. For other samples, the decrease in hematologic values is no more than 20% with respect to control.

For Sample 1, there is a decrease in hemoglobin and erythrocytes but an increase in the number of platelets and leukocytes, which may be due to a decrease in the aggregative ability of Sample 1. In blood contact with Sample 2, the number of white blood cells and platelets decrease. Sample 3 is characterized by a reduction of hemoglobin, erythrocytes and platelets, while the number of leukocytes with respect to control is much larger. In blood contact with Sample 4, there is a slight decrease in the number of white blood cells and platelets.

Table1. Changes in the hematologic and blood chemistry values after interaction with the synthesized nanostructures.

	Control	Sample 1	Sample 2	Sample 3	Sample 4
Hemoglobin	138	132	137	133	137
Erythrocytes	4.8	4.5	4.7	4.4	4.9
Color index	0.9	0.89	0.89	0.89	0.9
Leukocytes	5.6	5.9	5.4	6.4	5.3
Platelet count	117	128	109	102	114
Glucose, mmol/L	12.1	8.7	10.4	10.5	11.7
Bilirubin, μmol/L	48.3	8.7	10.4	10.5	11.7
Protein, g/L	3.9	8.7	10.4	10.5	11.7
Cholesterol, mmol/L	3.89	8.7	10.4	10.5	11.7
a-Amylase, U/L	32	8.7	10.4	10.5	11.7
Albumin, L	27	8.7	10.4	10.5	11.7

4. Conclusion

In this study, different methods were used to synthesize porous nanostructured particles based on oxides and hydroxides of aluminum and iron. The effect of the synthesized nanostructures on the hematologic and blood chemistry values was investigated. It was found that the synthesized samples affect the hematological values in different ways. Boehmite produced by water oxidation of AlN/Al nanoparticles has the greatest influence on the blood chemistry values. The reported results are useful for a general understanding of how metal oxides and hydroxides with a developed surface interact with the blood components, and can be used in the development of new hemostatic agents on their basis with high hemostatic efficiency.

Acknowledgments

The work was financially supported by the Ministry of Education and Science of the Russian Federation (Agreement No. 14.604.21.0156, Project Identifier RFMEFI60417X0156).

References

[1] Wolberg A S and Campbell R A 2008 *Transfus. Apher. Sci.* **38** 15
[2] Nair C H, Shah D A and Dhall D P 1986 *Thromb. Res.* **42** 809
[3] Susree M and Anand M 2016 *Int. J. Res. Eng. Technol.* **5** 102
[4] Kuznetsova N R and Vodovozova E L 2014 *Biochemistry (Moscow)* **79** 797
[5] Lozhkomoev A S, Lerner M I, Tsukanov A A, Kazantsev S O, Bakina O V and Psakhie S G 2017 *Phys. Mesomech.* **20** 134
[6] Křížek M, Pechoušek J, Tuček J, Šafářová K, Medřík I and Machala L 2012 *AIP Conf. Proc.* **1489** 88
[7] Lozhkomoev A S, Glazkova E A, Bakina O V, Lerner M I, Gotman I, Gutmanas E Y and Psakhie S G 2016 *Nanotechnology* **27** 205603
[8] Ma P, Luo Q, Chen J, Gan Y, Du J, Ding S, Hi Z and Yang X 2012 *Int. J. Nanomed.* **7** 4809
[9] Tsukanov A A and Psakhie S G 2017 *Phys. Mesomech.* **20** 43
[10] He T Xiang L and Zhu S *Langmuir* 2008 **24** 8284
[11] Chitrakar R, Tezuka S, Sonoda A, Sakane K, Ooi K and Hirotsu T 2006 *J. Colloid Interface Sci.* **298** 602

IOP Conf. Series: Materials Science and Engineering **447** (2018) 012073 doi:10.1088/1757-899X/447/1/012073

Effect of structural-phase transformations on the microhardness of the Al-22%Si-3%Cu-1.7%Ni alloy after high-pressure torsion and annealing

G R Khalikova, G F Korznikova and V G Trifonov

Institute for Metals Superplasticity Problems, Russian Academy of Sciences, 39 Khalturin st., Ufa 450001, Russia

E-mail: gulnara.r.khalikova@gmail.com

Abstract. The squeeze casting Al-22%Si-3%Cu-1.7%Ni alloy was subjected to high-pressure torsion (HPT) and subsequent annealing for investigation of their effect on the microstructure and microhardness. Specimens with a diameter of 8 mm and a thickness of 0.45 mm, cut from a squeeze cast ingot (region of ingot without primary silicon particles), were severely deformed by HPT via 5 revolutions under a pressure of 4 GPa at room temperature. The edge parts of the deformed samples were annealed at temperatures of 300, 400 and 500 °C for 5 minutes. HPT resulted in the refinement and partial dissolution of the eutectic silicon and primary intermetallic particles in an aluminum matrix with the formation of abnormal supersaturated solid solution and fragmentation of the structure that caused an increase in microhardness from 149±10 HV for the squeeze casting state to 285±14 HV for the deformed state. Subsequent annealing leads to the decomposition of the abnormal supersaturated solid solution, which occurred simultaneously with the recovery and recrystallization of the fragmented structure, which led to a decrease in the microhardness to 183±5, 108±5 and 132±6 HV at temperatures of 300, 400 and 500 °C, respectively.

1. Introduction

Severe plastic deformation techniques are actively used to form ultrafine grained and nanocrystalline structures in various aluminum alloys, which significantly change their physical and mechanical properties [1–3]. At the same time, during severe plastic deformation, a change in the phase composition can occur simultaneously with a structural change of the matrix of alloys, caused by the accelerated decomposition of the supersaturated solid solution i. e. dynamic strain aging [4,5] or dissolution of the second phases, including transition metal aluminides [6,7].

It is known that in the course of the conventional heat treatment of strain ageing aluminum alloys, along with recovery and recrystallization, a sequential decomposition process of a preliminarily supersaturated solid solution can occur [8]. However, after severe plastic deformation, the supersaturated solid solutions are metastable and at heating, accelerated decomposition undergoes, in some cases, with a change of its kinetics. As a result, the structure may differ from the structure obtained by conventional treatment [4,9–10].

The aim of the study was to evaluate the influence of high-pressure torsion and annealing on the microstructure, phase composition and microhardness of the Al-22%Si-3%Cu-1.7%Ni alloy.

2. Experimental

The commercial casting heat-strengthening Al-22%Si-3%Cu-1.7%Ni (AK21 in Russian classification) alloy was used in the present study. The initial state represents squeeze casting (SQ). The ingot was

characterized by inhomogeneity of a distribution of primary silicon crystals. In this work, parts of an ingot without primary silicon crystals were investigated.
Samples in the form of a disc with a diameter of 8 mm were mechanically cut and polished to a final thickness of about ~0.45 mm and subjected to severe plastic deformation by means of high pressure torsion (HPT) for 5 revolutions under a pressure of 4 GPa at room temperature. Then, the processed samples were subjected to annealing at 300, 400 and 500 °C for 5 minutes. The edge parts of the deformed and annealed samples were examined.

Microstructural characterization in the initial condition, after HPT and annealing was performed using a Tescan Mira 3LMH scanning electron microscope (SEM) by means of secondary electrons (SE) and back-scattered electrons (BSE) detectors. The fine structure of the HPT alloy was analyzed by a JEOL-2000 EX transmission electron microscope (TEM). The chemical composition of aluminum solid solution and secondary phases was investigated using SEM Tescan VEGA 3 SBH with an energy dispersive X-ray spectroscopy (EDS) supplement. Electron Backskatter Difraction (EBSD) analysis of the annealed samples was performed by a Tescan Mira 3 LMH SEM.

A quantitative estimation of the average area (S) and the volume fraction (V) of the primary (Pr) and secondary (Sec) phases was performed using computer analysis techniques circling each particle around the perimeter; the average grain size was calculated by the line intercept method. The microhardness of the alloy in all conditions was evaluated by the Vickers hardness method. Statistical analysis was performed with a confidence level of 95%.

3. Results and discussion

The typical microstructure of the Al-22%Si-3%Cu-1.7%Ni alloy in the initial squeeze casted state is shown in figure 1 a. The structure of the alloy consisted of a solid solution of α-Al, double eutectic (α-Al)+Si^{Eu}, and particles of intermetallic phases (IPh^{Pr}): $Al_5Cu_2Mg_8Si_6$, Al_3CuNi, $(NiCu)_2Al_3$, $(FeMnCu)_3Si_2Al_{15}$ and $CuSi_4Mg_5Al_4$ (identified by EDS analysis).

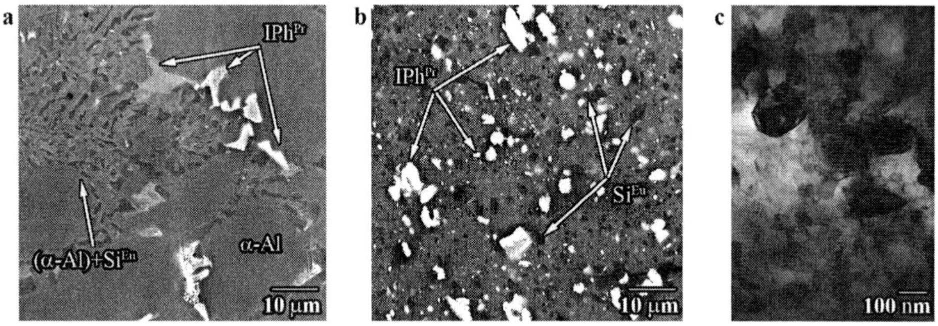

Figure 1. Typical structure of the Al-22%Si-3%Cu-1.7%Ni alloy in the initial squeeze casted state (a) and after HPT (b, c): a, b – SEM (SE imaging mode), c – TEM bright field.

High-pressure torsion resulted in fragmentation of excess phases of the alloy and a decrease in their volume fraction (figure 1 b). The quantitative characteristics of the structural constituents of the alloy are given in the table 1. The fine structure of the deformed alloy consisted mainly of individual fragments ranging in size from 20 to 200 nm, surrounded by nonequilibrium low and high angle boundaries (figure 1 c).

Since the Al-22%Si-3%Cu-1.7%Ni alloy was obtained by squeeze casting, during which the saturation of the solid solution by alloying elements did not occur due to the high cooling rate of the melt, deformation aging at HPT was suppressed. However, according to the results of the EDS analysis, HPT at room temperature resulted in an increase in the content of dopant and impurity elements in α-Al, contributing to the additional supersaturation of the solid solution (figure 2).

During the subsequent annealing, accelerated decomposition of the anomalously supersaturated solid solution occurred with the precipitation of particles of secondary phases (figure 3). Such rapid processes of nucleation and growth of the secondary phases during annealing of the deformed alloy were related to the nonequilibrium state of the alloy after HPT, which is confirmed by the results of fine structure analysis, and an increased concentration of deformation-induced defects in similar materials [11].

Table 1. Average area (S) and volume fraction (V) of silicon particles (Si), primary (Pr) and secondary (Sec) intermetallic phases (IPh) of the Al-22%Si-3%Cu-1.7%Ni alloy after various treatments.

Phase	Parameters	SQ	HPT	Annealing for 5 minutes at the temperature, °C		
				300	400	500
Si	S, μm^2	2.9±0.1	1.2±0.1	0.45±0.01	0.53±0.02	1.08±0.05
	V, %	14±1	8±1	11±1	13±1	13±1
IPh	S^{PR}, μm^2	5.6±0.5	0.49±0.05	0.46±0.04	0.53±0.04	0.50±0.04
	S^{SEC}, 10^{-3} μm^2			20.1±1.2	34.4±2.2	42.9±3.2
	V, %	11±1	6±1	9±1	10±1	14±1

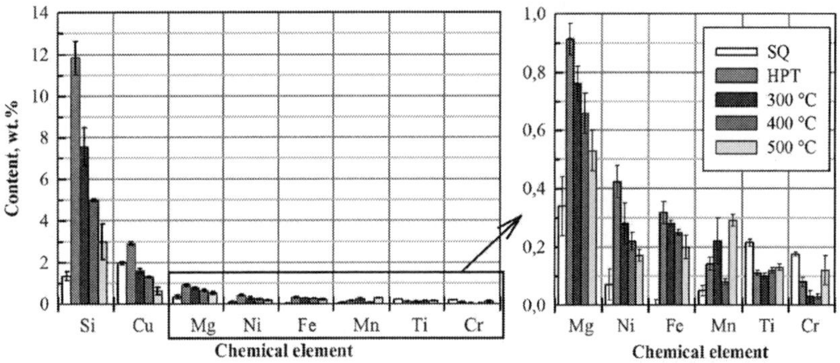

Figure 2. Distribution weight content of alloying and impurity elements in solid solution of Al-22%Si-3%Cu-1.7%Ni alloy in the initial squeeze casted (SQ) state, after the HPT and subsequent annealing at 300, 400, and 500 °C.

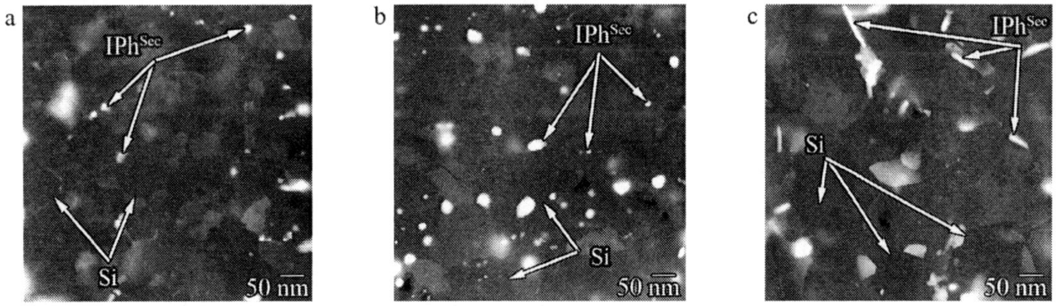

Figure 3. Typical structure of HPT Al-22%Si-3%Cu-1.7%Ni alloy after annealing for 5 minutes at 300 °C (a), 400 °C (b) and 500 °C (c) (SEM, BSE imaging mode). General view of strengthening secondary phases and (sub)grains.

The results of quantitative estimations showed that an increase of the annealing temperature resulted in an increase in the average area and volume fraction of silicon particles and intermetallic phases (table 1). This is due to the fact that the nucleation and growth of the dispersed secondary phases took place during the decomposition of an anomalously supersaturated solid solution.

The results of the quantitative estimation are in a good agreement with the results of the EDS analysis of the solid solution. After annealing of the deformed samples, the content of dissolved elements in the solid solution decreased (figure 2), and the higher the annealing temperature, the more intensive the decomposition of the solid solution.

During the post-deformation annealing of the alloy, the recovery and recrystallization of the fragmented nanostructure took place simultaneously with the decomposition of the anomalously supersaturated solid solution. This follows from the predominance of high-angle boundaries on the EBSD-maps of the alloy, for example, annealed at 400 °C (figure 4). With increasing annealing temperature, the average grain size increased up to 0.19±0.06, 0.43±0.05 and 0.67±0.07 μm for 300, 400 and 500 °C, respectively.

Structural changes of the alloy after HPT and subsequent annealing are in good agreement with the results of microhardness measurements. Thus, HPT led to an increase in microhardness from 149±10 HV (for SQ state) to 285±14 HV. Subsequent annealing of the HPT samples at 300, 400 and 500 °C reduced themicrohardness down to 183±5, 108±5 and 132±6 HV, respectively.

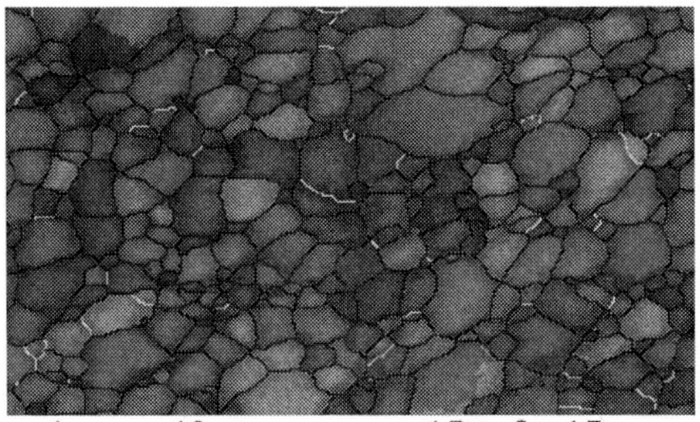

Figure 4. EBSD-map of HPT Al-22%Si-3%Cu-1.7%Ni alloy after annealing at 400 °C for 5 minutes.

4. Summary

During high pressure torsion of the Al-22%Si-3%Cu-1.7%Ni alloy, fragmentation of excess phases and their partial dissolution with the formation of an anomalously supersaturated aluminum solid solution, as well as the formation of a fragmented nanostructure with a fragment size from 20 to 200 nm took place. Subsequent annealing in the temperature range 300…500 °C led to intensive decomposition of anomalously supersaturated aluminum solid solution with the formation of secondary phase particles and simultaneous recovery and recrystallization of the fragmented nanostructure. With an increase in the annealing temperature, the average grain size raised from 0.19±0.06 to 0.67±0.07 μm. Such structural changes led to an increase in microhardness from 149±10 HV (for SQ state) to 285±14 HV (for HPT state), and then, upon subsequent annealing, – to a decrease to 183±5, 108±5 and 132±6 HV at 300, 400 and 500 °C, respectively.

References

[1] Burnhanettin A 2006 *Severe Plastic Deformation: Toward Bulk Production of Nanostructured Materials* (NY: Nova Science Publ. Inc.) p 612
[2] Zhilyaev A P and Langdon T G 2008 *Prog. Mat. Sci.* **53** 893
[3] Sabbaghianrad Sh and Langdon T G 2016 *Mater. Sci. Eng. A* **655** 36
[4] Murayama M, Horita Z and Hono K 2001 *Acta Mater.* **49** 21
[5] Mazilkin A A, Straumal B B, Rabkin E, Baretzky B, Enders S, Protasova S G, Kogtenkova O A and Valiev R Z 2006 *Acta Mater.* **54** 3933
[6] Senkov O N, Froes F H, Stolyarov V V and Valiev R Z 1998 *Scripta Mater.* **38**(10) 1511
[7] Mondolfo L F 1976 *Aluminum Alloys: Structure and Properties* (London: Butterworths) p 971
[8] Huang Y, Robson J D and Prangnell P B 2010 *Acta Mater.* **58** 1643
[9] Angella G, Bassani P, Tuissi A and Vedani M 2004 *Mater. Trans.* **45**(7) 2282
[10] Zhen L, Fei W D, Kang S B and Kim H W 1997 *Mater. Sci.* **32** 1895
[11] Korznikova E A 2011 *Letters on Materials* **1**(4) 208

IOP Conf. Series: Materials Science and Engineering **447** (2018) 012074 doi:10.1088/1757-899X/447/1/012074

Ultrafine powder of barium titanate for production of lead-free ceramic piezomaterials

A A Kholodkova[1,2], S G Ponomarev[1], A D Smirnov[1], Yu D Ivakin[2], M N Danchevskaya[1,2]

[1]Moscow Polytechnical University, B.Semenovskaya 38, Moscow, 115280, Russia
[2]Chemistry Department, Lomonosov Moscow State University, Leninskie Gory 1-3, Moscow, 119991, Russia

E-mail: s.ponomarev@mami.ru

Abstract. The paper is aimed to perform and compare the capabilities of the solid state technology and a new method of synthesis in supercritical water for the production of barium titanate powder. This material is of interest for applications such as dense piezoceramics. The new method of synthesis in an aqueous medium allowed obtaining barium titanate of higher phase purity and with a narrow crystal size distribution compared to the solid state method. The ceramics manufactured from the barium titanate powder synthesized in supercritical water possessed a consistently high dielectric constant (2400-2600) and a low loss tangent (~ 10^{-2}) at low and medium frequencies.

1. Introduction

Barium titanate is commonly used in a variety of applications in the microelectronic industry, such as the production of lead-free piezoelectric modules, electro-optic devices, ceramic capacitors, etc. [1]. This material owes its wide utilization to its high dielectric constant: at a temperature below 120 °C, $BaTiO_3$ exhibits ferroelectric properties. The functional characteristics of barium titanate ceramics depend on the properties of a starting powder. In this regard, it is recommended to use fine powders with high phase purity for the production of dense ceramic materials [2]. For barium titanate powder, it is also important to avoid using the metastable pseudo cubic modification that is paraelectric [3]. Traditionally, $BaTiO_3$ is synthesized by a simple and convenient solid state route [4]. However, this method is accompanied with high energy consumption and lacks the requirements for purity and morphological homogeneity of the product. A new method of synthesis of fine-crystalline $BaTiO_3$ in supercritical water was proposed in [5-6]. The method allows obtaining nano- and submicron powders with a low content of impurities, which was successfully applied for the production of dense dielectric ceramics. In this paper, we study the phase and morphological characteristics of $BaTiO_3$ powders synthesized by the traditional solid state route, and the new method, and compare the structural and functional properties of ceramics based on two kinds of synthesized powder.

2. Experimental

2.1. Materials

$BaCO_3$ (purity ≥ 99%) and TiO_2 (rutile, purity ≥ 99%) were used as starting reagents for the synthesis of $BaTiO_3$ *via* the solid state method. The stoichiometric mixture of reagents was milled in a planetary mill in isopropyl alcohol for 100 minutes, and then calcinated at 1300 °C in air for 1 hour. The synthesis in supercritical water (SCW) was carried out using BaO (purity ≥ 99%) and TiO_2 (rutile, purity ≥ 99%) as starting materials. The stoichiometric mixture of oxides was placed into a container in a stainless steel autoclave. The calculated amount of water was poured into the autoclave outside the container. The hermetically sealed autoclave was heated up to 400 °C and held at this temperature for 20 hours. The pressure of the SCW inside the autoclave was 26 MPa. At the end of the process, the autoclave was rapidly cooled down to room temperature, the product was removed from the container and dried in air at 50 °C for 12 hours. To manufacture ceramic samples, 5 mass. % of paraffin were added to the powders as a temporary technological binder, and discs of a diameter of 13 mm and of a thickness of 3-4 mm were pressed uniaxially at 150 MPa. The ceramics were sintered at 1300 °C for 1 hour. The obtained $BaTiO_3$ powder samples and the manufactured ceramics are listed in table 1. To prepare ceramic samples for electric measurements, their surfaces were polished and painted with silver paste as poles, and fired at 800 °C in for 20 minutes.

Table 1. Conditions of $BaTiO_3$ synthesis and corresponding ceramics

Sample of powder	Synthesis method	Parameters of synthesis	Conditions of sintering	Sample of ceramics
BT-S	Solid state	t = 1300 °C, 1 h.	t = 1300 °C, 1 h.	BTC-S
BT-F	In SCW medium	t = 400 °C, p = 26 MPa, 20 h.		BTC-F

2.2. Methods

Phase analysis of the synthesized powders was carried out using a STOE STADI P diffractometer with Cu-Kα radiation in the range of 20° ≤ 2θ ≤ 60°. The average crystallite size was calculated according to the Scherrer equation [7]. The particle size distribution of the powders was determined on an Analysette 22 NanoTec measuring unit. The morphology of the powders and the microstructure of the ceramic fractures were observed using a JSM-6390 LA scanning electron microscope. The apparent density of the ceramics was determined by the Archimedes method. The porosity of the ceramic samples was measured by the kerosene saturation method. The dielectric characteristics of the ceramics were measured with metallized samples on an Agilent E4980a LCR meter in the frequency range of 20Hz to 2MHz.

3. Results and discussion

XRD analysis of powders synthesized by the solid state method and in a medium of supercritical water (SCW) confirmed the formation of barium titanate in its tetragonal modification (figure 1 a). Along with the peaks corresponding to $BaTiO_3$, the weak reflections of other phases were presented in the XRD pattern of the BT-S sample in the range 25° ≤ 2θ ≤ 30°. These peaks indicated trace amounts of phases with a Ba:Ti mole ratio different from 1, which originated from the diffusion character of the solid state process. An XRD analysis of the BT-F sample obtained in the SCW medium revealed a pure $BaTiO_3$ phase within the sensitivity of the method. From the calculation of the average crystallite size, the BT-F sample possessed a well-ordered crystalline structure (D_{XRD} = 42 nm) compared to BT-S (D_{XRD} = 33 nm).

The particle size distribution of the BT-S powder is composed of a minor fraction consisting of submicron particles (0.07-0.2 μm, about 7%) and a major fraction of micron-sized particles (0.9-43 μm, about 93%). Particles of irregular shape and size of 0.5-6.9 μm fell within the SEM sight (figure 1 b). Round-shaped submicron crystals with a diameter from 120 to 740 nm were observed in the BT-F sample (figure 1 c) (mean size D_{SEM} = 310 ± 10 nm). From a laser diffraction analysis, the crystals in BT-F were considerably agglomerated. The sample exhibited a wide variation of particle sizes from 0.09 to 25 μm, as well as in a range of 107-409 μm (about 23% of the particles). Ultrasonic treatment might improve the dispersion of the powder.

The microstructure of the ceramic samples corresponded to the intermediate stage of sintering [8] (figure 2). On the fracture of BTC-S, the intergrowth of particles in the zones of their contact led to the formation of a grid of grains (figure 2, a). The size of the visible grains was 0.3-5.1 μm. In the BTC-F sample, the pore space had the shape of communicating channels, which formed isolated regions at the boundaries and in the body of grains (figure 2 b). The size of grains was estimated in the range of 1.2-3.1 μm. For both $BaTiO_3$ powders, a viscous flow contributed to the sintering mechanism. However, for BT-F this contribution was more significant due to the smaller crystal size and higher surface tension. In the BTC-S sample, the sintering bulk diffusion mechanism was implemented due to the higher defectiveness of the initial BT-S powder (see D_{XRD}).

Figure 1. XRD patterns (a) and SEM images of the $BaTiO_3$ powder samples: b – BT-S; c – BT-F.

Integral structural characteristics, such as density and porosity of the ceramic samples manufactured from BT-S and BT-F powders, possessed close values (table 2).

Table 2. Integral structural characteristics of the $BaTiO_3$ ceramics

Sample of ceramics	Apparent density (g cm^{-3})	Relative density (%)	Open porosity (%)
BTC-S	5.18	86	4.4
BTC-F	5.24	87	5.6

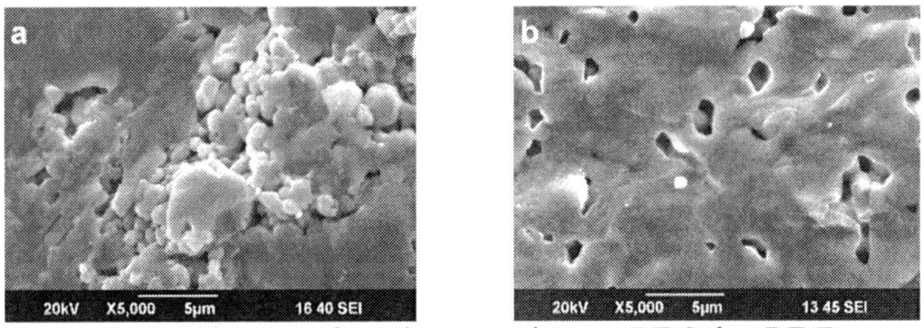

Figure 2. SEM images of $BaTiO_3$ ceramics: a – BT-S; b – BT-F.

Ceramics based on two types of $BaTiO_3$ powder possess dielectric characteristics that correspond to the previously reported data and requirements imposed by the electronic industry [9-11]. The BTC-F sample surpassed the BTC-S sample in dielectric properties. The ceramics manufactured from a powder synthesized in the medium of SCW had a higher dielectric constant (2400-2600) and a low loss tangent (~ 1%), and also showed a higher frequency stability of these characteristics compared to the ceramics from a powder obtained by the traditional solid-state technology (figure 3). This result is ensured by the structural homogeneity of BTC-F at similar density and porosity values of the two types of material.

4. Conclusions

$BaTiO_3$ powder, synthesized in the medium of SCW, excelled the powder obtained by the traditional solid state technology as a raw material for dense dielectric ceramics. The product of synthesis in an aqueous medium performed the phase purity, a well-ordered crystal structure, a narrow distribution of crystal sizes, and a regular shape of crystals. As a disadvantage of such a powder, it can be noted that submicron crystals agglomerated and formed micron particles. The phase purity and structural homogeneity of the powder synthesized in SCW contributed to the production of high-quality dielectric ceramics. These ceramics possessed a stably high dielectric permittivity and low dielectric losses in the low and medium frequency range. The method of $BaTiO_3$ synthesis in the medium of SCW is a promising method for obtaining raw materials for the microelectronic industry.

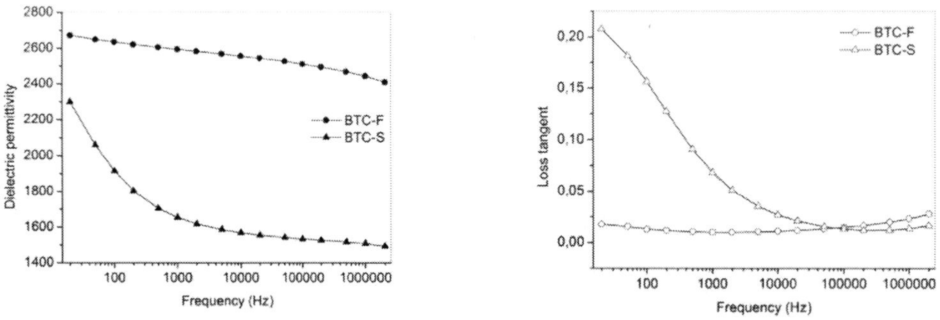

Figure 3. Frequency dependence of dielectric permittivity and loss tangent of $BaTiO_3$ ceramics.

Acknowledgments
The work was carried out with the financial support of The Ministry of Education and Science of the Russian Federation within the framework of the State Task "The Development of a Perspective Method for Obtaining of Ultra Dispersive Barium Titanate ($BaTiO_3$) as a Main Component in the Production of Ceramic Capacitors" № 10.2729.2017 from May, 31 2017 and with the use of equipment of the Centre of Collective Usage "High Technology in Engineering" of Moscow Polytech (Number for publications: 10.2729.2017/4.6.).

References
[1] Saravanan R 2018 *Titanate Based Ceramic Dielectric Materials* vol 25 (Materials Research Foundations)
[2] Somiya S, Komarneni S and Roy R 2013 *Ceramic Powders for Advanced Ceramics: What Are Ideal Ceramic Powders for Advanced Ceramics?* vol 35 (Elsevier Inc.)
[3] Gao L J, Liu X L, Zhang J Q, Wang S Q and Chen J F 2004 *Mater. Chem. Phys.* **88** 27
[4] Vijatovic M M, Bobic J D and Stojanovic B D 2008 *Sci. Sinter.* **40** 155
[5] Kholodkova A A, Danchevskaya M N, Ivakin Y D, Muravieva G P and Tyablikov A S 2016 *J. Supercrit. Fluids* **117** 194
[6] Kholodkova A A, Danchevskaya M N, Ivakin Y D, Muravieva G P, Smirnov A D, Tarasovskii V P, Ponomarev S G, Fionov A S and Kolesov V V. 2018 *Ceram. Int.* **44** 13129
[7] Nath A K, Jiten C and Singh K C 2010 *Phys. B* **405** 430
[8] De Jonghe L C and Rahaman M N 2003 *Handb. Adv. Ceram. Mater. Appl. Process. Prop.* **1–2** 187
[9] George C N, Thomas J K, Kumar H P, Suresh M K, Kumar V R, Wariar P R S, Jose R and Koshy J 2009 *Mater. Charact.* **60** 322
[10] Bocquet J F, Chhor K and Pommier C 1999 *Mater. Chem. Phys.* **57** 273
[11] Koripella C R 1992 *TechTopics* **2** 1

IOP Conf. Series: Materials Science and Engineering **447** (2018) 012075 doi:10.1088/1757-899X/447/1/012075

The structure of the surface layer in titanium VT1-0 after high-cycle fatigue tests

S. Konovalov[1,2], I. Komissarova[3], X. Chen[1], A. Feoktistov[3], D. Kosinov[3], V. Gromov[3], Yu. Ivanov[4]

[1] Wenzhou University, Wenzhou University Town, Wenzhou, 325035, China
[2] Samara National Research University, 34, Moskovskoye sh., Samara 443086, Russia
[3] Siberian State Industrial University, 42, Kirova str., Novokuznetsk, 654007, Russia
[4] Institute of High-Current Electronics of Siberian Branch of Russian Academy of Sciences, 2/3, Akademicheskii av., Tomsk, 634021, Russia

E-mail: ksv@ssau.ru

Abstract. Using the methods of optical and scanning electron microscopy, transformations that occur in the structure of commercially pure titanium VT1-0 subjected to electric-current pulse treatment and high-cycle fatigue testing to failure, were analyzed. The paper reports that as a result of the energy impact on the samples, its fatigue life increases by a factor of 1.3. It is revealed that the improvement of the fatigue life is possible due to the change in the size of crystallites in α-titanium, as well as to the decreased number of internal stress field concentrators in the surface layer.

1. Introduction

The fracture of metallic materials in conditions of any stresses is caused by the processes of plastic deformation, the formation and propagation of cracks. During fatigue testing it is associated with a sharply increasing temperature of a sample, leading probably, as a consequence, to processes of dynamic ageing, polygonization and recrystallization. The extent, to which these processes can be developed, significantly depends on both the nature of the external impact and the initial structural state of the material [1-7].

An efficient procedure to modify the structure and phase composition in a material is its electric-current pulse treatment [8-10]. A number of studies carried out on diverse structural classes of steel emphasizes that the impact of an electric-current on a material at a certain stage of fatigue stress improves the efficiency of samples by a factor of 1.5-2 [11-14].

Titanium alloys are widely applied industrial materials, since the unique combination of physical and mechanical characteristics such as high specific strength, sufficient technological plasticity and high corrosion resistance makes them rather competitive constructional materials. However, their use is restricted by low resistance to abrasive and erosive wear, low hardness, liability to seizure if applied in friction areas, and insufficient wear resistance.

This study is aimed at analyzing changes in the defect substructure, registered in samples of titanium alloy VT1-0 under regular stressing and after electric-current pulse treatment, as well as identifying the mechanisms relevant for improvement of fatigue life [15-17].

IOP Conf. Series: Materials Science and Engineering **447** (2018) 012075 doi:10.1088/1757-899X/447/1/012075

2. Test materials, equipment and research methods

Samples of titanium alloy VT1-0 (foreign analogue – alloy Grade 2) with chemical composition (up to 0.18 Fe; to 0.07 C; to 0.04 N; to 0.1 Si; to 0.12 O; to 0.004 H; 0.3 % - other impurities; balanced by Ti, mass %) was used as the material for research. Fatigue tests were carried out using a special device according to the scheme of asymmetric cantilever bending. Samples with a thickness of 4 mm and a width of 12 mm were made according to the Russian Standard (GOST 25.502-79) and had a symmetric stress concentrator with a minimal narrow spot in the center (5 mm). The temperature of tests was 300 K, the frequency of the bending stress of the sample was 10 Hz.

When fatigue testing, ultrasound monitoring was carried out using a universal unit to control the speed of ultrasound (Institute of Strength Physics and Materials Science, SB RAS), which was measured on a distressed sample. Electric-current treatment was carried out after a sharp change in the speed of ultrasound (180,000 cycles). The parameters of the electric-current treatment were as follows: the amplitude of the current pulses was 2 kA, the maximum current density in the zone of the stress concentrator was 100 A/mm^2, the frequency of the current pulses was 70 Hz, and the time of treatment was 2 minutes. After completion of the electric-current treatment a high-cycle fatigue testing to failure of the samples was carried out.

Using SEM methods (Tesla BS-301), the fractured surface was analyzed. The phase composition and state of the defect substructure were studied using transmission electron diffraction microscopy methods (JEM-2100). Foils for the purpose of research were prepared by ion thinning of plates cut of a solid sample, in parallel and as close as possible to the fractured surface using electric-spark method.

3. Results and Discussion

It is determined that the electric-current treatment causes an increase in cycles to failure from 281,333 to 359,532. This corresponds to an increase in fatigue life by a factor of 1.3.

The structure of the surface layer in as-delivered titanium fractured in the process of fatigue testing (281,333 cycles) has a relatively thin (maximum 4.5 μm) surface layer with a nano-crystalline grain and a sub-grain structure. Grains in this layer are on average 58.6 ± 21.5 nm. Grains form layers that are disoriented relative to each other and parallel to the surface of the sample. A ring structure is observed in the electron-diffraction micro-pattern, presented in figure 1 a. This fact demonstrates, firstly, the small size of crystallites, forming the electron-diffraction micro-pattern, secondly, large-angle disorientation of these crystallites, i.e. the formation of a grain and sub-grain structure in the surface layer of commercially pure titanium as a result of fatigue testing.

The analysis has revealed a lamellar structure of the material close to the nanostructured surface layer and a lot of flexural extinction contours in it. This is a good illustration of a large number of sources (concentrators) of internal stress fields in the surface layer, which cause bending and torsion in the crystal lattice of the material.

Indexing of the electron-diffraction micro-pattern shown in figure 1 b, demonstrates a multi-phase state of the surface layer in the material. A principal phase is the α-modification of titanium, the reflections of which form a ring structure. Therefore, nanocrystalline grains are formed by α-modification of titanium. An analysis of the dark-field of the surface layer in the reflections of the crystal lattice of α-titanium revealed a speckled contrast throughout crystallites (figure 1 c). This fact may be evidence of a defect substructure in nano-crystallites of α-titanium. Besides reflections of α-titanium, there are reflections of the titanium oxide phase detected in the electron-diffraction micro-pattern. Titanium oxides are located mainly along the edges of α-titanium crystallites; the particles are rounded with sizes varying in the range of 10 nm (figure 1 d). It should be noted that the particles of the oxide phase are found only in the nanostructured surface layer of the sample. Oxide phase particles were not detected throughout the sample. It can be assumed that particles of the oxide phase are formed, because atoms of oxygen penetrate along the small and large-angle edges of the crystallites during fatigue testing of the material.

It was found that after fatigue testing of samples subjected first to electric-current pulse treatment (359,532 cycles), the submicrocrystalline structure (average size of crystallites is 422.7 ± 400 nm) is formed in a relatively thin (maximum 4 μm) surface layer of commercially pure titanium VT1-0 (figure 2). Therefore, electric-current pulse treatment of commercially pure titanium VT1-0 during fatigue testing results in a multiple increase in dimensions of crystallites in the surface layer. The other special feature of the grain structure in the surface layer is the comparatively high dispersion of grain sizes. It is assumed that the reason for this is the recrystallization processes of the material that occur during electric-current pulse treatment.

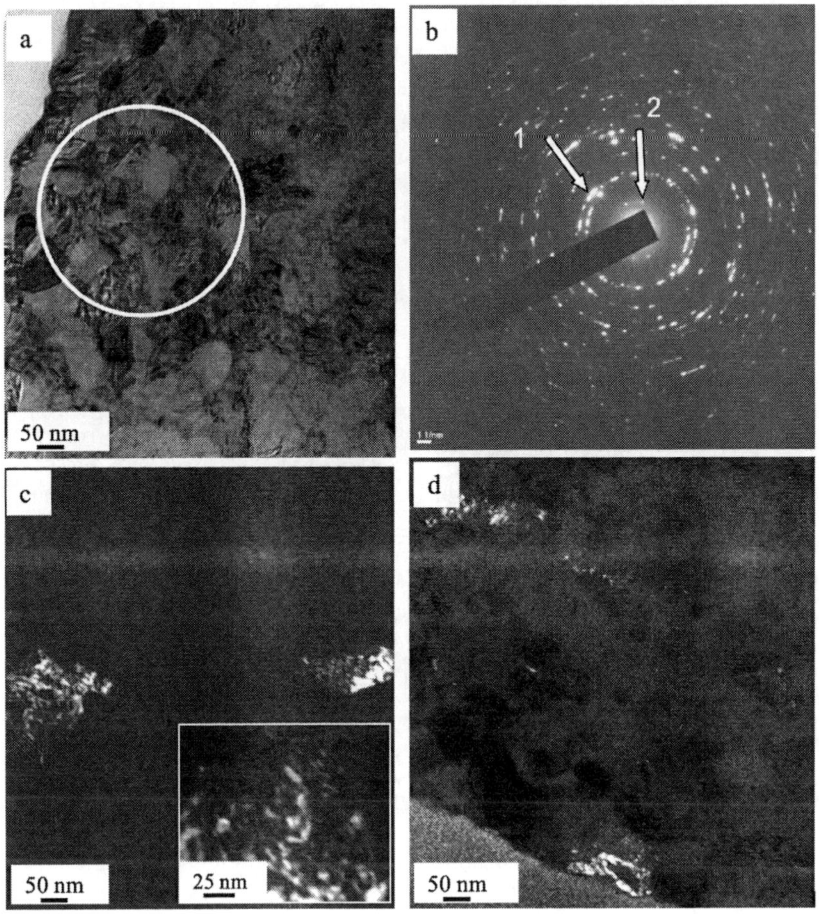

Figure 1. The cross section structure in a sample of commercially pure titanium VT1-0 fractured in the process of fatigue testing (281 333 cycles); a – bright-field image of the surface layer; b – electron-diffraction micro-pattern, taken on the zone in the oval (a); c – dark-field image of zone (a), taken in [002]α-Ti (reflection 1, shown by the arrow in b); d – dark-field image of zone (a), taken in [020]Ti_3O_5 (reflection 2, shown by the arrow in b).

Another feature of the samples of commercially pure titanium VT1-0, fractured during fatigue testing in conditions of electric-current treatment, is a rather low level of bending torsion in the crystal lattice of the material. Thus, this treatment of commercially pure titanium during fatigue testing decreases the number of sources (concentrators) of the internal stress fields in the surface layer of the material.

The principal phase in the layer of titanium fractured during fatigue testing with electric-current treatment is α-titanium, the second phases is the particles of titanium oxides located along the edges of α-titanium grains and subgrains. The oxide phase particles vary from 10 to 35 nm, significantly exceeding the size of the oxide phase particles formed in the surface layer of titanium subjected to fatigue testing without energy impact.

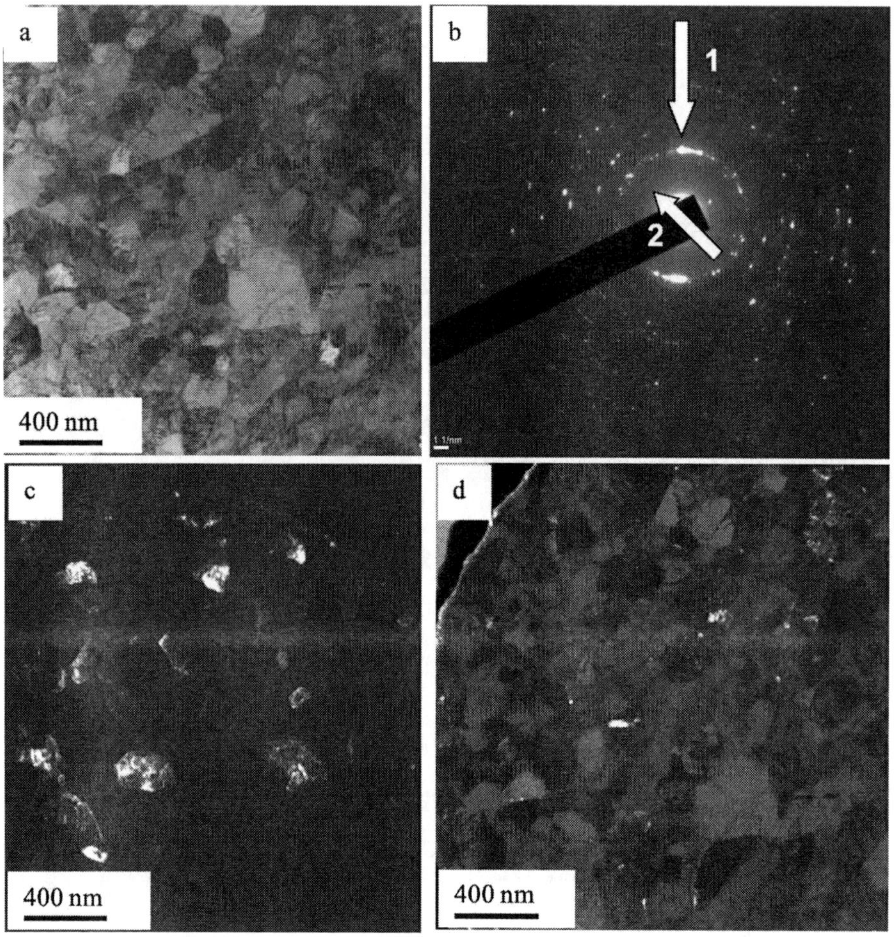

Figure 2. The cross section structure in a sample of commercially pure titanium VT1-0 fractured in the process of fatigue testing after electric-current treatment (359 532 cycles); a – bright-field image of the surface layer; b – electron diffraction micro-pattern of the foil section (a); c – dark-field image of the foil section (a), obtained in [011]α-Ti (reflex 1 indicated by the arrow in b); d – dark-field image of the foil section (a), obtained in [200]Ti_2O (reflex 2 indicated by the arrow in b).

4. Conclusions

To sum up, electric-current pulse treatment of commercially pure titanium VT1-0 results, firstly, in a multiple increase in sizes of α-titanium crystallites in the surface layer of the material due to recrystallization processes, secondly, reduces the number of sources (concentrators) of internal stress fields in the surface layer of the material, thus reducing the number of potentially dangerous spots of crack formation; thirdly, it causes a significant increase in size of the oxide phase particles formed in the surface layer of commercially pure titanium during fatigue testing.

Acknowledgments

The work is performed under the financial support of State Order Ministry of Education and Science Russian Federation [# 3.1283.2017/4.6] and Russian Foundation for Basic Research (project No 16-

32-60048_mol_a_dk).

References

[1] Leuders S, Thöne M, Riemer A, Niendorf T, Tröster T, Richard H A and Maier H J 2013 *Int. J. Fatigue* **48** 300-7.
[2] Williamson A, Lombardi D A, Folkard S, Stutts J, Courtney T K and Connor J L 2011 *Accid. Anal. Prevent.* **43** 498-515.
[3] Chan K S 2010 *Int. J. Fatigue* **32** 1428-47.
[4] Mughrabi H and Höppel H W 2010 *Int. J. Fatigue* **32** 1413-27.
[5] Yasniy P V, Okipnyi I B, Maruschak P O, Panin S V and Konovalenko I V 2013 *Theor. Appl. Fract. Mech.* **63-64** 63-68.
[6] Panin S V, Vlasov I V, Sergeev V P, Maruschak P O, Sunder R and Ovechkin B B 2015 *Int. J. Fatigue* **76** 3-10.
[7] Estrin Y and Vinogradov A 2010 *Int. J. Fatigue* **32** 898-907.
[8] Zhu Y H, To S, Lee W B, Liu X M, Jiang Y B and Tang G Y 2009 *Mater. Sci. Eng. A* **501** 125-32.
[9] Tang G, Zhang J, Yan Y, Zhou H and Fang W 2003 *J. Mater. Process. Technol.* **137** 96-99.
[10] Jiang Y, Tang G, Guan L, Wang S, Xu Z, Shek C and Zhu Y 2008 *J. Mater. Res.* **23** 2685-91.
[11] Sosnin O V, Gromova A V, Ivanov Yu F, Konovalov S V, Gromov V E and Kozlov E V 2005 *Int. J. Fatigue* **27** 1186-91.
[12] Petrunin V A, Sosnin O V, Konovalov S V, Tsellermaer V V and Gromov V E 2004 *Steel in Transl.* **34** 64-8.
[13] Leikina O S, Konovalov S V, Sosnin O V and Kozlov E V 2004 *Steel in Transl.* **34** 77-9.
[14] Veiga C, Davim J P and Loureiro A J R 2012 *Rev. Adv. Mater. Sci.* **32** 133-48.
[15] Shang G, Zhu Z, Chang H, Wang X, Kou H A and Li J 2011 *Xiyou Jinshu/Chinese Journal of Rare Metals* **35** 286-91.
[16] Agzamov R D, Tagirov A F and Ramazanov K N 2018 *Def. Diff. Forum* **383** 161-6.
[17] Pramanik A and Littlefair G 2015 *Machin. Sci. Technol.* **19** 1-49.

IOP Conf. Series: Materials Science and Engineering **447** (2018) 012076 doi:10.1088/1757-899X/447/1/012076

Estimation the surface roughness after mineralceramic smoothing of electrophysical coating

R R Dema[1], S I Platov[1], and D N Romanenko[2] O R Latypov[1]

[1]Department of Metal Forming Machines and Technologies, Nosov Magnitogorsk State Technical University, Lenin Avenue, 38, Magnitogorsk, 455000 Russia
[2]Laboratory of Structural and Special Steels, All-Russian-Scientific Research Institute of Aviation Materials. The street Radio, 17, Moscow, 105005 Russia

E-mail: demarr78@mail.ru

Abstract. Roughness is an important factor in the working properties of instruments. Roughness determines such working properties as durability, wear resistance from detritions, density of combinations, chemical fortitude and appearance. A method for evaluating the roughness of wear–resistant electric-spark coatings on a metal cutting instrument and a specimen of fast-cutting steel is a complex parameter of roughness (Δ). Complex parameter of roughness (Δ) allows more accurately estimating the surface roughness after smoothing the mineral ceramics of an electrophysical coating and others types of surface treatment.

1. Introduction

The development of modern mechanical engineering is connected with wear-resistance of cutting instruments to increase the efficiency of treatment processes. The surface layer of a detail exposes the strongest mechanical, thermal, chemical and some other types of impacts in the process of operation. The deprivation of the detail working capacity takes place, as a rule, from the surface as a result of wear, erosion, corrosion, thermal effect and etc. A significant resource for increasing efficiency consisted in the material from which the elements of constructions are made, as well as and the surface roughness of this material.

The aggregation of surface irregularities with relatively small steps on base length is the surface roughness (figure 1). Roughness concerns to microgeometry of a solid and determines its most important exploitative qualities. Wear resistance from detritions, durability, density of combinations, chemical resistance and appearance are the most important exploiting properties. There is also a connection between the ultimate deviation of the size and roughness [1, 3-5].

The altitudes parameters are as follows: *Ra* is the arithmetic mean deviation of the profile. $Ra = \frac{1}{n}\sum_{i=1}^{n} y_i$. *Rz* is the height of irregularities in the profile of 10 points. The step parameters are as follows: t_p is the relative support length of the profile, where p is the sectional values of the profile from the row: 10; 15; 20; 30; 40; 50; 60; 70; 80; 90%. $t_p = \frac{1}{l}\sum_{i=1}^{n} b_i$. *Ra*, *Rz* and *Rmax* are determined on base length l, which can accept values from the row: 0.01; 0.03; 0.08; 0.25; 0.80; 2.5; 8; 25 mm. The parameter *Ra* is preferred [2].

IOP Conf. Series: Materials Science and Engineering **447** (2018) 012076 doi:10.1088/1757-899X/447/1/012076

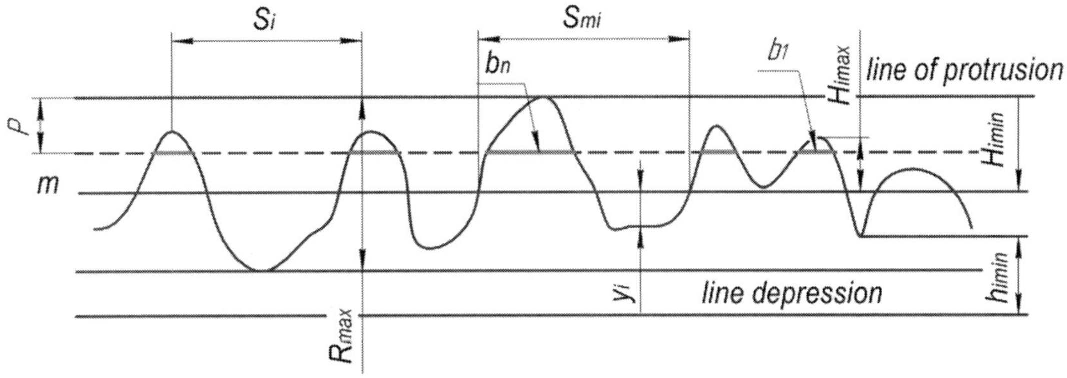

Figure 1. Normal profile and parameters of the surface roughness [2], where: l is the base length; m is the middle line of the profile; S_{mi} is the average pitch of surface irregularities; S_i is the average pitch of the local profile protrusions; H_{imax} is the deviation of the five highest maximums of the profile; H_{imin} is the deviation of the five highest minimums of the profile; h_{imax} is the distance from the highest points of the five largest maximums to the line, which is parallel to the middle line and does not cross the profile; h_{imin} is the distance from the lower points of the five largest maximums to the line, which is parallel to the middle line and does not cross the profile; Rmax is the largest height of the profile; y_i is the deviation of the profile from line m; p is the section level of the profile; b_n is the length of segments cut off at the level p.

2. Methods and procedure

Effective opportunities have been discovered to increase the operational capacity of the friction surface in connection with the development of new methods of the surface treatment of materials that provide a specified set of surface density and roughness.

Now, there are modern methods to increase the performance of blade metal cutting instrument due to changes in the physicochemical and operational properties of the working edges of metal surfaces in a given direction. These include: technological, thermal, chemical, chemicothermal, thermomechanical, electrophysical and mechanical methods.

Electric-spark coating has a row of advantages as compared to some other methods. There are several advantages of the electrospark coating:

1. The highest durability strength of the applied material with warp (due to mechanical mixing and the relative diffusion penetration of electrode materials in the formed layer);
2. The locality of the process;
3. The possibility of using pure metals and many alloys, metal-ceramic composition, refractory compound and etc. as alloying materials;
4. Lack of heating or minor heating of the detail in the treatment process, which may notice changes in its physical and mechanical attributes and geometry;
5. The technological process is simple, the equipment is compact and transportable;
6. Restoring the size of machine parts and instrument, decrease in inclination to grasp of surface in friction, holding of micro metallurgical process on the treatment surface for forming necessary connections.

3. Results

Hardening working surfaces of the metal cutting instrument made of powder fast-cutting steel of increased heat resistance R12M3K8F2-MP consist of the RF Specification TU 14-1-3647-83: C-1.0%; W-12.0%; Mo-3.1%; Cr-4.0%; V-2.1%; Co-8% realized on installation of «EIF-541» by local the electric spark method (LENP). Disk milling cutters, spiral drills, passing cutter, hard cutter, cutting cutter were subjected to the LENP method [3]. Cylindrical rods from self-fluxing powder alloy of the Ni-Cr-B-Si-C system on nickel chrome basis (%): Ni 80; Cr 15; B 2-3; Si 3-2 и C 0.1-0.7 with a diameter of 1…1.5 mm used as electrodes. This material was developed in the USA and was called «Colmonoy».

Local electric-spark coating on the working surfaces of the cutting instrument in places of intensive wear and tear on the rear and front surfaces is accompanied by the education of microrelief with rounded projections and hollows. Herewith, in contrast to mechanical treatment, the microrelief has not acute a scallop. However the quality of the instrument surface often does not match the technical conditions after LENP (figure 2). Ra roughness of instrument cutting edge after LENP is 4…6 μm under optimal conditions, but 0.2…0.4 μm is required. This significant drawback is decided by applying a combined treatment. This treatment consists of: LENP, adamantine and mineral–ceramic smoothing, LENP plus laser treatment.

(a) (b)

Figure 2. Photos of the drill: (a) with electric-spark coating,(b) surface with wear-resistant coating. Photos of the drill: (a) with electric spark coating, (b) with a surface of wear-resistant coating.

Significant contrast stress arises from the process of smoothing the surface in the place of contact of the deforming element and the workpiece. Plastic deforming of the surface layer occurs at a certain effort of smoothing, as a result of which microroughnesses crumble and the physic-mechanical properties of the surface layer change. Significant residual stress of compaction arises on the ironed surface. Details with a surface treated with smoothing have increased operating properties (high wear resistance and fatigue strength).

The technology of smoothing was realized on crew-cutting 1K62 lathe with the device described in works [4, 5]. Smother from mineral-ceramic BOK-60, belonging to the class of oxide-carbonate cutting ceramics of composition: Al_2O_3 – 76; TiC – 20; (WCo) C – 3…5; Mg<1 (at%) with a square cross-section 2×2 mm, a length of up to 15…20 mm and a working surface radius of 4.5…6.6 mm was used as an instrument. Optimal modes of processing conditions for smoothing with metal-ceramics were installed by an experiment with a target for ensuring maximum fortitude of smother from metal-ceramic BOK-60 and the required parameter of roughness. The roughness of processed surface depends on the moods of smoothing (effort of smoothing and innings). The number of approaches and the speed of smoothing to a lesser extent affect the roughness and microhardness of the processed surface.

The roughness was measured by an Outline-67 profilograph-protilometer by means of the complex parameter of roughness (Δ), proposed by the authors in work [6].

$$\Delta = \frac{R_{max}}{r \cdot b^{\frac{1}{v}}}, \qquad (1)$$

where R_{max} is the largest height of the profile; r is the middle radius of curvature of the vertices of the projection; b and v are the parameters of the initial section of the reference curve, which is built in absolute coordinates. These coordinates are determined by formulas (2) and (3) respectively.

$$b = t_{R_p} \left(\frac{R_{max}}{R_p} \right)^{\nu}, \quad (2)$$

$$\nu = 2 \cdot t_{R_p} \cdot \frac{R_p}{R_a} - 1, \quad (3)$$

where Ra is the arithmetic mean deviation of the profile; t_{Rp} is the relative support length of the profile on the level Rp; Rp is the deviation from the line of prominences to the line, which cuts the square into two equal parts.

4. Conclusions

The value of Ra amounted to 0.28…1.28 μm after smoothing the coatings with cermet. The complex parameter of roughness, which is defined by formula (1), was 1.55 before smoothing, but after smoothing it became 0.08…0.14.

The durability tests of the instrument were realized according to GOST 2034-80 «Twist drills. Specifications», 2M55 radial-drilling machine, workpiece material - steel 45GOST1050-88.

As a result of the tests, it is established: 1) Drill No. 1 with electric-spark alloying has wearing on the front surface no more than 0.3 mm; 2) DDrill No. 2 without electric-spark alloying has wearing on the front surface no more than 1.3 mm; 3) The actual persistence of the drill with electric-spark coating is 17.4 minutes to the first regrinding; 4) The actual persistence of the drill without electric-spark coating is 4.3 minutes to the first regrinding; 5) Industrial tests of hardened cutting instrument showed an increase in their persistence in 4 times.

Analysis of the obtained results shows that quality of the surface, which was formed after LENP, is insufficient by the roughness parameters Ra and Rmax. The complex roughness parameter (Δ) an order of magnitude more accurately to measure the roughness of surface after adamantine and cermet smoothing of electrophysical coatings, in particular LENP and some other surface treatment [7-9].

Acknowledgments

The work was supported by the Ministry of Education and Science of the Russian Federation under the project No. 11.2054.2017 / PH under the state assignment for 2017-2019 (Number for publication: 11.2054.2017 / 4.6).

References

[1] Anuriev V I 2001 *Handbook of the Designer-Machine Builder in Three Volumes* (Moscow: Machine-building industry) p 920
[2] State Standard 2789-73 2006 *Surface Roughness. Parameters and Characteristics* (Moscow: StandartInform) p 7
[3] Gadalov V N, Romanenko D N, Samoilov V V, Nikolaenko A V and Grigoriev S B 2012 *Russian J. Non-Ferr. Met.* **53** 348
[4] Gadalov V N, Gvozdev A E, Romanenko D N, Vornacheva I V and Kovalev S V 2016 *Russian Eng. Res.* **36** 731
[5] Gadalov V N, Romanenko D N, Markelov E A and Nikolaenko A V 2013 *Russian Eng. Res.* **33** 37980
[6] Burumkulov F X, Ivanov V I, Lyalakin V P, Cherkasov V V and Petrovskij S N 2005 *Technol. Metals* **6** 42
[7] Shcherenkova I S, Shkatov V V, Gadalov V N and Romanenko D N 2015 *Chem. Petrol. Eng.* **51** 277
[8] Kolmykov V I, Romanenko D N, Abyshev K I and Nasteka V V 2015 *Chem. Petrol. Eng.* **50** 610
[9] Kolmykov V I, Romanenko D N, Abyshev K I, Kolmykov D V and Bedin V V 2015 *Chem. Petrol. Eng.* **51** 58

IOP Conf. Series: Materials Science and Engineering **447** (2018) 012077 doi:10.1088/1757-899X/447/1/012077

Structure of SnO_2-Ag coating formed on copper by electroexplosion

D A Romanov, S V Moskovskii, K V Sosnin, V E Gromov and A D Filyakov

Prof. V. Finkel Chair of Natural Science, Siberian State Industrial University, 42 Kirov st., Novokuznetsk 654007, Russia

E-mail: romanov_da@physics.sibsiu.ru

Abstract. Studies of the phase and elemental composition of the surface layer of the copper electric contact of the KPV-604 contactor subjected to electro-explosive spraying of the composite SnO_2-Ag- coating were carried out by means of transmission electron microscopy. The scale of elemental structure of the coating surface varies within a very wide interval after electro-explosive spraying – from hundreds of micrometers to tens of hundreds of nanometers. According to the morphological structure, two layers can be distinguished in the coating volume: the coating itself and the heat affected layer smoothly transferring into the bulk of the sample. A nanocrystalline structure was detected. The main phases of the coating are SnO_2, Ag_3Sn, $Cu_{10}Sn_3$, Cu_3Sn, Cu_6Sn_5, Ag_4Sn and CuO. The volume of copper adjacent to the coating has a fragmented structure indicating a high level of deformation of the sample surface layer under the electro-explosive formation of the coating.

1. Introduction

According to the data of the International Association "Interelectromash", the share of failures in the operation of electrical equipment due to the failure of a contact instrument takes the first place among other problems and amounts to 26%. To restart the work, the contact must be replaced by the new one. A combination of different and incompatible requirements is typical for electric contact materials. These are high hardness, melting temperature, electrical and heat conductivity, electrocorrosion and corrosion resistance combined with the absence of welding and the formation of bridges. The application of powder metallurgy methods [1] enables to realize a diverse and contradictory set of properties that the electrocontact should have in one material. Nowadays, a large number of electrocontact materials are developed for application in various service conductions. Their composition includes, as a rule, a matrix of high electric conductivity and a high-melting component (filler) with high wear and electroerosion resistance [2]. Composite materials based on the silver, copper, copper-nickel, aluminum matrix and the high-melting filler are the most promising for the manufacturing of contacts. The following systems are among them: W-Cu, Mo-Cu, W-C-Cu, Mo-C-Cu, Mo-C-Cu, Ti-B-Cu, Ti-B-Cu, TiB_2-Cu, TiB_2-Al, W-Ni-Cu, Mo-Ni-Cu, Cr-C-Cu, Cr-Cu, CdO-Ag, SnO_2-Ag, W-Ag, Mo-Ag, W-C-Ag and Mo-C-Ag etc. [3-8].

One of the most advanced directions of condensed matter physics is the development of methods for enhancing the service characteristics of different materials. In this regard, the surface hardening of arc-resistant electric contacts is one of the most urgent problems in the development of modern technologies.

The aim of this work is to study the phase composition, the defect substructure and the phase morphology of the electroexplosion coating of the SnO_2-Ag system by means of transmission electron microscopy.

2. Materials and methods

The object of the study was the copper contacts of the KPV-605 contractor on the surfaces of which the coating of the SnO_2-Ag system was formed by the electro-explosive method [9]. 250 mg of silver foil was used as an electrical current conducting material, on the surface of which the charge of 50 mg of the SnO_2 powder was placed in the explosion area. The time of plasma impact on the sample surface was ~ 100 μs, the absorbed power density on the jet axis was ~5.5 W/m^2, the pressure in the shocked layer near the irradiated surface was ~ 12.5 MPa, the residual gas pressure in the working chamber was ~100 Pa; the plasma temperature on the nozzle cut was ~10^4K, the thickness of the thermal affected zone was ≈ 50 μm. Elemental and phase composition, morphology and defect structure of the coating were analyzed by means of transmission electron diffraction microscopy (JEM – 2100F device, JEOL). Foils for the material testing were manufactured from thin (≈100 μm) plates located in the sample cross section by the electron thinning method. It enabled to study the structure of the coating and the transition layer, which separates the coating from the substrate.

3. The results and discussion

A typical electron microscope image of a surface layer of copper subjected to an electro-explosive modification is shown in figure 1. It is clearly seen that the copper layer adjacent to the coating has a fragmentary structure, which indicates a high level of deformation of the sample surface during the formation of the coating.

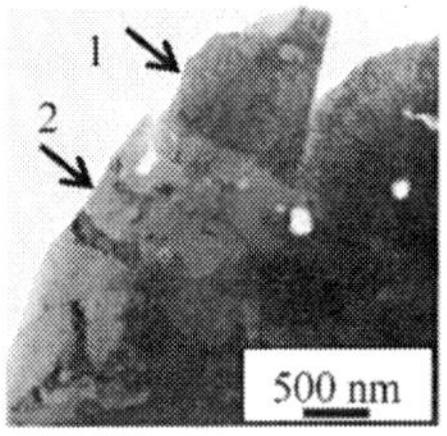

Figure 1. Electron microscope image of the structure of the "coating/substructure" system formed on copper by the electro-explosive method. The arrows indicate: a coating layer adjacent to the surface of copper (1) and a copper layer adjacent to the coating (2).

Figure 2 shows the electron microscope image of the coating formed under the electro-explosive modification of the copper contact of the KPV-605 contactor. The obtained images indicate that the coating was formed has a nanocrystalline structure with a size of crystallites varying from 20 to 40 nm.

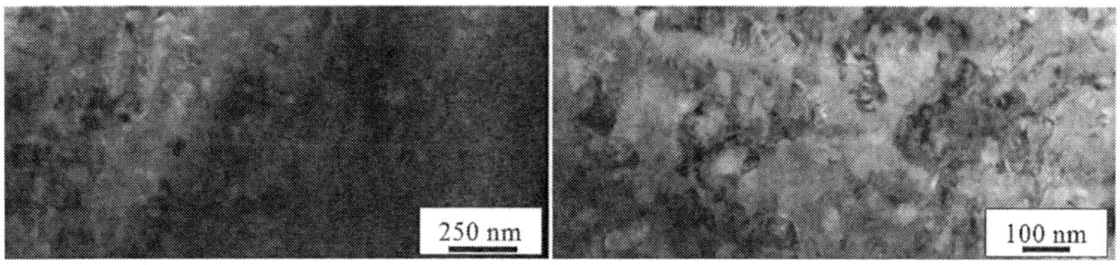

a *b*

Figure 2. Electron microscope image of the structure of the coating formed on the copper contact of the KPV-605 contactor by the electro-explosive method.

The elemental composition of a part of the coating surface layer (figure 3 b) was determined by means of the micro X-ray spectrum analysis. Energy spectra and conditions of the analysis are shown in figure 3 a. The results of the quantitative analysis of the elemental composition of the coating material are presented in table 1.

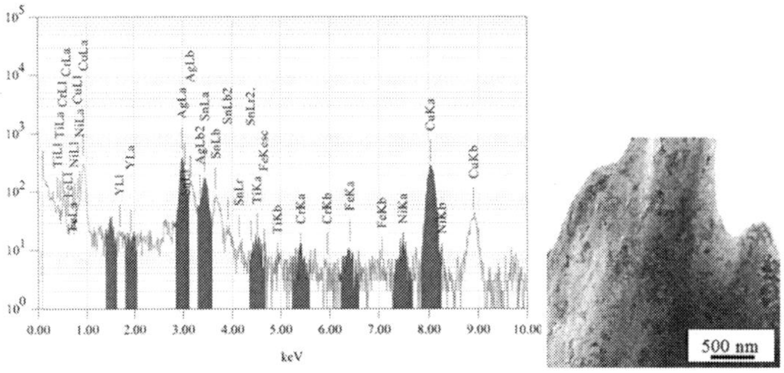

a b

Figure 3. Energy spectra (a) of a part of the coating (b) revealed by the micro X-ray spectrum analysis of the foil.

As a result of studies performed by this method, it was established that the analyzed coating appeared to be a multi-elemental material, the main metallic elements of which are copper, silver and tin. Consequently, a mix of elements of the substrate and the coating was observed as a result of electro-explosive coating.

Table 1. Results of the analysis of the micro X-ray spectrum of a part of the coating (figure 3 b) formed as a result of electro-explosive coating of modified copper.

Element	Excitation energy	Fractional content %
Al (K)	1.486	1.83
Ti (K)	4.508	0.8
Cr (K)	5.411	0.56
Fe (K)	6.398	0.34
Ni (K)	7.471	1.24
Cu (K)	8.040	39.3
Ag (L)	2.984	39.18
Sn (L)	3.443	16.76

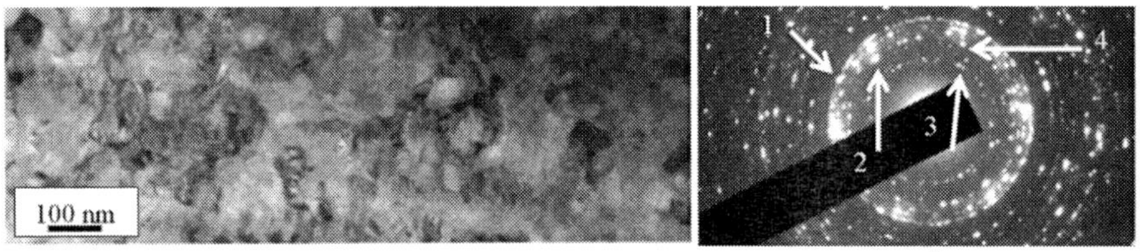

a *b*

Figure 4. Electron microscope image of a part of the coating formed on copper sample by the electro-explosive method (a) and the corresponding microelectron diffraction pattern (b). The arrows indicate the reflections in which the dark field images of the foil part (a) shown in figure 5 were obtained, namely reflection 1 – [004] Ag_3Sn; reflection 2 – [100] Ag_4Sn; reflection 3 – [002] Cu_4Sn_5; reflection 4 – [110] SnO_2.

The phase composition and the defect structure of the coating were analyzed by means of diffraction electron microscopy. Following the results shown before [10], it may be concluded that in the Ag-Sn-Cu system, the compounds based on Ag-Sn and Sn-Cu can be formed; the presence of silver, tin and copper based solid solutions is possible as well. The presence of oxide phases of the indicated elements in the coating should not be excluded.

A typical electron microscope image of the coating structure (figure 4 a) and the microelecton diffraction pattern corresponding to the structure (figure 4 b) are shown in figure 4. It is clearly seen that the microelecrton diffraction pattern has a circular structure, as evidenced by the essential quality of the coating as a class of nanocrystalline material. Taking into consideration the nonequilibrium conditions of transformation occurring during the electro-explosive formation of the coating and the possible presence of several phases in the analyzed part of the foil, the indication of microelectron diffraction patterns was done through determining the phase belonging of each of the five nearest diffraction rings separately.

The indication of the microelectron diffraction pattern shown in figure 4 b revealed reflections of the following phases: SnO_2, Ag_3Sn, $Cu_{10}Sn_3$, Cu_3Sn, Cu_6Sn_5, Ag_4Sn and CuO. When analyzing the results of studies of the phase composition of the coating, it may be noted that the particles of SnO_2 tin dioxide have a round shape and a size of particles varying within 15-40 nm. The particles of the Ag_3Sn, Ag_4Sn and Cu_6Sn_5 compounds are relatively coarser (20-50 nm) and they often form areas of 80-120 nm.

4. Conclusions

The study of the elemental and phase compositions, the defect substructure, and the morphology of the coating phases formed on the surface of copper samples was carried out by means of electron diffraction microscopy. The formation of a multi-elemental multi-phase coating of a nanostrystalline structure was determined. Through the indication of the microelectron diffraction pattern, it was stated that the main phases of the coating are SnO_2, Ag_3Sn, $Cu_{10}Sn_3$, Cu_3Sn, Cu_6Sn_5, Ag_4Sn and CuO. It was found that the volume of copper adjacent to the coating has a fragmentary structure, indicating a high level of deformation of the surface layer if the samples during the electro-explosive formation of the coating.

Acknowledgments

The investigation is financially supported by RFFI within the framework of scientific projects No. 16-32-60032 mol_a_dk, 18-32-00075 mol_a and the grant of the President of the Russian Federation for the state support of young Russian scientists – PHD MK – 1118.2017.2.

References

[1] Il'yushchenko A F 2010 *50 Year Anniversary of Belorussian Metallurgy. History, Chievements, Prospects* (Minsk: SRU "Institute of Powder Metallurgy") (In Russian)
[2] Gerard F 2005 *Advances in Condensed Matter and Materials Research* (New York: Nova Science Publisners)
[3] Taiasov K A, Isupov V P, Bokhonov B B, Gaponov Yu A, Tolochko B P, Sharafutdinov M R and Shatskaya S S 2000 *J. Mater. Synth. Process.* **8** 21
[4] Russell A M and Lee K L 2005 *Structure-Property Relations in Nonferrous Metals* (Hoboken: John Wiley & Sons)
[5] Talijan N M 2011 *Zastita Materijala* **52** 173
[6] Qureshi A H, Azhar S M and Hussain N 2010 *J. Therm. Anal. Calorimetry* **99** 203
[7] Bukhanovsky V, Rudnytsky M, Grechanyuk M, Minakova R and Zhang C 2016 *Mater. Technol.* **50** 523
[8] Timofeeva L A and Morozov V S 2015 *Information – control systems in railway transport* **6** 37 (In Russian)
[9] Panin V E, Gromov V E, Romanov D A, Budovskikh E A and Panin S V 2017 *Dokl. Phys.* **62** 67
[10] Kroupa A T, Dinsdale A, Watson A, Vrestal J and Zemanova A 2007 *J. Mining Metallurgy Sect. B Metallurgy* **43** 113

IOP Conf. Series: Materials Science and Engineering **447** (2018) 012078 doi:10.1088/1757-899X/447/1/012078

Gradient structure formation in the process of severe plastic deformation of carbon steels under the conditions of the effect of dynamic strain aging

G I Raab[1,2], I S Kodirov[1], G N Aleshin[1], A G Raab[1,2], N K Tsenev[3,4]

[1]Ufa State Aviation Technical University, 12 K. Marx str., Ufa, 450008 Russia
[2] Nosov Magnitogorsk State Technical University, 38 Lenin ave., Magnitogorsk, 455000 Russia
[3]Ufa Russia Transneft Research Institute, 144/3 pr. Oktyabrya, Ufa, 450055 Russia
[4]Ufa State Petroleum Technical University, Kosmonavtov st. 1, Ufa, 450062, Russia

E-mail: galioshin@mail.ru

Abstract. The formation of a gradient structure during the severe plastic deformation (SPD) by high-pressure torsion (HPT) of low- and medium-carbon steels under the influence of dynamic strain aging (DSA) is studied. Deformation mechanisms under various regimes of deformation processing are analyzed. The temperature ranges of the DSA effect during the ECAP processing of steel 10 and the fact of the formation of a gradient structure during the HPT processing under these conditions are established. It is shown that the deformation of carbon steels in the DSA temperature range enables controlling the structure of the deformed state.

1. Introduction

In recent years, a new field of research has emerged and is being developed – the study of gradient structural states in metallic materials. Research results demonstrate that the presence of a gradient structure enables rendering new, unprecedented properties to a material [1-5]. This structure combines a hard corrosion- and wear-resistant surface and a relatively «soft» core, which enables redistributing the loads and relaxing the stresses [5-6].

In this work, we make an attempt to study the features of the formation of a gradient structure under conditions of dynamic strain aging (DSA) during severe plastic deformation (SPD) of low- and medium-carbon steels by high-pressure torsion in the temperature range of DSA.

2. Material and experimental procedure

As materials for the study, we selected low-carbon steels 10 and 20, and medium-carbon steel 45, having standard chemical compositions [7]. Samples of the steels in the form of disks with a diameter of 10 and 20 mm and a thickness of 0.5 mm were subjected to high-pressure torsion (HPT) under a pressure of 6 GPa at temperatures of 20, 250 and 400 °C. For this purpose, a die-set similar to that described in [8, 9] was used,

HPT processing was performed for 5 revolutions, and the logarithmic strain in the middle of the radius was $\varepsilon \approx 3.45$. After processing, the microhardness of these samples was measured, and their microstructure was studied using light and electron microscopes. Samples of steel 10 were subjected to severe plastic deformation by HPT at 20, 250 and 400 °C, and samples of steels 20 and 45 were subjected to HPT at 20 and 400 °C.

After processing, the microhardness of all samples was measured, and their structure was studied using light and electron microscopes.

3. Experiment results

3.1. HPT processing of steel 10

The structure of steel 10 in the initial state represents ferrite grains with sizes of ~ 9-13 μm and pearlite grains with sizes of ~ 1=5 μm. Figure 1 shows the microstructure of steel 10 in the initial state and after 5 revolutions of HPT. As can be seen, the pearlite grains in the central region of the samples retain the equiaxed shape, but in the peripheral regions both the ferrite and pearlite grains are severely elongated along the direction of metal flow, and the strongest structure refinement takes place. Cementite plates in pearlite are, as a rule, fragmented in both the central and peripheral regions of the samples.

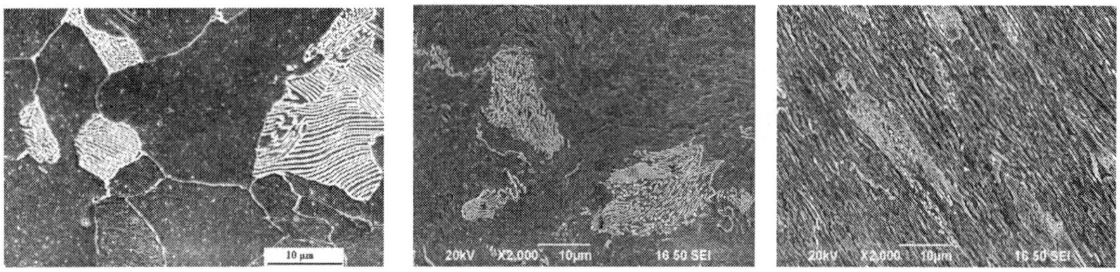

(a) (b) (c)

Figure 1. Microstructure of steel 10: a – initial state, b, c – 5 revolutions of HPT at 20 °C, b – central region of the samples, c – peripheral region. SEM.

Figure 2 shows the microhardness distribution along the diameter of the samples of steel 10 after HPT at different temperatures. It can be seen that the microhardness of the steel after HPT at 20 °C is lower than that after HPT at 250 and 400 °C. HPT processing at 250 °C leads to a more profound strengthening of the material in the peripheral region of the sample, while HPT processing at 400 °C results in a nearly uniform strengthening of the steel along the sample diameter. The microhardness in the central region of the sample (with a diameter not exceeding 2 mm), processed at 400 °C, is higher than that after processing at 250 °C.

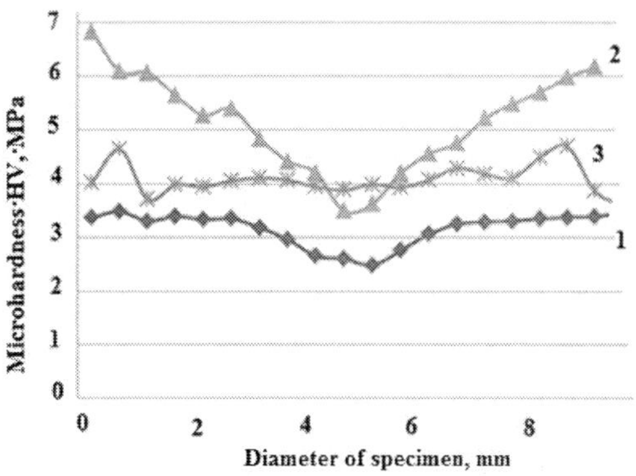

Figure 2. Microhardness of steel 10 along the diameter of disk-shaped specimens after 5 HPT revolutions: 1 – 20 °C, 2 – 250 °C, 3 – 400 °C.

This combination, observed in the studied samples, of the ultrafine-grained structure in the peripheral regions of the samples and the relatively coarse-grained structure in the central region after the HPT processing is in essence the gradient structure [10-14].

3.2. HPT processing of steel 20 and steel 45

The structure of the steels in the initial state consists of ferrite grains with a size of ~ 15-25 μm and pearlite grains with a size of ~ 0.5-20 μm. The volume fraction of a pearlite in the investigated steels 20 and 45 is, respectively, ~ 25% and 60%. The average microhardness value of steels 20 and 45 in the heat-treated condition was 2200 and 2450 MPa, respectively.

Figure 3 displays the HV microhardness values along the diameter of the steel 20 samples. A similar microhardness distribution along the sample diameter is also observed in steel 45.

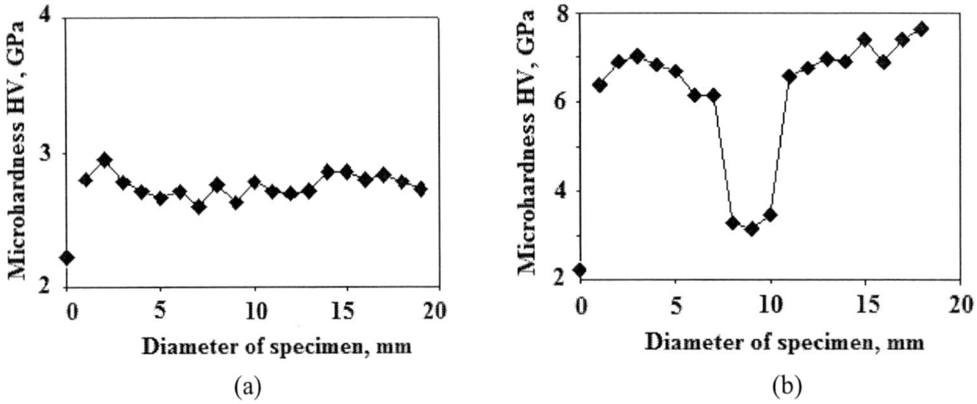

Figure 3. Microhardness distribution along the diameter of the steel 20 specimens after HPT at 20 °C (a) and at 400 °C (b).

Microstructural analysis shows that after the HPT processing of steels 20 and 45 at a temperature of 20 °C there is observed a slight grain refinement, which is more noticeable at the periphery of the samples. The shape of grains at the periphery is close to equiaxed. The microhardness distribution along the diameter of the samples is rather non-uniform and increases after HPT processing approximately by 20% as compared to the initial state, amounting to 2.7 GPa.

Thus, severe plastic deformation of low- and medium-carbon steels in the temperature range of dynamic strain aging results in an increase in the yield stress and the strain hardening parameters, primarily owing to the decrease in the size of structural elements, as well as the additional influence of the second-phase particles. This contributes to the formation of the gradient structure and properties in the investigated steels.

4. Conclusions

1. We considered the features of the gradient structure formation during the processing by high-pressure torsion of steel 10, steel 20 and steel 45 in the conditions of the effect of dynamic strain aging. We analyzed the role of activation of diffusion processes during the SPD processing.

2. We studied the structural changes occurring in the steels during the processing by high-pressure torsion in the temperature range of the effect of dynamic strain aging and, correspondingly, the formation of a gradient structure in the steels. The formation of the gradient structure is promoted by the activation of the DSA effect.

3. As a result of the DSA effect, in the process of severe plastic deformation, there occurred a further structure refinement in the alloys and, consequently, a larger increase in the yield stress and microhardness, as compared to the samples processed at room temperature.

4. We analyzed the mechanisms of plastic deformation leading to the formation of a gradient structure during the processing of steel 10 by high-pressure torsion – this is the shear strain leading to the activation of dislocation slip and the interaction of dislocations with impurity atoms, as well as the development of rotational deformation modes and the diffusion processes of carbon atoms redistribution.

Acknowledgements
The authors acknowledge gratefully the financial support from the Russian Foundation Basic Research through project № 17-08-00720. G I Raab and A G Raab acknowledge the financial support from the Ministry of Science and Higher Education of the Russian Federation (Grant 14.Z50.31.0043)

References

[1] Kozlov E V, Gromov V E, Kovalenko V V, Popova N A and Zhuleikin S G 2004 *Gradient Structures in Pearlitic Steel* (Novokuznetsk: Siberian State Industrial University) (in Russian)
[2] Ivanov Yu F, Kovalenko V V, Kozlov E V and Gromov V E 2006 *Gradient Structural and Phase States in Steels* (Novosibirsk: Nauka) (in Russian)
[3] Pashinskaya E H, Varyukhin V N, Maksakova A A, Maksakov A I, Tolpa A A and Makarova A V 2014 *Phys. Technol. High Pressure* **24** 120–126 (in Russian)
[4] Kovalenko V V, Kozlov E V, Ivanov Yu F and Gromov V E 2009 *Physical Nature of the Formation and Evolution of Gradient Structural and Phase States in Steels and Alloys* (Novokuznetsk: Poligrafist) (in Russian)
[5] Ivanov Yu F, Efimov O Yu, Popova N A, Kovalenko V V, Konovalov S V, Gromov V E and Kozlov E V 2008 *Fundamental'nye Problemy Sovremennogo Materialovedenia* **4** 55–58 (in Russian)
[6] Ivanov Yu F, Kovalenko V V, Ivakhin M P, Gromov V E and Kozlov E V 2004 *Fizicheskaya Mezomekhanika* **7(3)** 29–34 (in Russian)
[7] Sorokin V G, Volosnikova A V, Vyatkin S A, Gervas'ev M A, Greditor M A, Krylova K M, Kubachek V V and Mirmelstein V A 1989 *Grade Guide of Steels and Alloys* (Moscow: Mashinostroenie) (in Russian)
[8] Valiev R Z, Islamgaliev R K and Alexandrov I V 2000 *Prog. Mater. Sci.* **45** 103–189
[9] Valiev R Z, Zhilyaev A P and Langdon T G 2014 *Bulk Nanostructured Materials: Fundamentals and Applications* (Hoboken, New Jersey: John Wiley and Sons, Inc.)
[10] Lu K 2014 *Science* **345** 1455–56
[11] Fang T H, Li W L, Tao N R and Lu K 2011 *Science* **331** 1587–90
[12] Raab G I, Simonova L A and Aleshin G N 2016 *Metalurgija* **55** 177–180
[13] Raab G I, Gunderov D V, Shafigullin L N, Podrezov Yu M, Danylenko M I, Tsenev N K, Bakhtizin R N, Aleshin G N and Raab A G 2015 *Mater. Phys. and Mech.* **24** 242–252
[14] Raab G I, Podrezov Y M and Aleshin G N 2016 *Mater. Sci. Forum* **870** 253–258

IOP Conf. Series: Materials Science and Engineering **447** (2018) 012079 doi:10.1088/1757-899X/447/1/012079

Structural and mathematical modeling of Cosserat lattices composed of particles of finite size and with complex connections

A A Vasiliev[1], **I S Pavlov**[2]
[1] Tver State University, 35 Sadoviy per., Tver 170102, Russia
[2] Mechanical Engineering Research Institute of the Russian Academy of Sciences, 85 Belinsky St., Nizhny Novgorod 603024, Russia

E-mail: alvasiliev@yandex.ru

Abstract. This article provides a brief review of the work and discussion of the problems related to the discrete structural and generalized continuum models of the Cosserat lattices consisting of particles of finite size and with complex connections. A simple Cosserat lattice, a Cosserat lattice with auxetic properties, and a chiral microstructure are considered. Various approaches to constructing a hierarchy of multi-field models for materials having a square lattice are presented. These approaches are based on the introduction of macrocells of different types and, accordingly, on the use of a larger number of fields describing deformations. A combination of micropolar and multi-field theories with the development of models is discussed.

1. Introduction

In the ordinary theory of lattices and in classical models of the mechanics of continuous media, deformations of solids are characterized only by translational displacements of particles. However, in some cases, the correct description of the deformations requires a consideration of rotational degrees of freedom. In particular, this applies to solids composed of rigid rotating elements, as well as solids with a beam microstructure. The description of deformation of such bodies, as granular media, complex molecular lattices, nanomaterials, biomaterials, stone masonry, beam lattice constructions, liquid crystals etc, needs taking into account the rotation of their constituent structural elements.

Beam lattices and their models play an important role in the interpretation and application of the micropolar theory [1, 2]. Lattices with a beam-like microstructure are used in the mechanics of metamaterials for development of auxetic and chiral structures [3-5]. The models of lattices composed of particles of finite size and with complex connections [6-9] are discussed in this article.

2. Discrete and micropolar models of materials, taking into account the rotational degrees of freedom of particles

We consider the Cosserat square lattice, i.e., the lattice composed of particles with positions determined not only by the displacements u_n and v_n, but also by the rotations φ_n, see figure 1 a.

In the expression of kinetic energy, the inertia of rotation of particles is taken into account.

$$2E_{kin}^{n} = M\dot{u}_n^2 + M\dot{v}_n^2 + J\dot{\varphi}_n^2. \tag{1}$$

The potential energy of the elastic connections of the elements m and k can be written in the form

$$2E_{pot}^{k,m} = K_n^{k,m}\Delta u^2 + K_s^{k,m}\gamma^2 + G_r^{k,m}\Delta\varphi^2, \tag{2}$$

where the following notations are introduced:

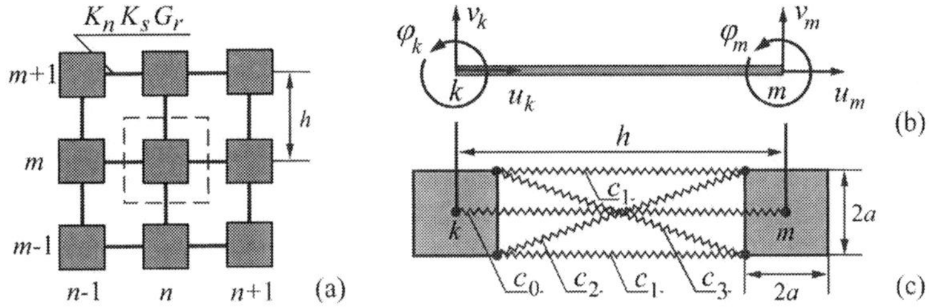

Figure 1. The Cosserat square lattice (a). The models of the two variants of connections in the Cosserat lattices: (b) a beam-type connection, (c) a complex connection of particles of finite size.

$$\Delta u = u_m - u_k ,\ \gamma = v_m - v_k - h(\varphi_m + \varphi_k)/2 ,\ \Delta\varphi = \varphi_m - \varphi_k .$$

This form of the potential of interaction was introduced in [10] for simulation of granular media. It should be noted that in the micropolar theory of elasticity, the potential also includes three similar terms. Particular cases of the potential (2) are the potential of a beam-type joint (figure 1b) that is used in modeling of materials with a beam microstructure and the potential of a complex symmetrical $(c_2 = c_3)$ connection of particles (figure 1 c), which is employed for development of continuum models of lattices consisting of particles of finite size [8, 9, 11].

On the basis of the Lagrangian, using the kinetic energy (1) and the potential of the connections (2), the equations of motion of a particle (n, m) of a unit cell are derived. Taylor series expansion up to the second order of the discrete equations with respect to the displacement components and rotations leads to continuum equations. They are analogs of the equations of the micropolar theory of elasticity. The derivation of continuum models from discrete models allows, in particular, establishing a relation between micro- and macroparameters.

3. Multi-field approach in modeling of microstructured solids. Combination of approaches of multi-field and micropolar mechanics

For construction of a micropolar model, in contrast to the classical theory of elasticity, the rotational degrees of freedom of the particles are taken into account. However, the drawback of the micropolar model is that it is not capable of describing rapidly changing short-wave-length deformations of the lattice.

The classical micropolar model is constructed on the basis of a primitive cell (figure 1 a), using the minimal number of fields corresponding to the degrees of freedom of the cell. Such an approach is a certain hypothesis, the rejection of which leads to the multi-field theory. The construction of multi-field models is based on the consideration of macrocells composed of several primitive cells, and,

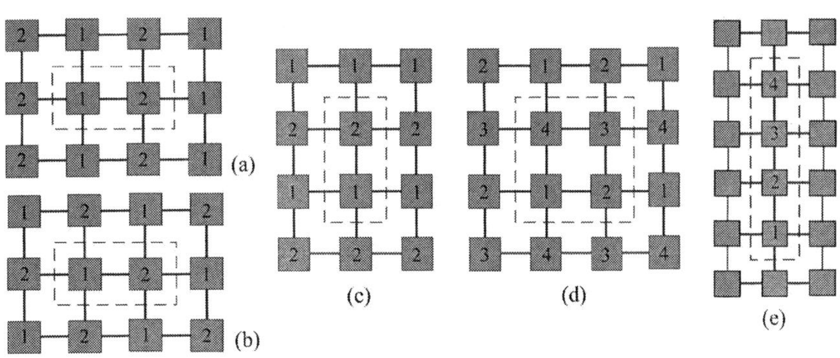

Figure 2. Variants of macrocells for construction of two-field (a-c) and four-field (d, e) models.

accordingly, more fields are used to describe deformations. Such an approach provides, particularly, a description of short-wavelength deformations. A review of the results on the development of the multi-field theory of bodies with a microstructure is given in [12].

Using various macrocells (figures 2 a-e) the multi-field approach enables one to construct a hierarchy of models describing the dynamic properties of the lattice with increasing accuracy [13, 14]. The classical single-field micropolar model is valid for approximation of lattices for low-frequency long-wavelength oscillations. Multi-field models improve the single-field model for high-frequency short-wavelength oscillations. These models can be useful, in particular, for modeling lattices as directed frequency filters. The area of the Brillouin zone in which the model is improved depends on the choice of a macrocell.

In [15,16], it was shown that there are systems with the long-wavelength and short-wavelength static boundary effects. The classical micropolar model describes long-wavelength boundary effects, but it is not suitable for systems with short-wavelength boundary effects, since for their description one function must rapidly vary, and the long-wavelength model containing only second-order derivatives is not applicable. Both long-wavelength and short-wavelength deformations are effectively described by two slowly varying functions of the two-field model. Therefore, the two-field model is applicable to describe the effects of both types. Consideration of such deformations can be important in the problems of destruction.

4. Cosserat lattices with special properties

4.1. Lattice of particles of finite size with auxetic properties

Materials with a negative Poisson's ratio are called auxetics. The property of auxetics to expand in the lateral direction under uniaxial tension makes them attractive for technological applications.

In [6], for the structural model presented in figure 3 a, discrete and micropolar models were developed. Further, the model of the ordinary elasticity theory is obtained by eliminating the rotational degree of freedom. This allows us to find the relationship between micro- and macroparameters, in particular, to express the Poisson's ratio in terms of microstructural lattice parameters and to find parameters for which the Poisson's ratio is negative. It is shown that the negative Poisson's ratio is realized only if the particle sizes are not equal to zero, i.e. the finiteness of the particle size plays an important role for the manifestation of the auxetic property. This result was confirmed in [11] for a similar lattice consisting of round particles with symmetric spring connections of three different types.

The short-wavelength rotation fields are typical for an auxetic lattice presented in figure 3 a. They are realized even in the case of simple stretching. However, as mentioned above, short-wavelength deformations are not described by the ordinary micropolar theory. In [7], two-field and four-field models were developed. It is shown that multi-field models describe short-wavelength oscillations and static short-wavelength boundary effects in an auxetic lattice.

4.2. The Cosserat lattice with a chiral microstructure

The potential (1) corresponds to the connection of particles shown in figure 1 c, in the case when $c_2 = c_3$. To study chiral structures, we consider the opposite case, when $c_2 \neq c_3$. In this case, the interaction potential takes the form

$$2E_{pot}^{k,m} = C_{11}^{k,m}\Delta u^2 + C_{22}^{k,m}\gamma^2 + C_{33}^{k,m}\Delta\varphi^2 + C_{12}^{k,m}\gamma\Delta u, \qquad (3)$$

where the following notations are used:

$$C_{11} = c_0/h^2 + 2c_1/r^2 + r^2(c_2 + c_3)/d^4\,,\ C_{22} = 4a^2(c_2 + c_3)/d^4\,,$$

$$C_{33} = 2a^2c_1/r^2\,,\ C_{12} = 4ar(c_2 - c_3)/d^4\,,\ r = h - 2a\,,\ d = \sqrt{h^2 + (2a)^2}\,.$$

It can be shown that the first three terms of the potential (3) can be obtained by choosing proper coefficients of the potential (1). The potential (3) contains the fourth additional component, which enables one to take chirality into account. Accounting for this component leads to changes in the

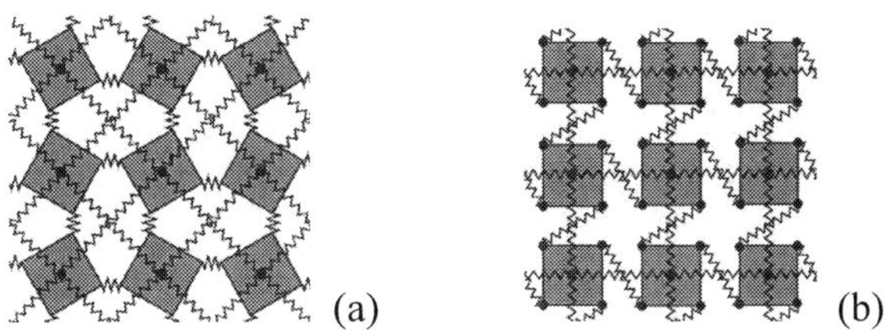

Figure 3. The Cosserat square lattices with particles of finite size: auxetic (a) and chiral (b).

equations of the micropolar theory. For a square lattice of finite size particles (figure 1 a) with complex asymmetric connections (figure 1 cc), they take the form

$$\rho u_{tt} = C_{11}u_{xx} + C_{22}(u_{yy} + \varphi_y) + C_{12}(v_{xx} - v_{yy} - \varphi_x)/2 + f_x(x,y) = 0,$$
$$\rho v_{tt} = C_{11}v_{yy} + C_{22}(v_{xx} - \varphi_x) + C_{12}(u_{xx} - u_{yy} - \varphi_y)/2 + f_y(x,y) = 0, \quad (4)$$
$$j\varphi_{tt} = (C_{33} - h^2C_{22}/4)(\varphi_{xx} + \varphi_{yy}) + C_{22}(v_x - u_y - 2\varphi) + C_{12}(u_x + u_y)/2 + f_\varphi(x,y) = 0.$$

Here $f_x(x,y) = F_x^{(m,k)}/h^2$, $f_y(x,y) = F_y^{(m,k)}/h^2$, and $f_\varphi(x,y) = F_\varphi^{(m,k)}/h^2$ are the components of the vector of distributed forces and moments, $\rho = M/h^2$ is the density of the medium, $j = J/h^2$ is the moment of inertia of a particle per unit cell, M is the mass of the particle, J is its moment of inertia. For the chiral lattice with $c_2 \neq c_3$ the elasticity modulus is $C_{12} \neq 0$. If $C_{12} = 0$, i.e. for symmetric connections with $c_2 = c_3$, equations (4) are structurally similar to the equations of the micropolar elasticity theory. If $C_{12} \neq 0$, the equations (4) contain additional components determined by chirality.

The chiral lattice possesses some interesting properties. For example, when a concentrated force acts on an ordinary lattice, the deformation field is symmetric with respect to the axis of the force action. In the case of a chiral medium, the axis of the force application is displaced and the deformation field is not symmetric with respect to this axis. The load is transmitted in the lateral direction from the force action axis. Moreover, the particles of the chiral lattice deviate in different directions, depending on the type of force acting - compressive or tensile.

Acknowledgements
The research was carried out under the financial support of the Russian Foundation for Basic Research (project N 16-08-00971-a) and the Ministry of Education and Science of Russian Federation within the framework of the basic part of State Work for scientific activity (Work No. 9.7446.2017/8.9).

References
[1] Noor A K 1988 *Appl. Mech. Rev.* **41**(7) 285
[2] Ostoja-Starzewski M 2002 *Appl. Mech. Rev.* **55**(1), 35
[3] Smith C W, Grima J N and Evans K E 2000 *Acta mater.* **48** (17) 4349
[4] Spadoni A and Ruzzene M 2012 *J. Mech. Phys. Solids* **60**(1) 156
[5] Li T, Chen Y, Hua X, Li Y and Wang L 2018 *Materials and Design* **142** 247
[6] Vasiliev A A, Dmitriev S V, Ishibashi Y and Shigenari T 2002 *Phys. Rev.* B **65** 094101
[7] Vasiliev A A, Dmitriev S V and Miroshnichenko A E 2005 *Int. J. Solids Struct.* **42** 6245
[8] Pavlov I S, Potapov A I and Maugin G A 2006 *Int. J. Solids Struct.* **43** 6194
[9] Potapov A I, Pavlov I S and Lisina S A 2010 *Acoust. Phys.* **56** 588
[10] Suiker A S J, Metrikine A V and de Borst R 2001 *Int. J. Solids Struct.* **38** 1563
[11] Erofeev V I and Pavlov I S 2015 *J. Appl. Mech. and Tech. Phys.* **56** (6) 1015
[12] Vasiliev A A, Dmitriev S V and Miroshnichenko A E 2010 *Int. J. Solids Struct.* **47** 510

[13] Vasiliev A A and Miroshnichenko A E 2005 *J. Mech. Behavior Mater.* **16**(6) 379
[14] Vasiliev A A 2010 *Acoust. Phys.* **56**(6) 831
[15] Vasiliev A A, Miroshnichenko A E and Ruzzene M 2010 *Mech. Res. Commun.* **37**(2) 225
[16] Vasiliev A A, Miroshnichenko A E and Dmitriev S V 2014 *Eur. J. Mech.* A *Solids* **46** 96

Study of plastic deformation and microstructure formation in bismuth telluride based thermoelectric materials during equal-channel angular pressing

A I Prostomolotov[1], N A Verezub[1], D I Bogomolov[2], V T Bublic[2], N Yu Tabachkova[2] and V B Osvenskii[3]

[1]Ishlinskii Institute for Problems in Mechanics, Russian Academy of Sciences, 101-1 Vernadskii ave., Moscow 119526, Russia
[2]National University of Science and Technology "MISIS", 4 Leninskii ave., Moscow 119049, Russia
[3]State Scientific-Research and Design Institute of Rare-Metal Industry "Giredmet" JSC, 5-1 B. Tolmachevsky lane, Moscow 119017, Russia

E-mail: prosto@ipmnet.ru

Abstract. This work is a complex experimental and theoretical study of the extrusion process by equal-channel angular pressing in the production of chalcogenides based on bismuth telluride, which are of interest for the creation of new functional thermoelectrics.

1. Introduction

A significant role in the analysis of equal-channel angular pressing (ECAP) is played by mathematical modeling [1,2], which allows describing the thermally stressed state of a sample during a plastic deformation in special geometry of ECAP (for example, for different angles and smoothing between channels etc.), in set conditions of thermal heating and for the set rate of the punch. The basis of the methodical approach is finite element method (FEM), implemented in commercial codes: *ADINA, DEFORM, ABAQUS, MARC*. Many studies are devoted to ECAP modeling in the production of high-quality aluminum alloys by using the elastic-plastic FEM model. The application of the elastic-viscoplastic model for the preparation of polymers is also considered. However, analogous studies of the ECAP process for the production of thermoelectrics (TE) based on bismuth telluride are absent. The theoretical approach is based on the mathematical model of the ECAP process using the elastic-plastic body approach. This model is supplemented by calculations of grain sizes, formed as a result of plastic deformation. Experimental studies were aimed at obtaining specific data on the distribution and size of grains in extruded materials, which are necessary for a verification of the mathematical model. A furnace for the production of TE materials based on solid solutions of chalcogenides was developed in the National University of Science and Technology (MISIS). Its functional diagram is shown in figure 1 a. The location of three channels with the angles of their rounded corners $\varphi_{12} = \varphi_{23}$, $\psi_{12} = \psi_{23}$ is shown in the plane (x, y) of this drawing. Dimensions of the 1st channel: 2 cm in x and 8.5 cm in y directions; 2nd channel: 1.6 cm in x and 1 cm in y; 3rd channel: 1 cm in x and 3 cm in y. In the direction z, perpendicular to the plane of the drawing, the thickness of the channels and the initial sample is 2 cm. The essence of the ECAP process consist in pressing the initial powder through three mutually perpendicular channels, which are located in a closed chamber heated to a predetermined temperature.

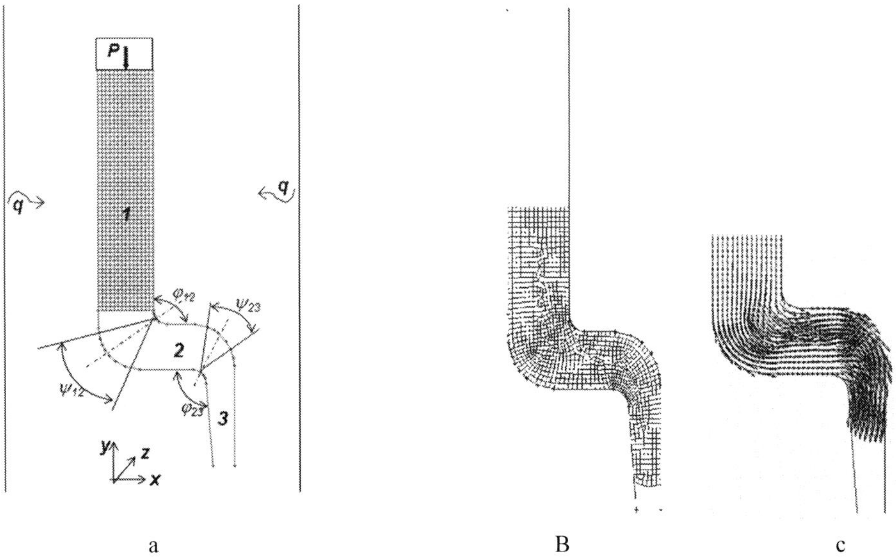

Figure 1. Press-die of the ECAP process (a): *P* – punch; the initial calculation grid for the sample in channel *1*, channels *2* and *3* with the angles of rounded corners $\varphi_{12} = \varphi_{23}$, $\psi_{12} = \psi_{23}$; *q* – heat flux from the heater. Calculation grid of the ECAP process (b). Vector image of plastic flow (V_{max} = 0.028 cm/s at V_P = 0.018 cm/s) at time t = 250 s (c).

2. The calculations of plastic deformation and grain formation

The methodical approach is based on the approximation of an elastic-plastic body. For the ECAP process, the thermal stresses may be neglected. The calculations were carried out by the FEM method using the *MSC Marc*® code, where the equivalent plastic deformation ε is used as a quantitative characteristic of plastic deformation. The maximum value of ε was verified by using of the well-known analytical formula (1) from [2]. The transition of bismuth telluride from an elastic to a plastic state was experimentally determined: σ_o= 102 MPa is the critical stress. Other parameters are known from the literature data: E = 40 GPa is the Young's modulus, ν = 0.3 is the Poisson's ratio. In this process, a graphite gasket is used to ensure the sample slipping along the die walls. Therefore, friction was not taken into an account. The punch rate was set equal to V_p = 0.018 cm/s. The process was modeled by taking into an account the heating of the die and the sample to 420 ÷ 515 °C.

Calculations of the thermoplastic state of this material in the ECAP process were carried out using a moving FEM grid (figure 1 b), which during this process was adapted to the geometry of the die. The number of grid nodes was increased or decreased depending on the magnitude of plastic deformation for a satisfaction of the specified calculation accuracy and for the convergence of the iterative process. The calculated picture of plastic motion is illustrated by vector lines, which show its layered character (figure 1 c). The same layering nature of the extruded material can be seen on the longitudinal sections of the samples after the ECAP process. A detailed methodological approach to the calculation of grain sizes is presented in [3]. It is assumed that the initial grain size d_o is set in the initial sample, and the formation of grains is limited by the magnitude of critical deformation, which is estimated as follows:

$$\varepsilon_C = \alpha\, exp(\tfrac{T_C}{T}). \qquad (1)$$

Here α = 4.76×10^{-4}, T_C = 773 K. It is assumed that at $\varepsilon < \varepsilon_C$ the initial grain size is preserved. Otherwise, the grain size *d* after recrystallization is calculated in depending on the strain rate and temperature according to the following formula:

$$d = \chi_1 \dot{\varepsilon}^{-\chi_2} exp(\frac{-\chi_3 Q}{RT}). \qquad (2)$$

Here: χ_1 = 11.3, χ_2 = 0.14 ÷ 0.24, χ_3 = 0.014, and Q = 267×10^3 J/mol is the activation energy of the formation of grains, R = 8.314 J/mol×K is the universal gas constant.

The criterion for the transition to plasticity is the specified value of the von Mises stress. The analysis of the data in figure 2 a shows that in the 1st channel the sample moves straight from top to bottom and in initial contact with the 1st fillet the excess of the critical stress is insignificant. Therefore, small plastic deformations occur. However, after turning, the area of plastic deformations expands in the 2nd channel and is significantly strengthened in the entrance to the 3rd channel.

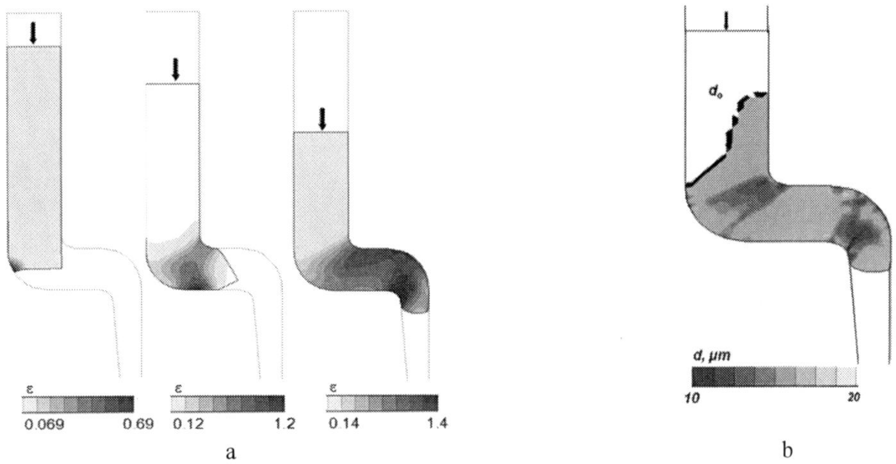

Figure 2. Contours of plastic deformation ε at different times of the ECAP process at 470 °C: 70, 150 and 250 s (a); contours of grain sizes *d* [μm] at d_o = 500 μm, 470 °C, *t* = 250 s (b).

The magnitude of plastic deformation and its rate play a decisive role in the formation of the microstructure of the material. In particular, they determine the process of grain formation. Quantitative data on changes in plastic deformation during the process served as the basis for calculating the distributions and sizes of the formed grains in extruded sample. The contours of grain sizes in figure 2 b show that with small values of plastic deformation ($\varepsilon < 1$), the grain sizes are larger. Near the die rounding, the plastic deformation becomes significant ($\varepsilon > 1$), that causes the reduction of grain sizes to 10 μm. In this macro picture, the grain size ~ 20 μm is the predominant one, which is formed in the 1st channel and is determined by the parameters of the process.

3. The measured data

ECAP processes were carried out with different percentage composition of components and grain sizes of the initial powder. The design press-die was optimized and the experimental processes were carried out in different temperature modes to find the ways to control the microstructure, composition and defects in the extruded TE material. The microstructure of the samples was investigated by X-ray diffractometry, scanning and transmission electron microscopy (SEM and TEM), and their thermoelectric characteristics were measured by the Harman's method [4]. Microstructure studies of the samples showed that their grains are fragmented, and the grain fragments themselves are separated by dislocation grids. It is noted that as the grain sizes of the initial powder increases, the volume density of dislocations decreases. The microstructure was studied by a scanning electron microscope

IOP Conf. Series: Materials Science and Engineering **447** (2018) 012080 doi:10.1088/1757-899X/447/1/012080

for ECAP samples extruded for different grain sizes of the initial powder. The obtained TE material had a homogeneous and finely dispersed microstructure regardless of the different granule composition of the initial powders. The slots of the samples basically were as plates, which are characteristic for the layered structures of solid solutions.

Despite the different initial granule composition, after ECAP process the grain sizes in the slots of all samples did not exceed 5 μm. This is much smaller than the grain sizes of the initial powder. It is noted that with an increase in the grain size of the original powder, the texture improves, and the volume density of dislocations in the extruded material decreases. The largest number of grains with a favorable texture (from the point of view of electric anisotropy) is observed for samples with initial grain sizes that are ~ 500 μm. There are no grains with split planes that are parallel to the surface of the TE samples. The larger grain size of the original powder corresponds to the finer favorable texture. Thus, the anisotropy of the electric characteristics may be used in cutting the TE material as perpendicularly to the extrusion direction. TEM studies of the microstructure showed that within the grains there is a sufficiently high density of chaotically located dislocations (figure 3 a). The measured fraction of grains (%) of their size at an ECAP temperature 470 °C is shown in figure 3 b.

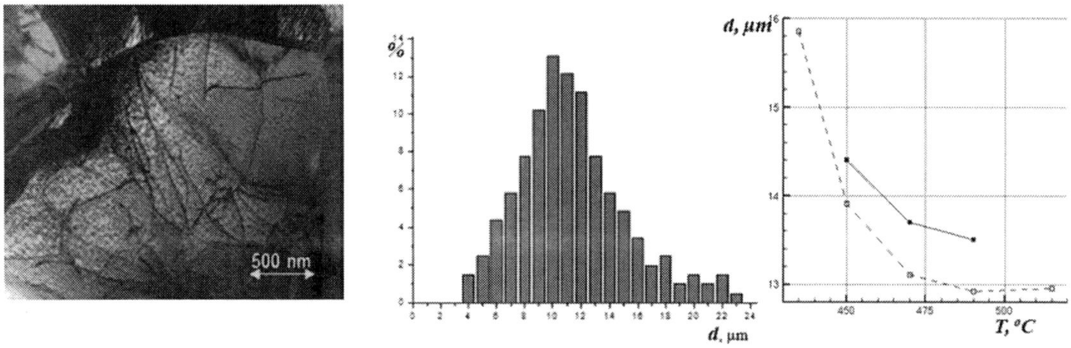

a b c

Figure 3. TEM image of the internal structure of grains with a resolution of 500 nm (a); measured grain fractions (%) upon grain sizes *d* for the ECAP process at 470 °C (b); dependences of maximum grain sizes *d* upon temperature *T* (c): dashed line – measurements, solid line – calculations.

4. Conclusions

Measurement data became the basis for verifying the calculated grain sizes. A reduction of grain sizes with the increasing ECAP temperature is consistent with satisfactory accuracy with measurement data (figure 3 c). The increase in the discrepancy between the calculated and experimental data at higher temperatures (475 and 500 °C) may be explained by the lower compliance of the elastic-plastic model. The drawback of the model is its static nature, since the sizes and distribution of the grains are calculated at a specific time, using only the current values of plastic deformation.

Acknowledgments

This work was carried out in framework of IPMech RAS project No. AAAA-A17-117021310373-3 and at financial support by the RFBR (grants: 18-02-00036, 17-08-00078, 16-29-11785).

References

[1] Aour B and Mitsak A 2016 *J. Theor.Appl. Mech.* **54** 263
[2] Krállics G, Széles Z and Malgyn D 2003 *Mater. Sci. Forum* **414-415** 439
[3] Prostomolotov A I and Verezub N A 2016 *Tambov University Rev.: Ser. Nat. Techn. Sci.* **21** 818 (In Russian)
[4] Bogomolov D I, Bublik V T, Tabachkova N Yu and Tarasova I V 2016 *J. Electron. Mater.* **45** 403

IOP Conf. Series: Materials Science and Engineering **447** (2018) 012081 doi:10.1088/1757-899X/447/1/012081

Structural studies of plastically deformed cobalt

R Almukhametov and L Gabdrakhmanova

Bashkir State University, 32 Zaki Validi st., 450076, Ufa, Russia

E-mail: AlmukhametovRF@rambler.ru

Abstract. This paper is devoted to the study of the fcc-hcp phase transition delay during cooling in nanocrystalline cobalt obtained by severe plastic deformation. Using the methods of X-ray and electron microscopy, it was revealed that this delay in the phase transition is due to the small size of the crystallites and the stressed state of the crystal lattice.

1. Introduction

Cobalt is one of the main components of alloys applied in magnetic recording devices. It is widely used as a catalyst for chemical reaction, as well as a material for thermoelectric devices. Due to the unusual physical properties, cobalt with ultrafine-grained and nanocrystalline (NC) structures presents great scientific and practical interest. It is known that NC cobalt can have both an hcp structure and a fcc structure. The physical properties of NC cobalt depend on the ratio of these phases. In turn, the ratio of fcc and hcp phases depends on the method of obtaining NC cobalt, the strain of deformation and the size of the crystallites, and a number of other factors. However, the results of research available in the literature are often contradictory [1-9]. We have already mentioned that the stability of fcc and hcp phases in NC cobalt in our opinion has not been sufficiently studied so far and requires more systematic research. This paper presents the results of studies of the structure and phase transformations in NC cobalt, obtained by severe plastic deformation (SPD) and subjected to subsequent annealing.

2. Experimental procedures and results

The NC samples were produced by SPD on Bridgman anvils under a pressure of ~ 7 GPa at room temperature by rotating the striker at 5 turns. Samples in different structural states were obtained by annealing at various temperatures. X-ray studies were carried out with a DRON-7 diffractometer on Co-radiation at room temperature. The microstructure was studied using a JEM-2000EX electron microscope.

According to the results of our electron microscopic studies, the microstructure of the samples after SPD is uniform, the grain boundaries are fuzzy and blurred. The average size of the crystallites is difficult to determine and it is about 20-50 nm. After annealing at 300 °C and 400 °C, partial recovery is observed in the samples, the grain boundaries remain blurred. After annealing at 500 °C, the primary recrystallization is completed and the grain sizes are approximately 200-500 nm. Electron microscope studies did not allow us to accurately determine the crystallite size. Therefore, to determine the phase composition in the samples and the crystallite size, we carried out X-ray studies. The diffractograms of SPD cobalt annealed at different temperatures are shown in figure 1. As can be seen from the figures, non-annealed SPD cobalt has an hcp structure at room temperature like non-deformed coarse-grained cobalt. There are no clearly defined data on the stability of hcp and fcc phases of NC-cobalt. Thus, our results confirm that in NC-cobalt obtained by SPD, the hcp structure is preserved at room temperature. Samples annealed at temperatures up to 450 °C retain the hcp structure.

According to literature data for coarse-grained non-deformed cobalt, the temperature of the hcp-fcc phase transition lies in the interval from 387 °C to 440 °C. According to the point of view of a number of authors [7,9], the temperature of the hcp-fcc phase transition of cobalt decreases with decreasing the crystallite size. The results of our investigations did not reveal a noticeable decrease in the temperature of the fcc-hcp phase transition with a decrease in the size of crystallites to ~ 20 nm.

One can see from figure 1c that the samples annealed at 500 °C consist mainly of the high-temperature fcc phase. Samples annealed at temperatures from 600 °C to 900 °C consist of a mixture of the low-temperature hcp phase and the high-temperature fcc phase (figure 1d).

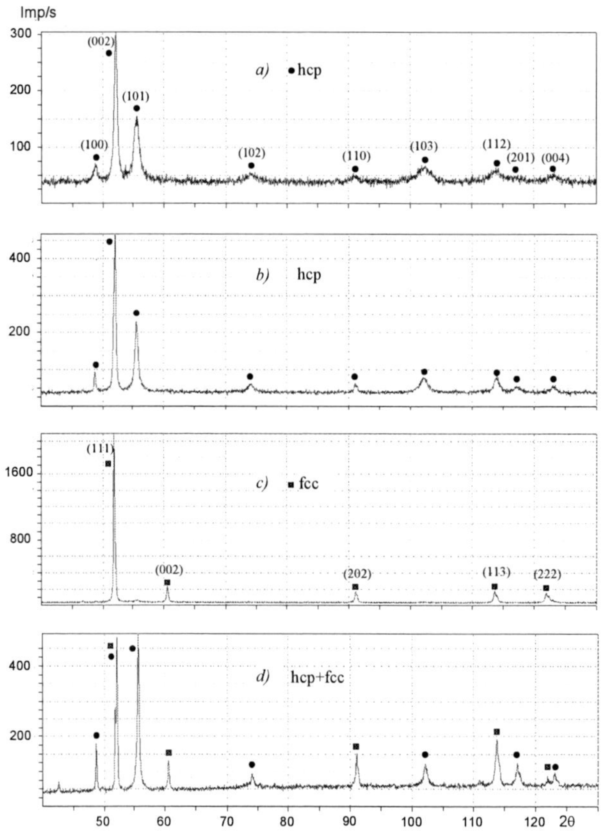

Figure 1. Diffractograms of SPD cobalt, annealed at different temperatures: *a*-non-annealed sample; b-T=450 °C; c-T=500 °C; d-T=900 °C.

Thus, according to the results of our research, the high-temperature fcc phase of NC-cobalt is retained after cooling to room temperature. We believe that the NC-structure of cobalt interferes with the transition of the high-temperature fcc phase to the low-temperature hcp phase upon cooling. We have already mentioned that this delay in the phase transition has not been studied sufficiently so far. We believe that the delay in the phase transition may be due to the small dimensions of the crystallites, the distortions and the stressed state of the crystal lattice. Since the distorted regions are the grain boundaries and the surface of the crystallites, the relative contribution of these regions to the elastic energy increases with decreasing the crystallite size. Therefore, for the fcc-hcp transition there must be some critical size. The existence of a critical size of crystallites under which the martensitic transformation is blocked has also been discussed in the literature.

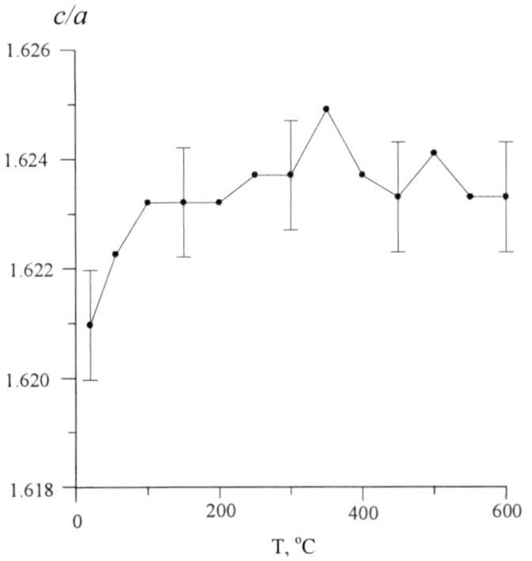

Figure 2. Dependence of the ratio of the hexagonal unit cell parameters c/*a* of the SPD cobalt on the annealing temperature

To confirm our assumptions about the stability of the hcp and fcc phases of NC cobalt, we investigated the dependence of the ratio of the hexagonal unit cell parameters c/*a* on the annealing temperature (figure 2). As can be seen from figure 2 for the non-annealed NC cobalt, the ratio is c/*a* = 1.621, that is noticeably smaller than this parameter for coarse-crystalline undeformed cobalt that equals to c/*a* = 1.632 and the theoretical value of this parameter is c/*a* = 1.633, which corresponds to the most stable state of the hcp structure. Consequently, the hcp phase of NC cobalt obtained by SPD is less stable than the hcp phase of coarse crystalline undeformed cobalt. When increasing the annealing temperature, the ratio of parameters c/*a* for NC cobalt increases to c/*a* = 1.6233. This indicates an increase in the stability of the hcp phase of NC cobalt with an increase in the annealing temperature.

To confirm our assumptions about the existence of the critical size for a fcc-hcp phase transition, we conducted studies of the coherent domain sizes (CDS) and the microstrains of the crystalline lattice as a function of the annealing temperature. The CDS and microstrain were determined by measuring the broadening of diffraction lines. The CDS were calculated with the help of the Sherrer formula using lines located at small diffraction angles:

$$D = \frac{\lambda}{W\cos\theta} \tag{1}$$

Here λ is the wavelength of the X ray radiation; W is the physical broadening of the diffraction line; θ is the diffraction angle. The values of the microstrain $<\varepsilon^2>^{1/2}=\Delta a/a$ were determined by the diffraction lines at large angles according to the formula:

$$W=4<\varepsilon^2>^{1/2} tg\theta \tag{2}$$

Here a is the average unit cell parameter and Δa is the average unit cell parameter devation from the a value. Microstrains were also determined taking into account the contribution of CDS to the broadening of X-ray lines by the formula:

$$<\varepsilon^2>^{1/2} = \left(W - \frac{\lambda}{D\cos\theta}\right)(4tg\theta)^{-1} \tag{3}$$

In addition, the average CDS and the microstrains $<\varepsilon^2>^{1/2}$ were determined from the Williamson-Hall relations

$$W \times \cos\theta = \frac{\lambda}{D} + 4 < \varepsilon^2 >^{1/2} \sin\theta . \tag{4}$$

Figure 3 shows that the CDS, determined by the Scherrer formula and the Williamson-Hall curves, are close. The values of D up to the annealing temperature T~300 °C vary insignificantly in the interval from ~ 250 Å to ~ 500 Å. These results agree well with our electron microscopic data. At annealing temperatures above 450 °C, a sharp increase in the CDS is observed. Thus, the obtained

IOP Conf. Series: Materials Science and Engineering **447** (2018) 012081 doi:10.1088/1757-899X/447/1/012081

results confirm our assumptions that the delay of the fcc-hcp phase transition in SPD cobalt is related to the smallness of the CDS. At annealing temperatures above 500 °C, some crystallites of the fcc phase grow to a critical size and, upon cooling, pass to the hcp phase. According to our results, the critical size for the fcc-hcp transition is approximately D ~ 1000 Å.

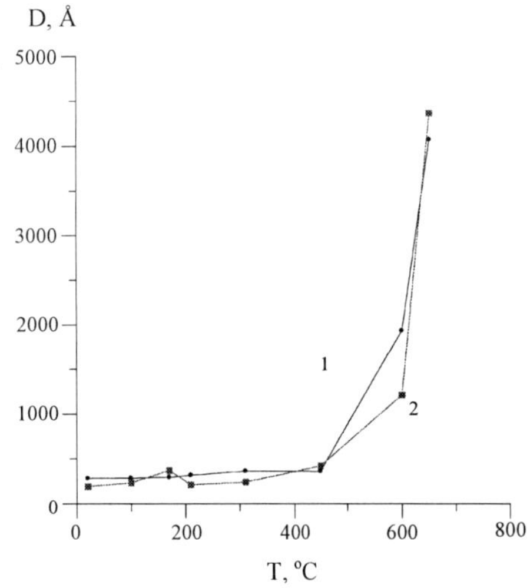

Figure 3. Dependence of the coherent domain sizes on the annealing temperature, determined: 1 - by the Scherrer formula, 2 - from the Williamson-Hall curves.

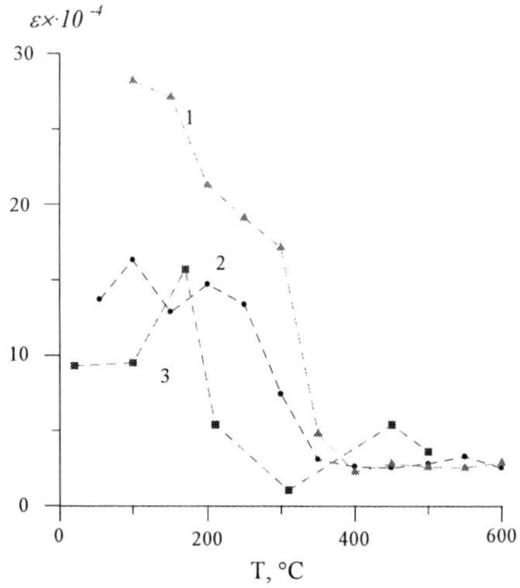

Figure 4. Dependence of the microstrains in SPD cobalt on the annealing temperature, determined: 1 - by formula 2; 2 - by formula 3; 3 - from the Williamson-Hall curves.

In the calculations of $<\varepsilon^2>^{1/2}$, using formula (3), the values of D determined from the line (100) were used, and their dependence on the annealing temperature was taken into account. As can be seen from figure 4, the values of microstrains determined by various methods differ only for annealing temperatures below 300 °C for the recovery region. Microstrains decrease with an increase in the annealing temperature and at annealing temperatures above 400 °C take a value equal to $\sim 2\times 10^{-4}$. This temperature is close to the temperature of a sharp increase in the CDS. This makes it possible to assume that the reasons for the delay of the fcc-hcp phase transition in SPD cobalt are the small dimensions of the crystallites and the strained state of the crystal lattice.

3. Conclusions

Thus, we have shown that the delay of the fcc-hcp phase transition in NC cobalt, obtained by SPD is due to the small crystallite size and the strained state of the crystal lattice.

References

[1] Betteridge W 1980 *Progr. Mater. Sci.* **24** 51
[2] Bolokang A S, Phasha M J, Motaung D E and Bhero S 2012 *J. Metallurgy* **2012** 290873
[3] Cardellinia F and Mazzonea G 1993 *Phil. Mag. A* **67** 1289
[4] Huang J Y, Wu Y K, Ye H Q and Lu K 1995 *Nanostr. Mater.* **6** 723
[5] Zhao J C and Notis M R 1995 *Scripta Metall. Mater.* **32** 1671
[6] Matsumoto H 1995 *J. All. Comp.* **223** L1
[7] Owen E A and Jones D M 1954 *Proc. Phys. Soc.* **67** 456
[8] Sorta J, Noguésa J, Suriñacha S and Baróa M 2003 *Phil. Mag.* **83** 439.
[9] Kajiwaraa S, Ohnoa S, Honmaa K and Udaa M 1987 *Phil. Mag. Lett.* **55** 215

IOP Conf. Series: Materials Science and Engineering **447** (2018) 012082 doi:10.1088/1757-899X/447/1/012082

Structure and phase changes in nanostructured compound coatings

M Sh Migranov [1], V R Mukhamadeev [1], A M Migranov [1], I R Mukhamadeev [1], A A Khazgalieva [2,3]

[1]Ufa State Aviation Technical University, 12 Karl Marx st., Ufa 450008, Russia.
[2]Institute for Metals Superplasticity Problems, Russian Academy of Sciences, 39 Khalturin st., Ufa 450001, Russia
[3] Bashkir State University, Zaki Validi St. 32, Ufa 450076, Russia

E-mail: vener_muhamadeev@mail.ru

Abstract. The paper deals with an experimental study of the influence of temperature and plastic deformation on the structural-phase changes occurring at the border of the "nanostructured coating-supporting plate".

1. Introduction

It is possible to increase the wear-resistance of the cutting tool by changing the state of near surface properties, under which the most efficient cutting tool resists wear, especially at high temperatures. Moreover, the material of the tool in the body volume of the cutting tool should simultaneously have sufficient strength at compression and bending, shock loading and reversed stresses. Wide industrial employing of cutting tools with coatings allows [1–3]: to increase the resistance and reliability of tools; reduce the rate of consumption of expensive tool materials; extend the field of application and reduce the list of applied solid-alloy tools.

Today, a large number of experimental studies are carried out in the field of designing innovative multilayer multiplex coatings to improve the workability and performance properties of cutting tools with various applications, different composition, structure, architecture and properties of coatings. These tasks can be solved by employing multilayer composite nanostructured coatings. One of the most promising trends is alloying the coatings by different elements and combinations (aluminum, yttrium [1-2], etc.) considering the phenomena arising while operating.

2. Theoretical part

The reduction of particles up to 100 nm and less leads to an increase in the parameters of the crystal lattice [4]. The coatings applied to cutting tools allow increasing their wear-resistance by reducing the friction force [3], reducing deformation and cutting forces [3], changing the direction and intensity of heat flows, meanwhile the increase of wear-resistance of the tool with coatings is also explained by the screening effect of contact surfaces from external interferences [1].

At present, an increase in wear-resistance of the cutting tool can be achieved through the use of titanium and aluminum nitrides (Ti, Al)N coatings. Coatings such as (Ti, Al)N are characterized by unique combinations of properties: high hardness at high temperatures in the contact area together with thermal and chemical resistance and low thermal conductance. An important advantage of (Ti, Al)N coatings is their tendency to oxidation and the formation of relatively stable surface oxide layers. A further increase in wear-resistance of (Ti, Al)N coatings during cutting may be achieved by bringing the grain refining to the nanoscale level (size of a grain less than 100 nm). This can be realized by alloying the (Ti, Al)N coatings on the tool material by means of magnetic-arc filtration (MAF).

3. Experimental studies

Studies [5-7] testify that the main influence on the hardness, elastic modulus and other physical and chemical properties of the coating is brought by parameters such as a grain size, constituents of the coating components, temperature of the supporting plate, ways of applying and possibility to form an amorphous phase from nitrides.

For experimental research, fast-changing multifaceted plates from hard-alloy titanium-cobalt compounds HF10, preliminary subjected to nitriding (for better adhesion of the coating and supporting plate) and having the (Ti, Al)N coating applied on them, were used. The coating thickness was measured by a Calotest device. The coating thickness was equal to $h_{coating}$ =6.124 µm.

The coating (Ti, Al)N was applied through filtration by a separator at a modified unit, such as NNV 6.6 – I1. The parameters of applying the coating with MAF are as follows: the arc current is 200 A, the nitrogen pressure is $5*10^{-3}$ Pa, the time is 30 minutes, the coil current is 20 A, the separator current is 20 A, the separator voltage is 16 V, the temperature of application is 340-370 °C. Heating of the supporting plate to the temperature of the coating application was performed by argon ions during their landing at a grid bias voltage of the supporting plate of 1 kV.

Nanocrystal coatings facilitate the formation of a protective layer of aluminum oxide on the surface. Coatings with a small grain structure have a greater length of grain boundaries. In this connection, a longer use of atoms is necessary for the outer diffusion of aluminum Al and inner diffusion of oxygen.

To study the chemical composition, the specimens were produced in the form of angle laps at an angle 5° to the upper part of the cutting plates. Before conducting the research, the patterns were skimmed of the fat, then placed into the preparation chamber of the spectrometer and subjected to 5-minute pickling in the medium of argon ions. The pickling was performed under a gas pressure of 10^{-4} Pa and with a velocity of pickling of approximately 20 monolayers per minute. Such preparation of samples excluded the impact of accidental pollution on the results of analysis. The surfaces of the samples were studied at different depths, starting from the surface: either by stationary electron and ionic beams, or by scanning them in a chosen direction.

A set of samples was annealed at 700 °C under a pressure of 0.5 hPa for 20 minutes in ambient conditions to determine the possible impact of the atmosphere on the cutting tool.

These samples allowed us to study the heterogeneity of the composition at a shallow depth under the surface, including the developed films. In connection with this, the width of the studied area is increased by 9.57 times, i.e. h_{res}=9.57* $h_{coating}$ =58.61 µm.

At the next stage of the experimental study, micro-hardness was measured according to the GOST 9450-76 standard on a Struers Duramin microhardness tester with the load of 5 H and a holding time of 10 s.

The values of microhardness were obtained at equidistance from the edge of the coating on a oblique loop. 12 values in the area of the coating and 12 more at the same distance were performed in the basis. According to the obtained microhardness values, graphs of the dependence HV=f(h_{res}) were plotted.

It is possible to see in the graph that a jump in hardness takes place on the border of the coating-supporting plate on the original sample. We can suppose that it comes from an increased concentration of nitrogen. Overviewing the graph of the sample hardness after annealing under pressure we may see the toughness annealing, however, therewith these values are higher than the hardness of the supporting plate. So as penetration into the base increases, the hardness value reaches the base hardness. This is apparently connected with structural changes occurring on the supporting plate.

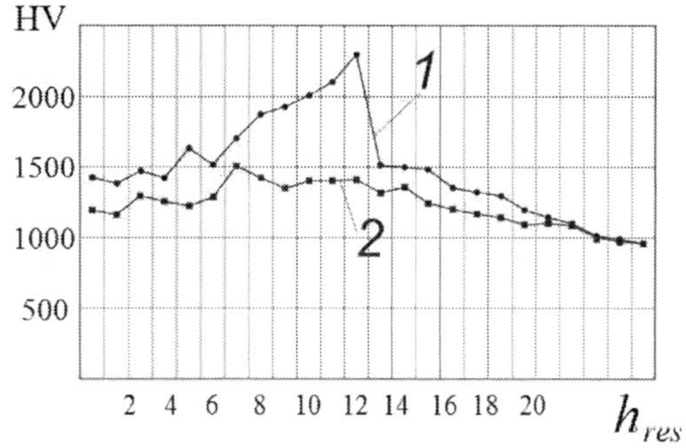

Figure 1. Dependence of the microhardness depending on the depth of measurement: 1 – original sample; 2 – annealed sample.

At the next stage of the study, the angle laps were examined by a Tescan Vega3 scanning electron microscope at an accelerating stress of 20 kV and magnification by 5000 and 20000 times. Images were obtained with the help of a BSE detector (BSE – backscattered electrons), designed for obtaining images which contain information on variations in composition depending on the opposition in the charge number. Figures 3, 4 represent images of the original sample and the annealed one.

4. Conclusions

On the basis of the conducted research it is possible to make the following conclusions:

- an increase in the microhardness in the original sample on the "border" "coating-supporting plate" is probably connected with a high nitrogen content obtained by the nitrogenation of the surface before coating (figure 1);
- the release of stresses leads to leveling the coating properties in near-surface layers of the tool thereby increasing the workability of the cutting tool happens (figure 2).
- the change in microhardness occurs due to the deliberate implementation of the structural elements of the coating into the cobalt bond, which leads to an increase in the average value of microhardness in near surface layers (figures 3, 4).

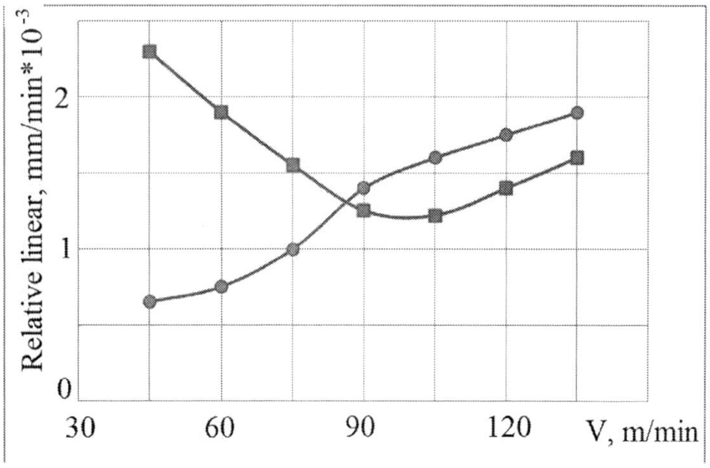

Figure 2. Relative linear wear: ● - HF10; ■ - HF10 with thermal treatment.

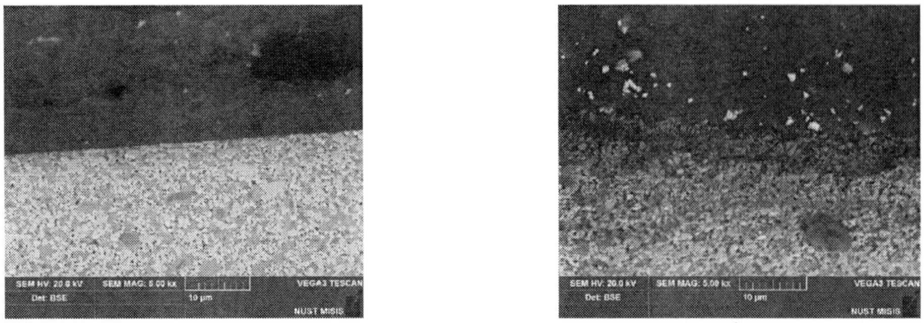

a b

Figure 3. The specimen (20 kV, BSE, 5000-fold increase): a – initial product; b – final product

a b

Figure 4. The specimen (20 kV, BSE, 20000-fold increase): a – initial product; b – final product

References

[1] Migranov M S 2007 *The Increase of the Tool Wear-Resistance on the Basis of Predicting the Surface Friction Processes at Metal Cutting,* Doc. Sc. Thesis (RUDN University, Moscow) p 433 (In Russian)

[2] Vereschaka A S 1993 *The Working Capacity of the Cutting Tool with Wear-Resistant Coatings* (Mechanical engineering, Moscow) p 336 (In Russian)

[3] Krioni N K, Migranov M S, Dementeva M S and Mukhamadeev V R 2016 *Vestnik UGATU* **72** 29 (In Russian)

[4] Maier J, Prill S and Reichert B 1988 *Solid State Ionics* **2** 28

[5] Ovchinnikov S V 2008 *9 Inter. Conference on Modification on Materials with Particles Beams and Plasma Flows* **1** 472 (In Russian)

[6] Mitterer C, Mayrhofer P and Musil J 2005 *Vacuum* **71** 279

[7] Zgalat-Lozinskii O B, Ragulya A V and Dub N V 2010 *Nanoscale Mater. Sciю* **1** 30

IOP Conf. Series: Materials Science and Engineering **447** (2018) 012083 doi:10.1088/1757-899X/447/1/012083

The improvement of the tribotechnical properties of materials and coatings for metal cutting tool

M Sh Migranov [1], V R Mukhamadeev [1], A M Migranov [1], I R Mukhamadeev [1], A A Khazgalieva [2,3]

[1]Ufa State Aviation Technical University, 12 Karl Marx st., Ufa 450008, Russia.
[2]Institute for Metals Superplasticity Problems, Russian Academy of Sciences, 39 Khalturin st., Ufa, 450001, Russia
[3] Bashkir State University, Zaki Validi St. 32, Ufa, 450076, Russia

E-mail: vener_muhamadeev@mail.ru

Abstract. The paper deals with experimental studies of tribotechnical properties (strength of adhesive bindings, wear-resistance etc.) of sintered cutting tools alloyed by the TiC, BN, Al_2O_3 additives and multilayer wear-resistant TiAlN, TiN, TiCrN coatings and others, which allow significantly increasing the performance properties of the cutting tool for edge cutting machining by forming secondary structures in friction.

1. Introduction

Cutting tool appears to be the most efficient element in the system of activities to improve the cutting process [1-6]. Thus, the problem of improving the cutting tool is considered to be urgent. At the same time, in mechanical engineering tool materials and coatings with controlled adaptation of friction surfaces (CAFS), which, taking into account the conditions of cutting (controlled variability of cutting mode elements, lubricant cooling agent, etc.), allow increasing wear-resistance of the tool due to the course of non-equilibrium processes in friction during metal cutting, are not widely used today [2-5]. It is well-known [1-6] that the major phenomena in friction are concentrated on a thin near surface layer. In the works of B.I. Kostetskiy and L.I. Bershadskiy it is shown that any tribosystem should be considered as an open non equilibrium thermodynamic system, which makes all types of bodies interconnection concentrate in a thin-film object – secondary structures (SS).

2. Experimental studies

To study the characteristic representatives of the processed materials, which possess distinguishing physical and mechanical properties, chemical and structured-phase compositions: constructional steels C45 (HB 180-200) and 41Cr4 (HB 200-220); corrosion-resistant steels – 431S29, 301S21, X6CrNiTi18-10, as well as refractory and high-temperature resistant steels and alloys – 15Cr18Ni12Si4TiAl, CrNi73MoNbTiAl (vacuum-arc remelt), were chosen.

High-speed steels were chosen from tool materials: HS6-5-2, HS6-5-2-5, T12008, sintered powder tool materials (SPTM) on the basis of high-speed steel HS6-5-2; hard alloys: HG30, HV10, K10, HT01, HF10, commonly used in machinery production. At the full-scale tests four-sided plates (12x12 mm) and solid bit tools, mills and drills were used. Circular milling was performed by the end mills from high-speed steel as well as from a hard-alloy material (d = 12; 13 mm; z = 4; 5), and face milling - by single-cutter mills (d = 90 mm), equipped with four-sided plates.

It has been established that SPTMs based on high-speed steel alloyed by titanium carbide possess high wear-resistant properties and can be classified as a new class of self-organizing tool materials. In particular, these may include SPTMs containing titanium carbide as a solid base and high-speed steel (HS6-5-2) – as a binder (SPTM + 20% TiC). With the help of AES (Auger-electron spectroscopy) and SIMS (Secondary-Ion Mass Spectrometry) it was found out that during the cutting process titanium carbides are transformed into thin surface films in the form of a titanium-oxygen composition (figures 1 and 2). This significantly improves frictional properties at operation temperatures of cutting (figure 3) and increases the wear-resistance of the cutting tool (figure 4). The study showed that the wear-resistance of such a tool is 2-3.5 times higher than the wear-resistance of conventional tools from high-speed steel.

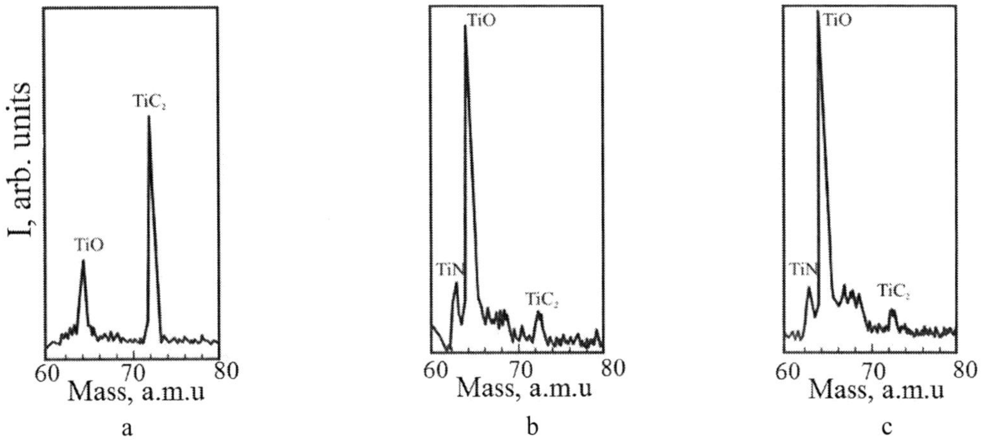

Figure 1. The spectra of the SIMS surface of SPTM tool (HS6-5-2 + 20% TiC) depending on the cutting time (steel turning 45: V = 60 m/min; t = 0.5 mm; S = 0.28 mm/r): a – after 4 minutes; b – after 20 minutes; c – after 24 minutes.

The essence of the second alloying principle is in the increase of the screening effect by a stable high-strength secondary structure (simple and complex oxygen-containing phases based on titanium and boron), appearing on the surface of the tool, which is achieved, for example, by adding 2% of BN. The addition of 5% of Al_2O_3 has not practically influenced the composition of secondary structures. The alloying was performed not by adding one or another element, but by adding compounds of the required density and instability at operation temperatures, that allowed using the compounds in relatively small amounts with a minimum possible impact on the volume quality. The application of both principles gives an opportunity to considerably increase the wear-resistance of the tool (for example, by adding 20% of TiCN) (figure 5).

Further improvement of the cutting tools includes applying the multilayer wear-resistant coatings on their working surfaces. Meanwhile, each layer of such a coating must be formed considering changes of the wear mechanism during the periods of running-in, normal (stable) and extremely gross wear.

It is established that the compromise between high wear-resistance and reliability is reached in a multilayer coating applied to the indium-enriched supporting plate. Indium is present in the lower coating layer in the metal as well as in the bound state (In – N). While warming during the cutting process indium partially turns into a liquid state and is partially oxidized (this is testified by the data of

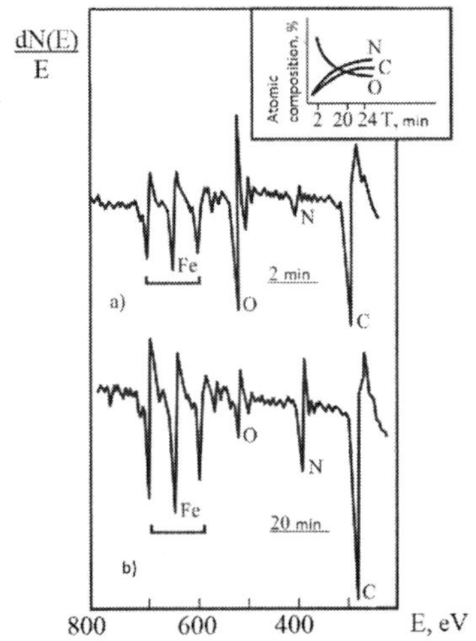

Figure 2. Auger spectra of the surface of the processed material (C45): a) – after 2 minutes of cutting; b) – after 20 minutes of cutting

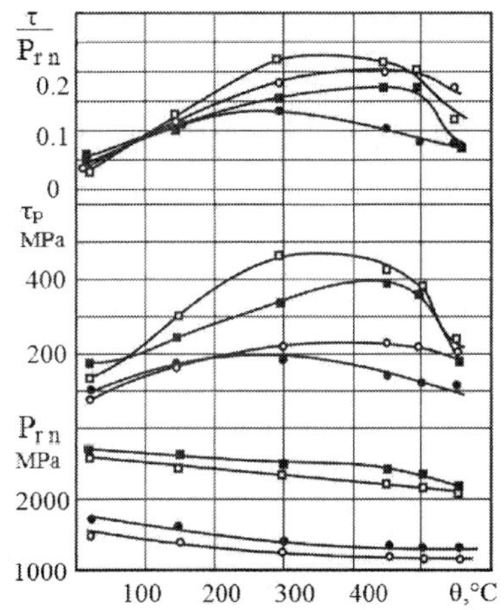

Figure 3. The influence of the temperature on tribotechnical properties of materials: ● ■ – SPTM (HS6-5-2 + 20% TiC); ○ ● – C45 (HB 180…200); ○ □ – HS6-5-2; □ ■ – C45 (HRC 30..32)

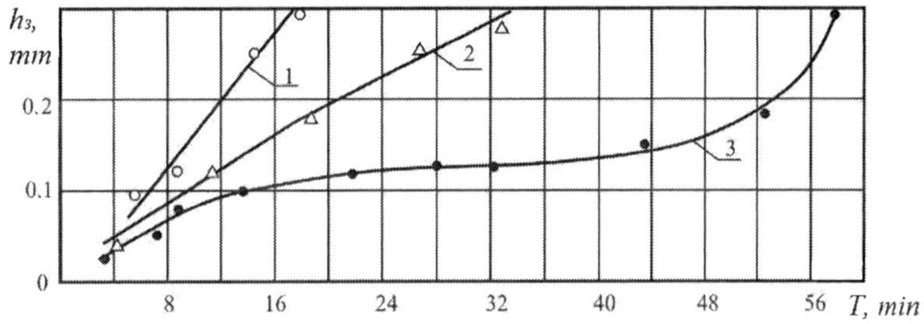

Figure 4. The influence of the processing time on cutting tool wear-resistance on the back surface of the tool at turning C45 (HB 180…200), V = 60 m/min; t = 0.5 mm; S = 0.2 mm/r: 1 – HS6-5-2; 2 – T12008; 3 – SPTM (HS6-5-2 + 20% TiC).

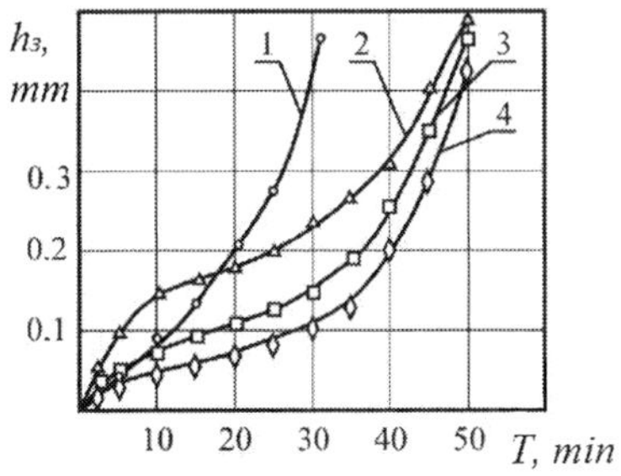

Figure 5. The influence of the processing time on the wear of the end mill Ø12 mm on back surface: 1 – SPTM (HS6-5-2 + 20% TiC); 2 – SPTM (HS6-5-2 + 20% TiC + 2% BN); 3 – SPTM (HS6-5-2 + 15% TiC + 5% Al_2O_3); 4 – SPTM (HS6-5-2 + 20% TiCN) (the processed material 41Cr4; V = 65 m/min; S_m = 63 mm/min; t = 3 mm; b = 10 mm).

metallographic studies). The liquid state as a lubricant reduces the friction coefficient. Oxygen-containing phases protecting the surface facilitate the prolongation of a normal wear stage, considerably increasing the tool wear-resistance.

It is established that, along with this, the following phenomena simultaneously occur: a – tribosplitting (decomposition) of a thin surface layer; b – formation of a stable amorphous-like protecting layer (of the type Ti-O and Ti-F) on the tool surface. Self-organization processes facilitate the improvement of wear-resistance during the stage of running-in wear, which leads to an increase in wear-resistance of cutters and end mills in 2 or 3 times.

3. Conclusions

The following conclusions can be made for the present studies:

- SPTMs based on high-speed steel, additionally alloyed by titanium carbide, possess high wear-resistant properties and can be considered as a new class of self-organizing tool materials.
- The optimal combination of strength and reliability (characterized by high adhesion of the coating (TiCr)N to the supporting plate) manifests itself in a multilayer coating with a lower layer, enriched by indium (implanted into the supporting plate surface), which facilitates the formation of amorphous-like structures In – N and In – O on friction surfaces and the prolongation of the normal wear stage, increasing the tool wear-resistance in 2.1 to 2.4 times.
- Multilayer coatings including ion nitriding of the lower layer from high-speed steel with further applying the PDC TiN coating, the upper layer of which is modified by PFPE (perfluoropolyether), provide high wear-resistance of the cutting tool.
- The main advantage of magnetic-arc filtration at alloying the coating by the PDC technique is a considerable (TiAl)N grain reduction, which leads to the formation of a surface layer with a grain size from 60 to 80 nm, i.e. within the nanoscale, and provides the possibility to use the cutting tool with such a coating under the conditions of high-speed processing by cutting with an increase in wear-resistance of the cutting tool in 3 or 4 times.

References

[1] Fox-Rabinovich G S 1997 *Wear* **206** 214
[2] Krioni N K and Migranov M S 2016 *Coatings and Lubricants for High Temperature Tribocouplings* (Innovationnoe Mashinostroenie, Moscow) 327 pp (In Russian)
[3] Migranov M S and Migranova R M 2012 *Key Eng. Mater.* **496** 75
[4] Vereschaka A A and Migranov M S 2014 *Adv. Mater. Res.* **871** 159
[5] Vereschaka A A, Migranov M S,Vereschaka A S and Sharipov O A 2015 *ARPN J. Eng. Appl. Sci.* **10** 9282
[6] Vereschaka A A, Volosova M A, Grigoriev S N and Vereschaka A S 2013 *Procedia CIRP* **9** 8

IOP Conf. Series: Materials Science and Engineering **447** (2018) 012084 doi:10.1088/1757-899X/447/1/012084

Evolution of the structure and properties of pure aluminum under severe plastic deformation

B K Rakhadilov, M K Kylyshkanov, M S Zhaparova

S. Amanzholov East Kazakhstan State University, 55 Kazakhstan st., Ust-Kamenogorsk 070004, Kazakhstan

E-mail: rakhadilovb@mail.ru

Abstract. The influence of severe plastic deformation (SPD) by equal-channel angular pressing (ECAP) on the formation of the structure and the mechanical properties of pure aluminum is studied. It is established that an ultrafine-grained structure is formed with an average size of structural elements of ~ 1.5 μm. X-ray diffraction studies have shown that SPD results in a broadening of interference lines with a decrease in their intensity, which indicates an increase in the imperfection of the structure and its high dispersion. SPD also results in a significant enhancement of microhardness, yield stress and ultimate tensile strength in aluminum.

1. Introduction

The study of the impact of the structure on properties of nanomaterials is of great relevance for materials science today [1]. The properties, inhered by the material in its usual coarse-grained state, undergo fundamental changes when a nanostructure is formed there. These materials are of primary interest in this regard [2]. For the last few years, material scientists have been showing their interest in ultrafine-grained (UFG) aluminum materials, where SPD is used for the structure refinement [3]. The formation of the UFG structure in these materials considerably improves their mechanical and physical properties, which can be widely applied. Currently, ECAP is one of the dominant approaches in the implementation of SPD. This method makes it possible to produce massive pore-free wrought semi products with ultra-fine grains with sizes ranging from tenths of microns for pure metals to dozens of nanometers for alloys and intermetallic compounds. The study of mechanical properties and structural-phase features influencing their formation in ECAP has been the subject of numerous studies [4].

In this view, the purpose of the study is to understand how SPD, performed by ECAP, affects the pure aluminium structure and its mechanical properties.

2. Experimental procedure

Pure aluminium rods (99.99 %) were employed as the test material. Aluminum rods were subjected to SPD by the ECAP method.

ECAP was carried out in a press mold at an angle of 90° for the intersection of working and output channels. Metal rods with a diameter of 20 mm and a length of 100 mm were pressed at a rate of 0.4 mm/s. The rods were subjected to ECAP at room temperature with a varying number of passes (from 1 to 12) rotating around their axis by 90° or 180° between the passes.

The structure of aluminium before and after ECAP was examined by optical microscopy using a Neophth-21 microscope and scanning electron microscopy using a JSM-6390LV microscope. Aluminum sheets have been polished in an electrolyte consisting of 20% HCl and 80% C_2H_5OH, at 20 °C and 25-35V prior to the study.

X-ray diffraction analysis was carried out on an X'Pert Pro diffractometer in Cu Kα- radiation to study the phase composition and crystal structure of the specimens.

Mechanical tests were carried out at room temperature on a Polyany type machine in accordance with the requirements of GOST 1497-87. During the tests, the sheet specimens were subjected to unaxial quasistatic tension till their rupture, meanwhile measuring the relative yield stress $\sigma_{0.2}$, the ultimate tensile strength σ_{UTS}, and the elongation to failure δ. Specimens with a gauge size of $0.5\times4\times10$ mm^3 were fabricated using electric spark cutting for these tests. Vickers microhardness was measured on a PMT-3 type machine at a load of 0.5 N for 10 seconds. The error of measurement was no more than 8%. A layer ~ 50 μm thick was removed from the surface of the specimen by mechanical grinding followed by electrolytic polishing.

3. Results and discussion

It was found out that in its initial state, aluminum consists of large grains with an average size of 80 μm (figure 1 a). Metallographic and electron-microscopic analyses have shown that ECAP results in a substantial refinement of pure aluminium. Subgrained fragments of ~ 4 and 1.5 μm were formed in the transverse and longitudinal sections after 8 and 12 ECAP passes (figure 1 b, c).

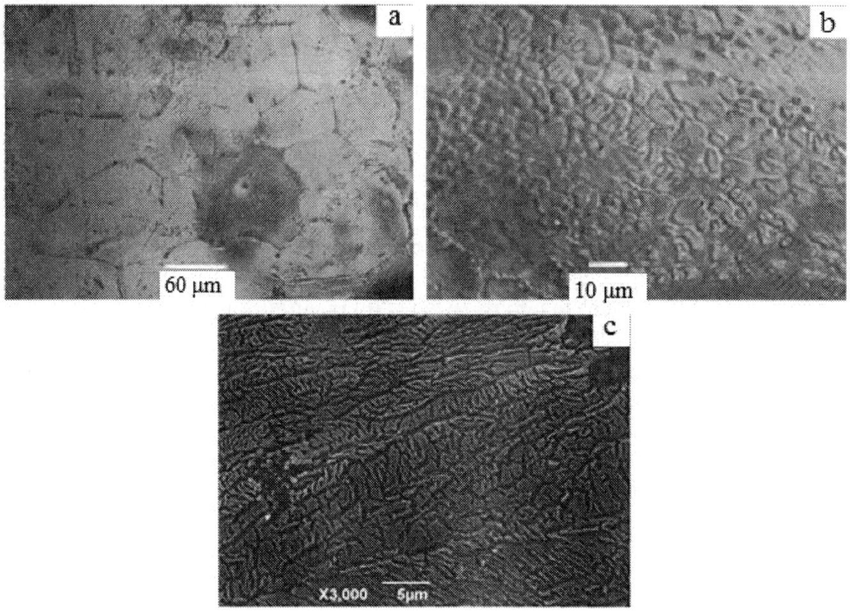

Figure 1. Microstructure of pure aluminum a - in the initial state; b - after ECAP-8 passes, c -after ECAP-12 passes

X-ray diffraction of coarse-grained and submicrocrystalline aluminum has shown considerable changes in the crystal structure after ECAP. As is known from [5], with increasing strain, along with a decrease in the grain size, the degree of imperfection of the structure also increases and the crystallographic deformation texture is formed. Figure 2 illustrates diffraction patterns of aluminum before and after 12 passes of ECAP. As can be seen, the reflexes (200) and (311) broaden, and their intensity decreases. Broadening of the lines and a decrease in their intensity indicate an increase in the imperfection of the structure and its high dispersion.

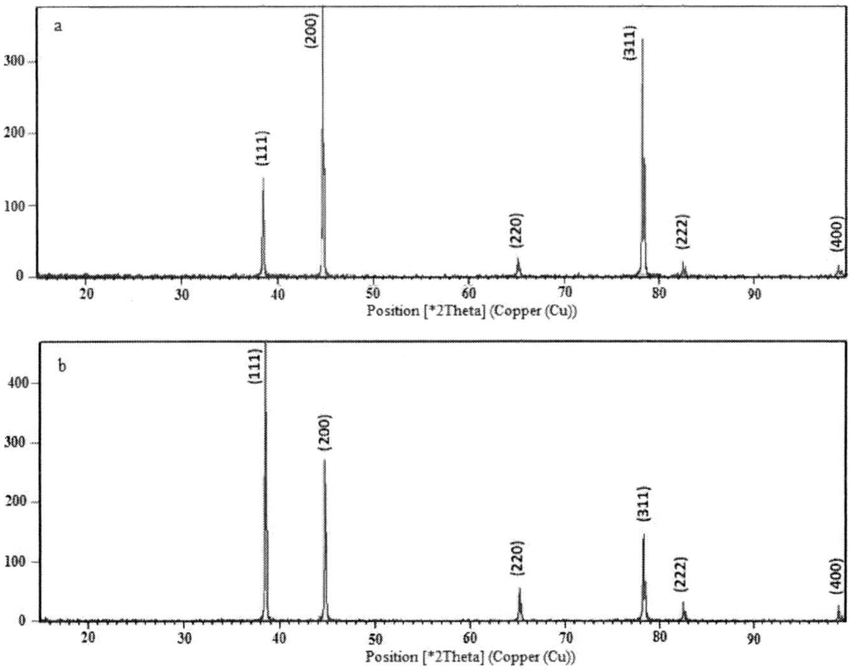

Figure 2. Diffraction patterns of aluminum before ECAP (a) and after 12 passes of ECAP (b).

The formation of ultrafine-grained structure of aluminum provides enhanced mechanical properties of the material. Figure 3 shows dependence of microhardness of aluminium on the number of steps of ECAP carried out with a rotation of the samples to 90° and 180° between the passes. As one can see, the ECAP results in an enhancement of the microhardness of pure aluminum depending on the number of steps, i.e. on the degree of deformation. One also can see that the specimens pressed with a rotation to 90° demonstrate a lower microhardness in comparison to that of the samples pressed with a rotation to 180°.

Thus, it is established that ECAP makes the microhardness of pure aluminum three times stronger.

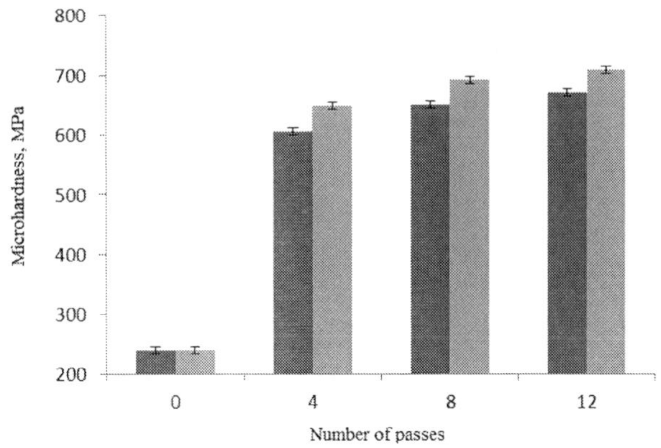

Figure 3. Dependence of the microhardness of aluminum on the number of ECAP passes with a rotation of samples to 90° (green) and 180° (brown).

After ECAP aluminum has improved yield limit and tensile strength (table 1). With an increase of the number of ECAP steps these properties increase too.

Table 1. Mechanical properties of aluminum before and after ECAP

ECAP mode	$\sigma_{0.2}$, MPa	σ_y MPa	δ, %
Before ECAP	10	45	50
ECAP 4-passes	105	137	7.5
ECAP 8-passes	170	191	12.5
ECAP 12-passes	183	214	14

The elongation to failure of the specimens drops down to 7.5% after 4 passes of ECAP and increase back up to 12.5% after 8 passes of ECAP and to 14% after 12 passes of ECAP. These changes are typical to UFG materials.

Thus, aluminium demonstrates mechanical properties which are typical of ultrafine-grained metals obtained by ECAP, i.e. enhanced strength properties, i.e. yield stress, ultimate strength and lower elongation to failure.

4. Conclusions

It was established that severe plastic deformation of pure aluminium by equal-channel angular pressing at room temperature leads to the formation of an ultrafine-grained structure with an average (sub)grain size of 1.5 μm.

It was revealed that as a result of ECAP, a significant hardening of pure aluminium occurs. As a result, the microhardness increased by a factor of 3, the UTS and YS increased by 4.7 and 18 times, respectively, with a simultaneous decrease in the elongation to failure up to 3.5 times.

References

[1] Andrievsky R A, Ragulya A V 2005 *Nanostructured Materials* (Moscow: Publishing Center «Academy») p 192 (In Russian)
[2] Gusev A I 1998 *Nanocrystalline Materials: Methods of Preparation and Properties* (Ekaterinburg: UB RAS) p 199 (In Russian)
[3] Kolobov Yu R, Valiev R Z, Grabovetskaya G P 2001 *Grain-Boundary Diffusion and Properties of Nanostructured Materials* (Novosibirsk: Science) p 232 (In Russian)
[4] Valiev R Z, Aleksandrov I V 2007 *Bulk Nanostructured Metallic Materials* (Moscow: ICC «Akademkniga») p 398 (In Russian)
[5] Kaibyshev O A, Utiashev F Z 2002 *Superplasticity, Refinement of Structure and Processing of Hard-Deformed Alloys* (Moscow: Nauka) p 210 (In Russian)

IOP Conf. Series: Materials Science and Engineering **447** (2018) 012085 doi:10.1088/1757-899X/447/1/012085

Electrical conductivity and strength of ultrafine-grained copper containing deformation vacancies, alloying atoms and dislocations

R G Chembarisova

Ufa State Aviation Technical University, 12 Karl Marx St., Ufa 450008, Russia

E-mail: chroza@yandex.ru

Abstract. The processes of electron scattering by crystal lattice defects such as vacancies, impurity atoms and dislocations are considered. The refined values of the sizes of the regions of interaction of electrons with the defects are obtained. The contribution of these defects to the resistivity and strength of the nanostructured Cu-0.5wt.% Cr alloy is estimated.

1. Introduction

In order to increase the strength properties of metallic materials, they are subjected to severe plastic deformation (SPD), which allows introducing various crystal lattice defects into the material. The influence of defects on the mechanical strength of ultrafine-grained materials obtained by SPD has been thoroughly studied. To analyze the effect of structural defects on physical properties and to predict their changes depending on the factors to which they are sensitive, the appropriate analytical dependencies between the quantities characterizing them are necessary. In particular, materials with high strength and sufficient electrical conductivity are in demand [1]. The introduction of a large number of dislocations into the material strengthens the material. The calculation of their contribution to the resistivity, performed in [2], establishes its dependence on the density of dislocations and the size of the region of interaction of an electron with a dislocation (size of the dislocation core). The estimates, based on the obtained analytical dependence [2] showed the correspondence with the known experimental data in order of magnitude. The purpose of this paper was to calculate the average scattering probability per unit time of an electron on vacancies and impurity atoms, which makes it possible to determine the contribution to the resistivity of the specified point defects. Comparison of the results with known experimental data will allow us to obtain refined data on the dimensions of the regions of interaction of an electron with the lattice imperfections under consideration.

2. Specific resistance and strength of metals containing vacancies

The contribution to the resistivity of metals of defects in the crystal lattice depends on the probability P of the scattering of electrons by them per unit time, which is determined by the dimensions of the scattering region. To calculate the value of P, it is necessary to specify the form of the scattering potential U.

Consider a system of N_e electrons-non-interacting fermions with spin ½, enclosed in a volume V at a temperature T=0. The solutions of the Schrödinger equation for a free particle have the form of plane waves: $\psi_k=\exp(i\vec{k}\cdot\vec{r})$, where $\vec{r}$ is the radius vector, $\vec{k}$ is the wave vector. The density of electronic states in a metal (the density of allowed electron states in a single energy E interval) is equal to

$$g(E)=\frac{\mathrm{d}(N_e/V)}{\mathrm{d}E}=\frac{\mathrm{d}n_e}{\mathrm{d}E}=\frac{3n_e}{2E_F}\sqrt{\frac{E}{E_F}}. \tag{1}$$

At the Fermi level $g(E_F)=3n_e/2E_F$, where n_e is the bulk density of electrons, E_F is the Fermi energy. The potential of the vacant region is defined as $U=n_e/g(E_F)=(2/3)E_F$ [2]. If the probability P of the fact that an electron per unit of time experiences a collision with a vacancy is known, then the relaxation time is $\tau_k \sim 1/P$. Only the electrons in the vicinity of the Fermi energy are scattered. We represent the density of quantum states in the form $g'(E)=\mathrm{d}N_e/\mathrm{d}E=km^*V/(\pi^2\cdot\hbar^2)=p^2V/\pi^2\hbar^3u_F$, where p is the electron momentum. The Fermi momentum is equal to $p_F=m^*u_F$, where u_F is the Fermi velocity, m^* is the effective mass. Then the probability of scattering into a solid angle dw per unit time is determined by the expression

$$P=\frac{2\pi}{\hbar}\left\langle |H_{kk'}|^2\right\rangle\frac{\mathrm{d}w}{4\pi}g'(E_k), \tag{2}$$

where $H_{kk'}$ is the scattering matrix element (transition from the state k to the state with k' in the k-space of the wave numbers):

$$H_{kk'}=\int \Psi_0^{k'*}\hat{H}\Psi_0^k \mathrm{d}\tau. \tag{3}$$

The perturbation of the Hamiltonian operator $\hat{H}$ is assumed to be equal to the potential of the vacant region U. For scattering, the particle energy is conserved: $|\vec{p}|=|\vec{p}'|$. Accordingly, the stationary solution of the Schrödinger equation is written in the form $\Psi_0^k=\left(\exp(i\vec{k}\cdot\vec{r})\right)/\sqrt{V}$. The electron scattering region τ by the vacancy is representable in the form of a cube, one of the faces of which is parallel to the vector $\vec{k}'-\vec{k}$. The volume of the cube is l^3. The integration in formula (3) with respect to the selected scattering region leads to the value

$$H_{kk'}=\frac{2}{3}\frac{E_F}{V}\frac{\sin|\vec{k}'-\vec{k}|l/2}{|\vec{k}'-\vec{k}|l/2}l^3. \tag{4}$$

The modulus of the vector $\vec{k}'-\vec{k}$ is $|\vec{k}'-\vec{k}|=2k\cdot\sin(\Theta/2)$. Then the scattering probability (2) takes the form

$$P=A\frac{\sin^2(kl\sin(\Theta/2))}{k^2l^2\sin^2(\Theta/2)}\frac{p^2V}{4\pi^3\hbar^3u_F}\mathrm{d}w. \tag{5}$$

where $A=(2\pi/\hbar)(Ul^3/V)^2$. The averaged probability of scattering per unit time $\langle P_N\rangle$ over all directions of the vector $\vec{k}$ in a sample with $N_\upsilon=C_{\upsilon V}V=C_{\upsilon a}n_aV$ vacancies, where n_a is the bulk density of atoms, $C_{\upsilon V}$ and $C_{\upsilon a}$ are the volume and atomic concentrations of vacancies, respectively, is

$$\langle P_N\rangle=\frac{1}{\tau_k}=\frac{k^3\hbar l^4C_{\upsilon V}}{9m^*\pi}\left[1-\frac{\sin 2kl}{2kl}\right]. \tag{6}$$

Then the contribution to the resistivity of vacancies is determined by the expression:

$$\rho_\upsilon=\frac{m^*}{n_ee^2}\frac{1}{\tau_k}=\frac{k^3\hbar l^4n_aC_{\upsilon a}}{9e^2\pi n_e}\left[1-\frac{\sin 2kl}{2kl}\right]. \tag{7}$$

According to well-known theoretical estimates [3 - 5], the contribution to the resistivity of noble metals per one percent of vacancies is equal to ρ_υ=1.5 μΩ·cm/at%, which coincides with the experimental values. Also known are the values obtained by a number of authors for copper, equal to 1.64 [6], 1.70 [7] and 1.73 μΩ·cm/at.% [8]. The contribution of vacancies to the resistivity of copper, according to formula (7), is equal to $\rho_\upsilon \approx 1.0C_{\upsilon a}$ μΩ·m, or 1.0 μΩ·cm/at.%, if we assume that the volume of the interaction region is equal to $l^3=a^3/4$, where a is the lattice parameter. The obtained estimate is lower than the known theoretical estimates. The result obtained in this paper using the potential U [2] depends on the size of the interaction region l and corresponds to the known data in order of magnitude. At l=1.06b, the increase in the resistivity of copper is $\rho_\upsilon \approx 1.73$ μΩ·cm/at.%. In the SPD process, many deformation vacancies are formed. Their concentration can reach values of $C_{\upsilon a}=10^{-5}$, the value ρ_υ is respectively $1.73\cdot10^{-3}$ μΩ·cm. At the same time, the resistivity of pure copper at room temperature is equal to $\rho_0 \approx 1.724$ μΩ·cm [1]. Thus, the contribution of vacancies to the electrical conductivity is very small.

The effect of vacancies on the strength and ductility of the metal is small and can be neglected.

3. Specific resistance and strength of metals containing impurity atoms

The introduction of an impurity atom into the crystal lattice leads to a change in its potential, which can be represented as the appearance of a Coulomb field of a point charge having the order of the elementary charge e. Scattering on a dissolved atom occurs as a result of the interaction of a free electron with this potential. The dependence on the distance r of the potential φ of the point charge q surrounded by an electron Fermi gas has the form $\varphi = qe^{-\delta\cdot r}/r$ [9]. The Coulomb potential of a point charge is screened at distances $r>1/\delta$, where $1/\delta\approx0.554\cdot10^{-10}$ m for copper. The contribution to the resistivity of impurity atoms the volume concentration of which is C_{aV}, is calculated by the formula $\rho_a = m_e u_F C_{aV} S_a / n_a e^2$. The scattering cross section is $S_a=\pi r^2$. If the screening length is assumed to be $r=3/\delta$, then at an atomic concentration of $C_{aa}=C_{aV}/n_a=6.1\cdot10^{-3}$ (0.5 wt%), the increase in the resistivity of copper is ρ_a=2.96 μΩ·cm (ρ_a=4.85 μΩ·cm/at.%).

The resistivity due to scattering of electrons by alloying atoms can also be calculated from formula (7) using the atomic concentration C_{aa} of atoms. The results of the calculations lead to the value of $\rho_a \approx 3.14$ μΩ·cm for l=1.35b and $C_{aa}=6.1\cdot10^{-3}$ (ρ_a=5.15 μΩ·cm/at.%).

The solid-solution hardening is proportional to the value of $C_{aa}^{1/2}$. Practically, the dependence of hardening on concentration does not differ from linear [10]. In the case of the Cu-0.5wt.% Cr alloy, its contribution can be neglected.

4. Specific resistance and strength of metals containing dislocations

To calculate the resistivity due to the presence of dislocations in the material, we can use the method described above to calculate the same for vacancies. It is assumed that electron scattering takes place at the dislocation core. The scattering potential is equal to U. The interaction region is represented in the form of a rectangular parallelepiped whose volume is equal to Ll^2. The averaged probability of scattering per unit time for an electron with a wave number k on $N_{disl.}=\rho_{tot}L^2$ dislocations is determined by the formula

$$\langle P_N \rangle = \frac{1}{\tau_k} = \frac{k^2\hbar l^2 L}{9m^* V}\rho_{tot}L^2\left[1-\frac{\sin 2kl}{2kl}\right]. \tag{8}$$

This formula leads to the value of the resistivity per unit dislocation density in copper $\rho_{disl.}/\rho_{tot}=5.24\cdot10^{-26}$ Ω·m³ for the size of the interaction region $l = a/\sqrt[3]{4}$ [2]. The values of $0.78\cdot10^{-25}$ and $1.9\cdot10^{-25}$ Ω·m³ were obtained in [11, 12], respectively. At the same time, experimental data are known that are equal to $(1.6\pm0.2)\cdot10^{-25}$ [13], $1.7\cdot10^{-25}$ [14], $(1.8\text{-}2.3)\cdot10^{-25}$ Ω·m³ [15, 16]. The result, consistent with the experimental data $\rho_{disl.}/\rho_{tot}=1.7\cdot10^{-25}$ Ω·m³, can be obtained using formula (8) for the scattering area sizes equal to $l\approx1.55b$.

The electrical conductivity of the Cu-0.5wt.% Cr alloy subjected to 4 passes of equal-channel angular pressing is 35% IACS ($\rho=4.93\cdot10^{-8}$ Ω·m) [17]. The characteristic value of the dislocation density is $\rho_{tot}= 8.33\cdot10^{14}$ m^{-2}, whose contribution to the resistivity of the alloy is $\rho_{disl.}=1.42\cdot10^{-10}$ Ω·m. Taking into account the contribution of vacancies ($C_{\upsilon a}=10^{-5}$, $\rho_\upsilon=1.73\cdot10^{-3}$ μΩ·cm) and the resistivity of pure copper, it is possible to determine the resistivity due to the alloying chromium atoms: $\rho_a=\rho-\rho_0-\rho_\upsilon-$

$\rho_{disl.}$=3.19 ·10^{-8} Ω·m, which corresponds to the dimensions of the interaction region of atoms with electrons, equal to l≈1.354b.

The dislocation hardening is described by the dependence $\sigma_{disl.} = M\alpha Gb\sqrt{\rho_{tot}}$ [18], where α is the coefficient [17], α=0.37. M is the Taylor factor, M=3.06. The shear modulus is G=42.1 GPa. In the specified state of the alloy, Cu-0.5wt.% Cr σ_{disl}≈352 MPa.

5. Summary and conclusions

The results obtained by the formulas presented in the paper agree with the experimental data in order of magnitude. From the known experimental data, the dimensions of the region of interaction of electrons with crystal lattice defects can be refined, which makes it possible to improve the accuracy of estimates. Vacancies have the smallest area of scattering (l≈1.06b), dislocations - the largest (l≈1.55b). The region of interaction of atoms is l≈1.35b. As follows from the calculations, the resistivity of the Cu-0.5wt.% Cr alloy is substantially increased by chromium atoms dissolved in the matrix, which do not exert a noticeable effect on its strength. The presence of vacancies and dislocations does not have any noticeable effect on the electrical conductivity of the material. At the same time, dislocations strengthen the material.

Acknowledgments

The work was carried out with the financial support of the Ministry of Education and Science of Russia within the framework of the project № 16.1969.2017 / 4.6

References

[1] Osintsev O E 2004 *Copper and Copper Alloys* (Moscow: Mashinostroyeniye) p 336 (In Russian)
[2] Harrison W A 1958 *J. Phys. Chem. Solids* **5** 44
[3] Blatt F J 1956 *Phys. Rev.* **103** 1905
[4] Abeles M F 1953 *Compt. Rend. Acad. Sci.* **237** 796
[5] Seeger A and Stehle H 1956 *Z. Phys.* **146** 242
[6] Seeger A 1962 *J. Phys. Rad.* **146** 242
[7] Polác J 1967 *Czech. J. Phys.* B **17** 171
[8] Fischer K 1963 *Phys. Status Solidi* **3** 2035
[9] Abrikosov A A 1987 *Fundamentals of the Theory of Metals* (Moscow: Nauka) p 520 (In Russian)
[10] Stremel M A 1997 *Strength of the Alloys. Part II. Deformation* (Moscow: MISIS) p. 527 (In Russian)
[11] Watts B R 1988 *J. Phys. F: Met. Phys.* **18** 1197
[12] Karolik A S and Luhvich A A 1994 *J. Phys.: Condens. Matter* **6** 873
[13] Brown R A 1977 *J. Phys. F: Met. Phys.* **7** 1283
[14] Fonteyn D and Pitsi G 1987 *Solid State Commun.* **64** 1113
[15] Kopetskii C V, Kulesko G I, Kokhanchik L S and Zharikov O V 1974 *Phys. Status Solidi a* **22** 185
[16] Blewitt T H, Coltman R R and Redman J K 1955 *Defects in Crystalline Solids* (London: Physical Society) p 309
[17] Wei K X, Wei W., Wang F, Du Q B, Alexandrov I V and Hu J 2011 *Mater. Sci. Eng.* A **528** 1478
[18] Hughes D A and Hansen N 2000 *Acta Mater.* **48** 2985

IOP Conf. Series: Materials Science and Engineering **447** (2018) 012086 doi:10.1088/1757-899X/447/1/012086

Development of a neural network model for predicting the physical and chemical properties of materials from the technological parameters of their formation

T A Bednaya[1], S P Konovalenko[2]

[1] Don State Technical University, 1 Gagarin square, Rostov-on-Don 344010, Russia
[2]Rostov State University of Economics, 69 Bolshaya Sadovaya st., Rostov-on-Don 344002, Russia

E-mail: bednayat@mail.ru

Abstract. A new approach for predicting the physical and chemical properties of materials based on the technological regimes of their formation is proposed. The developed method is applied to predict the response of the sensitive layer of a chlorine sensor based on thepolyacrylonitrile film containing silver. The artificial neural network is constructed in the form of multilayer perceptron for predicting the gas sensitivity coefficient based on the technological parameters used for the material formation (mass fraction of the additives, wt.%, the temperature and time of the 1st and 2nd IR annealing steps). The adequacy of the developed modelis proved: the root-mean-square error iss_t = 0.07, the correlation coefficient is R = 0.927, and the mean-square error of the forecast iss_v = 0.11. The sensitivity of the model for the mean square error of all the data is estimated to be s = 0.09.

1. Introduction

The prospect of using films of electrically conductive organometallic polymeric materials in chemical and biological sensors is the reason for intensive study of these properties of materials. Films of metal-containing polyacrylonitrile (PAN) are successfully used as a gas-sensitive layer of gas sensors operating at a temperature in the range of 17 – 32 °C [1]. The usage of PAN films is based on the adsorption-resistive effect, which consists in changing the resistance of the film material with selective absorption of gas molecules. A feature of PAN films is the ability to form polymer structures with a system of conjugated double bonds along the chain of macromolecules during heat treatment, which are characterized by increased electrical conductivity [2]. Thus, along with surface processes, structural changes in the linear polymer of PAN play an important part in the sensory response formation.

One of the problems that make it difficult to determine the criteria of the targeted synthesis of materials for solid-state gas sensors is the insufficient study of the effect of technological parameters of the material formation of PAN films obtained by the pyrolysis method under the influence of incoherent IR radiation on their electric, physical and gas sensitive properties. The development of technological bases for the controlled formation of nanocomposite films of metal-containing PAN is a laborious process for identifying the non-linear dependencies of the gas sensitive properties of PAN films on the given technological parameters [3-5].

2. Methodology

Films of the composition of Ag-containing PAN were formed by pyrolysis under the influence of incoherent IR radiation [6]. IR annealing was carried out using different temperature regimes in two stages. The first stage serves for the preliminary structuring of the PAN with the formation of a system of C = N-conjugated bonds at a temperature of 250 °C and 300 °C. The second stage of annealing, passing at a temperature of 350 and 450 °C, leads to the formation of also -C = C-bonds [7]. The time of exposure to IR radiation at each temperature varied (2 ÷ 20 minutes). The mass fraction of the modifying additive varied in the range of 0 ÷ 3 wt. %

The construction of an analytical model of a complex object is problematic, and sometimes impossible, and the relationship and dependence of the physicochemical properties of objects on the given parameters is complex and nonlinear. The neural network approach is used to describe complex relationships between numerous factors or parameters of obtaining objects and their physicochemical properties [8-9]. The prediction of properties of objects using structural data by the neural network approach has been successfully applied in various fields of physical research. For example, to predict the temperature in a storage tank for liquids using solar energy [10], the neural network model was used in the form of a multilayer perceptron with back propagation of the error.

To predict the properties of materials using neural network modeling, one can single out the general sequence of steps required for any modeled object (figure 1).

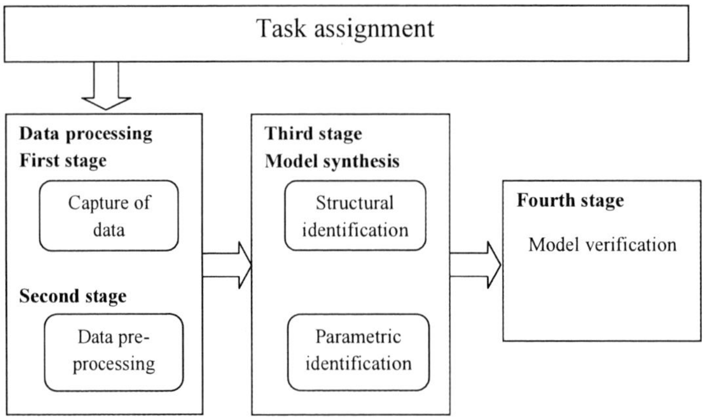

Figure 1. Algorithm for modeling of physical and chemical materials properties using neural networks.

3. Results

As a method of a neural network constructing for analyzing the experimental data, the method of constructing a neural network in the form of a two-layer perceptron is used (figure 2). The neural network contains 5 input neurons and 1 output. The neural network was trained using an algorithm for back propagation of the error.

Each input neuron corresponded to one of the technological parameters (mass fraction of the additive (mass%), temperature and time of the 1st and 2nd stages of IR annealing). The output neuron corresponded to the predicted property (S, rel.). The adequacy of the synthesized models was proved: the standard error of s_t = 0.07, the correlation coefficient R = 0.927, and the mean square error of the forecast s_v = 0.11. The sensitivity of the model was estimated for a standard data error of s = 0.09.

The correlation coefficient R determines the attributable fraction of the spread, R^2 is the fraction of the total spread relative to the gas sensitivity S, due to regression (R^2 = 0.86). The fact that the model explains by 86% the spread of data relative to the average, allows us to speak about the reliability of the model.

The root-mean-square errors of training $s_t = 0.07$ and test, $s_v = 0.11$ samples are approximately the same, which excludes retraining of the network. The accuracy of the calculations in the "learning" group and the "test" subgroup is the same (this is the forecast power), which indicates the stability of the model.

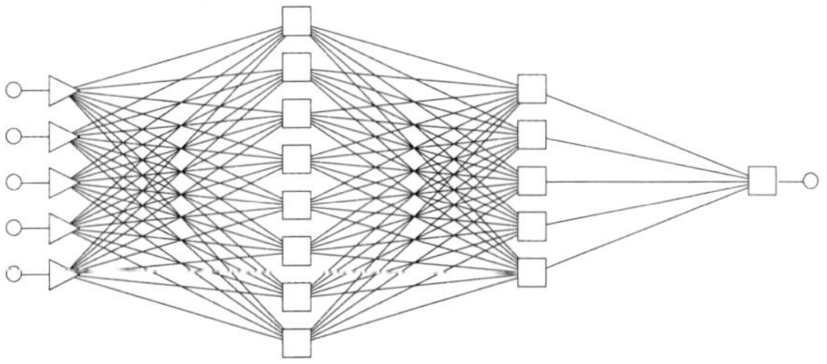

Figure 2. Structure of neural network.

The sensitivity of the model was estimated by means of the mean square error of all data $s = 0.09$.

An alternative way to verify the adequacy of the model is to test it.

The equation of the correlation dependence between the calculated and experimental values of the gas sensitivity of a silver-containing PAN is computed: $S_{pred} = 0.878S_{exp} + 0.078$. The visualization of the equation is shown in figure 3.

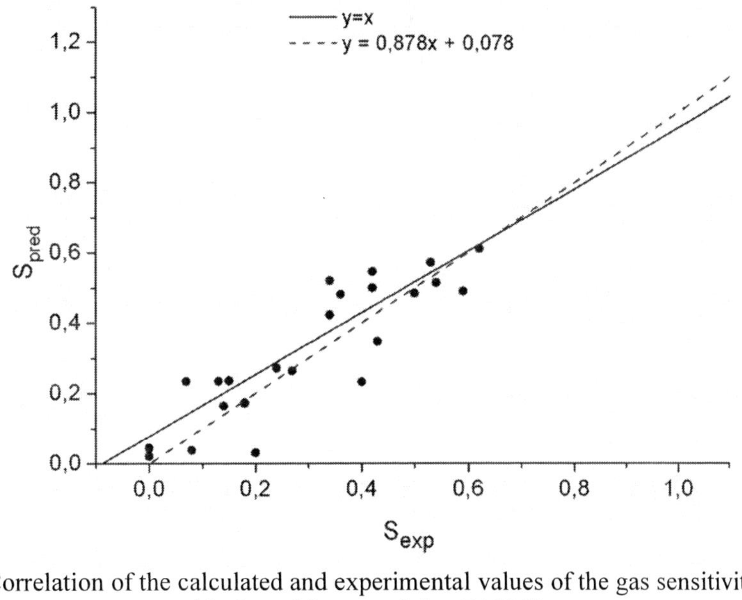

Figure 3. Correlation of the calculated and experimental values of the gas sensitivity coefficient to chlorine of Ag-containing PAN films

The optimum set of technological parameters for fabricating the Co-containing PAN films with the best values of gas-sensitivity to Cl_2 is established as follows: ω(Ag) = 0.05 mass. %, T_{drying}= 160 °C, t_{drying} = 30 minutes, $T_{IR\text{-}annealing1\ phase}$ = 250 °C, $t_{IR\text{-}annealing1\ phase}$ = 2 minutes, $T_{IR\text{-}annealing2\ phase}$ = 350 °C, $t_{IR\text{-}annealing\ 2\ phase}$ = 28 minutes.

Thus, a synthesized neural network can be used to predict the value of the gas sensitivity coefficient of Ag-containing PAN films in order to create an effective low-temperature gas sensor.

Experimental verification of the synthesized model was carried out (see table 1).

Table 1. Tests result of the synthesized model.

ω (Ag),. %	T_{drying} – t_{drying}, °C – min	$T_{IR\text{-}annealing1}$– $t_{IR\text{-}annealing1}$, °C – min	$T_{IR\text{-}annealing2}$– $t_{IR\text{-}annealing2}$, °C – min	Gas Sensitivity S, r.u.		Absolute and relative error	
				Prediction	Experiment	Δ	δ, %
0.05	160 – 30	250– 2	350 – 28	0.75	0.69	0.06	8
0.75	160 – 30	300 – 2	500 – 20	0	0	0	0
0.75	160 – 30	150– 3	400 – 2	0.13	0.15	0.02	13

4. Conclusion

A methodology for the synthesis of a neural network model has been developed for predicting the physical and chemical properties of materials taking into account the technological parameters of their formation. The developed method was used to predict the response of the sensitive layer of a chlorine sensor based on polyacrylonitrile containing silver.

It was shown that the use of the neural network approach allows reducing experimental work and establishing optimal technological parameters for obtaining gas sensitive materials to create efficient gas sensors.

References

[1] Bednaya T A, Konovalenko S P, Semenistaya T V, Perov V V and Korolev A N 2012 *Izvestiya Vysshikh Uchebnykh Zavedenii. Electronica.* **4** 66 (In Russian)
[2] Jing M, Wang C, Wang Q, Bai Y and Zhu B 2007 *Polym. Degrad. Stabil.* **92** 1737
[3] Semenistaya T V, Petrov V V, Kalazhokov K K, Kalazhokov Z K, Karamurzov B S, Kushkhov K V and Konovalenko S.P. 2015 *Surf. Eng. Appl. Electrochem.* **51** 9
[4] Konovalenko S P and Semenistaya T V 2013 *Izv. Sfedu. Eng. Sci.* **1** 178 (In Russian)
[5] Semenistaya T V, Petrov V V, Bednaya T A and Zaruba O A 2015 *Mater. Today: Proc.* **2** 77
[6] Konovalenko S P, Bednaya T A, Semenistaya T V, Perov V V and Maraeva E V 2012 *Inz. Vestn. Dona* **4** chapter 2 URL:ivdon.ru/magazine/archive/n4p2y2012/1356 (In Russian)
[7] Zemtsov L M and Karpacheva G P 1994 *Polym. Sci.* **36** 919 (In Russian)
[8] Kruglov V V and Borisov V V 2001 *Iskusstvennye Nejronnye Seti. Teorija i Praktika.* (Moscow: Gorjachaja linija) (In Russian)
[9] Himmel'blau D 1975 *Prikladnoe Nelinejnoe Programmirovanie* (Moscow: Mir) (In Russian)
[10] Géczy-Víg P and Farkas I 2010 *Solar Energy* **84** 801

IOP Conf. Series: Materials Science and Engineering **447** (2018) 012087 doi:10.1088/1757-899X/447/1/012087

Copper-containing nanocomposites in catalytic systems

I Yu Isaeva[1], G Yu Ostaeva[1], V V Grushina[1], A A Litmanovich[1] and O S Morozova[2]

[1] Moscow Automobile and Road Construction State Technical University (MADI), 64 Leningradskii pr., Moscow 125319, Russia
[2] Semenov Institute of Chemical Physics RAS, 4 Kosygin St., Moscow 119991, Russia

E-mail: irina-razumova-xim@yandex.ru

Abstract. Nanostructured systems, including the products of copper ions recovery – nanoparticles of copper, copper(I) oxide or complex copper particles and copper oxides, have been obtained. Copper ions were reduced by *tert*-butylamine borane in aqueous solutions of poly(N-vinylpyrrolidone) at room temperature under air. Different synthesis conditions allowed controlling both the size of the resulting particles and the final product of the synthesis. The composition, morphology and size characteristics of nanocomposites were studied by methods of electron microscopy and XRD. The efficiency of the Cu/CeO_2 catalytic system in the purification of hydrogen from CO was shown.

1. Introduction

The synthesis of copper sols by reducing Cu^{2+} ions in aqueous media, polymers or surfactants is used as size regulators of the forming metal nanoparticles. These compounds protect the nanoparticles from aggregation and oxidation [1-3]. The manufacturing procedure allows controlling the size of particles containing Cu, and to choose parameters for an optimal particle range of 2-15 nm.

The properties of a nanostructured system strongly depend on their dimensional characteristics. In our earlier work [4, 5], the formation of copper(I) oxide nanoparticles protected from aggregation by macromolecular screens polymeric protector of poly(N-vinylpyrrolidone) (PVP) was investigated.

Nanosized copper or copper(I) oxide are promising in many applications: solar energy conversion, catalysis and photocatalysis, the development of photovoltaic cells and nanofluids, etc. [6, 7]. They could also be effective substitutes for gold due to the abundance of copper and relatively low cost in catalytic systems.

For commercially and ecologically important processes, various Cu-containing catalysts are used, among which there is hydrogen purification from CO [8]. However, their preparation techniques are not very environmentally friendly due to the waste of water and energy expenditure. The main disadvantages are high costs and strict requirements for use and production. On the contrary, the mechanochemical synthesis applied for this purpose is fast and ecologically friendly. This article focuses on the preparation and morphology of nanosized particles of Cu/Cu_2O, as well as the redox properties of the composite $Cu\text{-}CeO_2$ manufactured by mechanochemical synthesis.

2. Experimental

Copper(II) sulfate pentahydrate (reagent grade, Reakhim, Russia), *tert*-butylamine borane (TBAB) (98% Aviabor, Russia) and PVP (BASF, Mn = 3.6×10^5) were used as the initial materials. The water was distilled twice. Cu nanoparticles were obtained via the reduction of copper ions with TBAB at room temperature in air. The calculated amount of a freshly prepared reducing agent solution was added to an aqueous solution containing a salt of copper (0.01 mol/L) and PVP (0.01 base-mol/L). The concentration of the reducing agent in the reaction mixture varied from 0.01 to 0.03 mol/L.

TEM images of the particles were obtained using a JEM-1011 microscope equipped with a side-mounted digital camera Gatan Erlangshen ES500W. The size distribution histograms were constructed from the obtained TEM images.

XRD analysis was performed using a DRON-3 diffractometer (Russia) with Cu-Kα radiation. The composition of the powder phase was determined using JCPDS files provided by the International Center for Diffraction Data.

The morphology of the samples was examined in a dual beam scanning electron microscope/ focused ion beam (SEM/FIB) Versa 3D (FEI, USA), equipped with a concentric backscatter detector (CBS). The microstructure of thin cross-sections of specimens prepared by Ga+ FIB was studied in a Titan 80-300 TEM/STEM (FEI, USA).

Nanocomposite catalysts were prepared by mechanochemical synthesis. Meanwhile, the mass ratio of Cu /CeO_2 was 8:92. Catalysts can be prepared for 0.5–1.5 hours at room temperature in air. Nanocomposite redox properties towards CO and H_2 were studied by temperature-programmed reduction (CO-TPR and H_2-TPR, respectively), which are given in detail. Earlier this method was tested on the system of CuO/CeO_2 and Cu/CeO_2 [9].

3. Results and Discussion

The formation of transparent stable dispersions of Cu_2O, Cu^0 in the presence of PVP indicates the small size of the particles of the dispersed phase and the shielding thereof by macromolecules. Figure 1 depicts the TEM image and the particle size distribution histogram of the particles.They have a spherical shape and a narrow size distribution with an average diameter of about 6.5 nm.

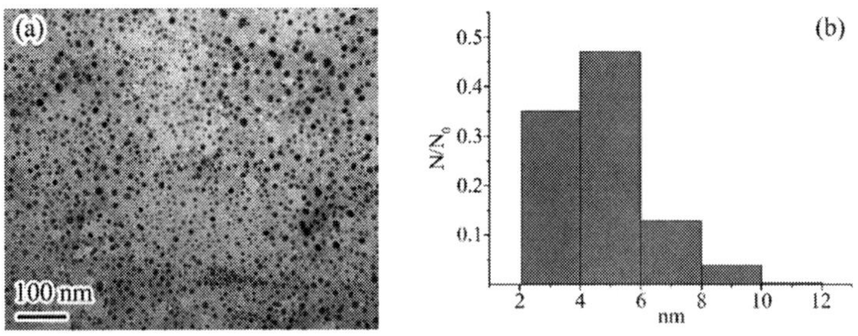

Figure 1. (a) TEM image and (b) histogram of numerical size distribution of particles in the Cu sol obtained by the reduction of copper ions in solution of PVP.

Weak cooperative non-covalent interactions with the participation of macromolecules and other macromolecules or the surface of species such as nanoparticles result in the formation of products the stability of which depends on the size of the smallest component of the complex formed. In particular, in the case of interaction between a long macromolecule and a nanoparticle, the stability of corresponding complex (MNC) representing a nanoparticle screened by the polymer chain increases exponentially with the surface area of nanoparticle. This explains the reason of essential consequences such as:

- In dilute MNC dispersions (sols), a very low polymer – nanoparticle interaction energy (-ΔG of order 10^{-3} - 10^{-4} J/m^2) is sufficient to ensure the stability of MNC, including nanoparticles with a diameter from 2 to 12 nm.

- A small size and a narrow size distribution of nanoparticles are typical of the soles obtained in processes of new phase formation proceeding in solutions of appropriate polymers (pseudo-template processes).

Figure 2 shows the XRD pattern of the final product obtained after reduction of $CuSO_4$. The phase composition is identified as metal Cu (~75 mass%) and Cu_2O (~25 mass%).

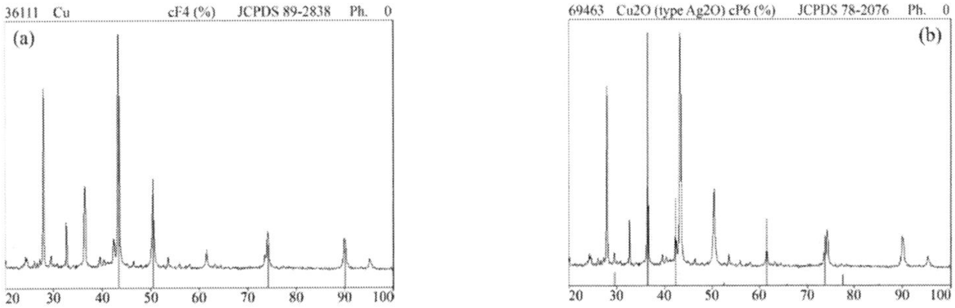

Figure 2. XRD patterns of $CuSO_4$. Characteristic peaks of Cu and Cu_2O, are marked in (a) and (b), respectively. The corresponding JCPDS phase numbers are given at the top of the figures.

The calculation of the lattice spacing indicates the presence of CeO_2 and Cu_2O, Cu^0 nanofragments. At the same time, the crystal lattice of the original Cu_2O and Cu metal contains defects and their particles are porous due to the special features of the synthesis.

Figure 3 shows images of BF STEM and HR TEM of Cu /CeO_2 particles. The characteristic structures of CeO_2 and Cu_2O, Cu^0 are shown by direct measurements of the lattice spacing.

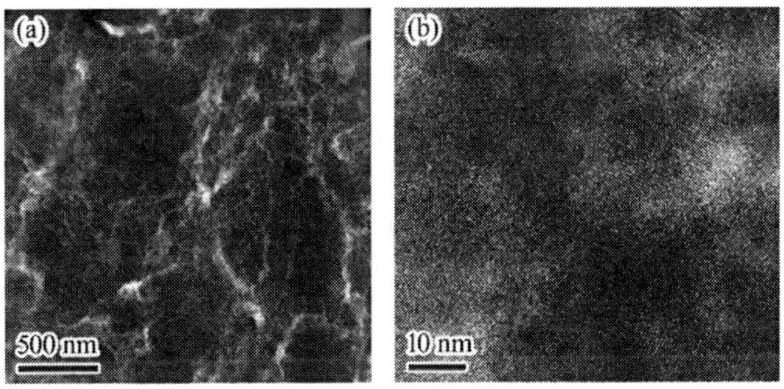

Figure 3. (a) TEM and (b) STEM images of Cu/CeO_2.

Nano-sized copper powder was used as an active component of the Cu-CeO_2 composite catalyst prepared by intermixing and mechanical activation of both components in a ball mill. High energy impact during the processing by ball-milling results in the adhesion of Cu particles to the surface of CeO_2. Figure 4 shows CO-TPR and H_2-TPR MS profiles (m/e = 44, CO_2 and m/e = 18, H_2O, respectively) recorded for 30 minute milled powder. In accordance with [9], low-temperature peaks were attributed to the reaction of both CO and H_2 with oxygen localized in the Cu/CeO_2 interface. This is "active" oxygen participating in the oxidation of CO in an excess of H_2. A large temperature loop between Tmax of low-temperature peaks of CO- and H_2-TPR point to the possibility of successful purification of H_2 from CO. Especially important are two features. (1) – Both CO- and H_2-TPR profiles are virtually identical for 30- and 60-minute milled powders (not shown) which indicate the extremely effective and rapid incorporation of nano-sized Cu particles in the CeO_2 matrix. (2) – The low-temperature peak fraction corresponding to the "active" oxygen fraction in the composite is ~ 40% after 30 minutes of milling instead of ~ 16.5% measured for a catalyst prepared from electrolytic (dendrite) Cu powder and CeO_2 [10] which also points to a significant effect of the size of Cu powder on the formation of the Cu/CeO_2 interface (figure 5).

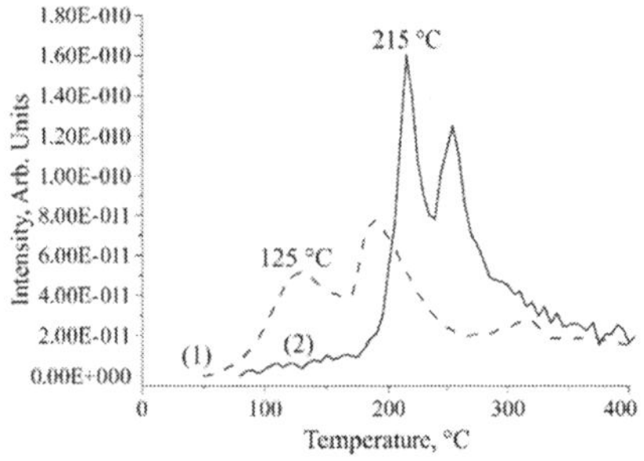

Figure 4. CO-TPR and H_2-TPR MS profiles recorded for 30 minute milled Cu-CeO_2 powder: 1 - m/e = 44, CO_2 and 2 - m/e = 18, H_2O.

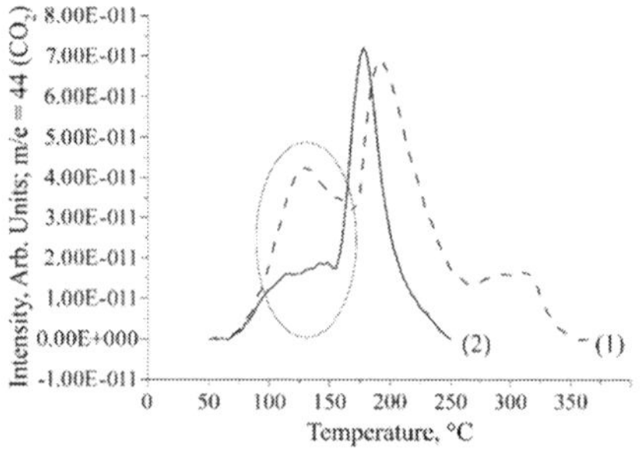

Figure 5. CO-TPR MS profiles (m/e = 44, CO_2) recorded for 30 minute milled Cu-CeO_2 powders prepared from: 1 – nanozised Cu and 2 – electrolytic Cu.

4. Conclusions

A new manufacturing procedure, including the reduction of copper ions by *tert*-butylamine borane in aqueous solutions of poly(N-vinylpyrrolidone) at room temperature under air is applied to synthesize copper, copper(I) oxide or complex copper nanoparticles. The size of Cu-containing particles of 2-15 nm is controlled by preparation parameters. The nanoscale particles obtained allow one to create new highly reactive oxygen states for CO oxidation in Cu-CeO_2 nanocomposites during a short time of joint mechanochemical treatment. This point on the promising potential of these systems in purification of hydrogen from CO.

Acknowledgment

This work was supported by the Ministry of Education and Science of the Russian Federation within the framework of the state assignment (project No. 16.11777.2018/11.12) and partially (XRD and DSC-DTG-MS measurements) by RFBR through research project No. 16-03-00330.

References

[1] Litmanovich O E 2008 *Polymer Sci. C* **50** 63
[2] Litmanovich O E, Ostaeva G Yu, Tatarinov V S, Bogdanov A G and Papisov I M 2010 *Polymer Sci. B* **52** 397
[3] Ostaeva G Yu, Bogdanov A G, Papisov I M 2006 *Polymer Sci. B* **48** 101
[4] Papisov I M, Isaeva I Yu, Ostaeva G Yu, Eliseeva E A, Papisova A I and Kozlovskii V F 2015 *Colloid J.* **77** 780
[5] Ostaeva G Yu, Isaeva I Yu, Grushina V V, Stuzhuk A N and Odinokova I V 2018 *Polymer Sci. B* **60** 455
[6] Gawande M B, Goswami A, Felpin F-X, Asefa T, Huang X, Silva R, Zou X, Zboril R and Varma R S 2016 *Chem. Rev.* **116** 3722
[7] Tamilvanan A, Balamurugan K, Ponappa K and Madhan Kumar B 2014 *Int. J. Nanosci.* **13** 1430001
[8] Guralskiy A V, Kosmambetova G R, Gritsenko V I, Strizhak P E, Kolko V P, Moroz E M, Gulyaev R V, Boronin A I and Ivashchenko T S 2009 *Nanosystems, Nanomaterials, Nanotechnologies* **7** 195
[9] Firsova A A, Morozova O S, Vorobjeva G A, Leonov A V, Kukharenko A I, Cholakh S O, Kurmaev E Z and Korchak V N 2018 *Kinetics and Catalysis* **59** 188
[10] Borchers Ch, Martin M L, Vorobjeva G A, Morozova O S, Firsova A A, Leonov A V, Kurmaev E Z, Kukharenko A I, Zhidkov I S and Cholakh S O 2016 *J. Nanopart. Res.* **18** 344

IOP Conf. Series: Materials Science and Engineering **447** (2018) 012088 doi:10.1088/1757-899X/447/1/012088

Development of severe plastic deformation methods for the production of contact wires from a Cu-Zr alloy for high-speed railways

A G Raab[2], D A Aksenov[1,2], R N Asfandiyarov[1,2], I S Kodirov[2] and G I Raab[2]

[1]Ufa State Aviation Technical University, 12 Karl Marx st., Ufa, 450008 Russia
[2]Ufa Federal Research Center RAS, 71 pr. Oktyabrya, Ufa, 450054 Russia

E-mail: agraab@mail.ru

Abstract. Zirconium is, like chromium, one of the most promising alloying elements. Microalloying with these elements enables increasing significantly not only the mechanical characteristics of copper alloys by cold deformation methods, while preserving a high electrical conductivity, but also imparting hot-strength properties. At room temperature, zirconium bronze is slightly inferior to chromium bronze in terms of strength characteristics, but it has a higher electrical conductivity and ductility. This set of properties determines the operational efficiency of wires. Another important parameter is the wear resistance of wires. Taking into account the operational restrictions related to wear, amounting to 20% of the wire's cross-sectional area, wires can be strengthened not across the whole section, but with a gradient, i.e. with a stronger surface layer and a weaker core. This will enable providing an enhanced wear resistance and strength of a wire and further increasing its ductility and bending fatigue limit. One of the methods to produce such properties is the method of rotary active alternating bending, which enables, in the conditions of multi-cycle processing, accumulating a high level of strain predominantly in the surface layers of wires. In this work, we present the study of this method with respect to the formation of a gradient-type structure in the samples of a Cu-Zr alloy, with the use of computer and full-scale modeling.

1. Introduction

Requirements to the properties of modern products and structural materials become higher and more specific. First of all, it refers to the responsible metal products operating under extreme conditions in aerospace, chemical, medical and other industries. In recent years, to increase the strength characteristics by the formation of ultrafine grained (UFG) and nanosized structures, methods of severe plastic deformation (SPD) have found a wide application [1]. Materials obtained by such processing, for example nanostructured aluminium and copper, can become harder than high-strength steels, but they have a very low ductility and become brittle.

One of the ways to improve the mechanical characteristics of products while maintaining their operational properties is to obtain metallic materials with a gradient structure. It was revealed that the gradient structure of a material, in which the grain size belongs to the nanoscale on a surface and to a coarse-grained scale in the center of a billet, is promising for the improvement of the physical and mechanical characteristics and operational properties of metals and alloys [2-5].

The presence of a gradient of the strain and structure can be of interest for controlling the local properties of materials, i.e. for obtaining surface areas with a noticeable increase of the strength properties, in long-length metallic semi-products. The combination of more durable surface layers with a plastic core in resulting materials can improve their wear, bending, fatigue and crack resistances [6,7].

The aim of the present work is to study the new method of rotational active alternating bending with the possibility of forming a gradient structure in the process of non-monotonic (alternating) plastic deformation. With this aim, the deformed state of billets obtained by rotational active alternating bending using computer simulation was investigated [8].

2. Material and experimental procedure

For the study we selected the Cu-Zr alloy having a standard chemical composition (table 1). Before the full scale modeling of deformation treatment of the alloy under study, computer simulation of rotational active alternating bending was made for an analysis of the investigated method (figure 1a). A simplified scheme, which includes one bend with the subsequent turning of the workpiece to 180° relative to its longitudinal axis, was used.

Table 1. Chemical composition and deformation treatment temperature of the investigated alloy

Alloy	The content of alloying elements, wt.%		Deformation treatment temperature, °C
	Cu	Zr	
Cu - Zr	99,82	0,18	20

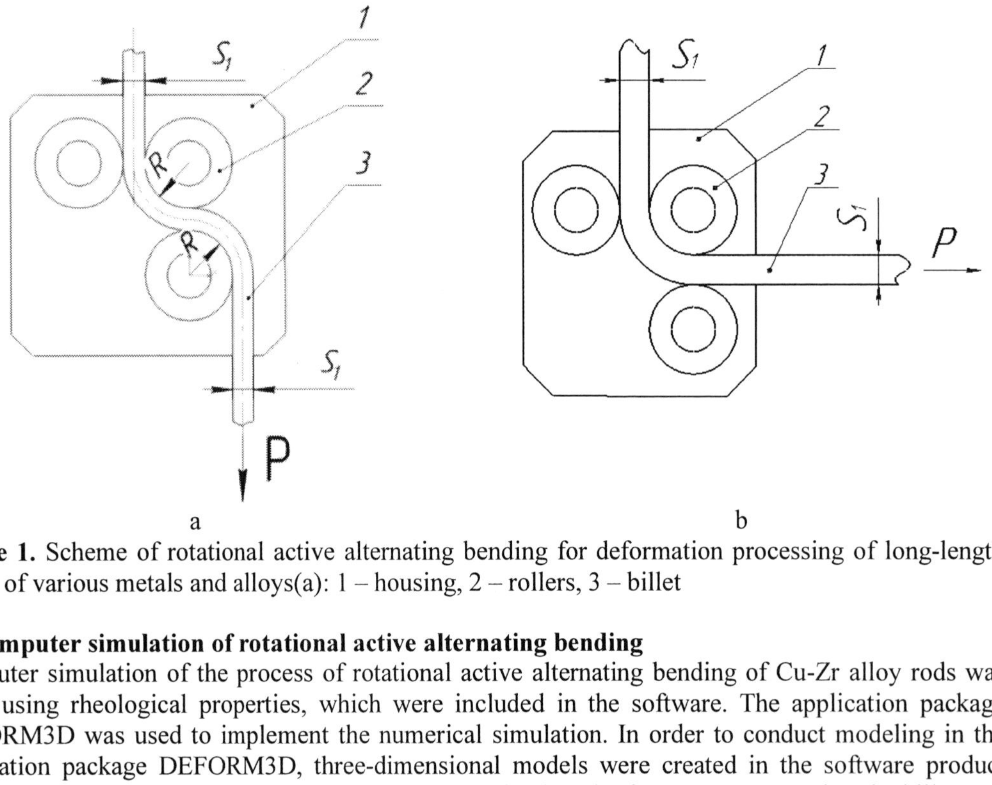

Figure 1. Scheme of rotational active alternating bending for deformation processing of long-length billets of various metals and alloys(a): 1 – housing, 2 – rollers, 3 – billet

3. Computer simulation of rotational active alternating bending

Computer simulation of the process of rotational active alternating bending of Cu-Zr alloy rods was made using rheological properties, which were included in the software. The application package DEFORM3D was used to implement the numerical simulation. In order to conduct modeling in the application package DEFORM3D, three-dimensional models were created in the software product Kompas-3D. The model of a 10x10 mm square rod with a length of 300 mm was used as the billet.

The following conditions were accepted for simulation:

1. The billet material in the initial state is isotropic and has no initial stresses.
2. The temperature is fixed at 20°C.
3. The tool is an absolutely rigid body.

IOP Conf. Series: Materials Science and Engineering **447** (2018) 012088 doi:10.1088/1757-899X/447/1/012088

4. In order to simulate the whole process of the passing of a workpiece through the rollers and obtain stable results of simulations, the number of simulation steps was taken equal to 150.
5. The number of finite elements is equal to 70000.
6. The friction coefficient between the tool and the billet is equal to 0.12.
7. The pull speed is assumed to be 60 m/min.
8. The rotation of rollers is free and occurs due only to the contact friction.

4. Results of simulation and analysis

Figure 2 shows the strain intensity after one and two operations of rotational active alternating bending. Analyzing the first picture, it should be noted that the deformation intensively increases already at the first stage of bending, and not only peripheral, but also central areas of the workpiece are processed. The graph of the distribution of strain intensity over the cross section of the sample has a hyperbolic shape. Moreover, in the region, where compressive stresses prevail (contact with the roller), the strain reaches approximately 1.4, and in the opposite region where tensile stresses occur, it amounts about 1.0, so the average value over cross section is equal to 0.4. After second bending with the change of deformation direction by 180°, the strain value reaches 1.95 in the compression zone and 1.76 in the opposite region, where tensile stresses occur, with an average value of about 1.0.

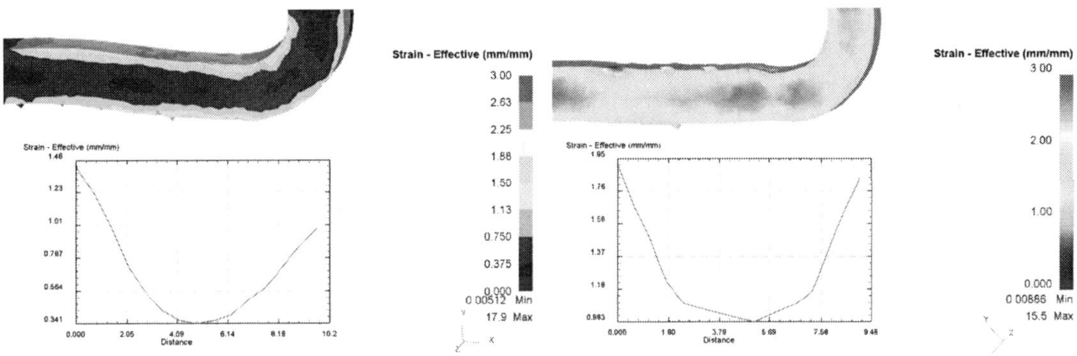

a b

Figure 2. Strain intensity distribution in longitudinal section after the deformation: a – first bend, b – second bend

Such an intensive deformation of the billet during processing by the method of rotational active alternating bending is associated with the arrangement of the rollers, their geometrical parameters and the conditions of their rotation about the axis of pulling. Thus, the mathematical model shows that by one cycle of processing by the method of rotational active alternating bending, the accumulation of a large strain intensity is possible in the conditions of high processing speeds of long-length products.

5. Conclusions

According to the results of experiments it was determined that the use of the method of rotational active alternating bending leads to a significant increase in the level of accumulated strain in the surface areas of the billet. So after one processing cycle the level of strain in the cross section of the billet changes is between e = 0.4 and e = 1.4 and after two cycles with a rotation of the billet to 180° before the second bend it changes in the interval from e = 1.0 to e = 1.9. This distribution indicates the gradient distribution of strain in the cross section of the billet with its maximum values in the peripheral regions of the deformed long-length square billets.

Acknowledgements

This research was supported by the Ministry of Science and Higher Education of the Russian Federation under the grant agreement No. 14.586.21.0059 (UIN: RFMEFI58618X0059).

References

[1] Valiev R Z, Zhilyaev A P and Langdon T G 2014 *Bulk Nanostructured Materials: Fundamentals and Applications* (New Jersey, Wiley) p 338

[2] Ivanov Yu F, Efimov O Yu and Kovalenko V V 2008 *Fundamental Problems of Modern Materials Science* **55–58** 4 (in Russian)

[3] Lu K 2014 *Science* **1455-1456** 345

[4] Lu K and Lu J 2004 *Mater. Sci. Eng. A* **373–377** 38

[5] Fang T H, Li W L, Tao N R and Lu K 2011 *Science* **1587-1590** 331

[6] Raab A G, Chukin M V, Aleshin G N and Raab G I 2014 *IOP Conf. Ser.: Mater. Sci. Eng. A* **63** 012008

[7] Raab G I, Gunderov D V, Shafigullin L N, Podrezov Yu M, Danylenko M I, Tsenev N K, R N Bakhtizin, Aleshin G N and Raab A G 2015 *Mater. Phys. Mech.* **242-252** 3

[8] Utyashev F Z and Raab G I 2013 *Deformation Methods for Fabrication and Processing of Ultrafine-Grained and Nanostructured Materials* (Gilem, Ufa) (in Russian)

AUTHOR INDEX

AUTHOR INDEX

AUTHOR INDEX